Methods in Cell Biology

VOLUME 109

Tetrahymena thermophila

Series Editors

Leslie Wilson
Department of Molecular, Cellular and Developmental Biology
University of California
Santa Barbara, California

Paul Matsudaira
Department of Biological Sciences
National University of Singapore
Singapore

Methods in Cell Biology

VOLUME 109

Tetrahymena thermophila

Edited by

Kathleen Collins
Department of Molecular and Cell Biology
University of California at Berkeley
Berkeley, CA, USA

ELSEVIER

AMSTERDAM • BOSTON • HEIDELBERG • LONDON
NEW YORK • OXFORD • PARIS • SAN DIEGO
SAN FRANCISCO • SINGAPORE • SYDNEY • TOKYO
Academic Press is an imprint of Elsevier

Academic Press is an imprint of Elsevier
225 Wyman Street, Waltham, MA 02451, USA
525 B Street, Suite 1900, San Diego, CA 92101-4495, USA
32, Jamestown Road, London NW1 7BY, UK
Linacre House, Jordan Hill, Oxford OX2 8DP, UK

First edition 2012

ISBN: 978-0-12-385967-9
ISSN: 0091-679X

For information on all Academic Press publications
visit our website at elsevierdirect.com

Printed and bound by CPI Group (UK) Ltd, Croydon, CR0 4YY

Transferred to Digital Print 2012

CONTENTS

CONTRIBUTORS

Numbers in parentheses indicate the pages on which the author's contributions begin.

Lydia J. Bright (141), Department of Biology, University of Indiana, Bloomington, 327 Jordan Hall, 1001 East 3rd Street Bloomington, IN 47405-7005, USA

Clifford Brunk (277), Department of Ecology and Evolutionary Biology, University of California, Los Angeles, 621 Charles E. Young Drive So., Los Angeles, CA, USA

Donna M. Cassidy-Hanley (239, 411), Department of Microbiology and Immunology, College of Veterinary Medicine, Cornell University, Ithaca, New York, USA

Douglas L. Chalker (327), Department of Biology, Washington University in St. Louis, St. Louis, MO, USA

Eric Cole (177), Biology Department St. Olaf College, Northfield, MN 55057, USA

Kathleen Collins (3, 347), Department of Molecular and Cell Biology, University of California at Berkeley, Berkeley, CA 94720-3200, USA

Mary T. Couvillion (347), Department of Molecular and Cell Biology, University of California at Berkeley, Berkeley, CA 94720-3200, USA

Robert S. Coyne (53), Department of Genomic Medicine, J. Craig Venter Institute, Rockville, MD, USA

F. Paul Doerder (11, 277), Department of Biological, Geological, and Environmental Sciences, Cleveland State University, 2121 Euclid Avenue, Cleveland, OH, USA

Joseph Frankel (83), Department of Biology, University of Iowa, Iowa City, IA, USA

Thomas H. Giddings Jr. (357), MCD Biology, University of Colorado at Boulder, 347 UCB, Boulder, Colorado 80309-0347, USA

Todd M. Hennessey (393), Department of Biological Sciences, University at Buffalo, Amherst, NY, USA

Jerry E. Honts (379), Department of Biology, Drake University, 2507 University Avenue, Des Moines, IA, USA

Kathleen M. Karrer (29), Department of Biological Sciences, Marquette University, Milwaukee, Wisconsin, USA

Thomas J. Lampert (393), Department of Biological Sciences, University at Buffalo, Amherst, NY, USA

Denis H. Lynn (11), Department of Zoology, University of British Columbia, 6270 University Blvd., Vancouver, BC, Canada V6T 1Z4

Wei Miao (53), Key Laboratory of Aquatic Biodiversity and Conservation, Institute of Hydrobiology, Chinese Academy of Sciences, Wuhan, China

Alejandro D. Nusblat (141), Cátedra de Biotecnología y Microbiología Industrial, Facultad de Farmacia y Bioquímica, Universidad de Buenos Aires, Junín 956, (C1113AAD) Buenos Aires, Argentina

Eduardo Orias (301), Department of Molecular, Cellular and Developmental Biology, University of California, Santa Barbara, Santa Barbara, CA 93106, USA

Chad G. Pearson (357), Department of Cell and Developmental Biology, University of Colorado – Denver, RC1 South, MS 8108, 12801 E. 17th Ave. Room 12104, Aurora, Colorado 80045, USA

Joshua J. Smith (411), Department of Biomedical Sciences, Missouri State University, Springfield, MO 65897, USA

Alexander J. Stemm-Wolf (357), MCD Biology, University of Colorado at Boulder, 347 UCB, Boulder, Colorado 80309-0347, USA

Nicholas A. Stover (53), Biology Department, Bradley University, Peoria, IL, USA

Toshiro Sugai (177), Department of Biology, College of Science, Ibaraki University, Mito, Ibaraki 310, Japan

Aaron P. Turkewitz (141), Department of Molecular Genetics and Cell Biology, The University of Chicago, 920 E. 58th Street, Chicago IL 60637, USA

Emily A. Wiley (411), W.M. Keck Science Department, Claremont McKenna, Pitzer, and Scripps Colleges, Claremont, CA 91711-5916, USA

Mark Winey (357), MCD Biology, University of Colorado at Boulder, 347 UCB, Boulder, Colorado 80309-0347, USA

Dorota Wloga (83), Department of Cell Biology, Nencki Institute of Experimental Biology, 3 Pasteur Street, 02-093, Warsaw, Poland

PREFACE

The goal of this volume is to provide current and future users of *Tetrahymena* with an enabling, enduring, inspiring guide. Chapter 1 introduces the model organism, its historical contributions, some of the more obvious compelling opportunities for high-impact discoveries, and community resources. The topical knowledge overviews in Chapters 2–7 explain organism features that are particularly useful and/or unique. Each of these chapters reaches through many decades of published studies, as well as unpublished work, to be comprehensive and not-soon-outdated in utility. The methodologically oriented Chapters 8–16 present experimental approaches, detailed protocols, literature references, and general forward-looking advice about handling *Tetrahymena* for purposes ranging from biochemistry to behavior and in contexts ranging from the classroom to the wild. For information beyond these chapters, anyone interested in the model organism will find a friendly welcome from members of the community, either individually or at the international ciliate meeting held every other year.

Past, present, and future contributions of *Tetrahymena* owe much to the strong community spirit of many researchers, who have invested their physical effort, training skills, and research careers in building invaluable intellectual and experimental resources. In particular, this volume is dedicated to the long-running record of visionary contributions by Peter Bruns, Marty Gorovksy, Ed Orias, and Meng-Chao Yao. Paul Matsudaira and Les Wilson also have been instrumentally supportive as *Methods in Cell Biology* series editors. Finally, there are thanks due to *Tetrahymena* itself: its many offerings as a useful experimental system are lovingly accounted in the following chapters, and no ode to the organism would be complete without acknowledging the simple, powerful physical appeal of watching the cells swim.

Kathleen Collins
University of California at Berkeley

PART I

Introduction

CHAPTER 1

Perspectives on the Ciliated Protozoan *Tetrahymena thermophila*

Kathleen Collins

Department of Molecular and Cell Biology, University of California at Berkeley, Berkeley, CA 94720-3200, USA

Abstract

In biology, scientific discoveries are often linked to technical innovations made possible by an inspired choice of model organism. Ciliate species, especially *Tetrahymena thermophila,* have had historically significant roles as uniquely enabling experimental systems. More importantly, as the chapters in this volume attest, ongoing efforts of the *T. thermophila* model organism community have created a knowledge and resource infrastructure for systems-level studies across a whole genome or proteome, setting the stage for understanding the fundamental biology underlying the sophisticated life cycle and environmentally responsive behaviors of this free-living, single-celled eukaryote. One hope is that these developments will stimulate the integration of ciliates into phylogenomic comparative analyses and also encourage the experimental use of *T. thermophila* by a broader scientific community. This early branching yet highly gene-rich eukaryote has much to offer for future studies of human-relevant basic biology.

0091-679X/10 $35.00
DOI 10.1016/B978-0-12-385967-9.00001-3

I. Introduction

In context of the eukaryotic family tree, ciliated protozoa launched their evolutionarily successful lineages long before the successful radiation of plants, fungi, or animals. What about the ciliates gives them the necessary advantages of growth, reproduction, and adaptability in the face of changing environmental conditions? Although these questions merit real answers from future studies, one can speculate based on features that are common to modern-day ciliates but are distinctive when ciliates are compared to other extant life. In addition to their defining characteristic of cilia-driven motility, ciliates share the properties of a large cell size, specialization of germline versus somatic nuclei within the same cytosol, and relatively high expressed gene content (comparable to mammals). Large cell size has been proposed to contribute to better feeding (ciliates are indiscriminate omnivores). The specialization of germline versus somatic nuclei allows remarkable *en masse* sampling of germline genotype allele combinations in the asexual phase of population growth and also an elegant mechanism for epigenetic influence of growth history on transmission of adaptive traits to sexual progeny. Streamlining of the expressed, somatic, macronuclear genome by large-scale elimination of repetitive DNA from the silent, germline, micronuclear genome could support the chronologically high rate of gene duplication and divergence that gives *T. thermophila* parity of gene number with animal genomes. These speculations have their origins in the literature summarized in the chapters that follow, which together provide an opportunity to integrate broadly across lessons learned from diverse research areas spanning from field studies of cell communities in their native habitat to reverse genetics of inbred strains maintained under controlled laboratory conditions.

II. Historical Contributions

Numerous useful features of *T. thermophila* account for its history of contributions as a model organism (Collins and Gorovsky, 2005; Orias *et al.*, 2011; Pearson and Winey, 2009; Turkewitz *et al.*, 2002). *T. thermophila* has a large size that is nonetheless modest among ciliates ($\sim 30 \times 50$ μm) and a rapid doubling time (about 2 h at 37 °C) made possible by a highly organized cortical architecture and a somatic nucleus streamlined for Herculean transcriptional output. Large size enables poking the cell with electrodes or with a needle to inject or ablate, the resolution of subcellular compartments by live or fixed whole-cell imaging, and obtaining lots of extract for biochemical studies from the more than 10^6 cells/mL that can be cultured in simple media. Combining conventional genetics (Chapter 10) and molecular genetics (Chapter 11) with the ease of biochemical analyses and purifications (Chapters 12 and 14) and advantages for cytology (Chapter 13), there is a wealth of opportunity for systems-level investigations to address complex mechanisms of cellular communication and behavior (Chapter 15). The ease of culture (Chapter 8), annotated genome contents (Chapter 4), and phenotypic and genotypic strain diversity (Chapters 2–5 and 9) lend themselves to student training opportunities that yield new findings and publications (Chapter 16).

Historical highlights of discoveries enabled by use of *T. thermophila* include the histone composition and modification differences between euchromatin and heterochromatin, which are readily detected by comparison of the macronucleus and micronucleus (Chapter 3). Much insight has also been gained about the process of DNA palindrome formation, which occurs during formation of the small macronuclear chromosome encoding large ribosomal RNAs (the rDNA chromosome) from its single-copy locus in the micronucleus (Tanaka and Yao, 2009). Some effort has been devoted to defining the principles of macronuclear chromosome counting (Donti *et al.*, 2009) and the conditionally essential, checkpoint-monitored processes of micronuclear mitosis and meiosis (Chapter 7). Pioneering discoveries exploiting *T. thermophila* also include the Nobel prize-winning self-splicing activity of the group I intron within the large ribosomal RNA precursor (Cech, 2004) and the simple-sequence repeat nature of chromosome telomeres and telomeric-repeat synthesis by telomerase (Blackburn, 2010; Greider, 2010), with additional seminal discoveries of microtubule motors, post-translational modifications, and dynamics (Chapter 5).

III. Compelling Opportunities

In addition to the established fields of study among researchers currently using *T. thermophila*, improved ciliate genome annotations and new methods (such as high-resolution imaging, deep sequencing, and quantitative proteomics) beg for expansion of ciliate model-system applications to new fields of study. Among these would be the biology of organellar biogenesis, remodeling, and function (see Chapter 5); membrane specialization, vesicle traffic, and regulated secretion (see Chapter 6); and different types of autophagy induced on massive scale to accomplish programmed nuclear death during sexual reproduction (Akematsu *et al.*, 2010) or to recycle cytosolic compartments and components of translation machinery when cells enter a state of growth arrest (Andersen and Collins, 2012; Nilsson, 1984). Also worthy of revisiting is the use of *T. thermophila* to characterize differential ribosome compositions that may reprogram translation (Hallberg and Sutton, 1977; McMullin and Hallberg, 1986) and to investigate stress-responsive regulation of translation in general (Calzone *et al.*, 1983).

Mechanisms that govern the selectivity of nuclear import have recently begun to be defined by directly exploiting nuclear dualism (Iwamoto *et al.*, 2009; Malone *et al.*, 2008; Orias *et al.*, 2011). There are also early hints that *T. thermophila* can provide new insights into principles of higher order chromatin organization, for example, the basis for clustering of rDNA chromosomes or condensin-dependent chromosome segregation (Cervantes *et al.*, 2006). Principles of developmentally induced genome remodeling are an obvious direction for continued study, including elucidation of the machinery that directs chromosome breakage and joining, chromosome breakage coupled to new telomere addition, and site-specific recombination (Chalker and Yao, 2011; Orias *et al.*, 2011). Also the roles of *T. thermophila* Piwi-protein RNPs in small RNA-mediated epigenetic regulation are just beginning to be understood (Couvillion *et al.*, 2009, 2010; Schoeberl and Mochizuki, 2011).

IV. Chapter Logic

The early Chapters 2–7 are knowledge summaries and systems perspectives. The later chapters 8–16 provide detailed methodological guidance, as well as general operating principles to enable extensions beyond established protocols. To supplement and update the compendium of this volume, which represents the cumulative expertise of the model organism community through 2011, the community is building updatable inventories of strains, plasmids, methods, and gene curations through the *Tetrahymena* genome database (http://ciliate.org/index.php/home/welcome), *Tetrahymena* functional genomics database (http://tfgd.ihb.ac.cn/), and *Tetrahymena* stock center (http://tetrahymena.vet.cornell.edu/). Beyond these resources, a ciliate list-serve allows queries for reagents and advice to be distributed across the model organism community (http://listserv.uga.edu/archives/ciliatemolbio-l.html).

Acknowledgments

The Collins laboratory research using *T. thermophila* has had long-term support from NIH R01 GM 54198. I thank my post-doctoral advisor Carol Greider for introduction to *T. thermophila*, my graduate advisor Paul Matsudaira for encouragement in providing this updated compendium of model organism information, Marty Gorovsky and Mary Couvillion for comments on this perspective, and Collins lab members for many years of vital inspiration and support. It should be clear from this volume that research using *T. thermophila* has benefited greatly from the model organism community commitment to collaborative generation of shared resources.

References

Akematsu, T., Pearlman, R. E., and Endoh, H. (2010). Gigantic macroautophagy in programmed nuclear death of *Tetrahymena thermophila*. *Autophagy* **6**, 901–911.

Andersen, K. L., and Collins, K. (2012). Several RNase T2 enzymes function in induced tRNA and rRNA turnover in the ciliate. *Tetrahymena. Mol. Biol. Cell.* **23**, 36–44.

Blackburn, E. H. (2010). Telomeres and telomerase: the means to the end (Nobel lecture). *Angew Chem. Int. Ed. Engl.* **49**, 7405–7421.

Calzone, F. J., Angerer, R. C., and Gorovsky, M. A. (1983). Regulation of protein synthesis in *Tetrahymena*. Quantitative estimates of the parameters determining the rates of protein synthesis in growing, starved, and starved-deciliated cells. *J. Biol. Chem.* **258**, 6887–6898.

Cech, T. R. (2004). Self-splicing and enzymatic activity of an intervening sequence RNA from *Tetrahymena*. *Biosci. Rep.* **24**, 362–385.

Cervantes, M. D., Coyne, R. S., Xi, X., and Yao, M. C. (2006). The condensin complex is essential for amitotic segregation of bulk chromosomes, but not nucleoli, in the ciliate *Tetrahymena thermophila*. *Mol. Cell. Biol.* **26**, 4690–4700.

Chalker, D., and Yao, M. C. (2011). DNA elimination in ciliates: transposon domestication and genome surveillance. *Annu. Rev. Genet.* **45**, 227–246.

Collins, K., and Gorovsky, M. A. (2005). *Tetrahymena thermophila. Curr. Biol.* **15**, R317–R318.

Couvillion, M. T., Lee, S. R., Hogstad, B., Malone, C. D., Tonkin, L. A., Sachidanandam, R., Hannon, G. J., and Collins, K. (2009). Sequence, biogenesis, and function of diverse small RNA classes bound to the Piwi family proteins of *Tetrahymena thermophila. Genes Dev.* **23**, 2016–2032.

Couvillion, M. T., Sachidanandam, R., and Collins, K. (2010). A growth-essential *Tetrahymena* Piwi protein carries tRNA fragment cargo. *Genes Dev.* **24**, 2742–2747.

Donti, T. R., Datta, S., Sandoval, P. Y., and Kapler, G. M. (2009). Differential targeting of *Tetrahymena* ORC to ribosomal DNA and non-rDNA replication origins. *EMBO J.* **28**, 223–233.

Greider, C. W. (2010). Telomerase discovery: the excitement of putting together pieces of the puzzle (Nobel lecture). *Angew Chem. Int. Ed. Engl.* **49**, 7422–7439.

Hallberg, R. L., and Sutton, C. A. (1977). Nonidentity of ribosomal structural proteins in growing and starved *Tetrahymena. J. Cell Biol.* **75**, 268–276.

Iwamoto, M., Mori, C., Kojidani, T., Bunai, F., Hori, T., Fukagawa, T., Hiraoka, Y., and Haraguchi, T. (2009). Two distinct repeat sequences of Nup98 nucleoporins characterize dual nuclei in the binucleated ciliate tetrahymena. *Curr. Biol.* **19**, 843–847.

Malone, C. D., Falkowska, K. A., Li, A. Y., Galanti, S. E., Kanuru, R. C., LaMont, E. G., Mazzarella, K. C., Micev, A. J., Osman, M. M., Piotrowski, N. K., Suszko, J. W., Timm, A. C., Xu, M. M., Liu, L., and Chalker, D. L. (2008). Nucleus-specific importin alpha proteins and nucleoporins regulate protein import and nuclear division in the binucleate *Tetrahymena thermophila. Eukaryot. Cell* **7**, 1487–1499.

McMullin, T. W., and Hallberg, R. L. (1986). Effect of heat shock on ribosome structure: appearance of a new ribosome-associated protein. *Mol Cell Biol.* **6**, 2527–2535.

Nilsson, J. T. (1984). On starvation-induced autophagy in *Tetrahymena. Carlsberg Res. Commun.* **49**, 323–340.

Orias, E., Cervantes, M. D., and Hamilton, E. P. (2011). *Tetrahymena thermophila*, a unicellular eukaryote with separate germline and somatic genomes. *Res. Microbiol.* **162**, 578–586.

Pearson, C. G., and Winey, M. (2009). Basal body assembly in ciliates: the power of numbers. *Traffic* **10**, 461–471.

Schoeberl, U. E., and Mochizuki, K. (2011). Keeping the soma free of transposons: programmed DNA elimination in ciliates. *J. Biol. Chem.* PMID:21914793.

Tanaka, H., and Yao, M. C. (2009). Palindromic gene amplification – an evolutionarily conserved role for DNA inverted repeats in the genome. *Nat. Rev. Cancer.* **9**, 216–224.

Turkewitz, A. P., Orias, E., and Kapler, G. (2002). Functional genomics: the coming of age for *Tetrahymena thermophila. Trends Genet.* **18**, 35–40.

PART II

Systems Perspectives

CHAPTER 2

The Life and Times of *Tetrahymena*

Denis H. Lynn* **and F. Paul Doerder**†

*Department of Zoology, University of British Columbia, 6270 University Blvd., Vancouver, BC, Canada V6T 1Z4

†Department of Biological, Geological, and Environmental Sciences, Cleveland State University, 2121 Euclid Avenue, Cleveland, OH, USA

Abstract

The genus *Tetrahymena* is defined on the basis of a four-part oral structure composed of an undulating membrane and three membranelles. It is a monophyletic genus with 41 named species and numerous unnamed species, many of which are morphologically indistinguishable. Nuclear small subunit rRNA and mitochondrial cytochrome *c* oxidase subunit 1 sequences indicate two major clades, a "borealis" clade of less closely related species and an "australis" clade of more closely related species that correlate to differences in mating-type determination and frequency of amicronucleates. Members of both clades show convergence for histophagy (primarily facultative), macrostome transformation, and (rare) cyst formation. Life cycle parameters of species are presented and problematic species discussed.

METHODS IN CELL BIOLOGY, VOL 109

0091-679X/10 $35.00
DOI 10.1016/B978-0-12-385967-9.00002-5

I. Introduction

Tetrahymena thermophila is the most widely known and best-studied member of the ciliate genus *Tetrahymena*. It is, however, but one of 41 recognized species and numerous unnamed species identified by DNA barcodes. In this chapter, we briefly review the history of *Tetrahymena* and update information on the named species, integrating data on life cycles and evolution. We also draw attention to areas in which further research is required to resolve ambiguities.

II. Brief History

Ciliates now assigned to the genus *Tetrahymena* were very probably seen in the 17th century by early microscopists, like Antoni von Leeuwenhoek. However, it was not until 1830 that Ehrenberg put a name to the "type" species, calling it *Leucophrys pyriformis* (Corliss and Dougherty, 1967). The genus *Tetrahymena* with its type species *Tetrahymena geleii* was established by Furgason (1940) for this same ciliate, which created a taxonomic and nomenclatural problem as two names cannot be used to identify the same organism, according to the International Code of Zoological Nomenclature. This was significantly more problematic for *Tetrahymena* because Lwoff (1923) had succeeded in culturing *Tetrahymena* on sterile medium, opening up a significant research opportunity in the biochemistry and physiology of protistan cells. In the intervening years from 1923 until 1967, when Corliss and Dougherty (1967) petitioned the International Commission on Zoological Nomenclature (ICZN), a significant body of literature had accumulated using *Tetrahymena* as the name. Thus, Corliss and Dougherty (1967) argued in a detailed submission to the Commission that it was essential to conserve the generic name *Tetrahymena* in the interests of stability and uniformity of nomenclatural usage. In 1970, the Commission agreed in their Opinion 915, preserving the generic name *Tetrahymena* for all future usage and setting as the type species *Tetrahymena pyriformis* Ehrenberg, 1830 (ICZN, 1970; Lwoff, 1947).

During the years since Lwoff (1923), an amicronucleate strain of *T. pyriformis*, Strain GL, served as the primary research model for biochemistry and physiology of *Tetrahymena*, and this research continued into the 1950s and beyond (Corliss and Daggett, 1983). In the 1950s, Elliott and Gruchy (1952) and Elliott and Nanney (1952) discovered mating types in micronucleate strains of *T. pyriformis*, opening up "this" ciliate as a genetical research model. Shortly thereafter Gruchy (1955) discovered that *T. pyriformis* was composed of eight varieties, actually true biological species, making it a cryptic species complex like *Paramecium aurelia*, whose cryptic nature had been discovered already by Sonneborn (1939). This period also saw *Tetrahymena* become an important model for cell cycle research when Zeuthen's lab devised a means of synchronizing mass cultures of this ciliate by repeated heat shocks (Scherbaum and Zeuthen, 1954; Zeuthen, 1953).

Corliss (1954, 1965, 1973a) has given overviews of the literature on *Tetrahymena*, demonstrating its importance as a model for research. In 1954, he claimed there were over 600 papers dealing with species in this genus; in 1973, this number had grown close to 3000 (Corliss, 1973a). A query on September 29, 2011 using "*Tetrahymena*" as a topic word in the "Web of Knowledge[SM]" for the period 1973 to the present provides 7814 titles, and from 1898 to the present 9305 titles! Joining the list of model organisms, *T. thermophila* has now had its genome sequenced and assembled, at least the macronuclear genome. This second edition of *Methods in Cell Biology* devoted to *Tetrahymena* confirms its place in the research community.

III. Description

Tetrahymena has been assigned without controversy to the family Tetrahymenidae in the hymenostome grouping of ciliates (Lynn, 2008). The genus is named for its four (i.e., tetra, Gr. = four) oral structures – the paroral or undulating membrane and three oral polykinetids or membranelles, which are "membrane"-like (i.e., hymen, L. = membrane) (Fig. 1) (see Chapter 5 for more details on structure). There are now 41 species assigned to the genus (Table I). However, this number is likely a significant underestimate of the genetic diversity within the genus (see section below).

Tetrahymena species, once the ciliate is confirmed as assignable to that genus, have been further characterized by four main categories of features (Corliss, 1970, 1973b): (1) the ciliature and infraciliature, particularly the number of somatic kineties or ciliary rows, and cortical features, particularly the patterning of the silver-line system following silver nitrate staining; (2) the life cycle characteristics, such as kinds of polymorphisms and the presence of a cyst; (3) their ecological habits, such as food preferences and kinds of relationships with host animals; and (4) physiological and biochemical properties, though these latter properties are not widely known for most species as this kind of research has concentrated primarily on *T. thermophila*.

For many years, the species were grouped into three infrageneric groupings, primarily based on life cycle characteristics (Corliss, 1970, 1973b). The *pyriformis* group contained bacterivorous species with a potential to facultatively parasitize invertebrates and vertebrates, both living and dying (Figs. 1 and 2). The *rostrata* group included larger-bodied species that can be strongly histophagous (i.e., tissue eating, primarily invertebrate) and/or parasitic and often divide in a cyst and may have a resting cyst stage (Fig. 2). The *patula* group comprised species that developed a huge cytopharyngeal pouch as a macrostome form, which preyed upon smaller ciliates, including its brothers and sisters who had not yet transformed into macrostomes (Fig. 2).

These infrageneric groupings are now most useful as a shorthand to identify the life cycle and general biology of the species as *pyriformis*-like, *rostrata*-like, or *patula*-like. They can no longer be considered to be phylogenetic groupings. Strüder-Kypke *et al.* (2001) argued that histophagy or the *rostrata*-like life style had evolved several times convergently within the genus, based on phylogenies derived from

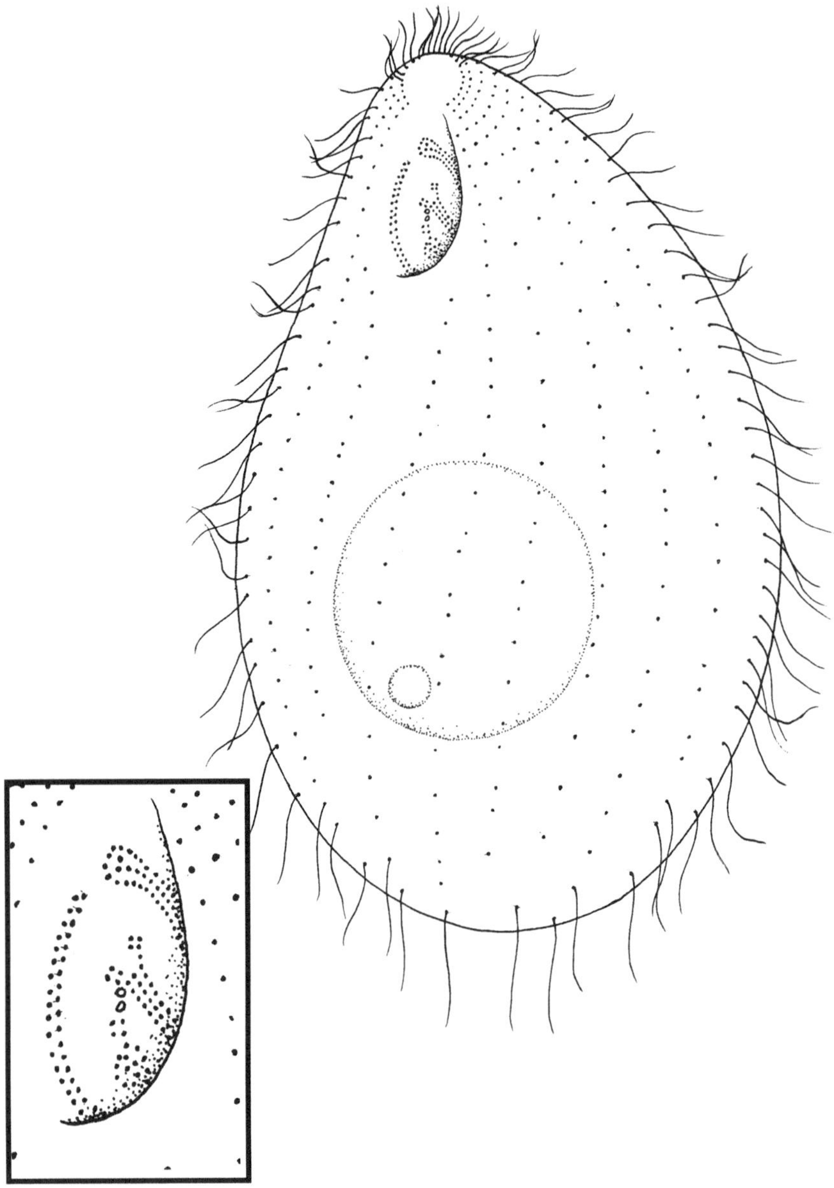

Fig. 1 Schematic drawing of *Tetrahymena pyriformis*. The ciliate's body is pear-shaped or pyriform. The inset shows a detail of the oral apparatus with the basal bodies or kinetosomes of its paroral or undulating membrane on the ciliate's right of the oral cavity and the three oral polykinetids or membranelles on the left (see Chapter 5 for more details). Adapted from Lynn (2008).

Table I
Brief characterization of valid species of *Tetrahymena* based on a summary of the literature.

Species name and taxonomic authorities	Ecological Habitus (if not specified, species can be considered bacterivorous but also facultative histophages and/or parasites	Cysts	Mating; Mating type determination	Micro-nucleus: Present (+)/ Absent (−)	Small subunit rRNA GenBank Accession Number	Cytochrome *c* oxidase subunit 1 GenBank Accession Number (Type Strain)	Reference
Tetrahymena americanis Nanney & McCoy, 1976		Resting – not viable	Conjugation; synclonal	+	EF070242	EF070267 (ATCC 205052)[a]	Nanney & McCoy, 1976
Tetrahymena asiatica Simon, Meyer & Preparata, 1985		No	Conjugation; synclonal	+	EF070243	EF070268 (ATCC 205167)	Simon, Meyer & Preparata, 1985
Tetrahymena australis Nanney & McCoy, 1976		No	Conjugation; synclonal	+/−	X56167	EF070269 (ATCC 30271)*[b]	Nanney & McCoy, 1976
Tetrahymena bergeri Roque, de Puytorac & Savoie, 1971	Facultative histophage	Reproductive	Cytogamy (Autogamy in pairs)	+	AF364039	EF070270 (ATCC 50985)*	Roque, de Puytorac & Savoie, 1971
Tetrahymena borealis Nanney & McCoy, 1976		No	Conjugation; karyonidal	+/−	M98020	EF070271 (ATCC 30317)	Nanney & McCoy, 1976
Tetrahymena canadensis Nanney & McCoy, 1976		No	Conjugation; karyonidal	+/−	X56170	EF070276 (ATCC 30368)	Nanney & McCoy, 1976
Tetrahymena capricornis Nanney & McCoy, 1976		No	Conjugation; synclonal	+	X56172	EF070277 (ATCC 30290)*	Nanney & McCoy, 1976
Tetrahymena caudata Simon, Meyer & Preparata, 1985	Macrostome	Resting cyst?	Unknown	+	EF070244	EF070278 (ATCC 50087)*	Simon, Meyer & Preparata, 1985
Tetrahymena chironomi Corliss, 1960	Parasite	No	Intraclonal conjugation? (selfing)	+	ND[c]	ND	Corliss, 1960
Tetrahymena corlissi Thompson, 1955	Facultative histophage	Resting, reproductive	Unknown	+	U17356	EF070279 (ATCC 50086)*	Thompson, 1955
Tetrahymena cosmopolitanis Nanney & McCoy, 1976		No	Conjugation; unknown	+	EF070245	EF070280 (ATCC 30324)	Nanney & McCoy, 1976
Tetrahymena dimorpha Batson, 1983	Facultative parasite	No	Intraclonal conjugation; unknown	+	ND	ND	Batson, 1983
Tetrahymena edaphoni Foissner, 1987		Resting cyst	Not observed	+	ND	ND	Foissner, 1987
Tetrahymena elliotti Nanney & McCoy, 1976		No	Intraclonal conjugation (selfing)	+/−	EF070246	EF070281 (ATCC 205065)*	Nanney & McCoy, 1976

(Continued)

Table I *(Continued)*

Species name and taxonomic authorities	Ecological Habitus (if not specified, species can be considered bacterivorous but also facultative histophages and/or parasites	Cysts	Mating; Mating type determination	Micro-nucleus: Present (+)/ Absent (−)	Small subunit rRNA GenBank Accession Number	Cytochrome *c* oxidase subunit 1 GenBank Accession Number (Type Strain)	Reference
Tetrahymena empidokyrea Jerome, Simon & Lynn, 1996	Obligate parasite?	No	Pairing observed	+	U36222	EF070282 (ATCC 50595)*	Jerome, Simon & Lynn, 1996
Tetrahymena farleyi Lynn, Gransden, Wright & Josephson, 2000	Obligate parasite?	No	Unknown	−?	AF184665	EF070283 (ATCC 50748)*	Lynn, Gransden, Wright & Josephson, 2000
Tetrahymena furgasoni Nanney & McCoy, 1976		No	None	−	EF070247	EF070284 (ATCC 30006)*	Nanney & McCoy, 1976
Tetrahymena hegewischi Nyberg, 1981		No	Conjugation; synclonal	+	M98019	GU439219 (ATCC 30832)	Nyberg, 1981
Tetrahymena hyperangularis Nanney & McCoy, 1976		No	Conjugation; synclonal	+	X56173	EF070286 (ATCC 30273)	Nanney & McCoy, 1976
Tetrahymena leucophrys Williams, Buhse & Smith, 1984	Macrostome	No	None	−	EF070248	EF070287 (ATCC 50069)*	Williams, Buhse & Smith, 1984
Tetrahymena limacis (Warren, 1932) Kozloff, 1946	Obligate parasite		Intraclonal conjugation (selfing)	+/−	EF070249	EF070288 (ATCC 30771)	Kozloff, 1946
Tetrahymena malaccensis Simon, Meyer & Preparata, 1985		No	Conjugation; karyonidal	+	M26360	EF070291 (ATCC 50065)*	Simon, Meyer & Preparata, 1985
Tetrahymena mimbres Meyer & Nanney, 1987		ND	None	−	EF070251	EF070292 (ATCC 30330)*	Meyer & Nanney, 1987
Tetrahymena mobilis (Kahl, 1926) n. comb. for *Sathrophilus mobilis* Kahl, 1926		No	Not observed	+	AF364040	GU439221 (CCAP 1630/22)[d]	This Chapter
Tetrahymena nanneyi Simon, Meyer & Preparata, 1985		No	Conjugation; synclonal	+	X56169	EF070294 (ATCC 50071)*	Simon, Meyer & Preparata, 1985
Tetrahymena nipissingi Nyberg, 1981		No	Conjugation; synclonal	+	EF070252	EF070295 (ATCC 30837)*	Nyberg, 1981
Tetrahymena paravorax Corliss, 1957	Macrostome		Intraclonal conjugation (selfing)	+/−	EF070253	EF070296 (ATCC 205177)*	Corliss, 1957
Tetrahymena patula (Ehrenberg, 1830) Corliss, 1951	Macrostome	Reproductive	Conjugation?	+/−	X56174	EF070297 (ATCC 50064)	Corliss, 1951

Tetrahymena pigmentosa Nanney & McCoy, 1976		No	Conjugation; synclonal	+	M26358	EF070299 (ATCC 30278)*	Nanney & McCoy, 1976
Tetrahymena pyriformis (Ehrenberg, 1830) Lwoff, 1947		No	None	-	X56171	EF070303 (ATCC 30327)	Lwoff, 1947
Tetrahymena rostrata (Kahl, 1926) Corliss, 1952	Facultative parasite	Resting	Autogamy in resting cyst	+/−	ND	GU439231 (ATCC PRA-326)	Corliss, 1952
Tetrahymena rotunda Lynn, Molloy & Lebrun, 1981	Obligate parasite?	Unknown	Unknown	Not observed	ND	ND	Lynn, Molloy & Lebrun, 1981
Tetrahymena setosa (Schewiakoff, 1893) McCoy, 1975		No	Conjugation?	+/−	AF364041	EF070306 (ATCC 30782)*	McCoy, 1975
Tetrahymena shanghaiensis Feng, Sun, Cao, Li & Chen, 1988		No?	Conjugation	+/−	EF070256	EF070307 (ATCC 205039)*	Feng, Sun, Cao, Li & Chen, 1988
Tetrahymena sialidos Batson, 1985	Facultative parasite?	No	Intraclonal conjugation?; unknown	+	ND	ND	Batson, 1985
Tetrahymena silvana Simon, Meyer & Preparata, 1985	Macrostome	Resting cyst?	Unknown	+	EF070257	EF070307 (ATCC 50084)*	Simon, Meyer & Preparata, 1985
Tetrahymena sonneborni Nyberg, 1981		No	Conjugation; synclonal	+	EF070258	GU439235 (ATCC 30834)	Nyberg, 1981
Tetrahymena stegomyiae (Keilin, 1921) Corliss, 1960	Obligate parasite	Resistant	Unknown	+	ND	ND	Corliss, 1960
Tetrahymena thermophila Nanney & McCoy, 1976		No	Conjugation; karyonidal	+/−	M10932	EF070310 (Strain B1975)	Nanney & McCoy, 1976
Tetrahymena tropicalis Nanney & McCoy, 1976		No	Conjugation; unknown	+/−	X56168	EF070314 (ATCC 30276)	Nanney & McCoy, 1976
Tetrahymena vorax (Kidder, Lilly & Claff, 1940) Kidder, 1941	Macrostome	Reproductive	None	−	AF364038	EF070319 (ATCC 30421)	Kidder, 1941

Note the following: *Tetrahymena setosa* has been considered a junior synonym of *Tetrahymena pyriformis* (Kher *et al.*, 2011); *Tetrahymena lwoffi* is considered a junior synonym of *Tetrahymena furgasoni* (see text; Meyer & Nanney, 1987); and herein we create the new combination *Tetrahymena mobilis* (Kahl, 1926) Lynn & Doerder, n. comb. given the identification of this strain to the genus *Tetrahymena* on the basis of its morphology (Schiftner & Foissner, 1998), submission of a type strain CCAP 1630/22 by Foissner, and its *COX*1 barcode (Kher *et al.*, 2011).

[a] ATCC = American Type Culture Collection

[b] * indicates species where only one strain has been sequenced

[c] ND = not determined

[d] CCAP = Culture Centre for Algae and Protozoa

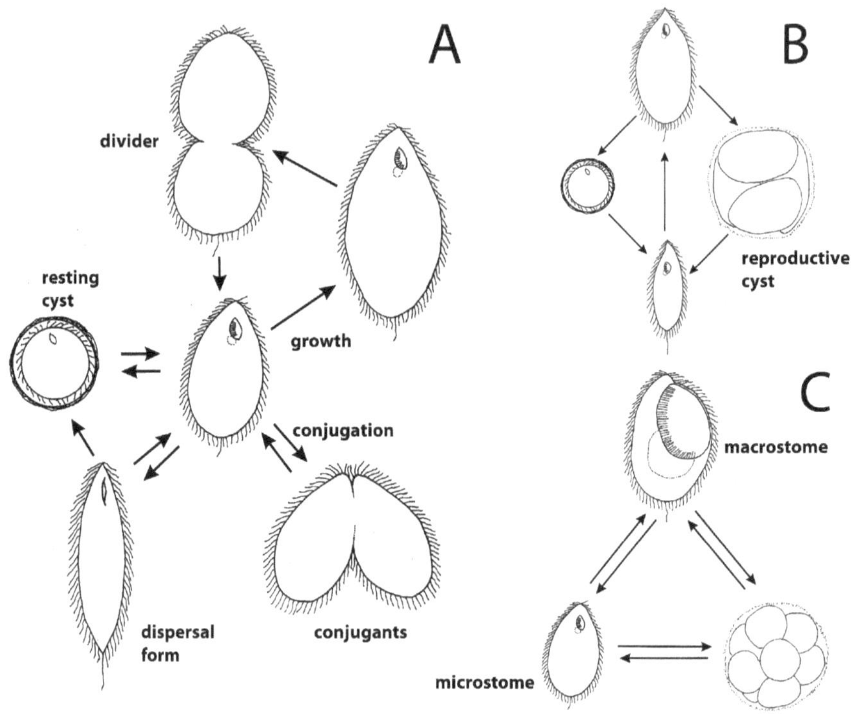

Fig. 2 Schematic representations of the life histories of *Tetrahymena* species that were assembled into three groups, which at one time were thought to be clades. (A) The *pyriformis* group, characterized here by a sexual species that undergoes both conjugation and forms resting cysts. The cell in the center can grow and divide when food is available (top right). When food, such as bacteria, is depleted, the cell can transform to a dispersal form (bottom left) or form a resting cyst (left). When food is encountered or reappears, the dispersal form can transform back or the cyst can excyst. Typically under nutrient deprivation, two cells undergo conjugation (bottom right), and the two cells separate afterward as exconjugants, which either begin growth if food is present or encyst or disperse if food is not; based on Lynn (2008). (B) The *rostrata* group, which includes larger bodied species that are strongly histophagous and/or parasitic. The cell at the bottom can grow into the cell at the top, which may divide in a reproductive cyst (right) or may form a resting cyst when conditions are unfavorable (left); based on Corliss (1973b). (C) The *patula* group comprised species that develop a huge cytopharyngeal pouch (stippled line) as a macrostome form (top) when bacterial food disappears. The macrostome preys upon smaller ciliates, including its brothers and sisters, such as the microstome (bottom left), which have not yet transformed into macrostomes. Division typically occurs in a reproductive cyst (bottom right). Macrostomes can reversibly transform to microstomes or enter a reproductive cyst to divide and produce more macrostomes or microstomes; based on Corliss (1973b).

small subunit rRNA gene sequences (SSUrRNA). This has now been confirmed for the histophages and extended to the *patula*-like life style, using the barcode region of the cytochrome *c* oxidase subunit 1 (*COX*1) gene (Fig. 3) (Chantangsi *et al.,* 2007; Kher *et al.,* 2011). These latter studies have also highlighted some areas for further

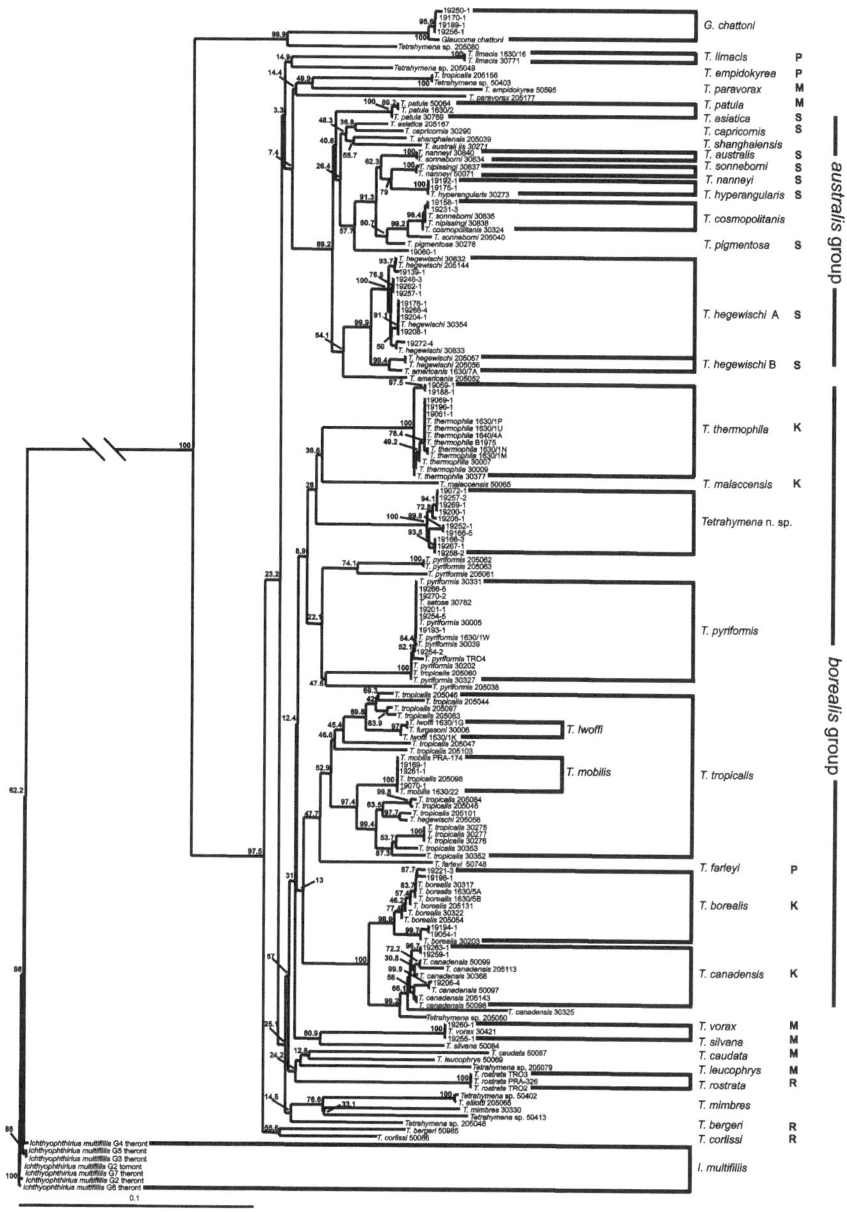

Fig. 3 A tree showing relationships among strains and isolates of the different species of *Tetrahymena*. The tree was generated using the neighbor-joining algorithm based on an 822-nucleotide stretch of the cytochrome *c* oxidase subunit 1 (*COX*1) gene. Genetic distances (scale bare = 0.1 nucleotide substitutions per site) were calculated using the Kimura two-parameter model and the data were bootstrap resampled 1000 times. **K** – karyonidal system of mating-type determination; **M** – macrostome; **P** – parasite; **S** – synclonal system of mating type determination; **R** – *rostrata*-like species. (Modified from Kher *et al.*, 2011.)

research and for taxonomic consideration, which we will briefly touch on in the remainder of this section.

Among the *pyriformis*-like species, it had long been recognized that it was problematic to use morphology to distinguish among them. In the 1970s, with the advent of isozyme electrophoresis, Allen and Weremiuk (1971) first demonstrated quantitative differences in esterase patterns between syngens of *T. pyriformis*, and this result was extensively broadened and strengthened to both sexual and asexual strains by Borden *et al.* (1973, 1977). Nanney and McCoy (1976) proposed to use these patterns in isozymes as operational means to identify any strain of *T. "pyriformis"*, sexual or asexual, using the following definition of an asexual species: "a population with approximately the same amount of molecular (genetic) heterogeneity as a sexual species and discontinuous in that heterogeneity from other sexual and asexual species" (Nanney and McCoy, 1976, pp. 671–672). Nanney and McCoy (1976) established four species for asexual groupings: their Phenoset A, which included Strain E, the neotype for the type species (Corliss, 1971), and Strain GL, the original type culture on which Lwoff (1947) based the name, became *T. pyriformis*; Phenoset B became *Tetrahymena elliotti*; Phenoset C became *Tetrahymena furgasoni*; and Phenoset E became *Tetrahymena lwoffi*. Isozyme patterns, while useful, were known to be problematic: significant biomass of cells was needed; marker standards always needed to be run and so reference cultures had to be maintained; and isozyme expression might vary with physiological state. Gene sequencing technologies (e.g., gene sequences of SSUrRNA and *COX*1) avoid these problems. Kher *et al.* (2011) suggested that strains differing by <1% on the *COX*1 barcode be assigned to the same species and that those diverging by >5% be assumed to be different species. Using these criteria, they were able to assign 98% (50/51) of environmental isolates to a species.

*COX*1 sequences have revealed problems with some species that are sufficiently important that we review them here. For instance, *T. furgasoni* and *T. lwoffi* were established as separate amicronucleate species based on isozyme mobilities (see above), but subsequent examination of cytoskeletal proteins (Williams *et al.*, 1984) and reexamination of strains using additional isozymes (Meyer and Nanney, 1987) failed to distinguish the species. Meyer and Nanney (1987) declared them to be synonymous and considered the species name *T. lwoffi* to be a junior synonym (i.e., a "younger" species name that does not have priority). Chantangsi *et al.* (2007) reported that both SSUrRNA and *COX*1 sequences of the two species are identical, essentially rendering it certain that they are identical. For these reasons, *T. lwoffi* is not included on the list of recognized species.

Problems also exist with micronucleate species. *Tetrahymena tropicalis* strains occupy several positions on both SSUrRNA and *COX*1 trees (Chantangsi *et al.*, 2007; Kher *et al.*, 2011). No differences were revealed by isozymes (Meyer and Nanney, 1987), though there are differences in the D2 region of the large subunit rRNA (LSUrRNA) sequences of *T. tropicalis* in GenBank. *COX*1 sequences differed by 6.2% (Kher *et al.*, 2011), a value inconsistent with other intraspecific values. While it is possible that some strains are mislabeled, which is a vexing problem in maintaining numerous isolates and species, differences among the 17 available *COX*1 sequences suggest it is not a major contributor. Rather, we suspect that there are cryptic species within this group. This is supported by the D2 variants and by *COX*1 sequences from additional wild isolates (Doerder, unpublished), which seem to fall into two groups. Further molecular investigations as well as breeding tests are needed to resolve the status of this species. The observation that degenerating macronuclei are uniquely anteriorly located in conjugating *T. tropicalis* (Simon and Doerder, 1981) may be useful in this regard. Strains of several other species established since 1976 on the basis of interbreeding criteria also appeared to be problematic according to Kher *et al.* (2011): strains of *Tetrahymena nanneyi* and *Tetrahymena nipissingi* appeared to be conspecific (i.e., belonging to the same species); strains of *T. nanneyi* and *Tetrahymena sonneborni* appeared to be conspecific; and strains of the latter two species appeared to be conspecific with strains of *Tetrahymena cosmopolitanis* (Fig. 3). More research will be needed on these matters.

With regard to *rostrata*-like species, Segade *et al.* (2009) isolated three *bona fide* strains of *Tetrahymena rostrata* from snails in Spain and confirmed these to be assignable to that species on the basis of their morphology and that they underwent autogamy (i.e., self-fertilization as a single cell) precisely as described many years earlier by Corliss (1952). However, these strains were not genetically similar to the *T. rostrata* strain(s) in the American Type Culture Collection for which information on autogamy and life history is not available. Segade *et al.* (2009) concluded that future research on *T. rostrata* use this newly authenticated strain (i.e., ATCC PRA-326, Table I) and not the other *T. rostrata* strain (i.e., ATCC 30770). Corliss (1971, Footnote, p. 244) expressed doubt about the distinctness of *Tetrahymena bergeri*, a patronym for one of his former students, Jacques Berger, from *T. rostrata*: *COX*1 barcoding demonstrates considerable distance between the two species (Fig. 3) (Chantangsi *et al.,* 2007; Kher *et al.,* 2011). The two strains of the other member of the former *rostrata* group, *Tetrahymena limacis*, isolated from slugs, are also confirmed as very different from other *rostrata* types (Fig. 3).

This leads us to comment on the *pyriformis*-like strains isolated from animals. The two species for which there are *COX*1 sequences – *Tetrahymena empidokyrea* isolated from adult mosquitoes (Jerome *et al.*, 1996) and *Tetrahymena farleyi* isolated from the urine of a dog named "Farley" (Lynn *et al.*, 2000) – are very different from each other and from adjacent sexual species: for the former from a strain of *T. tropicalis* and for the latter a larger grouping of *T. tropicalis* strains unrelated to the former (Fig. 3) (Kher *et al.,* 2011). Several other species have been isolated from the hemolymph of insects – *Tetrahymena chironomi* from European chironomid larvae, *Tetrahymena rotunda* from North American simuliid larvae, *Tetrahymena stegomyiae* from African *Aedes* mosquito larvae, *Tetrahymena sialidos* from alderfly larvae, and *Tetrahymena dimorpha* from simuliid larvae in Great Britain. These will all need to be reisolated from nature, although the latter two were kept in axenic culture but regrettably never submitted to a culture collection.

Finally attention should be drawn to strains that demonstrate a *patula*-like transformation to a macrostome. Since Corliss (1971, 1973b), three species could be added to this group: *Tetrahymena caudata*, *Tetrahymena leucophrys*, and *Tetrahymena silvana* (Table I). This brings to six, the number in this "group," but this is another life history strategy that appears to be convergent within the genus, given the significant genetic differences among these species (Fig. 3) (Chantangsi *et al.,* 2007; Kher *et al.,* 2011).

IV. Life Cycles and Breeding Systems

The majority of tetrahymenas possess the typical ciliate life cycle (Fig. 2A). In the absence of sufficient food for reproduction (i.e., through binary fission), cells engage in conjugation during which micronuclear meiosis and reciprocal fertilization occur. Additionally, the old macronucleus of each conjugant is destroyed, and new ones are assembled from mitotic products of the zygotic nuclei (see details elsewhere in this volume). Essentially, conjugation results in complete genome replacement in each exconjugant and genetic identity of both exconjugants. Such sex is evidently of considerable importance. Unlike the familiar *Paramecium* "*aurelia*" species, autogamy is rare in *Tetrahymena* (see Table I). In the laboratory, *T. thermophila* can be induced to undergo autogamy in pairs (i.e., cytogamy), a related process in which conjugants self-fertilize, but its occurrence in nature is doubtful. Genomic exclusion, another laboratory phenomenon of great genetic utility because it creates whole genome homozygotes, is also likely rare or absent in natural populations. Nevertheless, the occurrence of both cytogamy and genomic exclusion indicate that *T. thermophila* has evolved considerable flexibility regarding fertilization processes. See Chapter 10 for additional details on cytogamy and genomic exclusion.

Cells emerging from conjugation have two macronuclei, which are distributed to the two daughter cells at the first binary fission as karyonides. These karyonidal clones are immature, incapable of conjugation for many fissions. The length of the immaturity period is about 40–60 fissions in inbred *T. thermophila*, but is longer in descendants of wildcaught cells and in other species of *Tetrahymena* where it is poorly characterized. The length is under genetic control but is sensitive to environmental variables, such as temperature and nutrition (Nyberg and Bishop, 1981). A relatively long immaturity period is associated with an outbreeding economy (Sonneborn, 1957), suggesting that all tetrahymenas are primarily outbreeders, a conclusion supported by multiple mating types in most species. Once cells reach sexual maturity, they are capable of conjugation with cells of a complementary mating type, though initially they may mate with only a subset of testers (Rogers and Karrer, 1985).

The mating type that is expressed upon maturity is determined either at fertilization (i.e., is synclonal, Table I) or during macronuclear development (i.e., is karyonidal, Table I, and as discussed below). In synclonal systems, which include most of the "pyriformis" complex of species, descendants of a single pair have the same mating type determined by the common genotype (Fig. 3). This enforces outbreeding as descendants of the same pair of conjugants are not capable of mating among themselves. In the karyonidal system, each new macronucleus is independently determined during its development for a mating type specified by inherited *mat* alleles that specify arrays of possible mating types. In *T. thermophila* alleles at the *mat* locus typically specify four to six of the possible seven mating types (Arslanyolu and Doerder, 2000). This means that it is possible for descendants of a given pair of conjugants to mate among themselves. This has been useful in the laboratory, but how often it contributes to inbreeding in nature is unknown.

If a species possesses a micronucleus, it is theoretically capable of breeding, though breeding has not been observed in all species (Table I). There are, however, numerous amicronucleate species, and amicronucleate tetrahymenas are relatively common in nature (see chapter by Doerder and Brunk). Amicronucleates have never been observed to mate when brought into the laboratory, but can now be identified by *COX*1 barcodes (Chantangsi *et al.*, 2007; Kher, *et al.*, 2011). The barcodes indicate that some amicronucleates have micronucleate counterparts, whereas many, including the classical *T. pyriformis,* do not. The high frequency of *Tetrahymena* amicronucleates contrasts to the rarity of amicronucleates of other ciliate species (Ng, 1986) and raises questions as to their evolutionary success. With few exceptions, asexuals derived from sexual species are transitory (Schön *et al.*, 2009). Amicronucleate tetrahymenas would, however, be able to take advantage of new mutations through the phenomenon of macronuclear assortment, a poorly understood process similar to genetic drift (Doerder *et al.*, 1992). Such mutations in sexual lineages would be erased at conjugation by the construction of new macronuclei, but in the absence of sex, new mutations would be subject to selection as they increase in frequency in the assortment process. It would be an interesting project to examine the relative importance of sex and assortment in *Tetrahymena* evolution.

Some *Tetrahymena* species exhibit life-cycle traits that distinguish them from other members of the genus (Table I). These include, as mentioned above, macrostome formation, histophagy, parasitism, and cyst formation, all of which are poorly studied, especially recently. The lack of resting cysts in most species raises significant questions regarding mechanisms of dispersal and overwintering.

V. Evolution of *Tetrahymena*

As suggested by its morphology, the genus *Tetrahymena* appears to be monophyletic as determined by phylogenetic analysis of nuclear SSUrRNA and mitochondrial *COX*1 sequences (Chantangsi and Lynn, 2008; Strüder-Kypke *et al.*, 2001). Two major groups, "borealis" and "australis," originally suggested by various rRNA sequences and LSUrRNA sequences (Nanney *et al.*, 1998), were supported by the SSUrRNA and *COX*1 sequences (Chantangsi and Lynn, 2008). Moreover, *COX*1 sequences distinguished among members of the "australis" group that had identical SSUrRNA sequences (Chantangsi *et al.*, 2007). The average *COX*1 sequence divergence was $\sim$10% among *bona fide* species, with intraspecific variation generally

<2%. The utility of these sequences for species identification is described in the chapter by Doerder and Brunk.

The "australis" and "borealis" groups or clades are congruent with the mechanism of mating-type determination, suggesting a major evolutionary divergence. The "australis" group is uniformly synclonal in mating-type determination, whereas, the "borealis" clade is karyonidal in the species that have been studied (Fig. 3, Table I) (Meyer and Nanney, 1987). The identical SSUrRNA sequences of members of the "australis" clade and the greater similarity of their *COX*1 sequences indicate more recent divergence. It also may be significant that very few of the "australis" groups are associated with amicronucleates found in natural populations (Doerder, unpublished). Indeed, the most abundant amicronucleate is *Tetrahymena borealis* followed by *T. elliotti*, both in the "borealis" clade.

The list of species in Table I is by no means exhaustive. Both Chantangsi *et al.* (2007) and Kher *et al.* (2011) reported new species based on *COX*1 differences of >5%. Doerder (unpublished) has similar evidence of numerous other species: based on these unpublished data, as a crude estimate as to the abundance of new species, among 454 isolates, 36% had *COX*1 barcodes indicating new species. Of the ~30 species represented by these isolates, some species are represented by multiple isolates indicating that they are common, whereas others are represented by a single isolate. These include both micronucleate and amicronucleate forms. Many of the amicronucleates have no micronucleate counterpart, suggesting both that they might be ancient and pointing to a critical need for more sampling. Indeed, since the sampling was geographically restricted largely to areas of the northeast USA, there are likely hundreds of additional *Tetrahymena* species worldwide. Some *Tetrahymena*-like cells isolated from water samples do not grow in the laboratory, suggesting that there may be an additional pool of tetrahymenas in nature (see chapter by Doerder and Brunk). As indicated above, there has been little recent research on parasitic or symbiotic species, another potential pool of species.

Phylogenetic trees based on nuclear SSUrRNA and mitochondrial *COX*1 genes are in general agreement (see Chantangsi and Lynn, 2008), though there are differences in detail that remain to be resolved, perhaps by the use of additional nuclear and mitochondrial genes. One area of interest concerns the genus *Colpidium*, which both trees show is more closely related to *Tetrahymena* than to *Glaucoma* with which *Tetrahymena* is often associated. While the SSUrRNA tree places *Colpidium* outside of the *Tetrahymena* genus, the *COX*1 tree places both *Colpidium campylum* and *Colpidium colpoda* within the tetrahymenas. Further work is needed to resolve this potentially interesting conflict.

VI. Perspective

The genus *Tetrahymena* has a rich history of important contributions to genetics, cell biology, and modern molecular biology, mostly from the single species *T. thermophila*. The genus appears to be especially speciose and possesses considerable plasticity as evidenced by the multiple appearances of traits, such as histophagy and macrostomy. The diversity of the genus is such that significant discoveries are likely to be made using other species. Therefore, we welcome the news that sequences of the macronuclear genomes of *T. malaccensis*, *T. elliotti*, and *T. borealis* are becoming available.

References

Allen, S. L., and Weremiuk, S. L. (1971). Intersyngenic variations in the esterases and acid phosphatases of *Tetrahymena pyriformis*. *Biochem. Genet.* **5**, 119–133.

Arslanyolu, M., and Doerder, F. P. (2000). Genetic and environmental factors affecting mating type frequency in natural isolates of *Tetrahymena thermophila*. *J. Eukaryot. Microbiol.* **47**, 412–418.

Batson, B. S. (1983). *Tetrahymena dimorpha* sp. nov. (Hymenostomatida: Tetrahymenidae), a new ciliate parasite of Simuliidae (Diptera) with potential as a model for the study of ciliate morphogenesis. *Phil. Trans. R. Soc. Lond., Ser. B* **301**, 345–363.

Batson, B. S. (1985). A paradigm for the study of insect-ciliate relationships: *Tetrahymena sialidos* sp. nov. (Hymenostomatida: Tetrahymenidae), parasite of larval *Sialis lutaria* (Linn.) (Megaloptera: Sialidae). *Phil. Trans. R. Soc. Lond., Ser. B* **310**, 123–144.

Borden, D., Miller, E. T., Whitt, G. S., and Nanney, D. L. (1977). Electrophoretic analysis of evolutionary relationships in *Tetrahymena*. *Evolution* **31**, 91–102.

Borden, D., Whitt, G. S., and Nanney, D. L. (1973). Electrophoretic characterization of classical *Tetrahymena pyriformis* strains. *J. Protozool.* **20**, 693–700.

Chantangsi, C., and Lynn, D. H. (2008). Phylogenetic relationships within the genus *Tetrahymena* inferred from the cytochrome *c* oxidase subunit 1 and the small subunit ribosomal RNA genes. *Mol. Phylogenet. Evol.* **49**, 979–987.

Chantangsi, C., Lynn, D. H., Brandl, M. T., Cole, J. C., Netrick, N., and Ikonomi, P. (2007). Barcoding ciliates: a comprehensive study of 75 isolates of the genus *Tetrahymena*. *Int. J. Syst. Evol. Microbiol.* **57**, 2412–2425. doi: 10.1099/ijs.0.64865-0.

Corliss, J. O. (1952). Le cycle autogamique de *Tetrahymena rostrata*. *Compt. Rend. Acad. Sci.* **235**, 399–402.

Corliss, J. O. (1954). The literature on *Tetrahymena*: its history, growth, and recent trends. *J. Protozool.* **1**, 156–169.

Corliss, J. O. (1951). On nomenclature for congeneric ciliates in the *Colpidium-Glaucoma-Leucophrys-Tetrahymena* group. *Soc. Syst. Zool. News Letter, No.* **5**, 9–10 (Abstr.).

Corliss, J. O. (1957). *Tetrahymena paravorax* n. sp., the first caudal-ciliated member of the genus referable to the *vorax-patula* complex. *J. Protozool.* **4**(Suppl.), 13 (Abstr.).

Corliss, J. O. (1960). *Tetrahymena chironomi* sp. nov., a ciliate from midge larvae, and the current status of facultative parasitism in the genus *Tetrahymena*. *Parasitology* **50**, 111–153.

Corliss, J. O. (1965). *Tetrahymena*, a ciliate genus of unusual importance to modern biological research. *Acta Protozool.* **3**, 1–20.

Corliss, J. O. (1970). The comparative systematics of species comprising the hymenostome ciliate genus *Tetrahymena*. *J. Protozool.* **17**, 198–209.

Corliss, J. O. (1971). Type-specimens for the ten species of *Tetrahymena* (Protozoa: Ciliophora). *Trans. Amer. Microsc. Soc.* **90**, 243–252.

Corliss, J. O. (1973a). Guide to the literature on Tetrahymena: a companion piece to Elliott's "General Bibliography. *Trans. Amer. Miscrosc. Soc.* **92**, 468–491.

Corliss, J. O. (1973b). History, taxonomy, ecology, and evolution of species of *Tetrahymena*. *In* "Biology of Tetrahymena," (A. M. Elliott, ed.), pp. 1–55. Dowden, Hutchinson and Ross, Stroudsburg, PA.

Corliss, J. O., and Daggett, P. -M. (1983). *Paramecium aurelia*" and "*Tetrahymena pyriformis*": current status of the taxonomy and nomenclature of these popularly known and widely used ciliates. *Protistologica* **19**, 307–322.

Corliss, J. O., and Dougherty, E. C. (1967). An appeal for stabilization of certain names in the protozoan family Tetrahymenidae (subphylum Ciliophora, order Hymenostomatida), with special reference to the generic name *Tetrahymena* Furgason, 1940. *Bull. Zool. Nomencl.* **24**, 155–185.
Doerder, F. P., Deak, J. C., and Lief, J. H. (1992). Rate of phenotypic assortment in *Tetrahymena thermophila*. *Develop. Genet.* **13**, 126–132.
Elliott, A. M., and Gruchy, D. F. (1952). The occurrence of mating types in *Tetrahymena*. *Biol. Bull. (Abstr.)* **103**, 301.
Elliott, A. M., and Nanney, D. L. (1952). Conjugation in *Tetrahymena*. *Science* **116**, 33–34.
Feng, X. L., Sun, Q., Cao, T. G., Li, L. Y., and Chen, Y. Z. (1988). The S1 strain of *Tetrahymena* from Shanghai – *Tetrahymena shanghaiensis* sp. nov. *Acta Zool. Sinica* **34**, 42–51.
Furgason, W. H. (1940). The significant cytostomal pattern of the "*Glaucoma-Colpidium* group," and a proposed new genus and species, *Tetrahymena geleii*. *Arch. Protistenk.* **94**, 224–266.
Gruchy, D. G. (1955). The breeding system and distribution of *Tetrahymena pyriformis*. *J. Protozool.* **2**, 178–185.
ICZN (International Commission on Zoological Nomenclature)(1970). Opinion 915. *Tetrahymena* Furgason, 1940 (Ciliophora, Hymenostomatida): Preserved under the plenary powers and related matters. *Bull. Zool. Nomencl.* **27**, 33–38.
Jerome, C. A., Simon, E. M., and Lynn, D. H. (1996). Description of *Tetrahymena empidokyrea* n. sp., a new species in the *Tetrahymena pyriformis* sibling species complex (Ciliophora, Oligohymenophorea), and an assessment of its phylogenetic position using small-subunit rRNA sequences. *Can. J. Zool.* **74**, 1898–1906.
Kher, C. P., Doerder, F. P., Cooper, J., Ikonomi, P., Achilles-Day, U., Küpper, F. C., and Lynn, D. H. (2011). Barcoding *Tetrahymena*: discriminating species and identifying unknowns using the cytochrome *c* oxidase subunit 1 (*cox*-1) barcode. *Protist* **162**, 2–13.
Kidder, G. W. (1941). The technique and significance of control in protozoan culture. *In* "Protozoa in Biological Research," (G. N. Calkins, and F. M. Summers, eds.), pp. 448–474. Columbia University Press, New York.
Kozloff, E. N. (1946). The morphology and systematic position of a holotrichous ciliate parasitizing *Deroceras agreste* (L.). *J. Morph.* **79**, 445–465.
Lwoff, A. (1923). Sur la nutrition des infusoires. *C. R. Acad. Sci., Paris* **176**, 928–930.
Lwoff, A. (1947). Some aspects of the problem of growth factors for protozoa. *Ann. Rev. Microbiol.* **1**, 101–114.
Lynn, D. H. (2008). *The Ciliated Protozoa: Characterization, Classification, and Guide to the Literature.* Springer, Dordrecht.
Lynn, D. H., Gransden, S. G., Wright, A. -D. G., and Josephson, G. (2000). Characterization of a new species of the ciliate *Tetrahymena* (Ciliophora: Oligohymenophorea) isolated from the urine of a dog: first report of *Tetrahymena* from a mammal. *Acta Protozool.* **39**, 289–294.
Lynn, D. H., Molloy, D., and Lebrun, R. (1981). *Tetrahymena rotunda* n. sp. (Hymenostomatida, Tetrahymenidae), a ciliate parasite of the hemolymph of *Simulium* (Diptera, Simuliidae). *Trans. Amer. Microsc. Soc.* **100**, 134–141.
McCoy, J. W. (1975). Updating the tetrahymenids. III. Natural variation in *Tetrahymena* setosa nov. comb. *Acta Protozool.* **14**, 253–262.
Meyer, E. B., and Nanney, D. L. (1987). Isozymes in the ciliated protozoan Tetrahymena. *In* "Isozymes in the Biological and Medical Sciences," (M. C. Rattazzi, J. G. Scandalios, G. S. Whit, eds.), pp. 61–101. A. R. Liss, New York.
Nanney, D. L., and McCoy, J. W. (1976). Characterization of the species of the *Tetrahymena pyriformis* complex. *Trans. Amer. Microsc. Soc.* **95**, 664–682.
Nanney, D. L., Park, C., Preparata, R., and Simon, E. M. (1998). Comparison of sequence differences in a variable 23S rRNA domain among sets of cryptic species of ciliated protozoa. *J. Eukaryot. Microbiol.* **45**, 91–100.
Ng, S. F. (1986). The somatic function of the micronucleus of ciliated protozoa. *In* "Progress in Protislolology," (J. O. Corliss, and D. J. Patterson, eds.), pp. 215–286. Biopress Ltd., Bristol.

Nyberg, D. (1981). Three new "biological" species of *Tetrahymena* (*T. hegewischi* n. sp., *T. sonneborni* n. sp., *T. nipissingi* n. sp.) and temperature tolerance of members of the "pyriformis" complex. *J. Protozool.* **28**, 65–69.

Nyberg, D., and Bishop, P. (1981). The immaturity interval in *Tetrahymena*: genetic and environmental sources of variation. *Develop. Genet.* **2**, 159–170.

Roque, M., Puytorac, Pde., and Savoie, A. (1971). Caractéristiques morphologiques et biologlques de *Tetrahymena bergeri* sp. nov., cilié hyménostome tetrahyménien. *Protistologica* **6**(year 1970), 343–351.

Rogers, M. B., and Karrer, K. M. (1985). Adolescence in *Tetrahymena thermophila*. *Proc. Natl Acad. Sci. U. S. A.* **82**, 436–439.

Schiftner, U., and Foissner, W. (1998). Revision of *Tetrahymena* species with caudal cilium. *J. Protozool.* **45**(Suppl), 15A (Abstr.).

Scherbaum, O. H., and Zeuthen, E. (1954). Induction of synchronous cell division in mass cultures of *Tetrahymena pyriformis*. *Exptl. Cell Res.* **6**, 221–227.

Schön, I., Martens, K., and van Dijk, P. (2009). *Lost Sex.* Springer, Dordrecht, p. 615.

Segade, P., Kher, C. P., Lynn, D. H., and Iglesias, R. (2009). Morphological and molecular characterization of renal ciliates infecting farmed snails in Spain. *Parasitology* **136**, 771–782 [doi:10.1017/S0031182009006040].

Simon, E. M., Meyer, E. B., and Preparata, R. M. (1985). New wild *Tetrahymena* from Southeast Asia, China, and North America, including *T. malaccensis, T. asiatica, T. nanneyi, T. caudata,* and *T. silvana* n. spp. *J. Protozool.* **32**, 183–189.

Simon, E. M., and Doerder, F. P. (1981). The unique position of the degenerating macronucleus in *Tetrahymena tropicalis*. *J. Protozool.* **28**, 203–205.

Sonneborn, T. M. (1939). *Paramecium aurelia*: mating types and groups; lethal interactions; determination and inheritance. *Amer. Nat.* **73**, 390–413.

Sonneborn, T. M. (1957). Breeding systems, reproductive methods, and species problems in Protozoa. *In* "The Species Problem," (E. Mayr, ed.), pp. 155–234. AAAS Publ., Washington, D.C.

Strüder-Kypke, M. C., Wright, A. -D. G., Jerome, C. A., and Lynn, D. H. (2001). Parallel evolution of histophagy in ciliates of the genus *Tetrahymena*. *BMC Evol. Biol.* **1**, 5 Available at: http://www.biomedcentral.com/1471–2148/1/5.

Thompson Jr., J. C. (1955). Morphology of a new species of *Tetrahymena*. *J. Protozool.* **2**(Suppl), 12A (Abstr.).

Williams, N. E., Buhse Jr., H. E., and Smith, M. G. (1984). Protein similarities in the genus *Tetrahymena* and a description of *Tetrahymena leucophrys* n. sp. *J. Protozool.* **31**, 313–321.

Zeuthen, E. (1953). Growth as related to the cell cycle in single cell cultures of *Tetrahymena pyriformis*. *J. Embryol. Exptl. Cell Morphol.* **1**, 239–250.

CHAPTER 3

Nuclear Dualism

Kathleen M. Karrer

Department of Biological Sciences, Marquette University, Milwaukee, Wisconsin, USA

This chapter is dedicated to Dr. Martin A. Gorovsky and Jody Bowen LaRose who, together with the members of the Gorovsky laboratory, developed methods to separate Tetrahymena *micro- and macronuclei thereby allowing for the molecular analysis of nuclear dualism. That was the first of many seminal contributions to the field of* Tetrahymena *molecular biology from the Gorovsky lab, including the development of biolistic transformation, a major impetus for the* Tetrahymena *genome project and the development of the Tetrahymena Gene Expression Database. The members of the ciliate community are grateful for Marty's rigorous science, his generous mentoring, and his invaluable contributions to the field of ciliate molecular biology and to Jody for the essential part she played in making it all work.*

0091-679X/10 $35.00
DOI 10.1016/B978-0-12-385967-9.00003-7

Abstract

Nuclear dualism is a characteristic feature of the ciliated protozoa. *Tetrahymena* have two different nuclei in each cell. The larger, polyploid, somatic macronucleus (MAC) is the site of transcriptional activity in the vegetatively growing cell. The smaller, diploid micronucleus (MIC) is transcriptionally inactive in vegetative cells, but is transcriptionally active in mating cells and responsible for the genetic continuity during sexual reproduction. Although the MICs and MACs develop from mitotic products of a common progenitor and reside in a common cytoplasm, they are different from one another in almost every respect.

I. Vegetative Cell Division

MICs and MACs replicate their DNA at different points in the cell cycle. MIC DNA is replicated in late anaphase (Doerder and DeBault, 1975; Woodard *et al.*, 1972); thus, there is virtually no micronuclear G1, and the MIC has a DNA content of 4C for essentially all of the cell cycle. Macronuclear S phase, on the other hand, occurs midway through the cell cycle (Charret, 1969). The macronuclear ribosomal RNA genes (rDNA), which consists of pairs of genes on a palindromic minichromosome (Engberg *et al.*, 1976; Karrer and Gall, 1976), replicate early in MAC S phase. This was shown both by EM autoradiography (Charret, 1969), and by molecular analysis of DNA replication in cells that were starved to synchronize the cell cycle and then refed to initiate DNA replication (Engberg *et al.*, 1972).

In vegetatively growing cells, the nuclei divide without dissolution of the nuclear membranes (Jaeckel-Williams, 1978). Division of the MIC occurs first, and the chromosomes separate on a nuclear spindle that assembles within the nuclear membrane (LaFountain and Davidson, 1979, 1980) (for details, see Chapter 5). The MAC divides amitotically, without functional centromeres. Multiple copies of each macronuclear chromosome are randomly partitioned between the two daughter cells. As a result of the random segregation of alleles, vegetative progeny of a cell that is heterozygous following conjugation become pure for one allele or the other within approximately 100 cell fissions (Orias and Flacks, 1975). This is the molecular basis of the genetic phenomenon called phenotypic assortment in which a heterozygous cell expresses the dominant allele immediately following conjugation, but clones of the vegetative progeny can express either the dominant or the recessive allele (Sonneborn, 1974).

Molecular experiments are in accord with the genetic data and support the idea that the MAC lacks centromeres. All eukaryotes studied to date have a gene encoding a variant histone H3 that is specifically associated with the centromeres. The *Tetrahymena* homolog, CNA1, is essential for vegetative growth and for maintenance of the MIC. GFP-tagged Cna1p and antibodies to the protein produce patterns of staining consistent with association of the protein with centromeres. These include 10 dots at the periphery of the MIC, as expected for the five pairs of micronuclear chromosomes; alignment of the dots at the metaphase plate during micronuclear mitosis, and localization at the pole-proximal edge of the nuclei at anaphase. No staining was observed in the MACs of vegetatively growing cells (Cervantes *et al.*, 2006; Cui and Gorovsky, 2006).

II. Sexual Reproduction

The respective roles of the MICs and MACs are reflected in the nuclear events of sexual reproduction or conjugation (Chapter 7, Figs. 6 and 7). Mating pairs form between two cells of different mating types. The MIC (Fig. 1A), as the germ line nucleus, undergoes meiosis (Fig. 1B–D). It elongates to as much as 50 times the

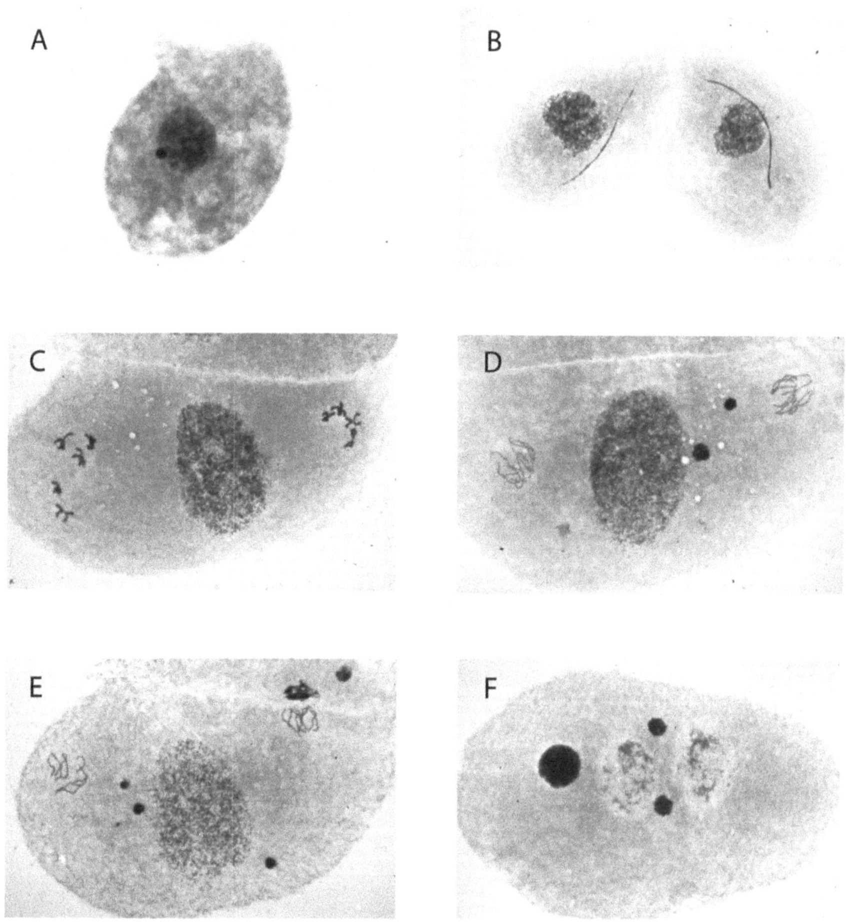

Fig. 1 Selected stages in the *Tetrahymena* life cycle. (A) Micro- and macronucleus in a vegetative cell. (B) Crescent micronucleus. (C) Meiosis I. (D) Meiosis II. (E) Prezygotic mitosis, just prior to nuclear exchange. (F) Early macronuclear anlagen development and condensation of the parental MAC. Figures (B–F) are courtesy of Joseph G. Gall. (For color version of this figure, the reader is referred to the web version of this book.)

usual diameter, to form a crescent MIC (Fig. 1B) that is analogous to the bouquet stage in multicellular organisms (Loidl and Mochizuki, 2009). The centromeres are at one end of the crescent MIC and the telomeres at the other end (Loidl and Scherthan, 2004; Mochizuki *et al.*, 2008). Chromosome pairing and recombination occur in the crescent MIC. Following the crescent stage, the chromosomes condense and meiotic divisions ensue (Fig. 1C and D). One of the four meiotic products is selected as the nucleus that will be inherited and the other three degenerate. The selected meiotic product undergoes mitosis to produce two identical pronuclei, one of which remains resident in the cell, and the other is transferred to the mating partner (Fig. 1E). The migratory and resident nuclei in each cell fuse, producing one zygotic MIC in each cell. The zygotic MIC undergoes two postzygotic mitoses. Two of the four nuclei develop into MICs and two develop into MACs (Fig. 1F). As the new MACs are developing, the parental MAC condenses and degenerates (Fig. 1F). How the destruction of the parental MAC is achieved along with the simultaneous preservation of the integrity of the MICs and the developing MACs is not well understood. Although a protein with homology to apoptosis inducing factor is associated with mitochondria and apparently plays a role in degradation of the old MAC (Akematsu and Endoh, 2010), most of the genes involved in apoptosis are absent from the *Tetrahymena* genome database (reviewed in detail in Chapter 5). Specific modifications to the nuclear envelope of the parental MAC have been detected, and it has been suggested that these may target an atypical lysosomal autophagy of that nucleus (Akematsu *et al.*, 2010).

Early events in conjugation are driven by transcription in the parental MAC. These include cell pairing, meiosis, exchange of nuclei, and the postzygotic mitoses. Several genes that are required for events at later stages of macronuclear development are also transcribed early in mating. As the new MAC, or macronuclear anlagen, develops it becomes transcriptionally active and produces transcripts required for the later stages of macronuclear development and sexual reproduction, including DNA replication and the transition from sexual reproduction to vegetative growth (Marsh *et al.*, 2001; Yin *et al.*., 2010).

III. Chromatin Structure

A. Structure of the Histone Genes

The chromatin of *Tetrahymena* is organized in nucleosomes that possess the standard complement of core and linker histones. In most eukaryotes, genes encoding the major histones H1, H2A, H2B, H3, and H4 are interspersed in cassettes that are tandemly repeated in the genome. In contrast, *Tetrahymena* histones are encoded by low-copy number genes that are dispersed in the micronuclear genome. They consist of two genes for linker histones, two for each of the major core histones, a single gene for the minor H2A variant, hv1, and one for each of the H3 variants, hv2 and Cnap1 (Cervantes *et al.*, 2006; Cui and Gorovsky, 2006; Liu and Gorovsky, 1996; Liu *et al.*, 1996; Thatcher *et al.*, 1994). Nuclear dualism and the structure of the histone genes in *Tetrahymena* have made the organism a particularly fertile ground for the study of the function of histone proteins. The structure and modification of histones in the MIC, which undergoes meiosis and mitosis, but is transcriptionally silent in vegetatively growing cells, can be compared to those in the amitotic, transcriptionally active MAC. In addition, the low-copy number of the histone genes has permitted genetic analysis by gene knockouts and gene replacements that are not possible in most eukaryotes.

B. Linker Histones

The average internucleosomal repeat length is 175 bp in the MIC and 202 bp in the MAC (Gorovsky *et al.*, 1977). This is likely to be due to differences in the constitution and modification of micronuclear and macronuclear histones. The most striking difference is between the linker histones in the MIC and MAC. Macronuclear H1 is encoded by the single copy gene, HHO. The protein of 163 amino acids is unusually small and lacks the central hydrophobic domain found in all other H1 proteins (Hayashi *et al.*, 1987; Wu *et al.*, 1986). Micronuclear linker histones consist of four proteins; α, β, γ, and δ, all encoded by the MLH gene (Wu *et al.*, 1994). The proteins are translated as a polypeptide, X, and the individual proteins are produced by specific proteolytic processing (Allis *et al.*, 1984). Both HHO and MLH are nonessential genes, since the respective knockout strains are viable and grow well (Shen *et al.*, 1995).

Although HHO and MLH are nonessential genes, DAPI staining showed that both micronuclear and macronuclear linker histones function in chromatin condensation (Shen *et al.*, 1995). A more surprising result relates to the role histone H1 plays in gene regulation. Although histone H1 acts as a general repressor of gene activity *in vitro* (Paranjape *et al.*, 1994), the overall transcriptional activity of the HHO knockout strain is not significantly different from that in the wild-type strain. However, histone H1 is required for both positive and negative regulation of specific inducible genes (Shen and Gorovsky, 1996), including a positive feedback regulation of CDC2, the gene encoding the kinase that phosphorylates histone H1 (Dou *et al.*, 2005; Song and Gorovsky, 2007).

C. Nucleosome core histones

Tetrahymena nucleosomes are composed of the conventional core histones, H2A, H2B, H3, and H4. However, the different biological roles of the MICs and MACs are reflected in their histone composition and modification.

Tetrahymena chromatin contains equimolar amounts of two major H2A proteins, H2A(1) and H2A(2). The proteins are slightly different from one another (Fusauchi and Iwai, 1983). However, neither of the genes encoding the major H2A histones is essential (Liu *et al.*, 1996), suggesting the two proteins may substitute for one another.

The chromatin of the transcriptionally active MAC contains a histone H2A variant, hv1, encoded by the HTA3 gene (White and Gorovsky, 1988; White *et al.*, 1988).

HTA3 is an essential gene; thus, hv1 performs a necessary function that cannot be supplied by either of the major H2A proteins. The hv1 protein is absent from the MIC except during the early stages of conjugation, when the MIC becomes transcriptionally active. It has been suggested that hv1 plays a role in establishing transcriptionally competent chromatin (Allis *et al.*, 1982; Stargell *et al.*, 1993).

There are three genes for histone H3 in *Tetrahymena*. HHT1 and HHT2 encode identical proteins (Horowitz *et al.*, 1987) and are transcribed only in growing cells. HHT3 encodes the minor histone variant, hv2, which differs in 16 amino acids from the major, abundant H3 proteins and is expressed constitutively (Bannon *et al.*, 1983). None of the genes encoding histone H3 are essential in *Tetrahymena*. However, in cells lacking HHT3, the HHT2 gene is transcriptionally activated in starved cells, where it would normally be downregulated. Furthermore, although the HHT1/HHT3 double knockout strains are viable, the HHT2/HHT3 double knockouts are not. This suggests that the constitutive expression of H3 replacement variants is more important than the differences in amino acid sequence, which may simply reflect an early evolutionary divergence (Yu and Gorovsky, 1997).

One difference between the core histones of the *Tetrahymena* MICs and MACs can be attributed to proteolytic processing. MICs contain two forms of histone H3: H3S, which is electrophoretically indistinguishable from macronuclear histone H3, and $H3^F$, which migrates more rapidly in an acid-urea polyacrylamide gel. $H3^F$ is derived from $H3^S$ by a specific proteolytic cleavage of six amino acid residues from the amino terminus of the protein (Allis and Wiggins, 1984). The physiological significance of $H3^F$ in the MIC is unknown.

As in most organisms, transcription of the histones is coordinated with S phase. As described above, the S phase of the MIC and the MAC occur at different points in the cell cycle. Since two unlinked genes encode the major variants of each of the *Tetrahymena* histones, it is tempting to speculate that the existence of gene pairs for each of the major histones is related to the specific DNA replication-associated deposition of the gene products in the two nuclei. This is true for the linker histones (Wu *et al.*, 1988); however, the situation is more complex in the case of histone H4. *Tetrahymena* histone H4 is encoded by two genes, HHF1 and HHF2, which produce identical proteins. Both genes are transcriptionally active in vegetatively growing cells, producing messages of different sizes with different flanking sequences (Bannon *et al.*, 1984; Horowitz *et al.*, 1987). *In situ* hybridization to RNA with gene-specific probes revealed that although HHF1 (formerly H4-1) transcription is limited to micronuclear S, HHF2 (formerly H4-II) is expressed during both micro- and macronuclear S (Yu *et al.*, 1987).

The earliest link between histone modification and transcriptional control was established in *Tetrahymena*. The core histones were shown to be acetylated in the transcriptionally active MAC, but not in the transcriptionally inert MIC (Vavra *et al.*, 1982). Histone acetyltransferase was purified from isolated MACs based on a gel activity assay (Brownell and Allis, 1995), and the gene was shown to have homology to the yeast transcriptional adaptor, Gcn5 (Brownell *et al.*, 1996). Subsequent *in vitro* and *in vivo* analysis demonstrated that yeast Gcn5 has HAT activity (Brownell *et al.*,

1996; Kuo *et al.*, 1998; Wang *et al.*, 1998). A histone deacetylase, encoded by the gene THD1, is also localized specifically to MACs (Wiley *et al.*, 2000). THD1 is an essential gene and plays a critical role in chromatin integrity in the MAC (Wiley *et al.*, 2005). Another histone modification that is associated with transcriptional activity is the methylation of histone H3 at lysine 4. This modification is also specific to MACs in *Tetrahymena* (Strahl *et al.*, 1999).

IV. Molecular Events of MAC Anlagen Development

As the macronuclear anlage develops, the genome undergoes a massive restructuring. The five micronuclear chromosomes are broken down into approximately 180 macronuclear chromosomes (E. Hamilton, personal communication). The single-copy ribosomal RNA gene (rDNA) is converted into a giant palindrome and amplified to a copy number of about 10,000. Approximately 6000 specific DNA elements, called IES (internal eliminated sequences) are removed from the genome, and the genome undergoes several rounds of endocycling, during which DNA replication occurs without cell division. Finally, the DNA in the MAC undergoes *de novo* DNA methylation. The molecular processes behind each of these events will be described in turn.

A. Chromosome Breakage

During macronuclear anlagen development, the five micronuclear chromosomes are broken down into about 180 macronuclear chromosomes ranging in size from 20 kb to over 1500 kb. Chromosome breakage is site specific. The macronuclear chromosomes, which are roughly comparable to yeast chromosomes in size, can be separated on agarose gels by pulsed field electrophoresis. The pattern of the fragmented chromosomes is constant, and specific DNA probes reproducibly recognize chromosomes of the same size (Altschuler and Yao, 1985; Conover and Brunk, 1986).

A 15-bp chromosome breakage sequence (CBS) has been identified that is necessary and sufficient for chromosome breakage. This was shown by *in vivo* analysis of DNA rearrangement on constructs (Yao *et al.*, 1990) and was confirmed by genetic analysis when a mutation in the CBS 3′ to the rRNA gene was shown to affect its excision during macronuclear anlagen development (Kapler and Blackburn, 1994). CBS sequences have a quite long and stringent sequence requirement (Fan and Yao, 2000), with a consensus sequence of TAAACCAACCTCTTT, but almost half of the CBS have some variation in the sequence, and only a 10-bp core sequence (underlined) is strictly conserved (Hamilton *et al.*, 2006).

B. rDNA Amplification

The micronuclear rDNA of *Tetrahymena*, unlike most organisms, is single copy (Yao and Gall, 1977). All of the macronuclear rDNA molecules are produced from a single germ-line gene. This affords a unique opportunity for mutation and genetic analysis that is not possible in most other organisms, where the rRNA genes are highly repeated in the germ line.

CBS are located on either side of the rDNA in the micronuclear chromosome. The gene is excised during macronuclear anlagen development and converted to a giant inverted repeat (Butler *et al.*, 1995; Engberg *et al.*, 1976; Karrer and Gall, 1976). The 20-kb rDNA minichromosome is amplified to a copy number of about 10,000 per cell (Yao and Gorovsky, 1974).

C. Telomeres

Following chromosome breakage, new telomeres are added to the ends of the macronuclear chromosomes. The first solution of the structure of eukaryotic telomeres was in *Tetrahymena* (Blackburn and Gall, 1978). Macronuclear telomeres are composed of 50–80 tandem repeats of the sequence 5′-G4T2-3′.

Micronuclear telomeres are somewhat more complex than their macronuclear counterparts in several respects (Kirk and Blackburn, 1995). The terminal GT tracts are about seven times longer than those of the macronuclear telomeres, and the centromere-proximal region of the tract is composed of about 0.5–1.0 kb of homogeneous G4T3 repeats. The telomere-associated sequences immediately adjacent to the G4T3 array are relatively GC rich and 55–87% identical to each other. The more complex structure of micronuclear telomeres may be related to a telomere function that is not required in the MAC, such as telomere localization or chromosome segregation.

One of the classic problems of modern molecular biology that was solved in *Tetrahymena* had to do with telomere synthesis. All of the known DNA polymerases synthesize DNA in a 5′ to 3′ direction and require a primer. How then, is the synthesis of the 5′ end of the DNA completed? Telomeric DNA is synthesized by telomerase (Greider and Blackburn, 1985), a ribonucleoprotein complex containing a 159 nucleotide RNA. Telomerase is essentially a reverse transcriptase in which the RNA component of telomerase serves as the template for synthesis of the GT rich strand of the telomere (Autexier and Greider, 1994; Greider and Blackburn, 1989; Yu *et al.*, 1990). There is a vast literature on the structure of telomeres and telomerase, which has recently been reviewed (Wyatt *et al.*, 2010).

D. IES Elimination

Macronuclear anlagen development involves the elimination of approximately 6000 specific elements (IES) from the genome (Yao *et al.*, 1984). The removal of these elements is interstitial and is accompanied by ligation of the flanking sequences. The majority of the IES are repeated in the micronuclear genome (Yao and Gorovsky, 1974), and they are completely eliminated from the macronuclear genome.

Many of the IES resemble transposable elements in structure. The largest to be described to date are the 22-kb TLR elements (for Tetrahymena long repeat) (Wells et al., 1994; Wuitschick et al., 2002). This was the first example of a group of elements called Maverick elements that have subsequently been found in various organisms including nematodes, zebrafish, and fungi (Pritham et al., 2007). Maverick elements are characterized by a 5–6 bp target site duplication, long subterminal inverted repeats and a number of conserved open reading frames.

Another family of IES, the REF elements, bears structural resemblance to non-LTR retrotransposons (Fillingham *et al.*, 2004). The *Tetrahymena* elements contain two open reading frames. One of them encodes a novel protein, but the other, ORF2, encodes a deduced protein with an apurinic/apyrimidinic endonuclease (APE) domain and a reverse transcriptase (RT) domain, typical for non-LTR elements in other species. Nucleotide substitutions in the APE and RT domains of the REF elements, and in several of the open reading frames of the TLR elements, including one encoding a putative integrase gene (Gershan and Karrer, 2000; Wuitschick *et al.*, 2002), are highly nonrandom. The vast majority of the nucleotide polymorphisms occur in the third nucleotide of the codon, suggesting that at some point the genes were under selective pressure to encode a functional enzyme.

Two additional classes of elements that resemble transposable elements in other systems are IES in *Tetrahymena*. Inserted in the TLR elements, there are several examples of an element that has a single open reading frame. The open reading frame has a domain that resembles the nuclease domain of HNH endonucleases and a putative DNA binding motif similar to the apetala2 domain found primarily in plants (Wuitschick *et al.*, 2004). The *Tetrahymena* genome project revealed the presence of elements resembling Tc1 elements of *Caenorhabditis elegans* (Eisen *et al.*, 2006).

Much of the molecular analysis of IES elimination has been done on short, single-copy elements of a few hundred base pairs that do not contain open reading frames. The primary subjects of these studies have been the M and R elements (Austerberry and Yao, 1988). Additional small elements include the mse2.9 element, and the C, H, RP, RR, and B elements (Chau and Orias, 1996; Hüvös *et al.*, 1998; Katoh *et al.*, 1993; Li and Pearlman, 1996.

Most IES differ from some transposable elements in that their excision from the genome is imprecise. Some IES display microheterogeneity of 10–20 bp at the sequence junction (Austerberry *et al.*, 1989; Heinonen and Pearlman, 1994; Patil *et al.*, 1997). For some elements, rearrangement can occur at alternative junctions over a range of a few hundred base pairs (Austerberry and Yao, 1988; Chau and Orias, 1996; Wells *et al.*, 1994).

Partial sequencing of the micronuclear genome revealed that IES are underrepresented in exon regions, as expected for elements that undergo imprecise excision. However, a novel class of small IES was discovered that undergo precise excision, and sequences from at least one of these IES are included in a biologically stable RNA that is transcribed during conjugation. Thus, there may be some cases where IES are protein coding (Fass *et al.*, 2011).

It is not known whether any of the eliminated elements are essential for life processes in *Tetrahymena*. However, it seems *Tetrahymena* has apparently evolved the ultimate method to silence transposable elements, which is to remove them from the somatic genome entirely (Fillingham and Pearlman, 2004; Yao *et al.*, 2003). Since the excision of most IES is imprecise, their elimination from the MAC does not solve the problem of invasion of transposons into coding sequences in the germ-line MIC. It does, however, preclude the spreading of these elements in the somatic genome.

Sequence-specific recognition between the IES and the elimination machinery is apparently not required, because foreign DNA introduced into the MIC can be recognized as MIC-specific and eliminated from the developing macronuclear genome. Although single copies of the Neo gene can be eliminated as IES, there is a position effect such that a Neo is not eliminated from other sites unless there are additional copies elsewhere in the micronuclear genome (Howard-Till and Yao, 2007; Liu *et al.*, 2005; Yao *et al.*, 2003). Thus, repetition of a sequence in the MIC promotes its elimination from the developing MAC (Liu *et al.*, 2005).

Elimination of IES is an RNA-mediated event, and many of the components are similar to those required for RNA interference in other organisms. The "Scan RNA" model has been proposed to describe the molecular processes (Mochizuki and Gorovsky, 2004c; Schoeberl and Mochizuki, 2011). Briefly, the model proposes that the entire micronuclear genome is bidirectionally transcribed early in sexual reproduction. The double-stranded RNA is digested by a Dicer-like enzyme to 28 bp scan RNAs (scRNA). These are exported to the cytoplasm where they are incorporated into a complex containing an argonaute family protein, Twi1p and a number of additional proteins. The complex migrates to the parental MAC, where the genome is "scanned" to determine which sequences are present. Complexes containing scRNAs with homology to macronuclear sequences are degraded. The remaining complexes, containing scRNAs complementary to MIC-limited sequences, are then transported the developing macronuclear anlagen, where they target the formation of specialized heterochromatin. The heterochromatic DNA is excised from the somatic genome and degraded.

The scan RNA model incorporates a wide variety of molecular data. Although the MIC is transcriptionally inactive during vegetative growth, micronuclear transcription occurs early in meiosis (Martindale *et al.*, 1985; Sugai and Hiwatashi, 1970). A subunit of RNA polymerase II is localized to the MIC in a developmental stage specific manner, suggesting that the transcription is catalyzed by RNA pol II (Mochizuki and Gorovsky, 2004b). Transcription of the M and R elements was shown to be bidirectional, suggesting that MIC transcription produces double-stranded RNA molecules (Chalker and Yao, 2001).

Transcripts of a Dicer-like gene (DCL1), transcribed in the parental MAC, appear early in mating. Dcl1p is localized in the MIC and is required for production of 28-bp scRNAs. In matings between somatic knockouts of DCL1, germ line limited transcripts accumulate, scRNAs are not produced, IES elimination fails and the progeny die (Malone *et al.*, 2005; Mochizuki and Gorovsky, 2005).

The scRNAs are exported to the cytoplasm, where they associate with the argonaute-like protein, Twi1p (Mochizuki *et al.*, 2002). Twi1p is required to stablilize the scRNAs. Another protein, Giw1p, associates with Twi1p and is required for transport of Twi1p into the parental MAC. Site-directed mutagenesis showed that a "slicer" activity of Twi1p is required for association with Giw1p. It has been proposed that the slicer activity removes the passenger strand of the double-stranded scRNAs, which converts the complex to a conformation that is recognized by Giw1p for binding and transport to the parental MAC (Noto *et al.*, 2010). Once transported to the parental MAC, the single-stranded scRNAs are stabilized by 2′-*O*-methylation at their 3′ termini, catalyzed by the *Tetrahymena* HEN1 homolog (Kurth and Mochizuki, 2009).

Initially, the population of scRNAs appears to contain sequences homologous to the entire micronuclear genome. As conjugation proceeds, the scRNAs are enriched for sequences homologous to IES (Aronica *et al.*, 2008; Mochizuki and Gorovsky, 2004a). This is likely due to the selective degradation of scRNAs with homology to sequences present in the parental MAC. A putative RNA helicase, encoded by the gene EMA1, is required for interaction of Twi1p with chromatin in the parental MAC. It has been postulated that this interaction is mediated by base pairing between scRNAs in the Twi1p complex and nascent RNAs in the parental MAC. In EMA1 knockouts, selective loss of scRNAs homologous to Mac-destined sequences (MDS) is abolished (Aronica *et al.*, 2008).

The scanning of the macronuclear genome by Twi1p complexes as proposed in the scan RNA model would explain a striking epigenetic phenomenon in IES elimination. The M and R elements are excised from the developing MAC of wild-type cells. However, these elements can be artificially introduced into the MAC. The presence of the normally MIC-limited sequences in the parental MAC of mating cells greatly reduces the efficiency of the elimination of the respective element from the developing MAC in the progeny (Chalker and Yao, 1996). Furthermore, IES elimination is blocked in a wild-type cell when it is mated to a cell with elements that are normally IES in the MAC, showing that the inhibition in transferred between mating cells. This is inconsistent with genetic imprinting models and suggests that the epigenetic effect is mediated by small molecules that can be transferred through the junction between mating cells (Chalker *et al.*, 2005). The most likely candidate is the Twi1p complexes.

The efficiency of IES elimination may depend on an appropriate ratio of scRNA to noncoding nascent RNAs in the parental MAC. Injection of dsRNA complementary to MAC-destined sequences results in the inappropriate elimination of those sequences from the developing MAC anlagen (Yao *et al.*, 2003). This might be explained if the dsRNA is converted to an excess of scRNAs, which cannot be efficiently removed during the scanning process. Abundance of scRNA might also explain the position effect on foreign DNA sequences in the MIC (Howard-Till and Yao, 2007; Liu *et al.*, 2005), if some regions are transcribed at a higher rate than others.

After the scanning process is complete, the remaining Twi1p complexes, with the scRNAs complementary to IES, are transported to the developing MAC (Mochizuki *et al.*, 2002). Here, as in the parental MAC, Twi1p complexes interact with noncoding transcripts in a manner that is dependent on the helicase Ema1p (Aronica *et al.*,

2008). The probable role of Twi1p complexes in the macronuclear anlagen is to target the formation of heterochromatin on IES. However, the mechanism of targeting and the interacting heterochromatin components are unknown.

An abundant phosphoprotein, Pdd1p (programmed DNA deletion) has a dynamic subcellular distribution similar to Twi1p, where it is first found in the parental MAC and subsequently transferred to the macronuclear anlagen (Madireddi *et al.*, 1994). Pdd1p has two chromodomains, often found in proteins associated with heterochromatin. Pdd1p and three other abundant proteins, the chromodomain protein Pdd3p, and the novel proteins Pdd2p and Lia1p are present in electron dense bodies that colocalize with IES in the developing macronuclear anlagen and are believed to be the site where IES elimination takes place (Nikiforov *et al.*, 2000; Rexer and Chalker, 2007; Smothers *et al.*, 1997). PDD1, PDD2, and LIA1 are all required in the parental MAC for IES elimination (Coyne *et al.*, 1999; Nikiforov *et al.*, 1999; Rexer and Chalker, 2007).

Once targeted to the IES by the Twi1p complex, chromodomain proteins and a histone methyltransferase may be responsible for propagating the spread of heterochromatin along the chromosome. One popular model to explain heterochromatin spreading in multicellular organisms is supported by a large body of experimental evidence (Bannister *et al.*, 2001). The model proposes that the chromodomain protein HP1 binds to histone H3 methylated at the lysine 9 residue (H3K9Me). HP1 recruits a histone methyltransferase, which methylates H3K9 on the adjacent nucleosome, providing in turn a new binding site for HP1. Thus, the heterochromatin structure is propagated along the chromosome until it reaches a barrier to heterochromatin spreading. A similar mechanism may be responsible for establishing heterochromatic structure over the IES in *Tetrahymena*.

Tethering of Pdd1p to a MAC-destined sequence in the macronuclear anlagen is sufficient to induce elimination of the target sequence, suggesting that, once localized to the IES, Pdd1p alone is sufficient to recruit all of the necessary machinery for IES elimination (Taverna *et al.*, 2002). Emerging evidence argues for a functional relationship between Pdd1p and H3K9Me.

In *Tetrahymena*, H3K9 methylation occurs specifically during conjugation, where it is localized with the electron dense DNA elimination structures in the macronuclear anlagen (Taverna *et al.*, 2002). Replacing the lysine 9 of histone H3 with glutamine prevents methylation and impairs IES elimination (Liu *et al.*, 2004). The chromodomain proteins Pdd1p and Pdd3p both bind H3K9Me *in vitro*, and loss of Pdd1p greatly reduces the level of H3K9 methylation (Taverna *et al.*, 2002), as would be expected if Pdd1p was required to recruit the histone methyltransferase.

The histone methyltransferase that catalyzes H3K9 methylation in *Tetrahymena* has not yet been identified, but some hints arise from a study of a different histone H3 modification. H3K27Me is another marker of heterochromatin in multicellular organisms. In Drosophila, this modification is achieved through the activity of E (z), a SET domain histone methyltransferase (Czermin *et al.*, 2002). There are three homologs of Drosophila E(z) in *Tetrahymena*. One of them, EZL1, is expressed specifically during conjugation and the gene product, Ezl1p is localized to the chromatin elimination bodies in macronuclear anlagen. EZL1 is required for both IES elimination and chromosome breakage. In matings between EZL1p somatic knockouts, both H3K9 and H3K27 methylation are abolished in the macronuclear anlagen. Thus, H3K27Me may be an upstream regulator of H3K9Me or, alternatively, Ezl1p may catalyze both modifications (Liu *et al.*, 2007). Possible interactions between Ezl1p and the chromodomain proteins Pdd1p or Pdd3p are under investigation (Y. Liu, personal communication).

One facet of IES elimination that is not well understood is the role of flanking sequences in the chromosomal DNA. Constructs introduced into the developing macronuclear anlagen undergo DNA rearrangement in a manner very similar to the chromosomal deletions. This has provided a useful tool to study DNA sequence requirements for IES elimination. For the M rearrangement, a 10-bp A_5G_5 sequence located in the flanking DNA at a distance of 45–50 bp from the rearrangement junction specifies the junction site (Godiska and Yao, 1990; Godiska *et al.*, 1993). Curiously, A_5G_5 repeats have not been found in the vicinity of any other IES, although less well-defined sequences in the flanking DNA also seem to direct deletion of the R (Chalker *et al.*, 1999; Fillingham *et al.*, 2001) and TLR elements (Patil and Karrer, 2000). These sequences also seem to act at a short distance from the rearrangement boundary. The interpretation of these data became more complicated when it was found that *Tetrahymena* IES can be eliminated from constructs without any natural flanking sequences (Wuitschick and Karrer, 2003) and foreign sequences can be eliminated from sites in the *Tetrahymena* genome where there are no endogenous IES, and thus no known flanking sequences that would normally promote deletion (Liu *et al.*, 2005; Yao *et al.*, 2003). In addition, there is a strain variant that has an additional 1.8 kb of DNA at one end of the HI IES. In this chromosome, approximately 1.5 kb of the additional sequence is deleted with the HI IES, but 300 bp is retained, meaning that the new deletion boundary is 300 bp from the boundary in the B strain cells (Hüvös *et al.*, 1998). One possible explanation of these observations is that flanking sequences delimit the boundaries of an IES through structural features rather than sequence specific protein binding. If the IES is incorporated into elimination structures via spreading of the heterochromatin along the chromosome, then perhaps that chromatin spreading is limited by structural features in the chromatin such as a bend or a kink in the DNA, or a relatively long distance between adjacent nucleosomes. Such features could result as a secondary effect of various sequences along the chromosome that have been identified as cis-acting sequences for IES elimination in the flanking DNA. Sequences that confer these features might occur at random along the chromosome, where they would limit the spreading of heterochromatin from newly inserted foreign sequences.

The IES appear to be excised from the chromosome as linear molecules. Sensitive experiments involving ligation-mediated PCR of DNA from mating cells detected developmental-stage specific, naturally occurring breaks at the junction of the M and R elements. The molecules have 4-bp staggered cuts with a recessed hydroxyl adenosine at the 3′ end (the A rule) and a 5′ phosphate on the protruding strand

(Saveliev and Cox, 1995, 1996). A model was proposed for IES excision by a mechanism involving a double-stranded break at one end of the element, and transesterification initiated by a 3′ adenosine. The branched intermediate is subsequently cleaved within 15–16 nucleotides of the IES boundary, releasing the linear IES (Saveliev and Cox, 2001).

A strong candidate has been identified for the enzyme that produces the double-stranded break for IES excision. A macronuclear gene in *Tetrahymena* encodes a nuclease that is similar to the transposase of the piggyBac transposons, found in a wide variety of phylogenetically diverse organisms. The gene, TBP2, encodes an endonuclease that produces double-stranded breaks with 4-bp 5′ protruding ends *in vitro*. The gene is transcribed specifically at the time of IES elimination, and the protein product, Tbp2p, localizes to the electron dense structures in the macronuclear anlagen. Knockdown of the TBP gene activity by hairpin RNA results in major defects in assembly of the DNA elimination structures, IES elimination, and chromosome breakage (Cheng *et al.*, 2010).

A number of additional genes have been identified that are required for IES elimination, although their roles are not yet understood. These were identified on the basis of association of the protein products with Twi1p, developmentally specific expression or localization in the macronuclear anlagen. Some encode deduced proteins that are related to Piwi-interacting proteins in other systems (Bednenko *et al.*, 2009) and some encode novel proteins (Matsuda *et al.*, 2010; Yao *et al.*, 2007).

E. Endocycling

Beginning at about the same time as IES elimination, the genome in the macronuclear anlagen undergoes multiple rounds of DNA replication without nuclear division, resulting in the increased ploidy of the MAC. Two rounds of DNA replication occur immediately after the postzygotic divisions of the MIC. Then there is a pause in DNA replication for about 4 h (Allis *et al.*, 1987) after which it resumes. Over the course of the first few fissions in the progeny cells, the DNA content rises to about 128°C (Doerder and DeBault, 1975; Marsh *et al.*, 2001), but it subsequently drops to the 50°C characteristic of vegetative growth.

Endoreplication of the *Tetrahymena* genome occurs in distinct rounds of DNA replication that are separated by gap phases (Yin *et al.*, 2010). Thus, it is an example of endocycling, a process that is conserved in evolution and occurs in specific tissues of multicellular organisms, including Drosophila, Arabidopsis, *C. elegans*, and mammals (reviewed in Edgar and Orr-Weaver, 2001).

In *Tetrahymena*, endocycling is controlled by the gene ASI2 (for anlagen stage-induced gene 2). Various domains in the deduced ASI2 protein are similar to those in bacterial signal transduction receptors. ASI2 is transcribed early in mating in the parental MAC, and at later stages in the macronuclear anlagen. The absence of ASI2 in the parental MAC results in delayed endocycling and reduced fertility (Yin *et al.*,

2010). Lack of ASI2 in the macronuclear anlagen causes the arrest of endocycling and lethality of the progeny (Li *et al.*, 2006).

F. DNA Methylation

A relatively late modification of the macronuclear genome, occurring shortly after IES elimination (Harrison and Karrer, 1985), is the *de novo* methylation of about 0.8% of the adenine residues to N6-methyladenine (Gorovsky *et al.*, 1973). In the *Tetrahymena* genome, which is approximately 75% AT, this amounts to about one methylated adenine per 165 bp of DNA. Methylation occurs at the sequence 5′-NAT-3′ (Bromberg *et al.*, 1982), and patterns of methylation are specific and reproducible. Some sites are methylated on 90% or more of the macronuclear DNA molecules (Harrison *et al.*, 1986). Other, partially methylated, sites are modified in a lower percentage of the molecules. The level of methylation is characteristic of the site and consistent between clonal cell lines. Although the molecules containing one partially methylated site have been shown to undergo phenotypic assortment, molecules that are unmethylated at that site do not assort to purity. Thus, it is likely that the maintenance methylase of *Tetrahymena* has a *do novo* activity in vegetatively growing cells (Capowski, 1989).

DNA methylation is related to chromatin structure in *Tetrahymena*. DNA sequence is not sufficient to determine methylation because sequences from a fully methylated chromosomal site were unmethylated when they were moved to the rDNA (Van Nuland *et al.*, 1995). Nucleosomes are phased over a significant portion of the *Tetrahymena* genome, and methyladenine is preferentially localized in linker DNA (Pratt and Hattman, 1981). However, both nucleosome phasing (Karrer and VanNuland, 1999) and DNA methylation are independent of histone H1 (Karrer and VanNuland, 2002).

Adenine methylation is very unusual in eukaryotes. Cytosine is the predominant methylated base in the genome of multicellular eukaryotes, where it plays an important role in gene regulation. Although methyladenine has been found in the genomes of various ciliates, its biological role is unknown. No change in methylation state was detected for sites in the vicinity of genes whose transcriptional activity varies with the physiological state of the cell (Karrer and Stein-Gavens, 1990). In prokaryotes, methyladenine plays a part in restriction-modification systems (Efimova *et al.*, 1988), in the discrimination between template and daughter strands for mismatch repair (Modrich, 1989), and in the control of DNA replication (Russell and Zinder, 1987). Genes encoding DNA methylases are difficult to identify by *in silico* analysis because they have short, poorly conserved domains, and the proteins are more conserved at the level of tertiary structure than primary amino acid sequence. Failure to identify the gene encoding *Tetrahymena* DNA methylase has impeded the investigation of the function of this modification. Whatever the function of adenine methylation in the *Tetrahymena* MAC, it must be one that is not required or is served in some other manner in the MIC.

V. Nuclear Transport

The structural and functional differences between the MIC and MAC necessitate the specific localization of macromolecules in the two nuclei. For example, the micronuclear and macronuclear linker histones are specifically localized to the respective nuclei (Dou *et al.*, 2002; Wu *et al.*, 1986), minor histone variants are specifically localized in transcriptionally active nuclei (Stargell *et al.*, 1993; Wenkert and Allis, 1984), centromeric histones are specific to the MIC (Cervantes *et al.*, 2006) and the various proteins involved in chromatin elimination are subject to elaborate developmental stage-specific trafficking between nuclei (Madireddi *et al.*, 1994; Mochizuki *et al.*, 2002; Rexer and Chalker, 2007). In some cases, the same protein will be localized to the MIC at one developmental stage and to the MAC at others. For example, a subunit of RNA polymerase II, encoded by RPB3, is localized in the MIC during meiotic prophase, and in the MAC during vegetative growth (Mochizuki and Gorovsky, 2004b). Emerging evidence suggests that much of the specific localization is due to the nuclear import apparatus.

The gateways to the nucleus are large nuclear pore complexes (NPC), a 44–60 MDa complex composed of approximately 30 proteins, nucleoporins. The micro- and macronuclear membranes have similar density of nuclear pore complexes, at about $45/mm^2$ (Iwamoto *et al.*, 2009). To investigate the molecular composition of *Tetrahymena* NPC, 13 candidate genes from the *Tetrahymena* genome database with structural domains and/or phenylalanine-glycine (FG) repeats characteristic of nucleoporins in other organisms were designated as nucleoporins based on the localization of GFP-tagged proteins around the periphery of the nuclei (Iwamoto *et al.*, 2009). Of the 13 proteins tested, nine localized to both the MICs and the MAC s. The Nup50 homolog was present in the nucleoplasm of both nuclei, consistent with its localization in other organisms.

The most striking finding was in the localization of the Nup98 homologs, two of which were exclusively found in the MIC and two in the MAC. The macronuclear Nup98 proteins contain GLFG repeats, which are characteristic of nucleoporins in mammals and yeast. However, the MIC Nup98 homologs have novel NIFN repeats. The GLFG or NIFN repeats are concentrated in the N-terminal half of the protein. These proteins also have C-terminal Nucleoporin2 domains, which are generally required for targeting to the NPC. In domain-swapping experiments, the N- and C-terminal halves were exchanged between the Nup98 homologs. The chimeric proteins were localized to the nucleus corresponding to the C terminal half of the protein, suggesting that this region of the protein is responsible for nucleus-specific targeting to the NPC.

The chimeric proteins had revealing effects on nuclear import. In cells with the "BigMic" gene, consisting of the macronuclear Nup98 N-terminus and the micronuclear Nup98 C-terminus, the level of the micronuclear linker histone was drastically reduced in the MIC. Similarly, localization of macronuclear linker histone H1 was dramatically reduced in the MACs of cells expressing the "BigMac" gene and, notably, micronuclear linker histone did not accumulate in those nuclei. These results suggest that the role of the GLFG/NIFN repeats may be to block the inappropriate import of proteins into the respective nuclei, rather than to facilitate the import of proteins (Iwamoto *et al.*, 2009).

Another component of the nuclear import apparatus are the importins. The importins generally consist of a family of a dozen or more genes which fall into two classes, imp α and imp β. Typically, most of the family members belong to the imp β class. There is only one imp α gene in yeast and three in humans. In *Tetrahymena*, however, the imp α family is expanded to at least 10 putative members. Nuclear proteins with a classical nuclear localization signal (NLS) associate with imp α and imp β in a ternary complex that binds through the imp β component to nucleoporins in the NPC for import of the nuclear protein. Alternatively, nuclear proteins with a nonclassical NLS can associate with imp β without the imp α carrier (reviewed in Malone *et al.*, 2008).

The intracellular localization of imp α and imp β homologs in *Tetrahymena* was determined by analysis of GFP fusion proteins. Most of the 11 imp β-like proteins examined localized to both the MIC and the MAC. In contrast, the imp a importins were nucleus specific. Nine of 13 proteins with homology to imp α localized specifically to the MIC. Ima1p localized strongly to the MAC, and may be the primary macronuclear imp α. The various micronuclear imp α proteins do not seem to be redundant in function, because the IMA10 gene is essential for micronuclear mitosis (Malone *et al.*, 2008).

Although a great deal of work is yet to be done before we fully understand the targeting of nuclear proteins to the MIC and the MAC, it is evident that multiple components of the nuclear import machinery contribute to specific nuclear targeting. Analysis of nucleus-specific protein import in *Tetrahymena* will undoubtedly lead to a better understanding of nuclear import in general.

References

Akematsu, T., and Endoh, H. (2010). Role of apoptosis-inducing factor (AIF) in programmed nuclear death during conjugation in *Tetrahymena thermophila*. *BMC Cell Biol.* **11**, 13.

Akematsu, T., Pearlman, R. E., and Endoh, H. (2010). Gigantic macroautophagy in programmed nuclear death of *Tetrahymena thermophila*. *Autophagy* **6**, 901–911.

Allis, C. D., Allen, R. L., Wiggins, J. C., Chicoine, L. C., and Richman, R. (1984). Proteolytic processing of H1-like histones in chromatin: a physiologically and developmentally regulated event in *Tetrahymena micronuclei*. *J. Cell Biol.* **99**, 1669–1677.

Allis, C. D., Colavito-Shepanski, M., and Gorovsky, M. A. (1987). Scheduled and unscheduled DNA synthesis during development in conjugating Tetrahymena. *Dev. Biol.* **124**, 469–480.

Allis, C. D., and Wiggins, J. C. (1984). Proteolytic processing of micronuclear H3 and histone phosphorylation during conjugation in *Tetrahymena thermophila*. *Exp. Cell Res.* **153**, 287–298.

Allis, C. D., Ziegler, Y. S., Gorovsky, M. A., and Olmsted, J. B. (1982). A conserved histone variant enriched in nucleoli of mammalian cells. *Cell* **31**, 131–136.

Altschuler, M. I., and Yao, M. -C. (1985). Macronuclear DNA of *Tetrahymena thermophila* exists as defined subchromosomal-sized molecules. *Nucl. Acids Res.* **13**, 5817–5831.

Aronica, L., Bednenko, J., Noto, T., DeSouza, L. V., Siu, K. W., Loidl, J., Pearlman, R. E., Gorovsky, M. A., and Mochizuki, K. (2008). Study of an RNA helicase implicates small RNA-noncoding RNA interactions in programmed DNA elimination in *Tetrahymena*. *Genes Dev.* **22**, 2228–2241.

Austerberry, C. F., Snyder, R. O., and Yao, M. -C. (1989). Sequence microheterogeneity is generated at junctions of programmed DNA deletions in *Tetrahymena thermophila*. *Nucl. Acids Res.* **17**, 7263–7272.

Austerberry, C. F., and Yao, M. -C. (1988). Sequence structures of two developmentally regulated, alternative DNA deletion junctions in *Tetrahymena thermophila*. *Mol. Cell. Biol.* **8**, 3947–3950.

Autexier, C., and Greider, C. W. (1994). Functional reconstitution of wild-type and mutant *Tetrahymena telomerase*. *Genes Dev.* **8**, 563–575.

Bannister, A. J., Zegerman, P., Partridge, J. F., Miska, E. A., Thomas, J. O., Allshire, R. C., and Kouzarides, T. (2001). Selective recognition of methylated lysine 9 on histone H3 by the HP1 chomo domain. *Nature* **410**, 120–124.

Bannon, G. A., Bowen, J. K., Yao, M. -C., and Gorovsky, M. A. (1984). *Tetrahymena* H4 genes: structure, evolution and organization in macro- and micronuclei. *Nucl. Acids Res.* **12**, 1961–1975.

Bannon, G. A., Calzone, F. J., Bowen, J. K., Allis, C. D., and Gorovsky, M. A. (1983). Multiple, independently regulated, polyadenylated messages for histone H3 and H4 in *Tetrahymena*. *Nucl. Acids Res.* **11**, 3903–3917.

Bednenko, J., Noto, T., DeSouza, L. V., Siu, K. W., Pearlman, R. E., Mochizuki, K., and Gorovsky, M. A. (2009). Two GW repeat proteins interact with *Tetrahymena thermophila* argonaute and promote genome rearrangement. *Mol. Cell. Biol.* **29**, 5020–5030.

Blackburn, E. H., and Gall, J. G. (1978). A tandemly repeated sequence at the termini of the extrachromosomal ribosomal RNA genes in *Tetrahymena*. *J. Mol. Biol.* **120**, 33–53.

Bromberg, S., Pratt, K., and Hattman, S. (1982). Sequence specificity of the DNA adenine methylase in the protozoan *Tetrahymena thermophila*. *J. Bacteriol.* **150**, 993–996.

Brownell, J. E., and Allis, C. D. (1995). An activity gel asssay detects a single, catalytically active histone acetyltransferase subunit in *Tetrahymena macronuclei*. *Proc. Natl. Acad. Sci. U.S.A.* **92**, 6364–6368.

Brownell, J. E., Zhou, J., Ranalli, T., Kobayashi, R., Edmondson, D. G., Roth, S. Y., and Allis, C. D. (1996). *Tetrahymena* histone acetyltransferase A: a homolog to yeast Gcn5p linking histone acetylation to gene activation. *Cell* **84**, 843–851.

Butler, D. K., Yasuda, L. E., and Yao, M. -C. (1995). An intramolecular recombination mechanism for the formation of the rRNA gene palindrome of *Tetrahymena thermophila*. *Mol. Cell. Biol.* **15**, 7117–7126.

Capowski, E. E. (1989). *Molecular Analysis of N6-Methyladenine Patterns in the Nuclear DNA of the Ciliate, Tetrahymena thermophila.* Brandeis University, Waltham, MA, USA.

Cervantes, M. D., Xi, X., Vermaak, D., Yao, M. C., and Malik, H. S. (2006). The CNA1 histone of the ciliate *Tetrahymena thermophila* is essential for chromosome segregation in the germline micronucleus. *Mol. Biol. Cell.* **17**, 485–497.

Chalker, D. L., Fuller, P., and Yao, M. C. (2005). Communication between parental and developing genomes during tetrahymena nuclear differentiation is likely mediated by homologous RNAs. *Genetics* **169**, 149–160.

Chalker, D. L., La Terza, A., Wilson, A., Kroenke, C. D., and Yao, M. C. (1999). Flanking regulatory sequences of the *Tetrahymena* R deletion element determine the boundaries of DNA rearrangement. *Mol. Cell. Biol.* **19**, 5631–5641.

Chalker, D. L., and Yao, M. -C. (1996). Non-Mendelian, heritable blocks to DNA rearrangement are induced by loading the somatic nucleus of *Tetrahymena* with germline-limited DNA. *Mol. Cell. Biol.* **16**, 3658–3667.

Chalker, D. L., and Yao, M. C. (2001). Nongenic, bidirectional transcription precedes and may promote developmental DNA deletion in *Tetrahymena thermophila*. *Genes Dev.* **15**, 1287–1298.

Charret, R. (1969). L'ADN nucleolaire chez *Tetrahymena pyriformis*: chronologie de sa replication. *Exptl. Cell Res.* **54**, 353–370.

Chau, M. F., and Orias, E. (1996). Developmentally programmed DNA rearrangement in *Tetrahymena thermophila*: isolation and sequence characterization of three new alternative deletion systems. *Biol. Cell* **86**, 111–120.

Cheng, C. Y., Vogt, A., Mochizuki, K., and Yao, M. C. (2010). A domesticated piggyBac transposase plays key roles in heterochromatin dynamics and DNA cleavage during programmed DNA deletion in *Tetrahymena thermophila*. *Mol. Biol. Cell* **21**, 1753–1762.

Conover, R. K., and Brunk, C. F. (1986). Macronuclear DNA molecules of *Tetrahymena thermophila*. *Mol. Cell. Biol.* **6**, 900–905.

Coyne, R. S., Nikiforov, M. A., Smothers, J. F., Allis, C. D., and Yao, M. C. (1999). Parental expression of the chromodomain protein Pdd1p is required for completion of programmed DNA elimination and nuclear differentiation. *Mol. Cell.* **4**, 865–872.

Cui, B., and Gorovsky, M. A. (2006). Centromeric histone H3 is essential for vegetative cell division and for DNA elimination during conjugation in *Tetrahymena thermophila*. *Mol. Cell. Biol.* **26**, 4499–4510.

Czermin, B., Melfi, R., McCabe, D., Seitz, V., Imhof, A., and Pirrotta, V. (2002). Drosophila enhancer of Zeste/ESC complexes have a histone H3 methyltransferase activity that marks chromosomal Polycomb sites. *Cell* **111**, 185–196.

Doerder, F. P., and DeBault, L. E. (1975). Cytoflourimetric analysis of nuclear DNA during meiosis, fertilization and macronuclear development in the ciliate *Tetrahymena pyriformis*, syngen 1. *J. Cell Sci.* **17**, 471–493.

Dou, Y., Bowen, J., Liu, Y., and Gorovsky, M. A. (2002). Phosphorylation and an ATP-dependent process increase the dynamic exchange of H1 in chromatin. *J. Cell Biol.* **158**, 1161–1170.

Dou, Y., Song, X., Liu, Y., and Gorovsky, M. A. (2005). The H1 phosphorylation state regulates expression of CDC2 and other genes in response to starvation in *Tetrahymena thermophila*. *Mol. Cell. Biol.* **25**, 3914–3922.

Edgar, B. A., and Orr-Weaver, T. L. (2001). Endoreplication cell cycles: more for less. *Cell* **105**, 297–306.

Efimova, E. P., Delver, E. P., and Belogurov, A. A. (1988). Alleviation of type I restriction in adenine methylase (dam) mutants of *Escherichia coli*. *Mol. Gen. Genet.* **214**, 313–316.

Eisen, J. A., Coyne, R. S., Wu, M., Wu, D., Thiagarajan, M., Wortman, J. R., Badger, J. H., Ren, Q., Amedeo, P., Jones, K. M., Tallon, L. J., Delcher, A. L., Salzberg, S. L., Silva, J. C., Haas, B. J., Majoros, W. H., Farzad, M., Carlton, J. M., Smith Jr., R. K., Garg, J., Pearlman, R. E., Karrer, K. M., Sun, L., Manning, G., Elde, N. C., Turkewitz, A. P., Asai, D. J., Wilkes, D. E., Wang, Y., Cai, H., Collins, K., Stewart, B. A., Lee, S. R., Wilamowska, K., Weinberg, Z., Ruzzo, W. L., Wloga, D., Gaertig, J., Frankel, J., Tsao, C. C., Gorovsky, M. A., Keeling, P. J., Waller, R. F., Patron, N. J., Cherry, J. M., Stover, N. A., Krieger, C. J., del Toro, C., Ryder, H. F., Williamson, S. C., Barbeau, R. A., Hamilton, E. P., and Orias, E. (2006). Macronuclear genome sequence of the ciliate *Tetrahymena thermophila*, a model eukaryote. *PLoS Biol.* **4**, e286.

Engberg, J., Anderson, P., and Leick, V. (1976). Free ribosomal DNA molecules from *Tetrahymena pyriformis* GL are giant palindromes. *J. Mol. Biol.* **104**, 455–470.

Engberg, J., Mowat, D., and Pearlman, R. E. (1972). Preferential replication of the ribosomal RNA genes during a nutritional shift-up in *Tetrahymena pyriformis*. *Biochim. Biophys. Acta* **272**, 312–320.

Fan, Q., and Yao, M. C. (2000). A long stringent sequence signal for programmed chromosome breakage in *Tetrahymena thermophila*. *Nucleic Acids Res.* **28**, 895–900.

Fass, J. N., Joshi, N. A., Couvillion, M. T., Bowen, J., Gorovsky, M. A., Hamilton, E. P., Orias, E., Hong, K., Coyne, R. S., Eisen, J. A., Chalker, D. L., Lin, D., and Collins, K. (2011). Genome-scale analysis of programmed DNA elimination sites in *Tetrahymena thermophila*. *G3: Genes| Genomes| Genetics* **1**, 515–522.

Fillingham, J. S., Bruno, D., and Pearlman, R. E. (2001). Cis-acting requirements in flanking DNA for the programmed elimination of mse2.9: a common mechanism for deletion of internal eliminated sequences from the developing macronucleus of *Tetrahymena thermophila*. *Nucleic Acids Res.* **29**, 488–498.

Fillingham, J. S., and Pearlman, R. E. (2004). Role of micronucleus-limited DNA in programmed deletion of mse2.9 during macronuclear development of *Tetrahymena thermophila*. *Eukaryot. Cell* **3**, 288–301.

Fillingham, J. S., Thing, T. A., Vythilingum, N., Keuroghlian, A., Bruno, D., Golding, G. B., and Pearlman, R. E. (2004). A non-LTR retrotransposon family is restricted to the germ-line micronucleus in the ciliated protozoan *Tetrahymena thermophila*. *Eukaryot. Cell* **3**, 157–169.

Fusauchi, Y., and Iwai, K. (1983). *Tetrahymena* histone H2A. Isolation and two variant sequences. *J. Biochem.* **93**, 1487–1497.

Gershan, J. A., and Karrer, K. M. (2000). A family of developmentally excised DNA elements in *Tetrahymena* is under selective pressure to maintain an open reading frame encoding an integrase-like protein. *Nucleic Acids Res.* **28**, 4105–4112.

Godiska, R., James, C., and Yao, M. -C. (1993). A distant 10-bp sequence specifies the boundaries of a programmed DNA deletion in *Tetrahymena. Genes Dev.* **7**, 2357–2365.

Godiska, R., and Yao, M. C. (1990). A programmed site-specific DNA rearrangement in *Tetrahymena thermophila* requires flanking polypurine tracts. *Cell* **61**, 1237–1246.

Gorovsky, M. A., Glover, C., Johmann, C. A., Keevert, J. B., Mathis, D. J., and Samuelson, M. (1977). Histones and chromatin structure in *Tetrahymena* macro- and micronuclei. *Cold Spring Harbor Symp. Quant. Biol.* **42**, 493–503.

Gorovsky, M. A., Hattman, S., and Pleger, G. L. (1973). (6 N)methyl adenine in the nuclear DNA of a eucaryote, Tetrahymena pyriformis. *J. Cell Biol.* **56**, 697–701.

Greider, C. W., and Blackburn, E. H. (1985). Identification of a specific telomere terminal transferase activity in *Tetrahymena extracts. Cell* **43**, 405–413.

Greider, C. W., and Blackburn, E. H. (1989). A telomeric sequence in the RNA of *Tetrahymena* telomerase required for telomere repeat synthesis. *Nature* **337**, 331–336.

Hamilton, E. P., Williamson, S., Dunn, S., Merriam, V., Lin, C., Vong, L., Russell-Colantonio, J., and Orias, E. (2006). The highly conserved family of *Tetrahymena thermophila* chromosome breakage elements contains an invariant 10-base-pair core. *Eukaryot. Cell* **5**, 771–780.

Harrison, G. S., Findly, R. C., and Karrer, K. M. (1986). Site-specific methylation of adenine in the nuclear genome of a eucaryote, *Tetrahymena thermophila. Mol. Cell. Biol.* **6**, 2364–2370.

Harrison, G. S., and Karrer, K. M. (1985). DNA synthesis, methylation and degradation during conjugation in *Tetrahymena thermophila. Nucleic Acids Res.* **13**, 73–87.

Hayashi, T., Hayashi, H., and Iwai, K. (1987). *Tetrahymena* histone H1. Isolation and amino acid sequence lacking the central hydrophobic domain conserved in other H1 histones. *J. Biochem.* **102**, 369–376.

Heinonen, T. Y. K., and Pearlman, R. E. (1994). A germ line-specific sequence element in an intron in *Tetrahymena thermophila. J. Biol. Chem.* **269**, 17428–17433.

Horowitz, S., Bowen, J. K., Bannon, G. A., and Gorovsky, M. A. (1987). Unusual features of transcribed and translated regions of the histone H4 gene family of *Tetrahymena thermophila. Nucl. Acids Res.* **15**, 141–160.

Howard-Till, R. A., and Yao, M. C. (2007). Tudor nuclease genes and programmed DNA rearrangements in *Tetrahymena thermophila. Eukaryot. Cell* **6**, 1795–1804.

Hüvös, P. E., Wu, M., and Gorovsky, M. A. (1998). A developmentally eliminated sequence in the flanking region of the histone HI gene in *Tetrahymena thermophila* contains short repeats. *J. Euk. Microbiol.* **45**, 189–197.

Iwamoto, M., Mori, C., Kojidani, T., Bunai, F., Hori, T., Fukagawa, T., Hiraoka, Y., and Haraguchi, T. (2009). Two distinct repeat sequences of Nup98 nucleoporins characterize dual nuclei in the binucleated ciliate tetrahymena. *Curr. Biol.* **19**, 843–847.

Jaeckel-Williams, R. (1978). Nuclear divisions with reduced numbers of microtubules in *Tetrahymena. J. Cell Sci.* **34**, 303–319.

Kapler, G. M., and Blackburn, E. H. (1994). A weak germ-line excision mutation blocks developmentally controlled amplification of the rDNA minichromosome of *Tetrahymena thermophila. Genes Dev.* **8**, 84–95.

Karrer, K. M., and Gall, J. G. (1976). The macronuclear ribosomal DNA of *Tetrahymena pyriformis* is a palindrome. *J. Mol. Biol.* **104**, 421–453.

Karrer, K. M., and Stein-Gavens, S. (1990). Constancy of adenine methylation in *Tetrahymena* macronuclear DNA. *J. Protozool.* **37**, 409–414.

Karrer, K. M., and VanNuland, T. A. (1999). Nucleosome positioning is independent of histone H1 in vivo. *J. Biol. Chem.* **274**, 33020–33024.

Karrer, K. M., and VanNuland, T. A. (2002). Methylation of adenine in the nuclear DNA of *Tetrahymena* is internucleosomal and independent of histone H1. *Nucleic Acids Res.* **30**, 1364–1370.

Katoh, M., Hirono, M., Takemasa, T., Kimura, M., and Watanabe, Y. (1993). A micronucleus-specific sequence exists in the 5′-upstream region of the calmodulin gene in *Tetrahymena thermophila*. *Nucl. Acids Res.* **21**, 2409–2414.

Kirk, K. E., and Blackburn, E. H. (1995). An unusual sequence arrangement in the telomeres of the germ-line micronucleus in *Tetrahymena thermophila*. *Genes Dev.* **9**, 59–71.

Kuo, M. -H., Zhou, J., Jambeck, P., Churchill, M. E. A., and Allis, C. D. (1998). Histone acetyltransferase activity of yeast Gcn5p is required for the activation of target genes *in vivo*. *Genes Dev.* **12**, 627–639.

Kurth, H. M., and Mochizuki, K. (2009). 2′-*O*-methylation stabilizes Piwi-associated small RNAs and ensures DNA elimination in *Tetrahymena*. *RNA* **15**, 675–685.

LaFountain Jr., J. R., and Davidson, L. A. (1979). An analysis of spindle ultrastructure during prometaphase and metaphase of micronuclear division in *Tetrahymena*. *Chromosoma* **75**, 293–308.

LaFountain Jr., J. R., and Davidson, L. A. (1980). An analysis of spindle ultrastructure during anaphase of micronuclear division in *Tetrahymena*. *Cell Motil.* **1**, 41–61.

Li, J., and Pearlman, R. E. (1996). Programmed DNA rearrangement from an intron during nuclear development in *Tetrahymena thermophila*: molecular analysis and identification of potential cis-acting sequences. *Nucleic Acids Res.* **24**, 1943–1949.

Li, S., Yin, L., Cole, E. S., Udani, R. A., and Karrer, K. M. (2006). Progeny of germ line knockouts of ASI2, a gene encoding a putative signal transduction receptor in *Tetrahymena thermophila*, fail to make the transition from sexual reproduction to vegetative growth. *Dev. Biol.* **295**, 633–646.

Liu, X., and Gorovsky, M. A. (1996). Cloning and characterization of the major histone H2A genes completes the cloning and sequencing of known histone genes of *Tetrahymena thermophila*. *Nucl. Acids Res.* **24**, 3023–3030.

Liu, X., Li, B., and Gorovsky, M. A. (1996). Essential and nonessential histone H2A variants in *Tetrahymena thermophila*. *Mol. Cell. Biol.* **16**, 4305–4311.

Liu, Y., Mochizuki, K., and Gorovsky, M. A. (2004). Histone H3 lysine 9 methylation is required for DNA elimination in developing macronuclei in *Tetrahymena*. *Proc. Natl. Acad. Sci. U.S.A.* **101**, 1679–1684.

Liu, Y., Song, X., Gorovsky, M. A., and Karrer, K. M. (2005). Elimination of foreign DNA during somatic differentiation in *Tetrahymena thermophila* shows position effect and is dosage dependent. *Eukaryot. Cell* **4**, 421–431.

Liu, Y., Taverna, S. D., Muratore, T. L., Shabanowitz, J., Hunt, D. F., and Allis, C. D. (2007). RNAi-dependent H3K27 methylation is required for heterochromatin formation and DNA elimination in *Tetrahymena*. *Genes Dev.* **21**, 1530–1545.

Loidl, J., and Mochizuki, K. (2009). *Tetrahymena* meiotic nuclear reorganization is induced by a checkpoint kinase-dependent response to DNA damage. *Mol. Biol. Cell* **20**, 2428–2437.

Loidl, J., and Scherthan, H. (2004). Organization and pairing of meiotic chromosomes in the ciliate *Tetrahymena thermophila*. *J. Cell. Sci.* **117**, 5791–5801.

Madireddi, M. T., Davis, M. C., and Allis, C. D. (1994). Identification of a novel polypeptide involved in the formation of DNA-containing vesicles during macronuclear development in *Tetrahymena*. *Dev. Biol.* **165**, 418–431.

Malone, C. D., Anderson, A. M., Motl, J. A., Rexer, C. H., and Chalker, D. L. (2005). Germ line transcripts are processed by a Dicer-like protein that is essential for developmentally programmed genome rearrangements of *Tetrahymena thermophila*. *Mol. Cell. Biol.* **25**, 9151–9164.

Malone, C. D., Falkowska, K. A., Li, A. Y., Galanti, S. E., Kanuru, R. C., LaMont, E. G., Mazzarella, K. C., Micev, A. J., Osman, M. M., Piotrowski, N. K., Suszko, J. W., Timm, A. C., Xu, M. M., Liu, L., and Chalker, D. L. (2008). Nucleus-specific importin alpha proteins and nucleoporins regulate protein import and nuclear division in the binucleate *Tetrahymena thermophila*. *Eukaryot. Cell* **7**, 1487–1499.

Marsh, T. C., Cole, E. S., and Romero, D. P. (2001). The transition from conjugal development to the first vegetative cell division is dependent on RAD51 expression in the ciliate *Tetrahymena thermophila*. *Genetics* **157**, 1591–1598.

Martindale, D. W., Allis, C. D., and Bruns, P. J. (1985). RNA and protein synthesis during meiotic prophase in *Tetrahymena thermophila*. *J. Protozool.* **32**, 644–649.

Matsuda, A., Shieh, A. W., Chalker, D. L., and Forney, J. D. (2010). The conjugation-specific Die5 protein is required for development of the somatic nucleus in both Paramecium and *Tetrahymena*. *Eukaryot. Cell* **9**, 1087–1099.
Mochizuki, K., Fine, N. A., Fujisawa, T., and Gorovsky, M. A. (2002). Analysis of a piwi-related gene implicates small RNAs in genome rearrangement in *Tetrahymena*. *Cell* **110**, 689–699.
Mochizuki, K., and Gorovsky, M. A. (2004a). Conjugation-specific small RNAs in Tetrahymena have predicted properties of scan (scn) RNAs involved in genome rearrangement. *Genes Dev.* **18**, 2068–2073.
Mochizuki, K., and Gorovsky, M. A. (2004b). RNA polymerase II localizes in Tetrahymena thermophila meiotic micronuclei when micronuclear transcription associated with genome rearrangement occurs. *Eukaryot. Cell* **3**, 1233–1240.
Mochizuki, K., and Gorovsky, M. A. (2004c). Small RNAs in genome rearrangement in Tetrahymena. *Curr Opin Genet. Dev.* **14**, 1–7.
Mochizuki, K., and Gorovsky, M. A. (2005). A Dicer-like protein in *Tetrahymena* has distinct functions in genome rearrangement, chromosome segregation, and meiotic prophase. *Genes Dev.* **19**, 77–89.
Mochizuki, K., Novatchkova, M., and Loidl, J. (2008). DNA double-strand breaks, but not crossovers, are required for the reorganization of meiotic nuclei in *Tetrahymena*. *J. Cell. Sci.* **121**, 2148–2158.
Modrich, P. (1989). Methyl-directed DNA mismatch correction. *J. Biol. Chem.* **264**, 6597–6600.
Nikiforov, M. A., Gorovsky, M. A., and Allis, C. D. (2000). A novel chromodomain protein, Pdd3p, associates with internal eliminated sequences during macronuclear development in Tetrahymena thermophila. *Mol. Cell. Biol.* **20**, 4128–4134.
Nikiforov, M. A., Smothers, J. F., Gorovsky, M. A., and Allis, C. D. (1999). Excision of micronuclear-specific DNA requires parental expression of Pdd2p and occurs independently from DNA replication in *Tetrahymena thermophila*. *Genes Dev.* **13**, 2852–2862.
Noto, T., Kurth, H. M., Kataoka, K., Aronica, L., DeSouza, L. V., Siu, K. W., Pearlman, R. E., Gorovsky, M. A., and Mochizuki, K. (2010). The *Tetrahymena* argonaute-binding protein Giw1p directs a mature argonaute-siRNA complex to the nucleus. *Cell* **140**, 692–703.
Orias, E., and Flacks, M. (1975). Macronuclear genetics of *Tetrahymena*: I. Random distribution of macronuclear genecopies in *T. pyriformis*, syngen 1. *Genetics* **79**, 187–206.
Paranjape, S. M., Kamakaka, R. T., and Kadonaga, J. T. (1994). Role of chromatin structure in the regulation of transcription by RNA polymerase II. *Annu. Rev. Biochem.* **63**, 265–297.
Patil, N. S., Hempen, P. M., Udani, R. A., and Karrer, K. M. (1997). Alternate junctions and microheterogeneity of Tlr1, a developmentally regulated DNA rearrangement in *Tetrahymena thermophila*. *J. Euk. Microbiol.* **44**, 518–522.
Patil, N., and Karrer, K. M. (2000). A developmentally regulated deletion element with long terminal repeats has cis-acting seuences in the flanking DNA. *Nucl. Acids Res.* **28**, 1465–1472.
Pratt, K., and Hattman, S. (1981). Deoxyribonucleic acid methylation and chromatin organization in *Tetrahymena thermophila*. *Mol. Cell. Biol.* **1**, 600–608.
Pritham, E. J., Putliwala, T., and Feschotte, C. (2007). Mavericks, a novel class of giant transposable elements widespread in eukaryotes and related to DNA viruses. *Gene* **390**, 3–17.
Rexer, C. H., and Chalker, D. L. (2007). Lia1p, a novel protein required during nuclear differentiation for genome-wide DNA rearrangements in *Tetrahymena thermophila*. *Eukaryot. Cell* **6**, 1320–1329.
Russell, D. W., and Zinder, N. D. (1987). Hemimethylation prevents DNA replication in *E. coli*. *Cell* **50**, 1071–1079.
Saveliev, S. V., and Cox, M. M. (1995). Transient DNA breaks associated with programmed genomic deletion events in conjugating cells of *Tetrahymena thermophila*. *Genes Dev.* **9**, 248–255.
Saveliev, S. V., and Cox, M. M. (1996). Developmentally programmed DNA deletion in *Tetrahymena thermophila* by a transposition-like reaction pathway. *EMBO J.* **15**, 2858–2869.
Saveliev, S. V., and Cox, M. M. (2001). Product analysis illuminates the final steps of IES deletion in *Tetrahymena thermophila*. *EMBO J.* **20**, 3251–3261.
Schoeberl, U. E., and Mochizuki, K. (2011). Keeping the soma free of transposons: Programmed DNA elimination in ciliates. *J. Biol. Chem.* **286**, 37045–37052.

Shen, X., and Gorovsky, M. A. (1996). Linker histone H1 regulates specific gene expression but not global transcription *in vivo*. *Cell* **86**, 475–483.

Shen, X., Yu, L., Weir, J. W., and Gorovsky, M. A. (1995). Linker histones are not essential and affect chromatin condensation *in vivo*. *Cell* **82**, 47–56.

Smothers, J. F., Mizzen, C. A., Tubbert, M. M., Cook, R. G., and Allis, C. D. (1997). Pdd1p associates with germline-restricted chromatin and a second novel anlagen-enriched protein in developmentally programmed DNA elimination structures. *Development* **124**, 4537–4545.

Song, X., and Gorovsky, M. A. (2007). Unphosphorylated H1 is enriched in a specific region of the promoter when CDC2 is down-regulated during starvation. *Mol. Cell. Biol.* **27**, 1925–1933.

Sonneborn, T. M. (1974). Tetrahymena pyriformis. *In* "Handbook of Genetics: Plants, Plant Viruses and Protists," (R. C. King, ed.), Vol. 2, pp. 433–467. Plenum Press, New York.

Stargell, L. A., Bowen, J., Dadd, C. A., Dedon, P. C., Davis, M., Cook, R. G., Allis, C. D., and Gorovsky, M. A. (1993). Temporal and spatial association of histone H2A variant hv1 with transcriptionally competent chromatin during nuclear development in *Tetrahymena thermophila*. *Genes Dev.* **7**, 2641–2651.

Strahl, B. D., Ohba, R., Cook, R. G., and Allis, C. D. (1999). Methylation of histone H3 at lysine 4 is highly conserved and correlates with transcriptionally active nuclei in *Tetrahymena*. *Proc. Natl. Acad. Sci. U.S. A.* **96**, 14967–14972.

Sugai, T., and Hiwatashi, K. (1970). Cytologic and autoradiographic studies of the micronucleus at meiotic prophase in *Tetrahymena pyriformis*. *J. Protozool.* **21**, 542–548.

Taverna, S. D., Coyne, R. S., and Allis, C. D. (2002). Methylation of histone H3 at lysine 9 targets programmed DNA elimination in *Tetrahymena*. *Cell* **110**, 701–711.

Thatcher, T. H., MacGaffey, J., Bowen, J., Horowitz, S., Shapiro, D. L., and Gorovsky, M. A. (1994). Independent evolutionary origin of histone H3.3-like variants of animals and *Tetrahymena*. *Nucl. Acids Res.* **22**, 180–186.

Van Nuland, T. A., Capowski, E. E., and Karrer, K. M. (1995). Position effect for adenine methylation in the macronuclear DNA of *Tetrahymena*. *Gene* **157**, 235–237.

Vavra, K. J., Allis, C. D., and Gorovsky, M. A. (1982). Regulation of histone acetylation in *Tetrahymena* macro- and micronuclei. *J. Biol. Chem.* **257**, 2591–2598.

Wang, L., Liu, L., and Berger, S. L. (1998). Critical residues for histone acetylation by Gcn5, functioning in Ada and SAGA complexes, are also required for transcriptional function in vivo. *Genes Dev.* **12**, 640–653.

Wells, J. M., Ellingson, J. L. E., Catt, D. M., Berger, P. J., and Karrer, K. M. (1994). A small family of elements with long inverted repeats is located near sites of developmentally regulated DNA rearrangement in *Tetrahymena thermophila*. *Mol. Cell. Biol.* **14**, 5939–5949.

Wenkert, D., and Allis, C. D. (1984). Timing of the appearance of macronuclear-specific histone variant hv1 and gene expression in developing new macronuclei of *Tetrahymena thermophila*. *J. Cell. Biol.* **98**, 2107–2117.

White, E. M., and Gorovsky, M. A. (1988). Localization and expression of mRNA for a macronuclear-specific histone H2A variant (hv1) during the cell cycle and conjugation of *Tetrahymena thermophila*. *Mol. Cell. Biol.* **8**, 4780–4786.

White, E. M., Shapiro, D. L., Allis, C. D., and Gorovsky, M. A. (1988). Sequence and properties of the message encoding *Tetrahymena* hv1, a highly evolutionarily conserved histone H2A variant that is associated with active genes. *Nucl. Acids Res.* **16**, 179–198.

Wiley, E. A., Myers, T., Parker, K., Braun, T., and Yao, M. C. (2005). Class I histone deacetylase Thd1p affects nuclear integrity in *Tetrahymena thermophila*. *Eukaryot. Cell.* **4**, 981–990.

Wiley, E. A., Ohba, R., Yao, M. -C., and Allis, C. D. (2000). Developmentally regulated Rpd3p homolog specific to the transcriptionally active macronucleus of vegetative *Tetrahymen thermophila*. *Mol. Cell. Biol.* **20**, 8319–8328.

Woodard, J., Kaneshiro, E., and Gorovsky, M. A. (1972). Cytochemical studies on the problem of macronuclear subnuclei in *Tetrahymena*. *Genetics* **70**, 251–260.

Wu, M., Allis, C. D., and Gorovsky, M. A. (1988). Cell-cycle regulation as a mechanism for targeting proteins to specific DNA sequences in *Tetrahymena thermophila*. *Proc. Natl. Acad. Sci. U.S.A.* **85**, 2205–2209.

Wu, M., Allis, C. D., Richman, R., Cook, R. G., and Gorovsky, M. A. (1986). An intervening sequence in an unusual histone H1 gene of *Tetrahymena thermophila*. *Proc. Natl. Acad. Sci. U.S.A.* **83**, 8674–8678.

Wu, M., Allis, C. D., Sweet, M. T., Cook, R. G., Thatcher, T. H., and Gorovsky, M. A. (1994). Four distinct and unusual linker proteins in a mitotically dividing nucleus are derived from a 71-kilodalton polyprotein, lack p34cdc2 sites, and contain protein kinase A sites. *Mol. Cell. Biol.* **14**, 10–20.

Wuitschick, J. D., Gershan, J. A., Lochowicz, A. J., Li, S., and Karrer, K. M. (2002). A novel family of mobile genetic elements is limited to the germline genome in *Tetrahymena thermophila*. *Nucl. Acids Res.* **30**, 2524–2537.

Wuitschick, J. D., and Karrer, K. M. (2003). Diverse sequences within Tlr elements target programmed DNA elimination in Tetrahymena thermophila. *Eukaryot. Cell.* **2**, 678–689.

Wuitschick, J. D., Lindstrom, P. R., Meyer, A. E., and Karrer, K. M. (2004). Homing endonucleases encoded by germ line-limited genes in *Tetrahymena thermophila* have APETELA2 DNA binding domains. *Eukaryot. Cell.* **3**, 685–694.

Wyatt, H. D., West, S. C., and Beattie, T. L. (2010). InTERTpreting telomerase structure and function. *Nucleic Acids Res.* **38**, 5609–5622.

Yao, M. -C., Yao, C. -H., and Monks, B. (1990). The controlling sequence for site-specific chromosome breakage in *Tetrahymena*. *Cell* **63**, 763–772.

Yao, M. -C., Choi, J., Yokoyama, S., Austerberry, C., and Yao, C. -H. (1984). DNA elimination in *Tetrahymena*: a developmental process involving extensive breakage and rejoining of DNA at defined sites. *Cell* **36**, 433–440.

Yao, M. -C., and Gall, J. G. (1977). A single integrated gene for ribosomal RNA in a eukeryote, Tetrahymena pyriformis. *Cell* **12**, 121–132.

Yao, M. C., Fuller, P., and Xi, X. (2003). Programmed DNA deletion as an RNA-guided system of genome defense. *Science* **300**, 1581–1584.

Yao, M. -C., and Gorovsky, M. (1974). Comparison of the sequences of macronuclear and micronuclear DNA of *Tetrahymena pyriformis*. *Chromosoma* **48**, 1–18.

Yao, M. C., Yao, C. H., Halasz, L. M., Fuller, P., Rexer, C. H., Wang, S. H., Jain, R., Coyne, R. S., and Chalker, D. L. (2007). Identification of novel chromatin-associated proteins involved in programmed genome rearrangements in *Tetrahymena*. *J. Cell. Sci.* **120**, 1978–1989.

Yin, L., Gater, S. T., and Karrer, K. M. (2010). A developmentally regulated gene, ASI2, is required for endocycling in the macronuclear anlagen of Tetrahymena. *Eukaryotic. Cell.* **9**, 1343–1353.

Yu, G. -L., Bradley, J. D., Attardi, L. D., and Blackburn, E. H. (1990). in vivo alteration of telomere sequences and senescence caused by mutated *Tetrahymena* telomerase RNAs. *Nature* **344**, 126–132.

Yu, L., and Gorovsky, M. A. (1997). Constitutive expression, not a particular primary sequence, is the important feature of the H3 replacement variant hv2 in *Tetrahymena thermophila*. *Mol. Cell. Biol.* **17**, 6303–6310.

Yu, S. -M., Horowitz, S., and Gorovsky, M. A. (1987). A novel approach for studying gene expression in the cell cycle reveals coordinate and independent regulation of members of the H4 multigene family in cycling and in nongrowing *Tetrahymena*. *Genes Dev.* **1**, 683–692.

CHAPTER 4

Whole Genome Studies of *Tetrahymena*

Robert S. Coyne*, Nicholas A. Stover† and Wei Miao‡

*Department of Genomic Medicine, J. Craig Venter Institute, Rockville, MD, USA

†Biology Department, Bradley University, Peoria, IL, USA

‡Key Laboratory of Aquatic Biodiversity and Conservation, Institute of Hydrobiology, Chinese Academy of Sciences, Wuhan, China

Abstract

Within the past decade, genomic studies have emerged as essential and highly productive tools to explore the biology of *Tetrahymena thermophila*. The current major resources, which have been extensively mined by the research community, are

0091-679X/10 $35.00
DOI 10.1016/B978-0-12-385967-9.00004-9

the annotated macronuclear genome assembly, transcriptomic data and the databases that house this information. Efforts in progress will soon improve these data sources and expand their scope, including providing annotated micronuclear and comparative genomic sequences. Future studies of *Tetrahymena* cell and molecular biology, development, physiology, evolution and ecology will benefit greatly from these resources and the advanced genomic technologies they enable.

I. Introduction

As other chapters in this volume attest, *Tetrahymena* has been a powerful model system for a wide variety of investigations; genomic resources have greatly expanded this utility and opened new avenues for research. In 2002, Turkewitz *et al.* titled a review "Functional genomics: the coming of age for *Tetrahymena thermophila*" (Turkewitz *et al.*, 2002). The intervening years have shown the aptness of this metaphor and validated the authors' forecast of "a bright future for research in this rising model eukaryote". In this chapter, we begin by briefly reviewing the major current *Tetrahymena* genomic resources and describing notable examples of how these tools have already been used to address questions of wide biological interest. As of this writing, the basic *Tetrahymena* genomic tools are in need of improvement; we next describe near-term objectives for an enhanced toolbox and the challenges faced in reaching them. Finally, we discuss longer term potential applications of genomic technologies to outstanding questions for which *Tetrahymena* is a well-suited model organism.

As described in other chapters, *Tetrahymena* carries in its cytoplasm two structurally and functionally distinct nuclei – the small, diploid, germline micronucleus (MIC) and the large, polyploid, somatic macronucleus (MAC) (Orias *et al.*, 2011). Because genic, and thus phenotypic, expression is confined to the MAC, initial genomic efforts focused on this nucleus. This was also the logical choice for technical reasons; methods to separately purify MACs, MICs, and their developmental intermediates based on their differential sedimentation properties have long been established (Allen, 2000b; Allis and Dennison, 1982; Chau and Orias, 1996; Gorovsky *et al.*, 1975), in the course of highly fruitful studies on their chromatin characteristics, but because MACs carry roughly 20-fold more weight in DNA, it is easier to minimize MIC contamination of MAC DNA preps than vice versa. Nevertheless, the challenge of MIC genomics is now also being met (see Section IV.B below).

II. To Make a Long Story Short: A Brief History of the *Tetrahymena* MAC Genome Project

From the time the first bacterial genome sequence was rolled out, it was obvious that genomics would transform how biology is done. As the technology advanced, tackling larger eukaryotic genomes, each research community was eager for its favorite model organism to be next in line. The *Tetrahymena* community, under the forward-thinking leadership of Ed Orias of the University of California at Santa

Barbara (UCSB), began plotting its genomic strategies in 1999 (Orias, 2000), leading in time to the joint NIH/NSF-funded MAC genome sequencing project, carried out at The Institute for Genomic Research (TIGR, later subsumed by the J. Craig Venter Institute, JCVI) and led by Jonathan Eisen (Eisen *et al.*, 2006). The Sanger paired-end shotgun assembly, scheduled to be completed in stages over three years, was actually finished early by accident! At a time when the sequencing facility had been expected to do some light test runs of its new libraries, they instead ran over two million clones. Fortunately, these reads turned out to be of high quality and allowed assembly of a good draft genome. Even more fortunately, the MAC genome turned out to be substantially smaller than originally estimated, resulting in higher sequence coverage.

Naturally, the most eagerly awaited result of this effort was a compendium of the genome's coding potential. Using a limited set of ESTs, homologies to other characterized genes and *ab initio* gene-finding algorithms, the TIGR team made an initial estimate of 27,424 protein-coding genes and performed automated annotation of this gene set. Although only a rough beginning, this annotation provided exciting insights into ciliate biology (Eisen *et al.*, 2006). The sequence data and preliminary gene models were freely released prior to publication, allowing researchers to make immediate use of these valuable resources. Continued efforts at JCVI and UCSB and further EST sequencing financed in part by Genome Canada led to significant improvements in the genome assembly, removal of most contamination from the MIC genome, refinement of the gene models, and a downward-revised estimate of 24,725 protein-coding genes (Coyne *et al.*, 2008).

Meanwhile, to make these data more accessible to the community, the *Tetrahymena* Genome Database (TGD) was established at Stanford University (Stover *et al.*, 2006). TGD includes the standard features of a model organism genome database, such as a genome browser, BLAST server, and separate informational pages for each predicted gene, including manual curation of the existing literature. TGD is currently housed at Bradley University and has been redeveloped there as a Wiki (http://ciliate.org). A separate, international effort used the preliminary gene models to design a microarray platform to evaluate genome-wide transcription patterns (Miao *et al.*, 2009). Focusing on the three nutritional/developmental conditions of widest general interest – growth in rich medium, starvation, and the sexual process of conjugation – this team provided clear evidence that most of the predicted genes are indeed transcribed. The data also revealed many distinct developmental patterns of expression and showed that strong correlation of such patterns within a group of genes is often predictive of shared biological function. To house these very valuable data, the *Tetrahymena* Gene Expression Database (TGED; http://tged.ihb.ac.cn/) was set up, linked to the corresponding gene model pages of TGD (Xiong *et al.*, 2011a).

III. Examples of Use

Availability of the whole genome sequence, annotation, and expression profiles has enabled a wide variety of analyses, including functional studies of entire gene families, proteomics, transcriptomics, and comparative genomics.

A. Homology-Aided Functional Genomics

In a preliminary survey of the predicted proteome, Eisen *et al.* (2006) reported that *Tetrahymena* retains many ancestral eukaryotic gene functions, boosting the case for its use as a general model organism. In fact, compared with the more common unicellular eukaryotic model *Saccharomyces cerevisiae*, *Tetrahymena* shares a greater number of orthologs with humans, including many associated with disease. Several expanded gene families were selected for detailed analyses, highlighting their remarkable diversity. For example, with 940 membrane transporters, *Tetrahymena* surpassed all sequenced metazoans, and, reflective of its highly elaborated cytoskeletal structures, *Tetrahymena* encodes a multitude of tubulins and associated microtubule components and regulators. As remarkable and interesting as this initial survey was, the number of genes considered was only a small portion of the total and only limited functional studies were involved. With the genome sequence in hand, several groups have performed more in-depth analyses of whole gene families of particular interest to them. We present three examples to highlight the potential of such genome-enabled functional analyses.

1. Nuclear Targeting

With two distinct nuclei inhabiting a single cytoplasm (and these undergoing dramatic developmental transformations at certain stages), it is perhaps not surprising that *Tetrahymena* contains a diverse set of proteins regulating the import of proteins into nuclei. Malone *et al.* (2008) found 11 importin α and 13 importin β homologs in the genome and fluorescently tagged each one, showing nuclear specificity for several. This study is also notable for its added value as an educational activity for the many undergraduates who took part. Several other ciliate labs have actively involved undergraduates in genome-enabled research, as for example the functional characterization of dyneins (Wilkes *et al.*, 2009). The Ciliate Genomics Consortium (http://tet.jsd.claremont.edu/) was set up as a web-based hub to coordinate such education/research efforts and make their results accessible to the full community.

2. Responding to the Environment

In order to survive in their daily and seasonally changing aquatic environment, ciliates must import, export, and metabolize many substances, from ions and nutrients to xenobiotic toxins. Fu *et al.* (2009) and Xiong *et al.* (2010) used the *Tetrahymena* genomic and transcriptomic resources to characterize two large families involved in such processes – the 165 ATP-Binding Cassette transporters and the 44 cytochrome P450 monooxygenases. First, manual sequence alignments and cDNA sequencing allowed correction of mis-annotated gene structures. Phylogenetic analyses (including not only sequence-based studies but also those based on conservation of intron positions) and transcriptional "heat map" clustering allowed the subclassification of these large families, providing insights into their evolution and potential functions, which will be investigated in future studies. These studies also supported the conclusion of Eisen *et al.* (2006) that most *Tetrahymena* gene family expansions occurred by local gene duplication, often resulting in tandem arrays that then diverged in function and expression patterns or sometimes resulted in pseudogenization of some duplicates.

3. Managing Membrane Compartments

Ciliates also possess elaborate, dynamic arrays of intracellular membrane-bound compartments. The formation, targeting, and fusion of these compartments are under the control of monomeric GTPases called Rabs. Bright *et al.* (2010) combined phylogenetics, expression analysis, and dynamic GFP localization in a massive study of the 56 *Tetrahymena* Rab genes, a number comparable to that found in mammals and plants. While a subset of *Tetrahymena* Rabs can be considered highly conserved, another group appears from current data to be restricted to the ciliates and, consistently, to localize to structures that have, at least, undergone significant elaboration in this lineage. The authors also proposed from their phylogenetic analysis the existence of a novel core ancestral Rab clade. A key finding was that some Rabs do not localize to the structures expected from their phylogenetic affinity, thus providing a cautionary note against inferring function by transferring annotation between organisms, especially distant ones.

B. Proteomics

In contrast to the above homology-based genome scans, several investigators have applied proteomics to the systematic identification of organellar components or the proteins associated with a particular cellular process, thus casting a wider net for functionally relevant players. This approach first requires a means to highly purify the source material free of contamination from general cellular proteins. Thus, it is not surprising that the first *Tetrahymena* proteomic study was conducted on cilia (Smith *et al.*, 2005), which can be readily separated from the cell body. As in typical proteomic studies, solubilized ciliary proteins were thoroughly digested with trypsin, the resulting fragments resolved chromatographically, and their precise masses determined by mass spectrometry. This study was performed prior to annotation of the genome sequence, so the draft assembly was translated in all six reading frames, the peptide sequences digested *in silico* with trypsin and the predicted fragment masses matched to the experimentally derived ones. This approach, born of necessity, has also proven very useful for detecting and correcting gene models that have been misannotated (see below). The *Tetrahymena* "ciliome" thus analyzed contained 223 proteins, 84 of which had no detectable similarity to proteins outside the ciliates.

A more ambitious study of the mitochondrial proteome followed (Smith *et al.*, 2007), identifying 573 unique proteins, most encoded by the nuclear genome and many having no known function or homologs outside the ciliates, showing the remarkable evolutionary flexibility of the mitochondrion. Other applications of proteomics to *Tetrahymena* have included studies of the phagosome (Jacobs *et al.*, 2006), basal body

(Kilburn *et al.*, 2007), nuclear exchange junction (Cole *et al.*, 2008), pellicle (Gould *et al.*, 2011), and ATP synthase complex (Balabaskaran Nina *et al.*, 2010). In each case, novel components have been detected that were unsuspected based on homology relationships, demonstrating the power of an unbiased proteomic approach.

A study of constitutive secretion (Madinger *et al.*, 2010) demonstrated the variability of the "secretome" across different strain backgrounds and, especially, different growth conditions. Variability of this sort (although perhaps not to as great a degree) will likely also be detectable even in organellar proteomes, raising caution against defining "the" proteome of any cellular body.

C. Studies of Small RNAs

In the past decade, researchers have uncovered an enormous diversity of ~20–30 nt RNAs and their protein partners that play distinct and essential roles in gene regulation and chromosome function (Farazi *et al.*, 2008). Studies in *Tetrahymena* have been at the forefront of this research. The first small RNA (sRNA) class discovered in this organism comprises the ~27–30 nt scan RNAs believed to guide developmental DNA elimination (Mochizuki and Gorovsky, 2004) (deep sequencing of scnRNAs has been performed [Collins and Malone, personal communication], but has been awaiting full assembly of the MIC genome for its most productive interpretation). This "genome cleansing" results in removal from the somatic MAC of most of the repetitive, transposon-related sequences that, in most other eukaryotes, would be natural targets for sRNA-induced heterochromatic gene silencing, a process also dependent on histone H3 K9 methylation, which occurs in *Tetrahymena* exclusively in association with the DNA elimination pathway (Taverna *et al.*, 2002). However streamlined the MAC genome becomes as a result, studies show that it still utilizes multiple sRNA-based silencing mechanisms independent of scnRNAs as further means of genome defense and regulation.

The first evidence for existence of such mechanisms came from studies of Lee and Collins (2006), who detected a novel size class (23–24 nt) of sRNAs present in all *Tetrahymena* life-cycle stages. A small number were cloned, sequenced, and aligned to the MAC genome assembly. Remarkably for such a small sample, most of the sequences clustered at a small number of genomic loci that, although annotated as putative protein-coding genes, showed little evidence of transcription and had structural features resembling mobile elements. Couvillion *et al.* (2009) applied to this problem the power of deep sequencing, not only of the total sRNA populations of wild-type cells, but also those specifically associated with each of the expressed PPD family proteins (TWIs in *Tetrahymena*) in wild type and a variety of mutant genetic backgrounds. This thorough approach allowed the authors to identify sRNAs of low abundance in the total pool and distinguish multiple sRNA pathways.

The sRNA sequences aligned, generally in a strand-specific manner, to a number of sites in the MAC genome, including several clusters of sequence-related pseudogenes, high- and low-copy repeat loci, telomeric repeats, "phased cluster" loci adjacent to potential hairpin-generating transcription units, and a number of EST-supported predicted protein-coding loci that generally exhibit features of potential hairpin formation, membership in paralogous families, or possible transcriptional interference or overlap. These results show that, despite jettisoning most repetitive DNA, the MAC genome still employs a diverse array of sRNA-based mechanisms for gene silencing and other as yet unclear purposes. The authors provocatively suggest these pathways may affect epigenetic inheritance of genome structure, allowing vegetative life history to influence sexual outcomes. Future studies will determine whether such novel mechanisms actually exist, as has also been suggested in certain other model systems (Koonin and Wolf, 2009). It will also be of great interest to compare the sRNA pathways of *T. thermophila* to amicronucleate species such as *Tetrahymena pyriformis* that have lost their "safe haven" for mobile elements.

D. Comparative Genomics

Tetrahymena is the most well-studied member of its phylum and, indeed, one of the most well-studied of all protozoa, which comprise most of the diversity of the eukaryotic kingdom. As such, its genome sequence is critical for addressing evolutionary questions regarding the deep origins of gene families (Parker *et al.*, 2007), the history of plastid acquisition and loss (Archibald, 2008; Coesel *et al.*, 2008; Reyes-Prieto *et al.*, 2008), and the consequences of alternative genetic code usage (Adachi and Cavalcanti, 2009; Ring and Cavalcanti, 2008).

The annotated *Tetrahymena* genome also serves as a guide for structural and functional annotation of other ciliate genomes and EST datasets. Of particular relevance is the recent whole genome (Coyne *et al.*, 2011) and EST (Cassidy-Hanley *et al.*, 2011) sequencing of *Ichthyophthirius multifiliis*, commonly known as Ich, a fish parasite causing significant losses to the aquaculture industry. As Ich's closest sequenced relative, *Tetrahymena*'s genomic data were invaluable in modeling gene structures, assigning probable gene functions, and reconstructing metabolic pathways. As the *Tetrahymena* annotation is improved (see Section IV.C below), this will allow concomitant improvements to Ich's annotation. Future ciliate genome projects (see below) will also draw on the *T. thermophila* genome for guidance in interpretation of results.

IV. Near-Term Goals for Improvement of Genomic Resources

As with other significant eukaryotic model organisms such as yeast, *Caenorhabditis elegans* and *Drosophila*, the initial *Tetrahymena* genome sequencing and annotation project should be viewed as only a beginning. Significant improvements and expansions must be made to existing resources to realize the full value of *Tetrahymena* as an experimental organism. Here, we outline an essentially modest set of goals, some of which are already in progress.

A. Putting the Pieces Together: Prospects for a Finished MAC Genome

The central problem of shotgun genome assembly is computational – finding a unique solution to the jigsaw puzzle presented by millions or billions of relatively short stretches of nucleotides (Pop *et al.*, 2002). What makes this a difficult problem is the presence in every genome of repetitive sequences that offer multiple solutions to local sub-assemblies. Obviously, genome assembly programs cannot tolerate multiple solutions and instead break the assembly into scaffolds, which may consist of multiple contigs separated from one another by intrascaffold gaps. Remarkably, despite prolonged efforts, only one eukaryotic genome has been reported as being 100% finished (Nozaki *et al.*, 2007). Often, highly repetitive centromeric and/or subtelomeric regions are the most recalcitrant. Because these and most other *Tetrahymena* repetitive DNA sequences are confined to the MIC, it may be possible that the *Tetrahymena* MAC genome could join this very exclusive club.

In this section, we review the current state of affairs and evaluate future directions for genome closure. But first, why is it important to go from 99.5% (or thereabouts) to 100% finished? One reason is that, without completely finishing the MAC, it would remain uncertain what proportion of the current assembly actually belongs to the MIC, thus complicating interpretation of some analyses. Secondly, an unknown, but perhaps significant, number of genes will be found within or spanning sequencing gaps, which may also contain uncharacterized sequence elements controlling gene expression or chromosome function. In addition, a complete MAC sequence will greatly facilitate MIC genome assembly, allow complete characterization of alternative genome rearrangements, and either confirm or refute the assumption of colinearity between MAC and MIC chromosomes. Finally, the results will provide the ultimate genetic map for this model organism well suited to genetic analysis.

At the time the *T. thermophila* MAC genome project commenced, Sanger paired-end sequencing was the technology of choice. Paired-end reads from libraries of defined insert size provide crucial linking information to position reads at unique places in the growing assembly. Ideally, long insert libraries, in vectors such as BACs or fosmids, are used to "jump" over repetitive regions and/or "anchor" such regions to adjacent unique sequences. Unfortunately, efforts to construct representative libraries from the *Tetrahymena* genome with inserts larger than about 6 kb have repeatedly failed, most likely due to instability of such low GC inserts during *Escherichia coli* propagation. Nevertheless, due to the MAC genome's low repetitiveness, the assembly produced from 2–4 and 4–6 kb insert plasmid libraries was very good (for a genome its size), and closure of intrascaffold gaps was straightforward (although labor- and thus cost-intensive) (Coyne *et al.*, 2008; Eisen *et al.*, 2006).

Through physical and genetic mapping conducted in the Orias lab, we now know that the MAC genome is contained in 181 chromosomes. Of these, 124, comprising 53% of the genome length, have been sequenced fully, from telomere to telomere. The remaining closure tasks are to connect about 1000 scaffolds in their correct order and orientation into the remaining 57 chromosomes, close the interscaffold gaps separating them, and also close about 650 intrascaffold gaps (with an average length of 271 bp). Most of the complex scaffold connecting work has been accomplished in the Orias and Dear laboratories, using HAPPY mapping, a method of identifying physical linkage between fragments by the coassortment of PCR product tags (Dear and Cook, 1993; Hamilton *et al.*, 2006). Despite this progress, closure of the remaining gaps by traditional, targeted methods would still represent a major undertaking. However, we expect that it will be possible to close many gaps by effectively "resequencing" the genome at higher coverage using Illumina or other emerging technologies, which have the added advantage of not relying on DNA propagation in *E. coli*. For example, high coverage sequencing of the MIC genome is well underway (see Section IV.B below), although gaps closed from the resulting assembly will need to be confirmed not to contain any MIC-limited sequences. Other sequencing efforts, such as the detection of genetic mutations (see Section V.D below), can provide the needed confirmation and supplement the MIC genome data.

Besides the gaps in the existing assembly, it is also fully to be expected that there are multitudes of minor errors that need to be corrected. As a point of comparison, the 72 Mb *Paramecium tetraurelia* genome was sequenced (Aury *et al.*, 2006) to higher (13X vs. 9X) coverage than *Tetrahymena*, but much higher Illumina sequencing coverage has since allowed the identification of over 25,000 errors in the original assembly, including both single base pair changes and indels (L. Sperling, O. Arnaiz; personal communication). The possibility of misassembly also exists and, indeed, one probable misassembly has been detected serendipitously (Fu *et al.*, 2009). Correcting such errors will be crucial for some future applications of genomics (see Section V.D below).

B. Sequencing the MIC Genome

Many researchers who study ciliate genome rearrangement have eagerly awaited the full sequence of the MIC. When complete, we will finally have a grasp of the full extent of DNA elimination, the range of mobile element families and other sequences represented, the arrangement of all the MAC's 181 chromosomes on the five MIC chromosomes, and potentially the structure of MIC telomeres and centromeres (depending on their size and complexity). The sequence will allow more direct testing of hypotheses concerning the mechanisms of genome rearrangement, including the roles of chromatin modification and bidirectional nongenic transcription (Chalker and Yao, 2001). From a genome sequencing perspective, the MIC presents a greater challenge than the MAC because of its much higher repetitive sequence content, which, as discussed above, tends to result in more fragmented assemblies and incomplete chromosome ends.

As described in Chapter 3, several internal eliminated sequences (IESs) have been sequenced and characterized, but ironically the first unbiased "genome-wide" sampling of MIC-limited genome content resulted from its (unavoidable) contamination of the shotgun libraries constructed for MAC genome sequencing. Because of low coverage and repetitiveness, these reads did not assemble well, but there was enough information to show the presence of a surprising diversity of transposable element coding regions (Eisen *et al.*, 2006). Clearly, multiple "invasions" of the

T. thermophila germline by mobile elements have occurred and the fact that many of their coding sequences have not completely degenerated indicates either that some invasions were recent, that the elements somehow remain active and/or even that they may play an active role in their own excision.

The first directed MIC genome project to get underway was led by Kathy Collins of the University of California at Berkeley. Sequencing of an 8 kb MIC genome plasmid library was conducted at the Joint Genome Institute through its Community Genome Sequencing Program. As noted above, large insert *Tetrahymena* libraries are notoriously unstable and unfortunately this one was no exception; the linkage information was unreliable, but the long Sanger reads did prove valuable in locating IES junction sites. Several important conclusions could be drawn from the results (Fass *et al.*, 2011). First, as proposed decades prior on the basis of a very limited sampling of random MIC clones (Yao *et al.*, 1984), there are most likely about 6000 IESs dispersed throughout the MAC-destined regions. Second, nearly all IESs are found in intergenic regions or the noncoding portions of genes, but third, some IESs do interrupt gene coding regions. Interestingly, one intergenic IES appears to provide a mechanism by which a functional gene product may only be expressed from the developing MAC after excision has occurred. A genome browser has been set up to visualize the data at http://www.gb.genomecenter.ucdavis.edu/cgi-bin/hgGateway?hgsid=2689&clade=alveolata&org=0&db=0.

While this information is valuable, it falls far short of the value of a complete MIC genome sequence. A separate effort, funded by the National Human Genome Research Institute (NHGRI) through a white paper submitted by RSC and Ed Orias, is now underway at the Broad Institute's genome sequencing center with the aim of producing a whole genome MIC assembly. Preliminary results suggest a MIC genome size of around 150 Mb, a higher estimate than arrived at through reassociation kinetics (Yao and Gorovsky, 1974). These results and updates to the assembly and annotation can now be accessed at: http://www.broadinstitute.org/annotation/genome/*Tetrahymena*/MultiHome.html and are available at NCBI's Genbank under the accession number AFSS00000000. The data will also be transferred to TGD when the assembly process is completed.

C. The Latest Models: Fixing Gene Structures

Ideally, researchers interested in functional gene characterization or evolutionary patterns of gene conservation (e.g., phylogenomics) should have access to fully accurate gene models. Furthermore, for the application on a genome-wide scale of many technologies, such as the mapping of chromatin structural elements and sRNA sequences relative to transcription units, it is also imperative to have accurate gene models. Planners of the massive ENCODE (Myers *et al.*, 2011) and modENCODE projects (Gerstein *et al.*, 2010; Roy *et al.*, 2010), which aim to map all functional elements in the genomes of humans, *Drosophila* and *C. elegans*, understood this well and have devoted substantial effort to this task. Their results illustrate the extent to which even the most well-studied genomes are still "works in progress" with respect to possessing a set of models completely free of errors and omissions in start and stop sites of transcription and translation, exon/intron junctions, and sites of alternative splicing. For example, despite ten years of extensive *Drosophila* annotation efforts (between initial genome sequencing and the first modENCODE publication), 74% of annotated genes still required at least one correction to their exon structures or displayed novel alternative splice forms (Roy *et al.*, 2010).

That said, the most current (2008) *Tetrahymena* gene models are in need of major improvement. Several recent in-depth studies of particular gene families have documented cases of inaccurate gene models, gene fusions, missing genes, and pseudogenes annotated as functional. Such expert corrections are very valuable and can be used to update database entries, but of course it would be unrealistic and inefficient to pursue such a gene-by-gene strategy on a genome-wide scale. The three most promising large-scale approaches to the improvement of structural gene annotation are, in order of utility, transcriptomics, comparative genomics, and proteomics. Application of these three methods, along with manual curation, to the *Tetrahymena* genome, will greatly benefit the research community, but, as the ENCODE results attest, needs to be an ongoing process.

1. Transcriptomics

Deep RNA sequencing (RNA-seq) has emerged as a powerful method to characterize transcriptomes and revealed unexpected levels of transcriptome complexity, including multiple transcription start sites and alternative splicing events (Ozsolak and Milos, 2011). To exploit this technology, Xiong *et al.* (2011; submitted) performed high-throughput RNA-seq on six polyA-purified RNA samples at five time points of three major physiological or developmental stages of *T. thermophila*: growth, starvation, and conjugation. They obtained about 94 million paired-end reads, with a total length of more than 14 Gb. About 65% of the reads could be uniquely mapped to the *T. thermophila* reference genome, covering 57 Mb of sequence, about 55% of the MAC genome. The six combined RNA-seq datasets detected 96% (23,770 of 24,725) of the previously annotated open reading frames in the genome. In 6633 cases, the gene models and RNA-seq data were in perfect agreement. However, the data indicate that at least 7300 predicted gene models require correction and identify 1474 potential alternative splicing (AS) events distributed over 5.2% of *T. thermophila* genes (this percentage represents a two orders of magnitude increase over previous EST-based estimates). Additionally, more than 1000 new transcribed regions were identified. These results significantly improve the genome annotation and indicate a larger *T. thermophila* transcriptome than previously predicted.

More specialized technologies, such as CAGE (Kanamori-Katayama *et al.*, 2011), 3P-Seq (Jan *et al.*, 2011), and strand-specific RNA-seq (Ozsolak and Milos, 2011) improve resolution of transcription start sites and polyA addition sites and allow detection of antisense transcripts. The application of these, as well as emerging technologies such as direct RNA sequencing (Ozsolak and Milos, 2011), will continue to augment and refine our understanding of the transcriptome of *Tetrahymena*.

2. Comparative Genomics

Studies of yeast, *Drosophila,* and other model organisms have demonstrated the enormous value of comparative genomic analysis in refining gene models (as well as in defining conserved noncoding sequences.; see Section V.A below) (Bergman *et al.*, 2002; Kellis *et al.*, 2003). Conservation of open reading frame structural features across species is a strong indication of their functional importance. In addition to the MIC genome sequencing described above, the Broad Institute is currently sequencing and annotating the MAC genomes of three new species of *Tetrahymena* – *T. malaccensis*, *T. elliotti,* and *T. borealis*. Two of these are the closest known relatives of *T. thermophila,* and the third is at an intermediate distance (see Fig. 1). The more distantly related *I. multifiliis* and *P. tetraurelia* (soon to be joined by several other *Paramecium* species) genomes are also available (Aury *et al.*, 2006; Coyne *et al.*, 2011). This range of distances should allow the detection of conserved sequence elements that diverged more or less rapidly from their ancestral states. Comparative genomic analysis of these data will serve to cross-check transcriptomic results and also validate structures of genes whose transcription is undetectable under the laboratory conditions thus far used.

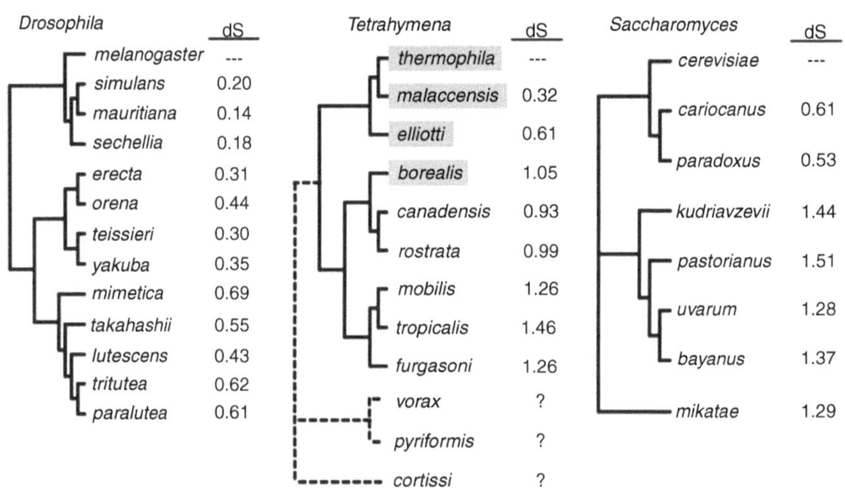

Fig. 1 Phylogeny of *Tetrahymena*, *Drosophila,* and *Saccharomyces* genera based on comparisons of the same gene, encoding the centromere-specific histone (*Tetrahymena* CNA1, *Drosophila* Cid, and *Saccharomyces* CSE4). dS values (ratio of synonymous nucleotide substitutions/total possible synonymous substitutions) were calculated for the 270 bp conserved histone fold domain. Figure generated by Nels Elde and used by permission (unpublished). *Tetrahymena* species sequenced or being sequenced are shaded.

3. Proteomics

Normally, a well-annotated genome is part of the input data for successful proteomic analysis. Several proteomic studies have been conducted on *Tetrahymena* organelles and protein complexes (see above). These studies have been very enlightening, but have also highlighted the shortcomings of *Tetrahymena* genome annotation. In fact, by comparing proteomic data not to predicted coding regions, but to sixfold translations of the entire genome, many corrections to gene models have been made (Smith *et al.*, 2005, 2007). This strategy will be continued in future proteomic studies.

D. Gene Identity: Functional Annotation

Beyond getting gene structures right, it is of critical importance to many downstream studies to characterize genes with respect to the predicted function of their protein products. Depending on the individual gene, this can be accomplished to varying degrees of specificity and confidence. Naturally, the highest confidence stems from direct experimental evidence regarding the gene product's localization, biochemical activity, interaction partners, and/or the phenotype resulting from its absence or impairment. This type of information is captured into databases by curators or, in the Wiki model currently in place at TGD, by the investigators themselves. At present, experimental evidence is available for relatively few *Tetrahymena* genes, though application of high-throughput methods will improve this situation (see Section V.D below). Meanwhile, what we can infer about the function of most *Tetrahymena* genes comes either from their developmental expression and coexpression profiles, proteomic association with a subcellular structure, or the protein sequence features they share with better characterized homologs and orthologs of other species. These sequence features were characterized by automated annotation methods on the 2005 predicted gene set (Eisen *et al.*, 2006). In light of improvements to existing databases and algorithms as well as to the gene models themselves, repeating these analyses will greatly improve the accuracy of functional annotation available in TGD and NCBI.

A highly valuable tool for defining probable gene function is orthology. Orthologs are genes in different species that evolved from a common ancestral gene by speciation and normally retain the same function. Two notable databases contain computed orthology relationships between the current *Tetrahymena* gene product set and those of certain other species. *Paramecium* researchers in France built CilDB (http://cildb.cgm.cnrs-gif.fr/)(Arnaiz *et al.*, 2009), a database focused on ciliary proteins, but which can also be mined for information on any gene. CilDB uses Inparanoid to calculate shared orthologs between any pair of 33 different eukaryotic species, including *T. thermophila*. The BioMart query tool can be used to filter output according to a number of user-selected criteria. Making use of a different ortholog classification algorithm, the OrthoMCL database (http://www.orthomcl.db)(Chen *et al.*, 2006) currently contains data on 138 prokaryotic and eukaryotic genomes, clustering over one million protein sequences into over 100,000 ortholog groups. The groups can be searched by domain, keyword, or phyletic pattern. This database was recently used to compare the proteomes of *Tetrahymena*, *Paramecium*, their parasitic relative Ich, and, as a representative host species, the zebrafish *Danio rerio* (Coyne *et al.*, 2011). A number of potentially ciliate-specific ortholog groups were identified that contain only representatives from one or more of these three ciliates. Enzyme Commission (EC) numbers were assigned to the proteomes based on their ortholog grouping, and these data used to reconstruct the main features of ciliate metabolism. In summary, the orthology relationships in these databases are useful in identifying probable gene function and interrogating pathways.

Despite their power, informatic sequence comparison tools such as orthology mapping and domain searches are inherently limited, not least because on the order of 50% of genes in a typical eukaryotic genome have no identifiable, functionally relevant sequence features. This accounts for the abundance of gene products annotated as simply "hypothetical protein". Comprehensive collections of knockouts and/or tagged genes (see Section V.D below) represent an alternative approach to assigning gene functions globally, but due to the technical challenges, these are long-term prospects. An immediately accessible approach taken by Xiong *et al.* (2011b) is to group genes by correlation of their expression patterns across multiple growth and development conditions (Hughes *et al.*, 2000; Kim *et al.*, 2001). This statistical method resulted in clustering of *Tetrahymena* genes into 55 "modules", providing potential clues to functional relatedness (see also Section IV.E.2, below).

E. Updates and Improvements to Database Resources

To maximize the value of genomic data to the *Tetrahymena* research community, it is vital to maintain them in user-friendly, web-accessible databases that are regularly updated in their content and upgraded in their functionalities. The two major existing *Tetrahymena*-specific data sources are described below. As the needs of the community evolve and new forms of data become available, the database resources will need to be expanded and adapted.

1. *Tetrahymena* Genome Database

The *T. thermophila* MAC genome sequence and annotations are provided by the *Tetrahymena* Genome Database (TGD) at www.ciliate.org (Stover *et al.*, 2006). This online resource was founded in 2004 at Stanford University on the same platform as the popular *Saccharomyces* Genome Database (SGD) (Dwight *et al.*, 2004) and includes a BLAST server, genome browser, and search and display functions for gene annotations. From 2004 to 2006, the initial functional annotations provided by TIGR were updated and expanded by curators for several hundred genes based on the published literature. After the backlog of papers was exhausted, the project was moved to Bradley University and reintroduced as a community-updatable Wiki. This format allows researchers to add functional annotations directly to the site based on their published or unpublished results. Registered users in the *Tetrahymena* community are able to add annotations to a variety of fields in the database, including standardized gene names (Allen, 2000a), Gene Ontology (GO) annotations (Ashburner *et al.*, 2000), free-text descriptions, and associated literature. Extensive guidelines detailing the annotation standards and practices for TGD Wiki have been written and can be accessed at http://ciliate.org/index.php/show/editguide.

Improvements to the hardware and software used to run TGD are made as new technology becomes available, as for example when new versions of Gbrowse and BLAST are released. New browser tracks offering comparative information will be uploaded as additional genomes and analyses become available (see Section IV.C.2 above). Programs will also be written to automatically update the genome browser and BLAST server with community annotations, and to update the database with the latest information from Pubmed and the Gene Ontology.

At this time, the annotations presented in TGD Wiki come from two primary sources: large-scale BLAST analyses of the putative proteins by JCVI and information from published articles. In both cases, it is relatively easy for researchers to identify the source of the annotation and to view or recreate the data underlying the annotation. However, many observations made about *Tetrahymena* genes are small in scope and are not published, even though they may be informative about one or more genes. To accommodate annotations made from unpublished data, while still maintaining transparency about their origins, a companion site to TGD Wiki is being developed. This site will act as a clearinghouse for unpublished studies, where researchers can upload and display abbreviated reports containing background, figures, methods, etc., that support annotations made at TGD Wiki. The new website will expand on the current Ciliate Genomics Consortium, which primarily houses student data collected during laboratory classes. These studies will continue to be a main source of information in the unpublished results database, though other researchers will be invited to submit data as well. Over time we expect this new, unique resource, combined with the ability of *Tetrahymena* researchers to edit the genome database directly, to significantly enhance the annotation of the *Tetrahymena* genome.

During the next few years, sequencing projects will produce a wealth of data of interest to *Tetrahymena* researchers. Completion of the MIC genome will be a major step in the study of MAC development, and it will be important to incorporate data from this effort into TGD Wiki in a meaningful way. The entire MIC genome sequence will be incorporated into the BLAST server and genome browser, and these sequences will be marked with sites of IES excision and chromosome breakage in the MAC genome browser. The genomes of *T. malaccensis*, *T. elliotti,* and *T. borealis* will be made available for search at TGD Wiki, though no plans are in place to create community annotation sites for these nonmodel species. The genomes of more distantly related ciliates, such as Ich and *Paramecium* species, will be used to augment the comparative genomics resources at TGD Wiki. Ich genomic data will be fully incorporated into TGD. *Paramecium* genome data are hosted independently at the full-service genome database website *Paramecium*DB (http://paramecium.cgm.cnrs-gif.fr/). Other advances described above, in particular closure and correction of the *T. thermophila* MAC genome sequence and reannotation of gene-coding regions, will also be incorporated. Alignment of MAC and MIC chromosomes will allow a more rational approach to numbering genes that reflects their natural arrangement in the genome.

TGD Wiki will continue its focus on improving the annotation of the *T. thermophila* genome throughout the coming years. It also maintains an intimate connection with the *Tetrahymena* Stock Center (http://*Tetrahymena*.vet.cornell.edu/), a comprehensive repository and distribution source for standard and mutant strains.

2. *Tetrahymena* Functional Genomics Database (TetraFGD)

Building on the foundation of the *Tetrahymena* Gene Expression Database (Xiong *et al.*, 2011a) (see Section II above), TetraFGD (http://tfgd.ihb.ac.cn/) has been established to house *Tetrahymena* microarray, RNA-seq, and gene network information (Miao *et al.*, 2009; Xiong *et al.*, 2011b). Gene expression profiles and candidate co-expressed genes can be retrieved using gene ID or gene description searches in TetraFGD. In addition, transcripts identified by RNA-seq can be accessed through Gbrowse or BLAST. TetraFGD will be expanded to develop and collect other functional genomics data (e.g., proteomics) as they become available and is intended to be a resource for all members of the *Tetrahymena* research community.

V. Looking Further Ahead

In contemplating the future direction of genome-wide studies in *Tetrahymena*, it is necessary to focus on the unusual strengths of this model organism and the ways in which it may make unique contributions to general understanding. Many of these strengths are more fully described in other chapters of this volume; we discuss them here in a genomics context.

A. Chromatin

Marty Gorovsky was one of the first to recognize the potential of *Tetrahymena*'s nuclear dualism and nuclear developmental program to reveal interesting features of chromatin (Gorovsky, 1973). Its robust biochemistry, powerful genetics (including the unusual ability to make histone gene replacements [Liu *et al.*, 1996]), flow cytometric methods (Allis and Dennison, 1982), etc., continue to recommend its use for this purpose. However, there are huge gaps in our basic understanding of many chromatin-associated functions in this and all ciliates. With very few exceptions, we are completely ignorant of the ciliate cis-acting DNA sequences controlling transcription and other basic chromosomal functions. Likewise, although many transcription factors can be identified by homology, their functions, activities, and network relationships are as yet almost entirely unstudied. The integration of a variety of whole genome studies represents a promising approach to remedying this state of affairs.

Comparative genomics is not only useful for gene structure annotation (see Section IV.C.2 above) but also to identify conserved noncoding elements such as promoters, enhancers, and potentially replication origins (Bergman *et al.*, 2002; Kellis *et al.*, 2003). Because the intergenic regions (in particular) of *Tetrahymena* and its relatives are extremely AT-rich, it may be challenging to define such DNA elements. The three additional species currently in the sequencing pipeline will not be fully sufficient, but with improved technology and reduced costs, we can expect to see yet more species in the future. And, because transcription of the MIC genome is critical for scnRNA-guided developmental genome rearrangement (see Section V.B below), it will be critical to sequence multiple additional MIC genomes to be able to compare their transcriptional control elements as well.

The number of genome-wide mapping studies of transcription factors and chromatin features in a variety of species has exploded in recent years with the use of chromatin immunoprecipitation, coupled now with deep sequencing (ChIP-Seq). Related technologies, such as DNase-Seq (Boyle *et al.*, 2008) and MNase-Seq (Schones *et al.*, 2008), can rapidly map open chromatin regions and nucleosome positions across the genome. Such knowledge could dramatically improve our understanding of *Tetrahymena*'s chromatin landscape and how it changes over the course of development or in response to environmental changes and also aid in the prediction of cis-acting DNA elements. Currently, full-scale mapping of most *Tetrahymena* transcription and replication factors is unrealistic (although building on the experience gained through the ENCODE projects, application of such technologies will become more readily available in other model systems). However, more focused applications of ChIP-Seq and related methods should soon be applied to *Tetrahymena*. Clearly, histone post-translational modification has been a fertile field for *Tetrahymena* that is now ripe for genome-wide studies. Availability of the MIC genome sequence will also open the genome rearrangement process to ChIP-Seq studies of histone modification, chromatin modifiers, and sRNA-associated factors as well as deep sequencing of chromatin-associated RNAs (Mondal *et al.*, 2010). Localization of RNA polymerase across the genome will help define the boundaries of genes and noncoding transcription units and perhaps reveal the existence of paused polymerases. Eventually, performing such studies on multiple divergent strains of *T. thermophila*, in particularly interesting mutant backgrounds or even in different species, will increase confidence in the results and improve spatial and temporal resolution.

B. Chromosome Structure and Rearrangement

Clearly, ciliates have evolved some unusual solutions to common challenges faced by eukaryotic genomes, but as so often seen in the past, investigating such oddball characteristics can provide universal insights. One obvious unusual feature of the

MAC genome is its paucity of repetitive DNA sequences, and in particular mobile elements. Whereas most eukaryotic chromosomes are divided into euchromatic and heterochromatic domains, gene mapping (Eisen *et al.*, 2006) and genome-wide expression studies (Miao *et al.*, 2009) of *Tetrahymena* have revealed no clear patterns of organization of the MAC genome that may reflect broad chromatin domains as, for example, associated with centromeres (absent in the MAC) or telomeres. Comparative genomics and genome-wide chromatin studies will help confirm or refute these initial indications. Comparing and contrasting the logic of chromosomal structure as it relates to gene regulation and DNA replication between *Tetrahymena*, other ciliates, and other eukaryotes will be enlightening.

Tetrahymena MAC chromosomes apparently lose their centromeres during development (Cervantes *et al.*, 2006; Cui and Gorovsky, 2006) and have no mechanism for equal segregation of their approximately 45 copies at each cell division. This results in the phenomenon of phenotypic assortment, in which initially heterozygous cells eventually become homozygous after repeated vegetative divisions (Orias and Flacks, 1975). The consistent rate of phenotypic assortment at different loci (Doerder *et al.*, 1992) suggests that each of the 181 MAC chromosomes maintains an equal copy number through an active control mechanism. Likewise, the equal depth of sequencing coverage observed in the genome project supports a generally equal copy number for each non-rDNA chromosome (at least at the population level) (Eisen *et al.*, 2006). However, especially under certain selective conditions, it may be that copy number variation is used adaptively by *Tetrahymena* and other ciliates. This could be evaluated using array or deep sequencing technologies. The mechanism of copy number control of nearly 200 chromosomes is mysterious (Larson *et al.*, 1991). This may become open to investigation using engineered whole chromosomes (see below) and/or with the development of technologies that allow determination of copy number in single cells and the associated changes in DNA replication control.

The study of DNA replication in *Tetrahymena* has also yielded interesting surprises, from the first eukaryotic origin to be mapped (on the rDNA) (Cech and Brehm, 1981) to the novel association of a small RNA with the origin recognition complex (ORC) (Mohammad *et al.*, 2007). Studies (Donti *et al.*, 2009) showing cell cycle-dependent chromosomal redistribution of ORC provide further justification for genome-wide studies of its localization. Another striking feature of *Tetrahymena* DNA replication is that the MAC and MIC go through S phase during different periods of the cell cycle (McDonald, 1962). It will be of great interest to compare on a genome-wide scale the replication origins of these two highly distinct nuclei.

As with the MAC, the silent germline MIC genome also presents an unusual adaptation. As an apparent safe haven for transposable elements, it is perhaps surprising that it has not expanded to greater size as, for example, the MIC genomes of spirotrichous ciliates (Prescott, 1994). It is clearly of immediate interest to characterize the bidirectional transcription that gives rise to scnRNAs (Chalker and Yao, 2001). Besides the obvious involvement of these transcripts in promoting DNA excision, it will be of interest to see if there are correlations between transcription and recombination rates or other features across the MIC genome. Comparative MIC genomics will also help reveal the selective pressures on MIC chromosome structure. It is intriguing that the only known amicronucleate *T. thermophila* mutant contains significant amounts of "MIC-limited" sequence in its MAC (Karrer *et al.*, 1984). Whole MAC genome and sRNA sequencing of several amicronucleate *Tetrahymena* species will help shed light on how they have learned to live without a germline and whether the absence of an epigenetic DNA elimination mechanism has left their MAC genomes open to recent invasion by mobile DNA elements.

The fact that mutations that perturb developmental DNA excision are lethal indicates that genome-wide retention of MIC-limited sequences in the MAC is incompatible with gene function. However, it is clear from studies of epigenetic interference of targeted excision events (Chalker and Yao, 1996) and from partial knockdown of the PDD1 gene (M.C. Yao; personal communication) that some MIC-limited elements can be tolerated. The number and variety of such elements may be identified by genome-wide interference studies and/or tiling array analysis or whole genome sequencing of survivors of partial knockdown matings.

C. Diversity

In the preceding sections, we have highlighted comparative genomics primarily as a means to better understand the workings of *T. thermophila*, and even more specifically, the inbred B strain, on which practically all research with this species is currently done. However, there is great value in studying diversity. For one thing, the torturous process of creating inbred strains, involving severe population bottlenecks, may have resulted in unknown genomic polymorphisms in comparison to the original "wild" state (Nanney and Simon, 2000). As sequencing costs drop, it may behoove us to compare the genomes and transcriptomes of the various inbred strains to each other and to stocks recently isolated from nature. The natural strain-to-strain variation in gene expression and co-expression patterns can also be used to more accurately model transcriptional networks (Wessel *et al.*, 2007; Zhu *et al.*, 2008).

Whole genome analyses are also increasingly being applied to questions of ecological and evolutionary interest, giving rise to the field of population genomics (Nadeau and Jiggins, 2010). One of the principal goals of this field is to define, by unbiased whole genome sampling, the molecular basis of adaptive phenotypes, which may be difficult to define *a priori*, particularly in microbes (Ellison *et al.*, 2011). As bacterial grazers, ciliates occupy an important ecological niche. *T. thermophila* is distributed widely in the eastern United States (and perhaps beyond) and other species have even wider distributions. It is expected that subpopulations will have undergone genetic selection to adapt to their local environments. Because of the phenomena of phenotypic assortment (Orias and Flacks, 1975) and epigenetic inheritance, it may even be possible for *Tetrahymena* populations to adapt rapidly, for example, to seasonal fluctuations, by selection for advantageous allelic ratios and/or alternative genome rearrangement patterns in the MAC, even in the absence of germline selection. Such questions are amenable to the unbiased "reverse ecology" methods of population genomics. Once candidate loci for adaptive traits are identified, the powerful genetic methods of *Tetrahymena* may be applied to their functional characterization.

Besides population diversity, there is considerable species diversity within the *Tetrahymena* genus. For example, there have been several apparently independent adoptions of parasitic lifestyles (Struder-Kypke *et al.*, 2001). Whole genome sequencing and annotation of the related fish parasite Ich revealed extensive gene loss compared to *T. thermophila*, but the retention of overall similar metabolic capabilities (Coyne *et al.*, 2011). It would be interesting to compare how the genomes of other, independently evolved oligohymenophoran ciliate parasites have been altered by this lifestyle adaptation. The opportunity to conduct such an analysis is not available in many taxa. Other examples of phenotypic diversity within the *Tetrahymena* genus that could be amenable to comparative genomics include (as already mentioned) the presence or absence of the germline micronucleus, the varied systems of mating-type determination (Simon and Orias, 1987), and the remarkable morphological transformation of some species from a bacterivorous form to a "macrostome" form that preys on other ciliates (Ryals *et al.*, 2002).

D. Genetics

The ciliates *Paramecium* and *Tetrahymena* have long been subjects of forward genetic research. A rich collection of *T. thermophila* mutant strains defective in pattern formation, secretion, phagocytosis, and other functions is stored at the *Tetrahymena* Stock Center. However, molecular and cellular analysis of these mutants awaits the identification of the responsible genes. Cloning by complementation has been achieved in *Paramecium* (in which any injected DNA is maintained in the MAC) (Haynes *et al.*, 1998; Keller and Cohen, 2000), but not yet in *Tetrahymena*. With reductions in costs, whole genome sequencing of mutant strains has emerged as a highly attractive alternative strategy. First demonstrated in eukaryotes in *C. elegans* (100 Mb genome) (Sarin *et al.*, 2008), it has since been applied to *Drosophila* (123 Mb genome) (Blumenstiel *et al.*, 2009) and recently *Paramecium* (72 Mb genome; O. Arnaiz, L. Sperling; personal communication). To achieve this goal in *Tetrahymena*, the first required step is to correct the many sequencing errors that are common in low-coverage genome assemblies (see Section IV.A above) so that deep sequencing of mutants does not return an unreasonable number of false positive SNPs and indels. Basic genetic mapping of mutants using simple tricks of *Tetrahymena* genetics such as mating to nullisomic or deletion strains and co-assortment mapping can rapidly narrow down the genome region of interest. Candidate genes can then be confirmed by rescue transformation. It is hoped that this technology will not only reveal the genes behind the mutant phenotypes of existing strains, but also generate renewed interest in conducting genetic screens in this productive system.

While forward genetics is a powerful approach, even in extensively studied organisms such as *Drosophila,* so-called saturation mutagenesis is known to miss numerous gene targets. Fortunately, because *Tetrahymena* undergoes transformation by homologous recombination, it is well adapted to "reverse genetic" analysis. The most comprehensive reverse genetic resource available in any eukaryotic organism is the bar-coded collection of yeast gene knockouts (Shoemaker *et al.*, 1996), which, besides being available individually to any investigator for focused studies, can also be used to simultaneously interrogate the competitive advantages or disadvantages of the entire collection of mutants under any experimental condition (Hillenmeyer *et al.*, 2008). Because of technical limitations and a fourfold larger gene number, it would currently be a great challenge to replicate this resource for *Tetrahymena*, but smaller scale high-throughput knockout production has now begun (RSC; unpublished).

An alternative genome-wide "knockdown" strategy would be to use a library of RNAi constructs (Howard-Till and Yao, 2006), but this approach would first require establishing conditions under which inhibition of gene function could be efficiently and reliably induced throughout development. The "antisense ribosome" method, invented in *Tetrahymena* (Sweeney *et al.*, 1996), has also been productively employed as a screen for factors associated with certain cellular functions (Chilcoat *et al.*, 2001) but has not been fully tested and would in any case not be applicable to essential genes.

In addition to knockouts, it is likewise feasible to envision genome-wide collections of GFP- and affinity-tagged constructs. Because these could be generated by *E. coli*-based cloning in an rDNA transformation vector, the technical challenge, while still great, is not as great as producing the full knockout collection. Each of these collections would be highly useful in terms of assigning potential functions to genes, especially those lacking informative sequence features. Affinity tagged constructs could be used for proteome-wide interactome mapping, as has been done in other model organisms (Gavin *et al.*, 2002; Ho *et al.*, 2002; Williamson and Sutcliffe, 2010).

Finally, we consider the prospects for the ultimate form of reverse genetics – the engineering of the *Tetrahymena* genome. The purpose of such an endeavor could either be related to basic research or to better adapting *Tetrahymena* for use in biotechnology. For example, proteomic studies of the secretome indicate an abundance of secreted proteases that could severely limit *Tetrahymena*'s potential for expression of foreign proteins (Herrmann *et al.*, 2006; Madinger *et al.*, 2010). Under controlled growth conditions, secretion of these enzymes could most likely be eliminated without adverse effects. The power to engineer "improved" versions of *Tetrahymena* in this and other ways may be within reach in the not too distant future.

Being transcriptionally silent, the MIC genome can be radically altered, even to the point of eliminating one or more chromosomes. Such nullisomic or unisomic strains (described in Chapter 10) are useful for genetic mapping. More fine-grained targeted deletions could be generated by a Cre-lox recombinase system currently under development (Cassidy-Hanley, Clark; personal communication). Altering the MIC genome would also be useful for mapping MIC-specific functional elements or making MAC-destined deletions beyond the limits of homology-dependent gene replacement (which have not actually been tested). For example, many expanded paralogous gene families are arranged in tandem arrays (Eisen *et al.*, 2006) that could potentially be deleted *en masse*.

It may also be possible and useful to directly modify the MAC genome, either of *T. thermophila* or, to avoid the complications of nuclear dualism, of an amicronucleate species such as *T. pyriformis*. One possibility would be the introduction of engineered chromosomes. The recent complete chemical synthesis of a bacterial chromosome (of a size comparable to many *Tetrahymena* MAC chromosomes) (Gibson *et al.*, 2010) even raises the possibility of a "designer" genome. Efforts are now well underway to chemically synthesize and replace the entire genome of *S. cerevisiae* (http://www.syntheticyeast.org) (Dymond *et al.*, 2011), which will likely make it the first fully "synthetic" eukaryote. The technology is advancing rapidly, with concomitantly decreasing costs. Current methods of chromosome assembly require propagation as circular Yeast Artificial Chromosomes (YACs) (Gibson, 2011). It has been demonstrated (RSC; unpublished results) that YACs can be maintained containing up to, at least, 180 kb of *Tetrahymena* genomic DNA, although successful reintroduction of such DNA into *Tetrahymena* has not yet been achieved. When such methods are developed, it will become possible to start building TACs – *Tetrahymena* Artificial Chromosomes – to test hypotheses about chromosome structure or genetic networks or even to begin the stepwise redesign and replacement of the MAC genome.

An alternative to fully synthetic genome remodeling is to apply methods of "accelerated evolution" such as Multiplex Automated Genome Engineering (MAGE) (Wang and Church, 2011). To date, this method has been applied only to bacteria, but its developers envision its use in a wide variety of organisms. With *Tetrahymena*'s rapid cell cycle, growth to high density, and precise homology-dependent recombination machinery, it may be well suited to such technology.

VI. Concluding Remarks

Clearly, genomic resources have made a major impact on *Tetrahymena* research in the few years they have been available. These resources are set to expand and be improved on in the near future, enabling more detailed studies in research areas for which *Tetrahymena* is well adapted. Substantial effort and funding will be required to maintain and further expand these resources to keep up with developments in the field. Genomics is a rapidly advancing field and we cannot predict what novel technologies may emerge at any time, but *Tetrahymena*'s experimental strengths will allow this model system to take advantage of them.

Acknowledgments

RSC is grateful for the support of the City of Paris through the program Research in Paris 2010 and of the National Institutes of Health (Grant 1R24GM089662). NAS would like to thank the National Institutes of Health for their support of the Tetrahymena Genome Database by a subaward of NIH grant 5P40RR019688-06 to Ted Clark at Cornell University and the members of the Ciliates in the Classroom working group for their support and their dedication to undergraduate research. W.M. is grateful for the support of the National Natural Science Foundation of China (No. 31071993), and Knowledge Innovation Program of the Chinese Academy of Sciences (No. KSCX2-YWG-072 and KSCX2-EW-G-6-4).

References

Adachi, M., and Cavalcanti, A. R. (2009). Tandem stop codons in ciliates that reassign stop codons. *J. Mol. Evol.* **68**, 424–431.

Allen, S. L. (2000a). Genetic nomenclature rules for Tetrahymena thermophila. *Methods Cell. Biol.* **62**, 561–563.

Allen, S. L. (2000b). Isolation of micronuclear and macronuclear DNA. *Methods Cell. Biol.* **62**, 241–252.

Allis, C. D., and Dennison, D. K. (1982). Identification and purification of young macronuclear anlagen from conjugating cells of *Tetrahymena thermophila*. *Dev. Biol.* **93**, 519–533.

Archibald, J. M. (2008). Plastid evolution: remnant algal genes in ciliates. *Curr. Biol.* **18**, R663–R665.

Arnaiz, O., Malinowska, A., Klotz, C., Sperling, L., Dadlez, M., Koll, F., and Cohen, J. (2009). *Cildb: a knowledgebase for centrosomes and cilia.* Database (Oxford), bap022.

Ashburner, M., Ball, C. A., Blake, J. A., Botstein, D., Butler, H., Cherry, J. M., Davis, A. P., Dolinski, K., Dwight, S. S., Eppig, J. T., Harris, M. A., Hill, D. P., Issel-Tarver, L., Kasarskis, A., Lewis, S., Matese, J. C., Richardson, J. E., Ringwald, M., Rubin, G. M., and Sherlock, G. (2000). Gene ontology: tool for the unification of biology. The gene ontology consortium. *Nat. Genet.* **25**, 25–29.

Aury, J. M., Jaillon, O., Duret, L., Noel, B., Jubin, C., Porcel, B. M., Segurens, B., Daubin, V., Anthouard, V., Aiach, N., Arnaiz, O., Billaut, A., Beisson, J., Blanc, I., Bouhouche, K., Camara, F., Duharcourt, S., Guigo, R., Gogendeau, D., Katinka, M., Keller, A. M., Kissmehl, R., Klotz, C., Koll, F., Le Mouel, A., Lepere, G., Malinsky, S., Nowacki, M., Nowak, J. K., Plattner, H., Poulain, J., Ruiz, F., Serrano, V., Zagulski, M., Dessen, P., Betermier, M., Weissenbach, J., Scarpelli, C., Schachter, V., Sperling, L., Meyer, E., Cohen, J., and Wincker, P. (2006). Global trends of whole-genome duplications revealed by the ciliate Paramecium tetraurelia. *Nature* **444**, 171–178.

Balabaskaran Nina, P., Dudkina, N. V., Kane, L. A., van Eyk, J. E., Boekema, E. J., Mather, M. W., and Vaidya, A. B. (2010). Highly divergent mitochondrial ATP synthase complexes in *Tetrahymena thermophila*. *PLoS Biol.* **8**, e1000418.

Bergman, C. M., Pfeiffer, B. D., Rincon-Limas, D. E., Hoskins, R. A., Gnirke, A., Mungall, C. J., Wang, A. M., Kronmiller, B., Pacleb, J., Park, S., Stapleton, M., Wan, K., George, R. A., de Jong, P. J., Botas, J., Rubin, G. M., and Celniker, S. E. (2002). Assessing the impact of comparative genomic sequence data on the functional annotation of the Drosophila genome. *Genome Biol.* **3**, RESEARCH0086.

Blumenstiel, J. P., Noll, A. C., Griffiths, J. A., Perera, A. G., Walton, K. N., Gilliland, W. D., Hawley, R. S., and Staehling-Hampton, K. (2009). Identification of EMS-induced mutations in *Drosophila melanogaster* by whole-genome sequencing. *Genetics* **182**, 25–32.

Boyle, A. P., Davis, S., Shulha, H. P., Meltzer, P., Margulies, E. H., Weng, Z., Furey, T. S., and Crawford, G. E. (2008). High-resolution mapping and characterization of open chromatin across the genome. *Cell* **132**, 311–322.

Bright, L. J., Kambesis, N., Nelson, S. B., Jeong, B., and Turkewitz, A. P. (2010). Comprehensive analysis reveals dynamic and evolutionary plasticity of Rab GTPases and membrane traffic in *Tetrahymena thermophila*. *PLoS Genet.* **6**, e1001155.

Cassidy-Hanley, D. M., Cordonnier-Pratt, M. M., Pratt, L. H., Devine, C., Mozammal Hossain, M., Dickerson, H. W., and Clark, T. G. (2011). Transcriptional profiling of stage specific gene expression in the parasitic ciliate *Ichthyophthirius multifiliis*. *Mol. Biochem. Parasitol.* **178**, 29–39.

Cech, T. R., and Brehm, S. L. (1981). Replication of the extrachromosomal ribosomal RNA genes of *Tetrahymena thermophilia*. *Nucleic Acids Res.* **9**, 3531–3543.

Cervantes, M. D., Xi, X., Vermaak, D., Yao, M. C., and Malik, H. S. (2006). The CNA1 histone of the ciliate *Tetrahymena thermophila* is essential for chromosome segregation in the germline micronucleus. *Mol. Biol. Cell* **17**, 485–497.

Chalker, D. L., and Yao, M. C. (1996). Non-Mendelian, heritable blocks to DNA rearrangement are induced by loading the somatic nucleus of *Tetrahymena thermophila* with germ line-limited DNA. *Mol. Cell. Biol.* **16**, 3658–3667.

Chalker, D. L., and Yao, M. C. (2001). Nongenic, bidirectional transcription precedes and may promote developmental DNA deletion in *Tetrahymena thermophila*. *Genes Dev.* **15**, 1287–1298.

Chau, M. F., and Orias, E. (1996). An improved method to obtain high molecular weight DNA from purified micro- and macronuclei of *Tetrahymena thermophila*. *J. Eukaryot. Microbiol.* **43**, 198–202.

Chen, F., Mackey, A. J., Stoeckert, C. J., Jr., and Roos, D. S. (2006). OrthoMCL-DB: querying a comprehensive multi-species collection of ortholog groups. *Nucleic Acids Res.* **34**, D363–D368.

Chilcoat, N. D., Elde, N. C., and Turkewitz, A. P. (2001). An antisense approach to phenotype-based gene cloning in Tetrahymena. *Proc. Natl. Acad. Sci. U.S.A.* **98**, 8709–8713.

Coesel, S., Obornik, M., Varela, J., Falciatore, A., and Bowler, C. (2008). Evolutionary origins and functions of the carotenoid biosynthetic pathway in marine diatoms. *PLoS One* **3**, e2896.

Cole, E. S., Anderson, P. C., Fulton, R. B., Majerus, M. E., Rooney, M. G., Savage, J. M., Chalker, D., Honts, J., Welch, M. E., Wentland, A. L., Zweifel, E., and Beussman, D. J. (2008). A proteomics approach to cloning fenestrin from the nuclear exchange junction of Tetrahymena. *J. Eukaryot. Microbiol.* **55**, 245–256.

Couvillion, M. T., Lee, S. R., Hogstad, B., Malone, Cn. D., Tonkin, L. A., Sachidanandam, R., Hannon, G. J., and Collins, K. (2009). Sequence, biogenesis, and function of diverse small RNA classes bound to the Piwi family proteins of *Tetrahymena thermophila*. *Genes Dev.* **23**, 2016–2032.

Coyne, R. S., Hannick, L., Shanmugam, D., Hostetler, J. B., Brami, D., Joardar, V. S., Johnson, J., Radune, D., Singh, I., Kumar, U., Saier, M., Wang, Y., Cai, H., Gu, J., Mather, M. W., Vaidyan, A. B., Wilkes, D. E., Rajagopalan, V., Asai, D. J., Pearson, C. G., Findly, R. C., Dickerson, H. W., Badger, J. H., Wu, M., Martens, C., Van de Peer, Y., Roos, D. S., Cassidy-Hanley, D. M., and Clark, T. G. (2011). Comparative genomics of the pathogenic ciliate Ichthyophthirius multifiliis, its free-living relatives and a host species provide insights into adoption of a parasitic lifestyle and prospects for disease control. *Genome Biol.* **12**, R100.

Coyne, R. S., Thiagarajan, M., Jones, K. M., Wortman, J. R., Tallon, L. J., Haas, B. J., Cassidy-Hanley, D. M., Wiley, E. A., Smith, J. J., Collins, K., Lee, S. R., Couvillion, M. T., Liu, Y., Garg, J., Pearlman, R. E., Hamilton, E. P., Orias, E., Eisen, J. A., and Methe, B. A. (2008). Refined annotation and assembly of the Tetrahymena thermophila genome sequence through EST analysis, comparative genomic hybridization, and targeted gap closure. *BMC Genomics* **9**, 562.

Cui, B., and Gorovsky, M. A. (2006). Centromeric histone H3 is essential for vegetative cell division and for DNA elimination during conjugation in *Tetrahymena thermophila*. *Mol. Cell. Biol.* **26**, 4499–4510.

Dear, P. H., and Cook, P. R. (1993). Happy mapping: linkage mapping using a physical analogue of meiosis. *Nucleic Acids Res.* **21**, 13–20.

Doerder, F. P., Deak, J. C., and Lief, J. H. (1992). Rate of phenotypic assortment in *Tetrahymena thermophila*. *Dev. Genet.* **13**, 126–132.

Donti, T. R., Datta, S., Sandoval, P. Y., and Kapler, G. M. (2009). Differential targeting of Tetrahymena ORC to ribosomal DNA and non-rDNA replication origins. *EMBO J.* **28**, 223–233.

Dwight, S. S., Balakrishnan, R., Christie, K. R., Costanzo, M. C., Dolinski, K., Engel, S. R., Feierbach, B., Fisk, D. G., Hirschman, J., Hong, E. L., Issel-Tarver, L., Nash, R. S., Sethuraman, A., Starr, B., Theesfeld, C. L., Andrada, R., Binkley, G., Dong, Q., Lane, C., Schroeder, M., Weng, S., Botstein, D., and Cherry, J. M. (2004). Saccharomyces genome database: underlying principles and organisation. *Brief Bioinform.* **5**, 9–22.

Dymond, J. S., Richardson, S. M., Coombes, C. E., Babatz, T., Muller, H., Annaluru, N., Blake, W. J., Schwerzmann, J. W., Dai, J., Lindstrom, D. L., Boeke, A. C., Gottschling, D. E., Chandrasegaran, S., Bader, J. S., and Boeke, J. D. (2011). Synthetic chromosome arms function in yeast and generate phenotypic diversity by design. *Nature* **477**, 471–476.

Eisen, J. A., Coyne, R. S., Wu, M., Wu, D., Thiagarajan, M., Wortman, J. R., Badger, J. H., Ren, Q., Amedeo, P., Jones, K. M., Tallon, L. J., Delcher, A. L., Salzberg, S. L., Silva, J. C., Haas, B. J., Majoros, W. H., Farzad, M., Carlton, J. M., Smith, R. K., Garg, J., Pearlman, R. E., Karrer, K. M., Sun, L., Manning, G., Elde, N. C., Turkewitz, A. P., Asai, D. J., Wilkes, D. E., Wang, Y., Cai, H., Collins, K., Stewart, B. A., Lee, S. R., Wilamowska, K., Weinberg, Z., Ruzzo, W. L., Wloga, D., Gaertig, J., Frankel, J., Tsao, C. C., Gorovsky, M. A., Keeling, P. J., Waller, R. F., Patron, N. J., Cherry, J. M., Stover, N. A., Krieger, C. J., Del Toro, C., Ryder, H. F., Williamson, S. C., Barbeau, R. A., Hamilton, E. P., and Orias, E. (2006). Macronuclear genome sequence of the ciliate *Tetrahymena thermophila*, a model eukaryote. *PLoS Biol.* **4**, 1620–1642.

Ellison, C. E., Hall, C., Kowbel, D., Welch, J., Brem, R. B., Glass, N. L., and Taylor, J. W. (2011). Population genomics and local adaptation in wild isolates of a model microbial eukaryote. *Proc. Natl. Acad. Sci. U.S.A.* **108**, 2831–2836.

Farazi, T. A., Juranek, S. A., and Tuschl, T. (2008). The growing catalog of small RNAs and their association with distinct Argonaute/Piwi family members. *Development* **135**, 1201–1214.

Fass, J. N., Joshi, N. A., Couvillion, M. T., Bowen, J., Gorovsky, M., Hamilton, E. P., Orias, E., Hong, K., Coyne, R. S., Eisen, J. A., Chalker, D. L., Lin, D., and Collins, K. (2011). Genome-scale analysis of programmed DNA elimination sites in *Tetrahymena thermophila*. *G3: Genes, Genomes, Genetics* **1**, 515–522.

Fu, C., Xiong, J., and Miao, W. (2009). Genome-wide identification and characterization of cytochrome P450 monooxygenase genes in the ciliate *Tetrahymena thermophila*. *BMC Genomics* **10**, 208.

Gavin, A. C., Bosche, M., Krause, R., Grandi, P., Marzioch, M., Bauer, A., Schultz, J., Rick, J. M., Michon, A. M., Cruciat, C. M., Remor, M., Hofert, C., Schelder, M., Brajenovic, M., Ruffner, H., Merino, A., Klein, K., Hudak, M., Dickson, D., Rudi, T., Gnau, V., Bauch, A., Bastuck, S., Huhse, B., Leutwein, C., Heurtier, M. A., Copley, R. R., Edelmann, A., Querfurth, E., Rybin, V., Drewes, G., Raida, M., Bouwmeester, T., Bork, P., Seraphin, B., Kuster, B., Neubauer, G., and Superti-Furga, G. (2002). Functional organization of the yeast proteome by systematic analysis of protein complexes. *Nature* **415**, 141–147.

Gerstein, M. B., Lu, Z. J., Van Nostrand, E. L., Cheng, C., Arshinoff, B. I., Liu, T., Yip, K. Y., Robilotto, R., Rechtsteiner, A., Ikegami, K., Alves, P., Chateigner, A., Perry, M., Morris, M., Auerbach, R. K., Feng, X., Leng, J., Vielle, A., Niu, W., Rhrissorrakrai, K., Agarwal, A., Alexander, R. P., Barber, G., Brdlik, C. M., Brennan, J., Brouillet, J. J., Carr, A., Cheung, M. S., Clawson, H., Contrino, S., Dannenberg, L. O., Dernburg, A. F., Desai, A., Dick, L., Dose, A. C., Du, J., Egelhofer, T., Ercan, S., Euskirchen, G., Ewing, B., Feingold, E. A., Gassmann, R., Good, P. J., Green, P., Gullier, F., Gutwein, M., Guyer, M. S., Habegger, L., Han, T., Henikoff, J. G., Henz, S. R., Hinrichs, A., Holster, H., Hyman, T., Iniguez, A. L., Janette, J., Jensen, M., Kato, M., Kent, W. J., Kephart, E., Khivansara, V., Khurana, E., Kim, J. K., Kolasinska-Zwierz, P., Lai, E. C., Latorre, I., Leahey, A., Lewis, S., Lloyd, P., Lochovsky, L., Lowdon, R. F., Lubling, Y., Lyne, R., MacCoss, M., Mackowiak, S. D., Mangone, M., McKay, S., Mecenas, D., Merrihew, G., Miller, D. M., 3rd, Muroyama, A., Murray, J. I., Ooi, S. L., Pham, H., Phippen, T., Preston, E. A., Rajewsky, N., Ratsch, G., Rosenbaum, H., Rozowsky, J., Rutherford, K., Ruzanov, P., Sarov, M., Sasidharan, R., Sboner, A., Scheid, P., Segal, E., Shin, H., Shou, C., Slack, F. J., Slightam, C., Smith, R., Spencer, W. C., Stinson, E. O., Taing, S., Takasaki, T., Vafeados, D., Voronina, K., Wang, G., Washington, N. L., Whittle, C. M., Wu, B., Yan, K. K., Zeller, G., Zha, Z., Zhong, M., Zhou, X., Ahringer, J., Strome, S., Gunsalus, K. C., Micklem, G., Liu, X. S., Reinke, V., Kim, S. K., Hillier, L. W., Henikoff, S., Piano, F., Snyder, M., Stein, L., Lieb, J. D., and Waterston, R. H. (2010). Integrative analysis of the *Caenorhabditis elegans* genome by the modENCODE project. *Science* **330**, 1775–1787.

Gibson, D. G. (2011). Gene and genome construction in yeast. *Curr. Protoc. Mol. Biol.* Chapter 3, Unit3 22.

Gibson, D. G., Glass, J. I., Lartigue, C., Noskov, V. N., Chuang, R. Y., Algire, M. A., Benders, G. A., Montague, M. G., Ma, L., Moodie, M. M., Merryman, C., Vashee, S., Krishnakumar, R., Assad-Garcia, N., Andrews-Pfannkoch, C., Denisova, E. A., Young, L., Qi, Z. Q., Segall-Shapiro, T. H., Calvey, C. H.,

Parmar, P. P., Hutchison, C. A., 3rd, Smith, H. O., and Venter, J. C. (2010). Creation of a bacterial cell controlled by a chemically synthesized genome. *Science* **329**, 52–56.

Gorovsky, M. A. (1973). Macro- and micronuclei of *Tetrahymena pyriformis*: a model system for studying the structure and function of eukaryotic nuclei. *J. Protozool.* **20**, 19–25.

Gorovsky, M. A., Yao, M. C., Keevert, J. B., and Pleger, G. L. (1975). Isolation of micro- and macronuclei of *Tetrahymena pyriformis*. *Methods Cell Biol.* **9**, 311–327.

Gould, S. B., Kraft, L. G., van Dooren, G. G., Goodman, C. D., Ford, K. L., Cassin, A. M., Bacic, A., McFadden, G. I., and Waller, R. F. (2011). Ciliate pellicular proteome identifies novel protein families with characteristic repeat motifs that are common to alveolates. *Mol. Biol. Evol.* **28**, 1319–1331.

Hamilton, E. P., Dear, P. H., Rowland, T., Saks, K., Eisen, J. A., and Orias, E. (2006). Use of HAPPY mapping for the higher order assembly of the Tetrahymena genome. *Genomics* **88**, 443–451.

Haynes, W. J., Vaillant, B., Preston, R. R., Saimi, Y., and Kung, C. (1998). The cloning by complementation of the pawn-A gene in Paramecium. *Genetics* **149**, 947–957.

Herrmann, L., Erkelenz, M., Aldag, I., Tiedtke, A., and Hartmann, M. W. (2006). Biochemical and molecular characterisation of *Tetrahymena thermophila* extracellular cysteine proteases. *BMC Microbiol.* **6**, 19.

Hillenmeyer, M. E., Fung, E., Wildenhain, J., Pierce, S. E., Hoon, S., Lee, W., Proctor, M., St Onge, R. P., Tyers, M., Koller, D., Altman, R. B., Davis, R. W., Nislow, C., and Giaever, G. (2008). The chemical genomic portrait of yeast: uncovering a phenotype for all genes. *Science* **320**, 362–365.

Ho, Y., Gruhler, A., Heilbut, A., Bader, G. D., Moore, L., Adams, S. L., Millar, A., Taylor, P., Bennett, K., Boutilier, K., Yang, L., Wolting, C., Donaldson, I., Schandorff, S., Shewnarane, J., Vo, M., Taggart, J., Goudreault, M., Muskat, B., Alfarano, C., Dewar, D., Lin, Z., Michalickova, K., Willems, A. R., Sassi, H., Nielsen, P. A., Rasmussen, K. J., Andersen, J. R., Johansen, L. E., Hansen, L. H., Jespersen, H., Podtelejnikov, A., Nielsen, E., Crawford, J., Poulsen, V., Sorensen, B. D., Matthiesen, J., Hendrickson, R. C., Gleeson, F., Pawson, T., Moran, M. F., Durocher, D., Mann, M., Hogue, C. W., Figeys, D., and Tyers, M. (2002). Systematic identification of protein complexes in *Saccharomyces cerevisiae* by mass spectrometry. *Nature* **415**, 180–183.

Howard-Till, R. A., and Yao, M. C. (2006). Induction of gene silencing by hairpin RNA expression in *Tetrahymena thermophila* reveals a second small RNA pathway. *Mol. Cell Biol.* **26**, 8731–8742.

Hughes, T. R., Marton, M. J., Jones, A. R., Roberts, C. J., Stoughton, R., Armour, C. D., Bennett, H. A., Coffey, E., Dai, H., He, Y. D., Kidd, M. J., King, A. M., Meyer, M. R., Slade, D., Lum, P. Y., Stepaniants, S. B., Shoemaker, D. D., Gachotte, D., Chakraburtty, K., Simon, J., Bard, M., and Friend, S. H. (2000). Functional discovery via a compendium of expression profiles. *Cell* **102**, 109–126.

Jacobs, M. E., Desouza, L. V., Samaranayake, H., Pearlman, R. E., Michael Siu, K. W., and Klobutcher, L. A. (2006). The *Tetrahymena thermophila* phagosome proteome. *Eukaryot Cell* **5**, 1990–2000.

Jan, C. H., Friedman, R. C., Ruby, J. G., and Bartel, D. P. (2011). Formation, regulation and evolution of Caenorhabditis elegans 3′UTRs. *Nature* **469**, 97–101.

Kanamori-Katayama, M., Itoh, M., Kawaji, H., Lassmann, T., Katayama, S., Kojima, M., Bertin, N., Kaiho, A., Ninomiya, N., Daub, C. O., Carninci, P., Forrest, A. R., and Hayashizaki, Y. (2011). Unamplified cap analysis of gene expression on a single-molecule sequencer. *Genome Res.* **21**, 1150–1159.

Karrer, K., Stein-Gavens, S., and Allitto, B. A. (1984). Micronucleus-specific DNA sequences in an amicronucleate mutant of Tetrahymena. *Dev. Biol.* **105**, 121–129.

Keller, A. M., and Cohen, J. (2000). An indexed genomic library for Paramecium complementation cloning. *J. Eukaryot. Microbiol.* **47**, 1–6.

Kellis, M., Patterson, N., Endrizzi, M., Birren, B., and Lander, E. S. (2003). Sequencing and comparison of yeast species to identify genes and regulatory elements. *Nature* **423**, 241–254.

Kilburn, C. L., Pearson, C. G., Romijn, E. P., Meehl, J. B., Giddings, T. H., Jr., Culver, B. P., Yates, J. R., 3rd, and Winey, M. (2007). New Tetrahymena basal body protein components identify basal body domain structure. *J. Cell Biol.* **178**, 905–912.

Kim, S. K., Lund, J., Kiraly, M., Duke, K., Jiang, M., Stuart, J. M., Eizinger, A., Wylie, B. N., and Davidson, G. S. (2001). A gene expression map for *Caenorhabditis elegans*. *Science* **293**, 2087–2092.

Koonin, E. V., and Wolf, Y. I. (2009). Is evolution Darwinian or/and Lamarckian? *Biol. Direct* **4**, 42.
Larson, D. D., Umthun, A. R., and Shaiu, W. L. (1991). Copy number control in the Tetrahymena macronuclear genome. *J. Protozool.* **38**, 258–263.
Lee, S. R., and Collins, K. (2006). Two classes of endogenous small RNAs in *Tetrahymena thermophila*. *Genes Dev.* **20**, 28–33.
Liu, X., Li, B., and Gorovsky, M. A. (1996). Essential and nonessential histone H2A variants in *Tetrahymena thermophila*. *Mol. Cell. Biol.* **16**, 4305–4311.
Madinger, C. L., Collins, K., Fields, L. G., Taron, C. H., and Benner, J. S. (2010). Constitutive secretion in *Tetrahymena thermophila*. *Eukaryotic Cell* **9**, 674–681.
Malone, C. D., Falkowska, K. A., Li, A. Y., Galanti, S. E., Kanuru, R. C., LaMont, E. G., Mazzarella, K. C., Micev, A. J., Osman, M. M., Piotrowski, N. K., Suszko, J. W., Timm, A. C., Xu, M. M., Liu, L., and Chalker, D. L. (2008). Nucleus-specific importin alpha proteins and nucleoporins regulate protein import and nuclear division in the binucleate *Tetrahymena thermophila*. *Eukaryotic Cell* **7**, 1487–1499.
McDonald, B. B. (1962). Synthesis of deoxyribonucleic acid by micro- and macronuclei of *Tetrahymena pyriformis*. *J. Cell. Biol.* **13**, 193–203.
Miao, W., Xiong, J., Bowen, J., Wang, W., Liu, Y., Braguinets, O., Grigull, J., Pearlman, R. E., Orias, E., and Gorovsky, M. A. (2009). Microarray analyses of gene expression during the *Tetrahymena thermophila* life cycle. *PLoS One* **4**, e4429.
Mochizuki, K., and Gorovsky, M. A. (2004). Small RNAs in genome rearrangement in Tetrahymena. *Curr. Opin. Genet. Dev.* **14**, 181–187.
Mohammad, M. M., Donti, T. R., Sebastian Yakisich, J., Smith, A. G., and Kapler, G. M. (2007). Tetrahymena ORC contains a ribosomal RNA fragment that participates in rDNA origin recognition. *EMBO J* **26**, 5048–5060.
Mondal, T., Rasmussen, M., Pandey, G. K., Isaksson, A., and Kanduri, C. (2010). Characterization of the RNA content of chromatin. *Genome Res.* **20**, 899–907.
Myers, R. M., Stamatoyannopoulos, J., Snyder, M., Dunham, I., Hardison, R. C., Bernstein, B. E., Gingeras, T. R., Kent, W. J., Birney, E., Wold, B., and Crawford, G. E. (2011). A user's guide to the encyclopedia of DNA elements (ENCODE). *PLoS Biol.* **9**, e1001046.
Nadeau, N. J., and Jiggins, C. D. (2010). A golden age for evolutionary genetics? Genomic studies of adaptation in natural populations. *Trends Genet.: TIG* **26**, 484–492.
Nanney, D. L., and Simon, E. M. (2000). Laboratory and evolutionary history of *Tetrahymena thermophila*. *Methods Cell. Biol.* **62**, 3–25.
Nozaki, H., Takano, H., Misumi, O., Terasawa, K., Matsuzaki, M., Maruyama, S., Nishida, K., Yagisawa, F., Yoshida, Y., Fujiwara, T., Takio, S., Tamura, K., Chung, S. J., Nakamura, S., Kuroiwa, H., Tanaka, K., Sato, N., and Kuroiwa, T. (2007). A 100%-complete sequence reveals unusually simple genomic features in the hot-spring red alga *Cyanidioschyzon merolae*. *BMC Biol.* **5**, 28.
Orias, E. (2000). Toward sequencing the Tetrahymena genome: exploiting the gift of nuclear dimorphism. *J. Eukaryot. Microbiol.* **47**, 328–333.
Orias, E., Cervantes, M. D., and Hamilton, E. P. (2011). Tetrahymena thermophila, a unicellular eukaryote with separate germline and somatic genomes. *Res. Microbiol.* **162**, 578–586.
Orias, E., and Flacks, M. (1975). Macronuclear genetics of Tetrahymena: I. Random distribution of macronuclear genecopies in *T. pyriformis*, syngen 1. *Genetics* **79**, 187–206.
Ozsolak, F., and Milos, P. M. (2011). RNA sequencing: advances, challenges and opportunities. *Nat. Rev. Genet.* **12**, 87–98.
Parker, J. D., Bradley, B. A., Mooers, A. O., and Quarmby, L. M. (2007). Phylogenetic analysis of the Neks reveals early diversification of ciliary-cell cycle kinases. *PLoS One* **2**, e1076.
Pop, M., Salzberg, S. L., and Shumway, M. (2002). Genome sequence assembly: algorithms and issues. *Computer* **35**, 47–54.
Prescott, D. M. (1994). The DNA of ciliated protozoa. *Microbiol. Rev.* **58**, 233–267.
Reyes-Prieto, A., Moustafa, A., and Bhattacharya, D. (2008). Multiple genes of apparent algal origin suggest ciliates may once have been photosynthetic. *Curr. Biol.* **18**, 956–962.

Ring, K. L., and Cavalcanti, A. R. (2008). Consequences of stop codon reassignment on protein evolution in ciliates with alternative genetic codes. *Mol. Biol. Evol.* **25**, 179–186.

Roy, S., Ernst, J., Kharchenko, P. V., Kheradpour, P., Negre, N., Eaton, M. L., Landolin, J. M., Bristow, C. A., Ma, L., Lin, M. F., Washietl, S., Arshinoff, B. I., Ay, F., Meyer, P. E., Robine, N., Washington, N. L., Di Stefano, L., Berezikov, E., Brown, C. D., Candeias, R., Carlson, J. W., Carr, A., Jungreis, I., Marbach, D., Sealfon, R., Tolstorukov, M. Y., Will, S., Alekseyenko, A. A., Artieri, C., Booth, B. W., Brooks, A. N., Dai, Q., Davis, C. A., Duff, M. O., Feng, X., Gorchakov, A. A., Gu, T., Henikoff, J. G., Kapranov, P., Li, R., MacAlpine, H. K., Malone, J., Minoda, A., Nordman, J., Okamura, K., Perry, M., Powell, S. K., Riddle, N. C., Sakai, A., Samsonova, A., Sandler, J. E., Schwartz, Y. B., Sher, N., Spokony, R., Sturgill, D., van Baren, M., Wan, K. H., Yang, L., Yu, C., Feingold, E., Good, P., Guyer, M., Lowdon, R., Ahmad, K., Andrews, J., Berger, B., Brenner, S. E., Brent, M. R., Cherbas, L., Elgin, S. C., Gingeras, T. R., Grossman, R., Hoskins, R. A., Kaufman, T. C., Kent, W., Kuroda, M. I., Orr-Weaver, T., Perrimon, N., Pirrotta, V., Posakony, J. W., Ren, B., Russell, S., Cherbas, P., Graveley, B. R., Lewis, S., Micklem, G., Oliver, B., Park, P. J., Celniker, S. E., Henikoff, S., Karpen, G. H., Lai, E. C., MacAlpine, D. M., Stein, L. D., White, K. P., and Kellis, M. (2010). Identification of functional elements and regulatory circuits by Drosophila modENCODE. *Science* **330**, 1787–1797.

Ryals, P. E., Smith-Somerville, H. E., and Buhse, H. E., Jr. (2002). Phenotype switching in polymorphic Tetrahymena: a single-cell Jekyll and Hyde. *Int. Rev. Cytol.* **212**, 209–238.

Sarin, S., Prabhu, S., O'Meara, M. M., Pe'er, I., and Hobert, O. (2008). Caenorhabditis elegans mutant allele identification by whole-genome sequencing. *Nat. Methods* **5**, 865–867.

Schones, D. E., Cui, K., Cuddapah, S., Roh, T. Y., Barski, A., Wang, Z., Wei, G., and Zhao, K. (2008). Dynamic regulation of nucleosome positioning in the human genome. *Cell* **132**, 887–898.

Shoemaker, D. D., Lashkari, D. A., Morris, D., Mittmann, M., and Davis, R. W. (1996). Quantitative phenotypic analysis of yeast deletion mutants using a highly parallel molecular bar-coding strategy. *Nat. Genet.* **14**, 450–456.

Simon, E. M., and Orias, E. (1987). Genetic instability in the mating type system of Tetrahymena pigmentosa. *Genetics* **117**, 437–449.

Smith, D. G., Gawryluk, R. M., Spencer, D. F., Pearlman, R. E., Siu, K. W., and Gray, M. W. (2007). Exploring the mitochondrial proteome of the ciliate protozoon *Tetrahymena thermophila*: direct analysis by tandem mass spectrometry. *J. Mol. Biol.* **374**, 837–863.

Smith, J. C., Northey, J. G., Garg, J., Pearlman, R. E., and Siu, K. W. (2005). Robust method for proteome analysis by MS/MS using an entire translated genome: demonstration on the ciliome of *Tetrahymena thermophila*. *J. Proteome. Res.* **4**, 909–919.

Stover, N. A., Krieger, C. J., Binkley, G., Dong, Q., Fisk, D. G., Nash, R., Sethuraman, A., Weng, S., and Cherry, J. M. (2006). Tetrahymena Genome Database (TGD): a new genomic resource for *Tetrahymena thermophila* research. *Nucleic Acids Res.* **34**, D500–D503.

Struder-Kypke, M. C., Wright, A. D., Jerome, C. A., and Lynn, D. H. (2001). Parallel evolution of histophagy in ciliates of the genus Tetrahymena. *BMC Evol. Biol.* **1**, 5.

Sweeney, R., Fan, Q., and Yao, M. C. (1996). Antisense ribosomes: rRNA as a vehicle for antisense RNAs. *Proc. Natl Acad. Sci. U.S.A.* **93**, 8518–8523.

Taverna, S. D., Coyne, R. S., and Allis, C. D. (2002). Methylation of histone h3 at lysine 9 targets programmed DNA elimination in tetrahymena. *Cell* **110**, 701–711.

Turkewitz, A. P., Orias, E., and Kapler, G. (2002). Functional genomics: the coming of age for *Tetrahymena thermophila*. *Trends Genet.* **18**, 35–40.

Wang, H. H., and Church, G. M. (2011). Multiplexed genome engineering and genotyping methods applications for synthetic biology and metabolic engineering. *Methods Enzymol.* **498**, 409–426.

Wessel, J., Zapala, M. A., and Schork, N. J. (2007). Accommodating pathway information in expression quantitative trait locus analysis. *Genomics* **90**, 132–142.

Wilkes, D. E., Bennardo, N., Chan, C. W., Chang, Y. L., Corpuz, E. O., DuMond, J., Eboreime, J. A., Erickson, J., Hetzel, J., Heyer, E. E., Hubenschmidt, M. J., Kniazeva, E., Kuhn, H., Lum, M., Sand, A., Schep, A., Sergeeva, O., Supab, N., Townsend, C. R., Ryswyk, L. V., Watson, H. E., Wiedeman, A. E., Rajagopalan, V., and Asai, D. J. (2009). Identification and characterization of dynein genes in Tetrahymena. *Methods Cell. Biol.* **92**, 11–30.

Williamson, M. P., and Sutcliffe, M. J. (2010). Protein–protein interactions. *Biochem. Soc. Trans.* **38**, 875–878.

Xiong, J., Feng, L., Yuan, D., Fu, C., and Miao, W. (2010). Genome-wide identification and evolution of ATP-binding cassette transporters in the ciliate *Tetrahymena thermophila*: a case of functional divergence in a multigene family. *BMC Evol. Biol.* **10**, 330.

Xiong, J., Lu, X., Lu, Y., Zeng, H., Yuan, D., Feng, L., Chang, Y., Bowen, J., Gorovsky, M., Fu, C., and Miao, W. (2011a). Tetrahymena Gene Expression Database (TGED): a resource of microarray data and co-expression analyses for Tetrahymena. *Sci. China Life Sci.* **54**, 65–67.

Xiong, J., Yuan, D., Fillingham, J. S., Garg, J., Lu, X., Chang, Y., Liu, Y., Fu, C., Pearlman, R. E., and Miao, W. (2011b). Gene network landscape of the ciliate *Tetrahymena thermophila*. *PLoS One* **6**, e20124.

Yao, M. C., Choi, J., Yokoyama, S., Austerberry, C. F., and Yao, C. H. (1984). DNA elimination in Tetrahymena: a developmental process involving extensive breakage and rejoining of DNA at defined sites. *Cell* **36**, 433–440.

Yao, M. C., and Gorovsky, M. A. (1974). Comparison of the sequences of macro- and micronuclear DNA of *Tetrahymena pyriformis*. *Chromosoma* **48**, 1–18.

Zhu, J., Zhang, B., Smith, E. N., Drees, B., Brem, R. B., Kruglyak, L., Bumgarner, R. E., and Schadt, E. E. (2008). Integrating large-scale functional genomic data to dissect the complexity of yeast regulatory networks. *Nat. Genet.* **40**, 854–861.

CHAPTER 5

From Molecules to Morphology: Cellular Organization of *Tetrahymena thermophila*

Dorota Wloga* **and Joseph Frankel**†

*Department of Cell Biology, Nencki Institute of Experimental Biology, 3 Pasteur Street, 02-093, Warsaw, Poland

†Department of Biology, University of Iowa, Iowa City, IA, USA

0091-679X/10 $35.00
DOI 10.1016/B978-0-12-385967-9.00005-0

Abstract

Tetrahymena thermophila is both a cell and an organism, which combines great intracellular complexity with a remarkable accessibility to investigation using many different approaches. In this review, we start with a description of the elaborate cortical organization of the *Tetrahymena* cell, and then proceed inward to consider the mitochondria and then the nuclei. For each of these cellular organelles and organelle-systems, first we familiarize the reader with its location in the cell and its structure and ultrastructure, and then we analyze the molecular mechanisms associated with organelle assembly, function, and subdivision. This analysis includes a molecular inventory of the organelle or organelle system, as well as a review of the consequences of modification, disruption or overexpression of important molecular components of each structure or system. Relevant comparisons to results obtained with other well-studied organisms, from *Paramecium* to *Homo sapiens,* are also included. Our goal is to provide investigators, in particular those who are new to this organism, both the background and the motivation to work with this model system and achieve further insight into its organization and dynamics.

I. Overview

The free-living ciliate *Tetrahymena thermophila* is a small teardrop-shaped eukaryote, about 50 μm long. Its unicellular body is enclosed by a semi-rigid and structurally complex cortex containing multiple layers. The outermost layer, the plasma membrane, is underlain by flattened sub-membrane vesicles (the cortical alveoli), which in turn are underlain by a fibrogranular layer, the epiplasm, and a system of microtubular and non-microtubular arrays arranged into repeating units (Fig. 1A and C). Each typical cortical unit consists of a centrally located basal body (BB) accompanied by a set of appendages that anchor and position each BB. The kinetodesmal fiber (KF, Figs. 1C and 2F) and a transient anterior left filament (Jerka-Dziadosz, personal communication) are non-microtubular structures, while microtubule-based appendages include arrays of transverse (TM) and postciliary (PM) microtubules (MTs) (Allen, 1967). Outside of the oral area, the cortical units are arranged into 15–25 longitudinal rows (on average 17–21) (Corliss, 1973; Loefer *et al.*, 1966; Nanney, 1966a). An array of longitudinal (LM) MTs extends along and to the right of the BB row (observer's right; the position of the *Tetrahymena* cell structures is described as it would be seen by a hypothetical observer standing inside the cell and facing the cell surface) while the basal MTs run to the left (hypothetical viewer's left) of the BB row (Allen, 1967).

Anterior to each BB, the cell membrane forms an invagination called a coated pit or "parasomal sac" (Allen, 1967; Nilsson and van Deurs, 1983) (Fig. 2F), the site of endocytosis (pinocytosis) (Elde *et al.*, 2005). Multiple mitochondria, organized in one or two rows (see below) and accompanied by the endoplasmic reticulum, are located underneath the epiplasm and cortical MTs (Aufderheide, 1979; Elliott and

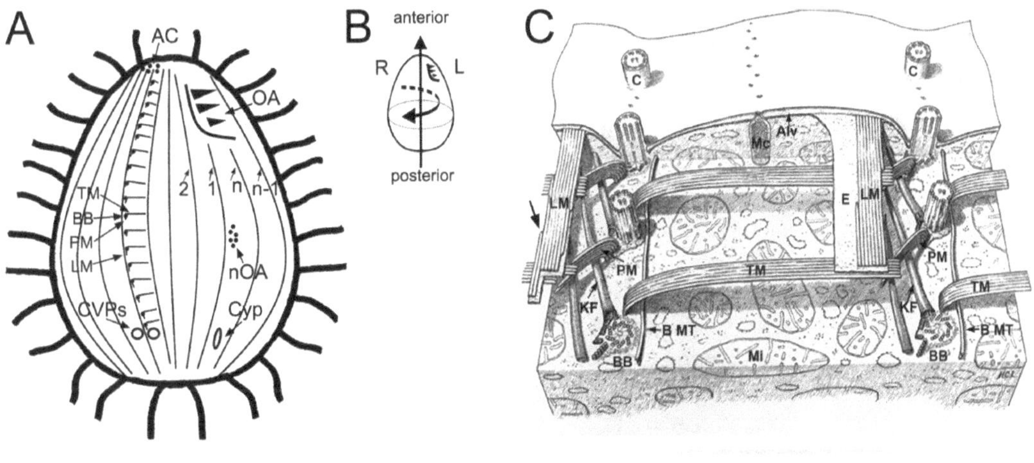

Fig. 1 The organization of the *Tetrahymena* cell surface. (A) A schematic overview of the *Tetrahymena* cell. The numbers inside the cell represent the conventional numbering of *Tetrahymena* somatic basal body (BB) rows (identical to ciliary rows) with the lowest number (1) and the highest number (n) assigned to the two postoral BB rows. Each typical BB within a ciliary row is associated with a postciliary (PM) and a transverse (TM) microtubule band, with a longitudinal microtubule band (LM) running to its right (observer's right for a hypothetical observer situated inside the cell and facing the cell surface). Cortical landmarks include an oral apparatus (OA) and apical crown (AC) near the anterior end of the cell and contractile vacuole pores (CVPs) and cytoproct (Cyp) near the posterior end of the cell; the newly assembling oral apparatus (nOA) is visible in the equatorial region, next to row number 1. (B) An illustration of *Tetrahymena* cell polarity in two orthogonal directions, antero-posterior and circumferential (L – left, R – right); (C) A cutaway diagram showing the organization of a portion of the *Tetrahymena* cell cortex. The anterior end of the cell is oriented upward in the diagram, receding from the viewer. The *Tetrahymena* cell is covered with rows of cilia (C) embedded in a cortex, which is composed of plasma membrane underlain by alveoli (Alv), which in turn are underlain by the epiplasm (E) and a system of kinetodesmal fibers (KF) and microtubular bands. Transverse (TM) and postciliary (PM) microtubule (MT) bands originate near each basal body (BB), while longitudinal MT bands (LM) and basal MTs (B MT) extend along the BB rows. A diagonal arrow indicates the anterior end of one MT at the right edge of an LM. The mucocysts (Mc) are arranged in two sets of longitudinal rows: Mucocysts of the first row are among the BBs of the ciliary rows, with the mucocysts positioned between two BBs (openings of mucocysts visible between last two BBs), and the mucocysts of the second row are between two adjacent BB rows. Numerous mitochondria (Mi) are located beneath the cell cortex. Fig. 1 (C) is taken from Fig. 2.6 of J. Frankel, "Pattern Formation: Ciliate Studies and Models", © 1989; slightly modified from the original, Fig. 22 of R. D. Allen (1967), *J. Protozool.* **14**, 553–565 © 1967, reproduced with permission of the authors, of Oxford University Press, and of the International Society of Protistologists, successor to the Society of Protozoologists.

Bak, 1964a). The Golgi compartment, which consists of dispersed dictyosomes, is located in proximity to mitochondria and near the base of the BBs and/or coated pits (Kurz and Tiedtke, 1993). The *Tetrahymena* cortex also contains multiple dense-core secretory granules (mucocysts), the membrane-limited organelles that are arranged in two sets of longitudinal rows – the first set within the BB rows with the mucocysts positioned between BBs, and the second parallel set of rows between two BB rows (Allen, 1967; Turkewitz and Kelly, 1992) (Fig. 1C). The *Tetrahymena* cell architecture is further complicated by the presence of asymmetrically located structures: an apical crown (AC) of BB couplets (Fig. 3A) underlain by a microfilamentous apical band (AB) (Fig. 3B and C), an oral apparatus (OA) (Fig. 4), contractile vacuole pores (CVPs), and a cytoproct (Cyp) (Fig. 1A). These asymmetrically located structures together with the directionally arranged kinetodesmal fibers and microtubular arrays create cell polarity in two orthogonal directions, antero-posterior and circumferential (Frankel, 2000a). The molecular mechanisms that control the formation and maintenance of this complicated cortical architecture are still not well understood.

Except for two so-called postoral BB rows (located posteriorly to the OA), all BB rows extend from the anterior to the posterior cell pole. The right-postoral BB row is referred to as row number 1 (Fig. 1A). By convention, the subsequently enumerated BB rows are positioned to the cell's right. Except for the two postoral BB rows (rows 1 and n, n being the row with the highest number) and rows number 2, 3, 4, and $n - 1$, all BB rows terminate at their anterior end with a pair of BBs (McCoy, 1974), known as the "apical crown" (Fig. 3A). The more anterior BBs of the apical crown are not ciliated (Jerka-Dziadosz, 1981a).

A detailed description of *Tetrahymena* cell ultrastructure and spatial patterning can be found in *Methods in Cell Biology*, Volume 62 (Frankel, 2000b). Here, we will briefly present the background information (including some early observations obtained during analysis of *Tetrahymena pyriformis*), and wherever possible discuss the more recently elucidated molecular mechanisms associated with the assembly and function of these cellular structures.

II. Basal Bodies

Non-dividing *Tetrahymena* cells have about 150 oral BBs (Bakowska *et al.*, 1982a; Nilsson and Williams, 1966) and approximately 500–600 somatic BBs that are generally arranged into 17 to 21 longitudinal rows (Frankel, 1980; Nanney, 1966a, 1971). Genetically identical *Tetrahymena* clones maintained under suboptimal conditions tend to be heterogeneous in the number of BB rows per cell. Although during division the progeny tends to maintain the number of BB rows of the "mother" cell, there is also a tendency to increase or decrease their number to 18–19 BB rows, the so-called "stability center" (Nanney, 1966a). With the exception of the two postoral BB rows, *Tetrahymena* has on average 30 BBs per row (Nanney, 1971). Interestingly, the number of BBs per row decreases as the number of BB rows per cell increases, thereby maintaining a rough constancy in the total number of BBs in cells grown in nutrient media (Nanney, 1971).

The number of BBs doubles prior to each cell division (Nanney, 1975; Nanney *et al.*, 1978). Within ciliary rows, new BBs assemble in proximity and anterior to pre-existing BBs. The formation of new BBs is more frequent in the mid and posterior part of a dividing cell (Frankel *et al.*, 1981; Kaczanowski, 1978; Nanney, 1975), and more than one round of BB duplication per cell cycle may occur (Allen, 1969; Nanney, 1975). A dramatic increase in the number of somatic BBs, which is independent of cell division, is observed in cells undergoing a morphogenetic transformation to a "rapid swimmer" phenotype that takes place after a shift to an inorganic medium (Nelsen and Debault, 1978; reviewed in Frankel, 2000b, pp. 88–90).

New BBs, visible initially as cartwheel structures with attached single MTs, are located perpendicular to the existing BBs. Each new BB assembles its additional MTs and gradually elongates, tilts up, and shifts away from the mature BB to finally dock at the cell surface. A detailed description of the changes at the ultrastructural level associated with the formation of new BBs was provided by Allen (1969).

The skeleton of the mature BB cylinder-like structure is built out of nine triplets of MTs (Fig. 2A). The diameter of the mature BB is about 180–220 μm, and its length is about 500–600 nm (Allen, 1969). The proximal end of the BB (positioned deeper in the cytoplasm) contains the cartwheel, a structure that assembles early during new BB formation and in *Tetrahymena* is maintained in the mature BB (Fig. 2A). The distal end of the BB is limited by a terminal plate (Fig. 2B), composed of two electron-opaque layers (Allen, 1969). The terminal plate marks the transition zone between the BB and the cilium. The center of the BB cylinder is occupied by an electron-dense core. The ultrastructure of the BB was described in detail (Allen, 1969), while the description of improved TEM visualization methods can be found in Giddings *et al.* (2010) and Meehl *et al.* (2009), see also Chapter 13 of this volume, by Winey Stemm-Wolf, Giddings, and Pearson.

The tubulin of the BB MTs is post-translationally modified by multiple mechanisms including glutamylation (Wloga *et al.*, 2008a). A decrease in the level of tubulin glutamylation, caused by deletion of enzymes that glutamylate tubulin (TTLL1- and TTLL9-type), is associated with defects in the docking of BBs at the cell surface (Wloga *et al.*, 2008a).

Efforts to establish the *Tetrahymena* BB proteome (Kilburn *et al.*, 2007) as well as localization and functional analysis of newly identified BB proteins have advanced our understanding of the molecular mechanisms that regulate BB assembly (Culver *et al.*, 2009; Kilburn *et al.*, 2007; Pearson *et al.*, 2009b; Pearson and Winey, 2009), see also Chapter 13 of this volume. BBs contain or are associated with at least 97 proteins (Kilburn *et al.*, 2007). Most of the proteins that build BBs are assumed to be incorporated during BB assembly, but some of these proteins are exchanged dynamically after the BB is formed (Pearson *et al.*, 2009a). Using antibodies or protein tagging, it was shown that in addition to α- and β-tubulin, the following proteins are present in or are closely associated with *Tetrahymena* BBs: γ-tubulin (Liang *et al.*, 1996; Shang *et al.*, 2002), actin (Hoey and Gavin, 1992), centrin1 and centrin2 (Pearson *et al.*, 2009a; Stemm-Wolf *et al.*, 2005), calmodulin (Ueno *et al.*, 2003), NIMA-related kinases (Nrks) (Wloga *et al.*, 2006), a Rab GTPase TtRabD23 (Bright *et al.*, 2010), Bbc14, 20, 23, 29, 30, 31, 52, 53, 57, 73, 78, 82, Spag6, PACRG, Eno1, Ftt18, Ftt49 (Kilburn *et al.*, 2007), Poc1 (Kilburn *et al.*, 2007; Pearson *et al.*, 2009b), and Sas6 (Culver *et al.*, 2009; Kilburn *et al.*, 2007; Pearson *et al.*, 2009a).

Functional studies were performed on a few of these proteins. γ-tubulin, a MT nucleating factor, is required for both the formation of new BBs and, interestingly, also for the maintenance of pre-existing BBs (Shang *et al.*, 2002). Moreover, the

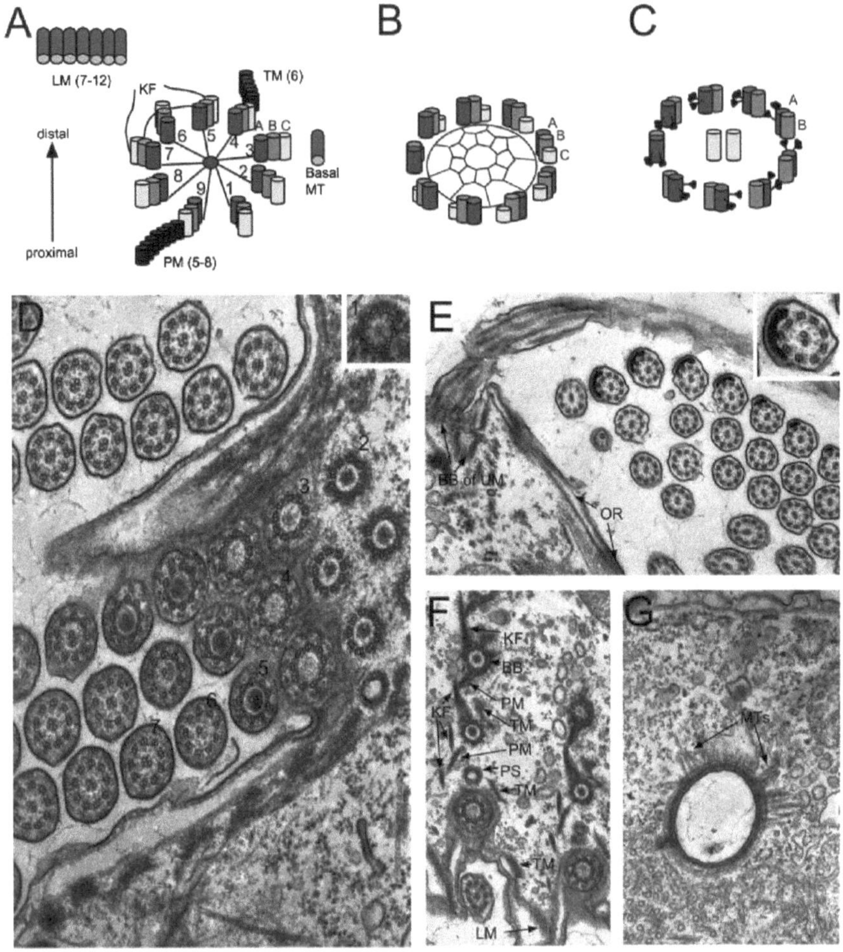

Fig. 2 Schematic representations (A-C) and transmission electron micrographs (D-G) of the cortical ultrastructure of *Tetrahymena*. (A) A basal body at its most proximal end, with triplets of MTs (tubules A, B and C) and the cartwheel structure; the MT triplets are numbered according to Beisson and Jerka-Dziadosz (1999) and Lynn (1981). The positions of the bands of transverse MTs (TM), postciliary MTs (PM) and the basal MT and kinetodesmal fiber (KF) are indicated in relation to the basal body. An array of longitudinal MTs (LM) runs distal to other microtubular and non-microtubular BB appendages (compare to Fig. 1C). The numbers in parentheses indicate the average number of MTs in each array. (B) A basal body at the distal end (terminal plate) showing the termination of the C tubule; (C) A cilium with A and B tubules of outer doublets marked, also showing MTs of central pairs and outer and inner dynein arms attached to the A tubules. (D-G) The ultrastructure of *Tetrahymena* cilia and basal bodies (D-F), and contractile vacuole pore (G), visualized by transmission electron microscopy. (D) A section through an oral membranelle with the details of cilia and basal body ultrastructure seen in cross-sections at different levels along the proximal-distal axis: 1. the proximal end of a basal body with MT triplets and cartwheel structure (inset in the upper right corner); 2. the central part of a basal body with electron-dense material inside the basal body cylinder; 3. a basal body beneath the terminal plate; 4. a basal body at the level of the terminal plate, note that in some positions the C tubule is no longer visible; 5. a cilium with outer doublets composed of an A and B tubule and one central MT, note the electron-dense material between the ciliary membrane and outer doublets; 6. a cilium with outer doublet MTs, one clearly visible central MT, and an indistinct second

GTP-binding domain of the γ-tubulin is essential and is involved in the regulation of the protein's function. Most mutations in the GTP-binding domain are lethal; however, two point mutations (A101G, T146 V) lead to over-duplication and mislocalization of new BBs (Shang *et al.*, 2005).

Tetrahymena has two homologs of the conserved cartwheel protein SAS-6 (reviewed by Schuldt (2011)), TtSas6Ap, and TtSas6Bp (Culver *et al.*, 2009). These homologs have nonidentical functions (Culver *et al.*, 2009); however, their functions may partly overlap. TtSas6Ap localizes to the cartwheel and electron-dense lumen in all BBs as well as near the base of the BBs (Culver *et al.*, 2009; Kilburn *et al.*, 2007; Vonderfecht *et al.*, 2011) and is involved in cartwheel assembly and consequently the assembly of new BBs, but not in maintenance of existing BBs. Interestingly, while pre-existing BBs persist in the absence of TtSas6Ap, the associated ciliary axonemes undergo shortening (Culver *et al.*, 2009). Thus, either SAS-6 has another function in the axoneme or somehow the structural state of the BB determined by SAS-6 is required for axoneme length regulation. In contrast to TtSas6Ap, TtSas6Bp is enriched in assembling and unciliated BBs. Cells lacking TtSas6Bp assemble cartweel structures but show defects in the organization of ciliary rows and oral membranelles. Thus, it seems that TtSas6Bp may play a role in the location of assembly of new BBs (Culver *et al.*, 2009). While the initial studies indicated that TtSas6Ap is essential (Culver 2009), recent data indicate that neither of the two SAS-6 homologs is essential (Winey, personal communication). Given that SAS-6 is the main component of the cartwheel and is required for BB assembly in other model organisms, it is likely that *Tetrahymena* needs either TtSas6Ap or TtSas6Bp for survival.

Tetrahymena also has a homolog of POC1, a conserved protein that in human cells is involved in centriole duplication and controls centriole length (Keller *et al.*, 2009). In *Tetrahymena,* Poc1p is required for BB assembly, and cells lacking this protein have fewer BBs, especially at elevated temperature. Moreover, this protein, which is associated with the cartwheel and BB MTs, affects the stability of BBs (Pearson *et al.*, 2009b).

Centrin is a conserved Ca^{2+} binding protein that is associated with centrioles and BBs (Andersen *et al.*, 2003; Keller *et al.*, 2005; Liu *et al.*, 2007). Among the four *Tetrahymena* centrins, Cen1p, a human centrin 2 homolog, is essential and has been localized: (1) at the base of existing BBs, where new BBs assemble, (2) in the mid-region of BBs, and (3) in the transition zones (Stemm-Wolf *et al.*, 2005). Deletion of

---

central MT; 7. a cilium with outer doublet MTs and both central pair MTs. (E) A cross section through the first oral membranelle revealing the most anterior ciliary row with a unique rod-like electron-dense structure positioned between the ciliary membrane and the three adjacent outer doublets. The inset shows such a cilium at higher magnification. Note also the fragment of oral ribs (OR) and two basal bodies (outer one ciliated) of the undulating membrane (UM). (F) A cross section of the *Tetrahymena* cell cortex showing two basal body (BB) rows with associated structures: kinetodesmal fiber (KF), transverse MTs (TM) and postciliary MTs (PM). Longitudinal MTs (LM) run parallel to basal body row, and a parasomal sac (PS) is located between two basal bodies of the same row. (G) A cross section though the contractile vacuole pore with apparent MTs radiating around the pore. Unpublished electron micrographs were kindly supplied by Dr. Maria Jerka-Dziadosz.

CEN1 results in ultrastructural defects in BBs and loss of BBs (Stemm-Wolf *et al.*, 2005). Centrin is a two-domain protein with two EF-hand Ca^{2+} binding motifs in each domain (Errabolu *et al.*, 1994). Data obtained in mutagenesis studies of EF-hand motifs of *Tetrahymena* centrin 1 suggest that different EF hand motifs bind Ca^{2+} with different affinity, and that; the N-terminal part of Cen1p plays a role in the separation of new and old BBs and in the proper spatial orientation of new BBs within BB rows, while the C-terminal part of Cen1p seems to play a role in protein localization within BBs and subsequently in BB stability (Vonderfecht *et al.*, 2011).

III. Cilia

A. Structure

In *Tetrahymena,* these hair-like MT-based organelles play an essential role in cell motility (Satir, 1984) and in accumulation of food particles in the OA funnel (phagocytosis) (Nilsson, 1979). Cilia also play an indirect role in the completion of cell division via "rotokinesis," a rotational motility of future daughter cells that facilitates scission of the cytoplasmic bridge (Brown *et al.*, 1999a). Furthermore, cells lacking cilia (e.g., IFT mutants) are unable to form conjugal pairs even with a ciliated partner (J. Gaertig, personal communication), presumably because either ciliary motility is required for pair formation or cilia have a signaling role in mating. Some of the methods used to evaluate function and to analyze properties of cilia in *Tetrahymena* cells were recently summarized by Rajagopalan *et al.* (2009a).

In growing cells, the ultrastructurally identical somatic (ciliary-row) cilia show some minor differences in their length, as the anterior cilia are slightly shorter than those present in the mid- and posterior region. This may suggest the existence of a spatial control of ciliary length within the cell (Wloga *et al.*, 2006). As in the flagellate *Chlamydomonas* (Bradley and Quarmby, 2005), the length of cilia is regulated by the activity of NIMA ("Never in Mitosis-A")-related serine/threonine Kinases (NRKs). Overexpression of specific GFP-tagged Nrks causes shortening or disassembly of subpopulations of cilia located either at the anterior or posterior cell pole (Wloga *et al.*, 2006). Thus, the length of different subsets of *Tetrahymena* cilia may be partly controlled by different members of the NRK family that in turn are subject to spatial control within this complex cell. The above studies indicate that despite similar morphology, cilia in different regions of the cell have distinct molecular composition. Also in other organisms, changes in the level of protein phosphorylation brought about by protein kinases are involved in the control of ciliary length and stability (for review, see Cao *et al.* (2009b)).

In cells from log-phase cultures, nearly 50% of the ciliary-row (somatic) BBs lack cilia, which suggests a significant delay between the initial assembly of new BBs and their subsequent ciliation (Frankel *et al.*, 1981; Nanney, 1975). New BBs are formed within the ciliary rows simultaneously with pre-division oral development; these BBs remain unciliated until the cell begins to divide (Frankel *et al.*, 1981) and largely account for the high overall proportion of unciliated somatic BBs in a growing population. In *Tetrahymena* cells that are not preparing to divide, the majority of BBs are ciliated. Exceptions are the more anterior BB of the paired BBs of the apical crown (Jerka-Dziadosz, 1981a), BBs of the inner row of the OA UM and some BBs of OA membranelles (Bakowska *et al.*, 1982a). Cells starved for several hours in inorganic medium assemble cilia on generally unciliated somatic BBs, including anterior BBs of the apical crown (Nelsen and Debault, 1978). Moreover, such cells form an exceptionally long cilium (15–20 μm) at the posterior pole of the cell (caudal cilium) as a part of their transformation into the "rapid swimmer" phenotype (Nelsen, 1978; Nelsen and Debault, 1978).

Tetrahymena assembles motile cilia with highly conserved 9 + 2 axonemes (Allen, 1968) (Fig. 2C). The MTs of peripheral doublets are extensions of two out of the three MTs of the BB triplets (A and B tubules). The central-pair (CP) MTs originate distal to the BBs terminal plate in an electron-dense granule (axosome) of unknown protein composition; more precisely, one of CP MTs originates from the axosome, while the other one is initiated slightly above the axosome (Allen, 1969) (Fig. 2D). The ciliary membrane is a continuation of the plasma membrane. A ciliary necklace, a structure that in other organisms, is suggested to act as a part of the functional barrier that ensures specific protein and lipid composition of the ciliary membrane (Rohatgi and Snell, 2010) is visible at the base of the cilium, between the plasma membrane and the peripheral MTs (Sattler and Staehelin, 1974). In the most distal segment of the axoneme, the B-tubules of the doublets terminate and only the A tubules and CP MTs extend to the tip of the cilium. The distal ends of both the CP MTs and the A-tubule extensions of the peripheral doublets are capped by complex structures known as the ciliary caps (Dentler, 1984; Suprenant and Dentler, 1988). The core molecular components of caps remain unknown.

Several oral cilia of the most anterior row of the first oral membranelle are ultrastructurally unique in having a rod-like electron-dense structure positioned between the ciliary membrane and three adjacent outer doublets (Sattler and Staehelin, 1974; Williams and Luft, 1968) (Fig. 2E).

B. Ciliary Proteins

Mass spectrometry has identified 223 proteins in *Tetrahymena* cilia (Smith *et al.*, 2005). The cross-species comparisons between model organisms with diverse axonemes (Ciliaproteome; http://www.ciliaproteome.org and CilDB; http://cildb.cgm.cnrs-gif.fr; Arnaiz *et al.*, 2009) suggest that the total number of proteins in *Tetrahymena* cilia can be higher. α- and β-tubulin subunits encoded by the *ATU1* and *BTU1* or *BTU2* genes, respectively, represent roughly 50% of the total protein mass of the cilia of *T. thermophila*. Tubulin in ciliary MTs, especially MTs of peripheral doublets, is extensively post-translationally modified including glutamylation, glycylation, and lysine acetylation (Gaertig *et al.*, 1995; Suryavanshi *et al.*, 2010; Wloga *et al.*, 2008a, 2009; Xia *et al.*, 2000). The functions of tubulin post-translational modifications (PTMs) including extensive studies in *Tetrahymena* have been recently reviewed (Wloga and Gaertig, 2010).

Tubulin PTMs may play an important role in ciliary assembly. Most of the available data implicate tubulin glutamylation and glycylation in this process. These two PTMs are based on the addition of glutamyl or glycyl side chains, respectively, made of one (monoglutamylation and monoglycylation) or multiple residues (polyglutamylation and polyglycylation) to the glutamic acids within the primary sequence of the tubulin C-terminal tail (for review, see Verhey and Gaertig, 2007; Wloga and Gaertig, 2010). Most likely these two polymeric PTMs share modification sites, based on the evidence of competition between enzymes that generate glutamylation (e.g., TTLL1 and TTLL6) and glycylation (TTLL3) (Wloga *et al.*, 2009, 2010).

The function of polymodification sites on α- and β-tubulin was initially studied by site-directed mutagenesis. The mutants in which all six potentially modifiable glutamic acids (E) in the α-tubulin tail domain are replaced by nonmodifiable aspartates (D) are viable and nearly normal morphologically (Wloga *et al.*, 2008a; Xia *et al.*, 2000). However, D substitution of three or more Es in the β-tubulin tail domain is lethal and strongly affects both cilia formation and cell division (see below) (Thazhath *et al.*, 2002; Xia *et al.*, 2000). The β-tubulin triple site mutant ($\beta DDDE_{440}$) assembles excessively short immotile cilia that lack central-pair MTs and have fewer or incomplete outer doublets (Thazhath *et al.*, 2002). Similar short immotile cilia without central pairs are also assembled in cells that overproduce the potent GFP-tagged TtTtll6Ap (*T. thermophila* tubulin-tyrosine-ligase-like 6A) glutamyl elongase that brings about hyper-elongation of tubulin glutamyl side chains (Janke *et al.*, 2005; Wloga *et al.*, 2010). Thus, both insufficient and excessive tubulin polymodifications affect axoneme assembly. Interestingly, nearly complete removal of tubulin glycylation by deletion of six TTLL3-type ligases that initiate glycyl side chains is associated with the assembly of only slightly shorter cilia of normal ultrastructure but with an apparent change in the dynamics of axonemal MTs (Wloga *et al.*, 2009). In contrast to these findings, *Tetrahymena* cells lacking acetylated α-tubulin assemble motile cilia of normal length (Akella *et al.*, 2010; Gaertig *et al.*, 1995).

Taken together, the data summarized here suggest that proper maintenance of cilia requires an optimal level of and possibly balance between the tubulin polymodifications, glycylation, and glutamylation [perhaps in combination with other PTMs that all together form a specific "tubulin code" (Verhey and Gaertig, 2007)].

In mammalian cells, it was shown that the level of tubulin glutamylation on cytoplasmic MTs affects the activity of MT severing proteins, spastin, and katanin (Lacroix *et al.*, 2010). Interestingly, *Tetrahymena* cells with a knocked-out katanin, either its p60 catalytic subunit (*KAT1*) or its p80 noncatalytic subunit (*KAT3*), also form abnormally short cilia without the central pair of MTs (Sharma *et al.*, 2007) and thus phenocopy the $\beta DDDE_{440}$ mutation. This suggests that there exists a potential link between the level of tubulin PTMs and the activity of katanin during *Tetrahymena* cilia assembly/maintenance. The role of katanin in cilia assembly can be conserved, as a mutation in the p80 katanin subunit in *Chlamydomonas reinhardtii* results in formation of nonmotile flagella that lack central-pair MTs (Dymek *et al.*, 2004).

Coordinated beating of the hundreds of cilia that cover the *Tetrahymena* cell enables directional cell motility. The level of PTMs of ciliary MTs, more precisely the level of tubulin glutamylation brought about by two TTLL6 paralogs, Ttll6A and 6F glutamyl elongases, affects not only the ultrastructure and length of cilia, but also ciliary motility. A decrease in the level of tubulin glutamylation in cells brought about by deletion of *TTLL6A* and *6F* is associated with extremely slow cell motility, reduced ciliary beat frequency, and an abnormal waveform. These effects may be mostly a result of disregulation of inner dynein arms. An *in vitro* analysis of the sliding of MTs of isolated wild-type and mutant axonemes that were exposed to ATP indicates that the tubulin glutamylation generated by Ttll6A and 6F glutamylases regulates the rate of sliding of inner dynein arms but not of outer dynein arms along the B-tubule wall of the neighboring outer doublets (Suryavanshi *et al.*, 2010). Similar observations linking tubulin glutamylation to inner dynein arm activity were made in *Chlamydomonas reinhardtii* with elimination of TTLL9 tubulin glutamyl elongase activity (Kubo *et al.*, 2010).

The axoneme has outer and inner dynein arms attached to the A-tubule of doublets. Analysis of the *Tetrahymena* genome database revealed 23 axonemal and 2 nonaxonemal (cytoplasmic, *DYH1, DYH2*) dynein heavy chain genes (Asai and Wilkes, 2004; Wood *et al.*, 2007). Among the 23 axonemal dynein heavy chains, three (*DYH3, DYH4,* and *DYH5*) "...are characterized as outer arm dyneins..." while "... a remaining 20 different genes are all tentatively annotated as inner arm heavy chains" (Wood *et al.*, 2007, pp 3081–3082). Single knockout of the genes that encode the inner arm dynein heavy chains *DYH6* (Angus *et al.*, 2001; Hennessey *et al.*, 2002), *DYH7* (Wood *et al.*, 2007), and *DYH8, 9* and *12* (Liu *et al.*, 2004) causes reduction in swimming rate, and, with the exception of *DYH8,* affects ciliary waveform, while elimination of *DYH8* results in reduced ciliary beat frequency. Moreover, the above analysis indicates that the dynein heavy chain encoded by the *DYH6* gene is essential for Ca^{2+}-induced ciliary reversal (Hennessey *et al.*, 2002).

The total number of kinesins in *Tetrahymena* cells is exceptionally high (over seventy) (Eisen *et al.*, 2006; Wickstead and Gull, 2006) and we have only started to uncover their function (Awan *et al.*, 2004, 2009; Brown *et al.*, 1999b).

The *Tetrahymena* macronuclear genome encodes four conventional actins and eight actin-related proteins, ARPs (Kuribara *et al.*, 2006). The conventional actin, Act1p, localizes to the cilia (Muto *et al.*, 1994; Williams *et al.*, 2006). Cells with either a somatic or a germ-line knockout of *ACT1* assemble full-length cilia with only occasional (1–5% of cilia) structural defects in outer doublets, namely separation of the one side of the B-tubule from the A-tubule (Williams *et al.*, 2006). However, the ciliary beating rate is significantly reduced, causing extremely slow cell motility, inhibition of phagocytosis, and lack of rotokinesis that in turn leads to reintegration of cells that had failed to separate after they had completed division constriction. Thus, TtAct1p is required for ciliary motility but not for ciliary assembly and stability. The ciliary roles (if any) of the other three *Tetrahymena* actins are unknown. Kuribara and co-authors (Kuribara *et al.*, 2006) reported on the presence of one of the ARPs (tArp) in cilia and an increase in the expression level of tArp after deciliation.

Immunolocalization studies or GFP-tagging revealed several other proteins that localize in cilia: the hsp70 family member hsp73 strongly and the hsp90 family-member hsp82 weakly (Williams and Nelsen, 1997), calmodulin and EF-1α (Ueno *et al.*, 2003), NIMA-related kinases (Wloga *et al.*, 2006), centrin (Guerra *et al.*, 2003), and homologs of radial spoke head proteins (Ueno *et al.*, 2006).

C. Intraflagellar Transport

The assembly and maintenance of cilia involve the intraflagellar transport (IFT) pathway, a bidirectional motility of multisubunit protein complexes (IFT particles) along outer doublet MTs. IFT is required for delivery of most if not all structural axoneme components from the BB area to the tip, where the components are incorporated during ciliary assembly and maintenance. Furthermore, IFT recycles components by moving them back to the BB. The anterograde transport of ciliary proteins from the BB to the distal end of the cilium is mediated by the microtubule-plus-end-directed kinesin-2-type motor proteins and involves a subset of IFT particles that form complex B. The retrograde transport of recycled ciliary proteins from the tip of cilia to the BB is carried out by the microtubule minus-end-directed cytoplasmic dynein, and requires IFT complex A subunits (reviewed by Hao and Scholey, 2009).

Assembly of cilia in *Tetrahymena* is crucially dependent upon anterograde intraflagellar transport. Double knockout of genes encoding kinesin-2 type motor proteins, Kin1 and Kin2 (Brown *et al.*, 1999b) causes a similar phenotype as a knockout of the genes that encode IFT complex B proteins, *IFT52* (Ift52p) (Brown *et al.*, 2003), *IFT80* (Ift80p) (Beales *et al.*, 2007), *IFT172* (Ift172p) (Tsao and Gorovsky, 2008a), and *DYF-1* (Dyf-1p) (Dave *et al.*, 2009). In each of these situations, the mutant cells are nearly immotile and assemble extremely short cilia containing remnants of outer doublets or lack cilia altogether. The cell's cytokinesis is inhibited at a very late stage, by failure of ciliary-dependent "rotokinesis," which leads to the regression of the cell division furrow and eventual formation of fused multisubcell "monster" cells with numerous nuclei. The GFP-tagged (or HA-tagged in case of Ift172p) IFT complex B proteins localize mainly to the BB region as well as into assembling cilia and weakly to full-length cilia; for details see descriptions of Kin1p (Brown *et al.*, 1999b), Ift52p (Brown *et al.*, 2003; Dave *et al.*, 2009), Ift80p (Beales *et al.*, 2007), Ift88p (Tsao and Gorovsky, 2008a, 2008b), Ift172p (Tsao and Gorovsky, 2008a), and Dyf-1p (Dave *et al.*, 2009). Ift172p may be a bifunctional protein, involved also in IFT turnaround or retrograde transport, as truncation of part of the C-terminal domain of Ift172p does not prevent ciliary assembly but causes the accumulation of truncated Ift172p and another complex B protein, Ift88p, at ciliary tips (Tsao and Gorovsky, 2008a).

Contrary to the dramatic ciliary phenotype observed in cells with eliminated kinesin-2 or IFT complex B proteins that are responsible for anterograde intraciliary transport, *Tetrahymena* cells with knocked out retrograde-transport components such as (1) the cytoplasmic dynein-2 heavy chain (*DYH2*) or (2) the cytoplasmic dynein-2 light intermediate chain (*D2LIC*) (Asai *et al.*, 2009; Rajagopalan *et al.*, 2009b), or (3) lacking a gene encoding *IFT122A*, a complex A protein (Tsao and Gorovsky, 2008b) display relatively mild changes in cilia. The assembled mutant cilia have normal axoneme ultrastructure (not investigated for *ΔIFT122A* cells), but there are fewer cilia per cell and those present are more variable in length when compared to the wild type. The cells in which the retrograde IFT component was knocked out swim significantly more slowly than wild-type cells and their swimming pathways are more winding (Rajagopalan *et al.*, 2009b). The population of dynein-2 knockout cells is heterogeneous, with both normal cells and some multinucleate monster cells, indicating a defect in cytokinesis (Rajagopalan *et al.*, 2009b). Knockdown of *IFT140*, another gene encoding complex A protein, also has only a mild effect on ciliogenesis (Rajagopalan *et al.*, 2009b, see discussion, p. 717). Strikingly, in *Chlamydomonas,* a mutation of dynein-2 heavy or light-intermediate chains results in the formation of short flagella with bulbs filled with accumulated IFT particles (Hou *et al.*, 2004; Pazour *et al.*, 1999). Such bulbs are not observed in *Tetrahymena* cells with either dynein-2 heavy chain (*ΔDYH2*) or light-intermediate chain (*ΔD2LIC*) knockout (Rajagopalan *et al.*, 2009b). However, cells with a knockout of *IFT122A* accumulate Ift172p and Ift88p proteins at the ciliary tip (in wild-type cells these complex B IFT proteins are distributed along the cilia and near the BBs) (Tsao and Gorovsky, 2008b). Thus also in *Tetrahymena* cilia, there is a retrograde-transport machinery and Ift122Ap is a part of it.

D. Ciliary Resorption

In *Tetrahymena*, under normal physiological conditions, somatic cilia once formed do not get resorbed. This is generally true of oral cilia as well, except in two special circumstances: the shedding of the outer row of cilia of the undulating membrane associated with cell division (Nelsen, 1981) and the resorption of the oral membranelles during oral replacement (see below). Ciliary resorption in the latter circumstance has been investigated ultrastructurally, and two patterns of cilia resorption were observed: (1). Regression *in situ*, and (2). Regression after incorporation into the cytoplasm (Williams and Frankel, 1973). The more frequently observed *in situ* regression starts with the partial loss of the wall of the B tubule and is followed by the loss of the A tubule. The central-pair MTs regress at different stages of outer doublet regression. Cilia that regress within the cytoplasm show a similar pattern of axoneme disorganization. Interestingly, these regressing cilia are not surrounded by autophagic vacuoles (Williams and Frankel, 1973). Nrks could be some of the factors that spatially and temporally regulate cilia resorption (Wloga *et al.*, 2006).

IV. Cortical Microtubular Arrays and Non-microtubular Fibers

The position of cortical microtubular arrays and non-microtubular fibers in relation to the BB triplets (Figs. 2A, F, and 4C) was described by Allen (1967). Please note, however, that the numbering of the BB triplets that we are using here (Fig. 2A) follows that described by Beisson and Jerka-Dziadosz (1999; Lynn, 1981) instead of that of Allen (1967).

A. Non-microtubular Cortical Fibers

Two types of non-microtubular structures originate near BBs: the kinetodesmal fiber (Allen, 1967) (KF in Figs. 2F and 4C) and a transient structure, the anterior-left fiber (Jerka-Dziadosz, personal communication). The periodically striated KF originates to the right of the base of the BB, next to triplets number 5, 6, and 7 (Allen, 1967; Beisson and Jerka-Dziadosz, 1999) and extends anteriorly and parallel to the BB row underneath the arrays of cortical MTs. The KF extends past at least two more anterior BBs. The anterior-left fiber originates to the BBs' left; however, unlike the KF, this filament is a transient structure visible only in dividing cells (Jerka-Dziadosz, personal communication).

In *Paramecium tetraurelia,* the KF is composed of a family of proteins with an estimated molecular mass of 30–36 kDa (Sperling *et al.*, 1991). Recently, J. Honts has detected a family of ~32 kDa KF-proteins in *T. thermophila* similar to the proteins reported by Sperling in *Paramecium* (Honts, personal communication). Interestingly, in green algae, the non-microtubular BB appendages called striated flagellar roots are predominantly composed of a 34 kDa protein, a striated fiber assemblin (SFA, Lechtreck and Melkonian, 1991). Two genes encoding proteins similar to SFA are present in the *Tetrahymena* genome database (Harper *et al.*, 2009).

B. Cortical Microtubular Arrays

The somatic BBs are accompanied by two bands of MTs that originate in their proximity: postciliary MTs (PM) and transverse MTs (TM) and also by arrays of MTs that extend parallel and to the right to the BB rows (longitudinal MTs, LM) as well as by basal MTs (B MT) positioned parallel and to the left of each BB row (Allen, 1967) (Figs. 1C and 2F). The transverse band (TM) originates anterior to each BB in proximity to triplet number 4 and is directed laterally toward the next BB row on its left, where it terminates. The TM is composed of on average of six MTs and extends between the epiplasmic layer and the mitochondria. The postciliary band (PM), composed of five to eight MTs, arises posterior to the BB next to triplet number 9 and extends obliquely toward the posterior cell pole underneath the epiplasmic layer. The mechanisms that regulate the length of the transverse and postciliary MT bands remain unknown.

Arrays of LMs arc positioned to the right of the BB rows and kinetodesmal fibers and run proximal to the cortical alveoli but distal to the epiplasm (Allen, 1967) (Fig. 1C). Although the LM bands extend from the anterior to the posterior end of the cell, the individual MTs within each bundle are much shorter (Ng, 1979; Pitelka, 1961). On the average, the LM array is composed of 7–12 partly overlapping unidirectionally organized MTs (Allen, 1967). Analysis of heteropolar *Tetrahymena* doublet cells resulting from fission blockage of a temperature-sensitive cytokinesis mutant revealed that after breakage of the LM arrays into two shorter bundles, the broken MTs elongated preferentially and more rapidly at the newly formed anterior ends of MTs that are posterior to the site of breakage than at the newly formed posterior ends of MTs that are anterior to the site of breakage. Such directionality of MT assembly suggests that (with the exceptions of inverted ciliary rows) the MTs of the LM are oriented within the cell with their plus ends directed toward the anterior and minus ends directed toward the posterior end of the cell (Ng, 1978, 1979).

The tubulin in cortical MTs is post-translationally modified by acetylation (Gaertig *et al.*, 1995), monoglycylation (Xia *et al.*, 2000), and monoglutamylation (Wloga *et al.*, 2008a) and weakly by polyglycylation (Thazhath *et al.*, 2002). In wild-type cells, polyglutamylation is not observed on cortical MTs (Wloga *et al.*, 2008a); however, polyglutamylation appears at least on some subcortical MTs in mutant cells with an increased activity of tubulin glutamylases (Janke *et al.*, 2005; Wloga *et al.*, 2010) and on some LMs in *KAT1* knockout cells (Sharma *et al.*, 2007). Since LM bands extend from the posterior to the anterior cell ends, *Tetrahymena* cell division requires breakage of LMs within the region of the fission furrow to allow daughter cells to separate. This phenomenon seems to depend upon the activity of the MT severing protein, katanin, and upon the levels of beta-tubulin polymodifications. A knockout of katanin, either of the catalytic subunit p60 or of the regulatory subunit p80, leads to inhibition of cytokinesis and formation of chains with three to four tandem subcells with LM bundles intact across adjacent subcells (Sharma *et al.*, 2007). The involvement of katanin in LM breakage is also supported by the immunolocalization studies. The GFP-tagged Kat1 p60 katanin localizes in lines parallel and to the right to the BB rows. Such localization is consistent with the position of the LMs (Sharma *et al.*, 2007). Formation of cell chains with unbroken LMs is also observed in mutagenized cells with elimination of three out of nine putative glycylation sites on the β-tubulin C-terminal tail (β $DDDE_{440}$) (Thazhath *et al.*, 2002, 2004). Since subsequent studies have shown that cells with an almost complete reduction of the level of tubulin glycylation (due to elimination of six Ttll3-type enzymes that function as tubulin monoglycylases) are viable and morphologically normal (Wloga *et al.*, 2009), it is unlikely that formation of cell-chains in β $DDDE_{440}$ mutant cells is caused exclusively by a decrease of the level of tubulin glycylation. However, since the mutated glycylation sites could also be required for tubulin glutamylation, it is possible that elimination of these putative glycylation/glutamylation sites could affect the level of tubulin glutamylation or the ratio of glycylation and glutamylation on β-tubulin, or the ratio of either of these modifications on β- versus α-tubulin, or the ratio of other tubulin PTMs affecting the entire "tubulin code" (Verhey and Gaertig, 2007).

The elimination of three putative modification sites in $\beta DDDE_{440}$ mutant cells affects not only breakage of LMs during cytokinesis but also causes hypertrophy of LMs (Thazhath *et al.*, 2004). This could be due to either excessive assembly of cortical MTs or to a delay in MT disassembly. Interestingly, *Tetrahymena* mutant cells with two knocked out genes encoding tubulin glutamylases, Ttll1 and Ttll9, have a slightly reduced number of MTs in their LM arrays (Wloga *et al.*, 2008a). Thus, the level of tubulin glutamylation or the balance mentioned above between tubulin modifications may be involved in the regulation of the number of MTs in the LM and possibly other cortical bands as well.

The two to three MTs called basal MTs (B MT in Fig. 1C) that usually run to the left of the BB row (Allen, 1967) have been little studied. In immunofluorescence images, they seem to overlap with the BB rows. These basal MTs become more conspicuous in cells that overexpress TtTtll6Ap glutamylase elongase (Wloga *et al.*, 2010).

While the cortical MTs described above are highly stable and not easily disaggregated by treatments (such as cold or colcemid) that normally disrupt MTs, the cytoplasmic MTs form a dynamic network that is sensitive both to cold and to antitubulin drugs (Kushida *et al.*, 2011; Stargell *et al.*, 1992). In wild-type cells, tubulin in cytoplasmic MTs is modified by the addition of a single glycyl (Xia *et al.*, 2000) or glutamyl residue (Wloga *et al.*, 2008a). However, overproduction of Ttll6Ap tubulin glutamylase elongase leads to hyperglutamylation and an increase in the number of cytoplasmic MTs (Wloga *et al.*, 2010). Moreover, stabilization of MTs in cells depleted of the MT severing enzyme, p60 KAT1 (Sharma *et al.*, 2007) also leads to an increase in the density of cytoplasmic MTs and to their apparent lysine acetylation (Sharma *et al.*, 2007), a modification typical of long-lived MTs (Verhey and Gaertig, 2007). The hyperglutamylated cytoplasmic MTs resulting from overexpression of TTLL6A tubulin glutamylase also accumulate a high level of acetylation and are more resistant to the tubulin depolymerizing drugs, oryzalin, and nocodazole (Wloga *et al.*, 2010). Thus, the level of tubulin modifications may have a direct or indirect stabilizing effect on cytoplasmic MTs, causing the observed differences in the stability of cortical and cytoplasmic MTs.

C. Invariant Zones

The formation of new cortical units in the somatic BBs rows during *Tetrahymena* cell growth is initiated by the assembly of new BBs anterior to the existing BBs followed by the assembly of their microtubular and non-microtubular appendages (Allen, 1969). At the whole-cell level, this proliferation of ciliary units and their subsequent ciliation is largely restricted to the equatorial and postequatorial regions of the cell (Frankel *et al.*, 1981; Kaczanowski, 1978; Nanney, 1975). An indirect consequence of this mainly centrally localized proliferation is that after cell division, the old BBs and cilia assembled in the previous cell generations are maintained for several generations at the most anterior region of the anterior daughter cell (proter) or the most posterior cell region of the posterior daughter cell (opisthe); these are the so-called invariant zones (Thazhath *et al.*, 2004). Similar regions postulated to be devoid of BB duplication in dividing *Paramecium* were named "invariant regions" (Iftode *et al.*, 1989). This concept of anterior and posterior "invariant regions" was subsequently applied to *Tetrahymena* (Thazhath *et al.*, 2004).

D. Tubulin Exchange

The microtubule-based structures present in invariant zones continue to slowly exchange components. In *Tetrahymena,* a slow exchange of tubulin subunits is observed in already assembled microtubular structures; however, the rate of the exchange is structure-specific. In cilia, the incorporation of new tubulin is especially prominent in their distal parts (Thazhath *et al.*, 2004). Over a period of 72 h, tubulin subunits are exchanged along the entire persisting cilia of the invariant zones (Thazhath *et al.*, 2004). It was estimated that the rate of tubulin turnover is 1–2.25% per hour (Nelsen, 1975; Thazhath *et al.*, 2004). The exchange of tubulin subunits is also observed in cortical MTs; in transverse MTs, the exchange of tubulin subunits is faster in the distal part of this array than it is near the BB, where the transverse MTs originate. In longitudinal MTs and MTs of contractile vacuoles pores, the tubulin exchange is faster than in transverse MTs (Thazhath *et al.*, 2004). The function of tubulin exchange is unknown.

V. Epiplasm (Membrane Skeleton)

Epiplasm, the fibrogranular layer that underlies the inner alveolar membranes, is proposed to anchor and coordinate positions of the cortical structures and is known to maintain cell shape. As in *Paramecium* (Williams *et al.*, 1989b) and *Euplotes* (Williams *et al.*, 1989a), depletion of MTs of *Tetrahymena* by extraction with a non-ionic detergent and high-salt solution leads to the formation of membrane skeleton "ghosts" in which the size and shape of intact cells is preserved (Williams *et al.*, 1990). Among roughly 35 proteins visible on a one-dimensional gel after separation of *Tetrahymena* membrane skeletons, the three most abundant proteins are named as epiplasmic band proteins A (EpiA), B (EpiB), and C (Epc1p, formerly known as EpiC) (Williams, 2004; Williams *et al.*, 1990).

Immunolocalization analysis shows that EpiA, EpiB, and Epc1p are distributed throughout the cell cortex in an overlapping but not identical pattern (Williams *et al.*, 1995) and are excluded from the regions of the OA, around somatic BBs, the cytoproct, and CVPs (Williams *et al.*, 1987, 1995). In dividing cells, the epiplasmic proteins are no longer detected in areas of new BB formation including the oral primordium (new OA formation), as well as in the region of the cleavage furrow (Kaczanowska *et al.*, 1993; Williams *et al.*, 1987, 1990). Thus the assembly of new cortical structures requires reorganization of the cell cortex, including spatial and temporal changes in the epiplasm that could be based on rearrangement or degradation. The molecular mechanism that controls such reorganization of the epiplasmic layer is unknown.

Cells with a knockout of *EPC1* are more rounded, show breaks in the BB rows, misplaced BBs, and increased variability in the number of BB rows, skewed toward an increase in their number, as well as branched and misaligned membranelles in the OA. However, the distribution of other epiplasmic proteins (EpiA, EpiB, TCBP-23, and TCBP-25) seems normal (Williams, 2004). The frequency of the more severe cortical abnormalities decreases with an increasing number of cell generations after the *EPC1* knockout.

The spatial pattern of the three epiplasm proteins is similar to that of two EF-hand Ca^{2+} binding proteins, TCBP-23 [whose localization seems to resemble that of Epc1p (Hanyu *et al.*, 1996)] and TCBP-25 [whose localization is similar to that of EpiB (Hanyu *et al.*, 1995)]. A knockdown of TCBP-25 results in some abnormally large cells and the presence of cell debris in the culture medium that suggests cell disruption (Nakagawa *et al.*, 2008). Such phenotypes could be caused by a weakened submembrane cytoskeleton. It would be interesting to see the spatial pattern of EpiA, EpiB, and Epc1p as well as the pattern of cortical MTs in the TCBP-25 knockdown cells. In mating cells, TCBP-25 is also detected in the junction area and around pronuclei (Hanyu *et al.*, 1995; Nakagawa *et al.*, 2008).

Fenestrin, another *Tetrahymena* cortical protein, is located beneath the epiplasmic layer (Nelsen *et al.*, 1994). The name of the protein reflects its localization, as fenestrin surrounds cell apertures including the cytoproct and CVPs, and is located around somatic and OA BBs (Cole *et al.*, 2008; Kaczanowska *et al.*, 2003; Nelsen *et al.*, 1994). In dividing cells, fenestrin is distributed asymmetrically, at the anterior pole and posterior to the cleavage furrow. In conjugating cells, fenestrin forms a filamentous network in the junction area and around pronuclei (Nelsen *et al.*, 1994). Functional studies on fenestrin have not been reported.

Epiplasmins are another family of cortical cytoskeleton proteins. They were first identified in *Paramecium* (Damaj *et al.*, 2009; Nahon *et al.*, 1993; Pomel *et al.*, 2006). Four epiplasmin proteins (EpiT1, EpiT2, EpiT3, and EpiT5) are also encoded by the *Tetrahymena* macronuclear genome (Damaj *et al.*, 2009). The localization of the epiplasmins has been studied in *Paramecium* (Damaj *et al.*, 2009; Nahon *et al.*, 1993) and the distribution and function of these proteins in *Tetrahymena* cells is starting to be investigated (Honts, personal communication). Recently, Gould and colleagues reported the presence of two genes encoding alveolin-like proteins (TtALV1 and TtALV2) in the *Tetrahymena* genome database (Gould *et al.*, 2008). Alveolins can be found in the Alveolata, and in *Paramecium* the alveolin epitopes were detected in the cell cortex (alveoli) (Gould *et al.*, 2008).

Despite the availability of a proteome of the "pellicle," which in this study also includes other cortical structures in addition to the membrane skeleton (epiplasm) (Gould *et al.*, 2010), our knowledge about the protein composition of the membrane skeleton, and about the associated pathways that reorganize the submembranous cytoskeleton and specifically the epiplasm, is rudimentary (for more details see

Chapter 14 by Honts). However, based on the data available so far, it is remarkable that different types of epiplasmic proteins appear to predominate in *Paramecium* and in *Tetrahymena*.

VI. Apical Crown, Apical Band, and Contractile Ring

The ciliary rows numbered 5 through $n - 2$ all terminate at their anterior end with a pair of specialized BBs ("apical couplets") that together make up an asymmetrical "apical crown" (McCoy, 1974) (Fig. 3A). The BBs of the couplets are linked together by dense material. While the posterior BB of each couplet is accompanied by regular microtubular and non-microtubular appendages, the anterior BB has only an atypically oriented transverse MT, and only the posterior BB of each BB couplet is ciliated (Jerka-Dziadosz, 1981a).

The paired BBs of the apical crown are underlain by a continuous ribbon of parallel microfilaments, originally called an "apical ring" (Jerka-Dziadosz, 1981b) and later re-named an "apical band" (Jerka-Dziadosz *et al.*, 1995) because

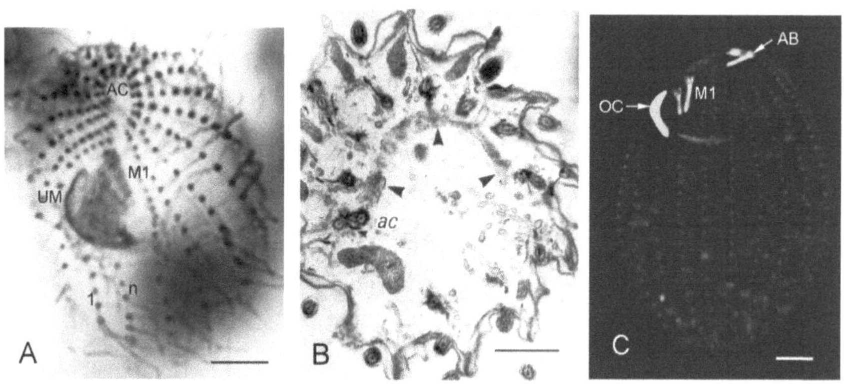

Fig. 3 The crown of apical BB couplets and the associated apical band of *Tetrahymena thermophila* as visualized by light microscopy (A), by transmission electron microscopy (B), and by immunofluorescence microscopy (C). (A) A protargol preparation, showing basal bodies as black dots. Ciliary rows 1 and n are labeled. The apical crown of basal-body couplets (AC) extends from ciliary rows 5 to n-2. The oral apparatus, including its undulating membrane (UM) and first membranelle (M1) is partly out of focus. (B) A cross section through the apical region. One apical basal-body couplet (ac) is visible. Large arrowheads point to the apical filament band that underlies the apical couplets. (C) Immunofluorescence signal from live cells expressing GFP-TtCen3. The image (originally green) is rendered in black and white, with brightness and contrast enhanced. The strong signals are from the oral crescent (OC) and from the apical band (AB). The signals from membranelles 1 (M1) and 2 (not marked) are considerably weaker than those from the OC and AB, and M3 is unlabeled in this and other similar preparations. Scale bars: 5 μm in A and C, 1 μm in B. (A) from Fig. 3B in Nelsen, E.M., Frankel, J., and Martel, E. (1981) *Dev. Biol.* **88**: 27–38. (B) from Fig. 1 of Jerka-Dziadosz, M. (1981) *J. Cell Sci* **51**:241–253, reproduced with permission of the author and of the Company of Biologists; (C) from an original unpublished image kindly supplied by Alexander Stemm-Wolf and Mark Winey of the University of Colorado at Boulder.

it does not encircle the entire cell. This horseshoe-shaped ribbon is a permanent structure located just posterior to the cell apex, adjacent to the anterior BB pairs (Fig. 3B). During cell division, the apical band of the posterior daughter cell is formed a short distance posterior to the site of constriction (Jerka-Dziadosz, 1981b). It is initiated from seeding-sites situated underneath the newly formed anterior BBs of the pairs located near the anterior end of the posterior daughter cell. Short segments of filaments extend laterally from these seeding sites and eventually join to form a continuous band (Jerka-Dziadosz *et al.*, 2001; Jerka-Dziadosz, personal communication). This new apical band is situated posterior to the contractile ring of microfilaments that is located at the narrowest circumference of the constricting cell (Jerka-Dziadosz, 1981b).

The apical band contains two of the four *Tetrahymena* centrins, TtCen3 and TtCen4 (Stemm-Wolf *et al.*, 2005), centrin 3 as a continuous band (Fig. 3C) and centrin 4 more punctate. TtCen3 and TtCen4 belong respectively to ciliate-specific and apicomplexan-specific centrin subfamilies (Gogendeau *et al.*, 2008) and "are similar to the centrins found in the infraciliary lattice of *Paramecium tetraurelia*" (Stemm-Wolf *et al.*, 2005 p. 3609; see Ruiz *et al.*, 2005). Structurally, the association of filaments of the developing apical band of *Tetrahymena* with the apical BB couplets (Jerka-Dziadosz, 1981b) strongly resembles the associations of similar fine filaments of the regenerating infraciliary lattice with BB pairs of *Paramecium* (Beisson *et al.*, 2001), which are distributed over the entire body surface (Iftode *et al.*, 1989) instead of being restricted to the anterior end of certain ciliary rows as in *Tetrahymena* (McCoy, 1974). Such BB-pairs, or "dikinetids" (Lynn, 1981) are widespread among ciliates and may be primitive (Lynn and Small, 1981). Thus, there are grounds for believing that the apical filament band of *Tetrahymena* might be an evolutionary vestige of the more extensive infraciliary lattice of a distant common ancestor of both *Paramecium* and *Tetrahymena.*

Additional proteins associated with the apical band of *T. thermophila* are fimbrin (Shirayama and Numata, 2003), "p85" (Gonda *et al.*, 1999a), calmodulin (Gonda *et al.*, 1999a), and elongation factor-1α (EF-1α, Numata *et al.*, 2000). There is also a strong localization of the antigen recognized by the 12G9 monoclonal antibody, which binds to cortical filaments in a variety of ciliates. Interestingly, the 12G9 antibody also decorates two transient structures observed only in dividing cells: the post-oral meridional filament and the fission line (Jerka-Dziadosz *et al.*, 1999, 2001; Strzyzewska-Jowko *et al.*, 2003). It was suggested that these structures might be involved in the regulation of cell polarity and patterning (Jerka-Dziadosz *et al.*, 2001).

p85 is a protein that was discovered through a mobility difference in 2D acrylamide gels between wild-type *T. thermophila* and a temperature-sensitive cell-division-arrest mutant (*cdaA-1*) (Ohba *et al.*, 1986); however, p85 may not be the product of *CDA1* gene (Gonda *et al.*, 1999b). p85 co-localizes with the apical band, starting with the nascent apical BB couplets that develop immediately posterior to the fission zone in cells preparing to divide (Numata and Gonda, 2001; Numata *et al.*, 1995a; Ohba *et al.*, 1986). Numata and colleagues hypothesize that a p85-Ca^{2+}-calmodulin (CaM) complex determines the division plane by inducing the polymerization of G-actin to F-actin (Gonda and Numata, 2002), with assistance of profilin (Edamatsu *et al.*, 1992; Numata and Gonda, 2001), EF-1α (Numata and Gonda, 2001), and fimbrin (Shirayama and Numata, 2003). However, p85 and CaM may not be contractile ring components as both were localized at the posterior side of the furrow while actin was seen in the center of the furrow (Gonda *et al.*, 1999a). Furthermore, when the contractile ring disappears at the end of constriction, p85 persists at the newly formed anterior apex of the new posterior daughter cell.

In animal and fungal cells, division furrowing is driven by a largely conserved mechanism involving an interaction between actin and myosin II (reviewed by Pollard, 2010). Myosin II is, however, present only in "unikonts", including amoebae, fungi, and animals, and is absent in "bikonts", including plants and many unicellular eukaryotic lineages, including the alveolates (Richards and Cavalier-Smith, 2005). Therefore, it is perhaps not surprising that a disruption of the gene encoding the major actin (Act1p) of *T. thermophila* has no effect on division furrowing (Williams *et al.*, 2006), indicating that Act1p is not essential for division furrowing. However, in *Paramecium tetraurelia*, the furrow is immunolabeled by an antibody against a member of one of the nine different *P. tetraurelia* actin subfamilies, actin 4–1 (Sehring *et al.*, 2007, 2010), which is not the closest ortholog of Act1p in *T. thermophila*. This discovery in *Paramecium* opens up the possibility that one of the three other *Tetrahymena* actin species (Williams *et al.*, 2006) might be involved in furrowing in *T. thermophila*. The *Tetrahymena* genome encodes 13 myosins, most of which belong to type XIV that is specific to Alveolata (Foth *et al.*, 2006; Sugita *et al.*, 2011). However, one of these myosins has a coiled-coil domain that is reminiscent of myosin II (Sugita *et al.*, 2011). The mechanism of cytokinesis in *Tetrahymena* is thus a fascinating area for future investigation.

VII. Oral Apparatus

The oral apparatus (OA) is a complex funnel-like structure located a short distance posterior to the anterior cell pole. The OA functions as the site of phagocytosis-based uptake of food particles (reviewed by Nilsson, 1979). In wild-type growing cells, the complex asymmetrical structure of the OA is generally invariant (Bakowska *et al.*, 1982a). Nearly 150 oral BBs are organized into four clusters that when ciliated give rise to four compound ciliary structures: three adoral-zone membranelles (M1, M2, and M3) and an undulating membrane (UM) (Bakowska *et al.*, 1982a; Nilsson and Williams, 1966) (Fig. 4). The parallel M1-M3 clusters that are located on the dorsal wall of the buccal cavity extend toward the left edge of the OA and are partly covered by the OA's rim (Kiersnowska and Golinska, 1996; Sattler and Staehelin, 1976). Each adoral membranelle contains 3 BB rows, whose right ends become "sculptured" into more irregular arrangements during the final stages of membranelle development (Bakowska *et al.*, 1982a, 1982b). M1, the most anterior

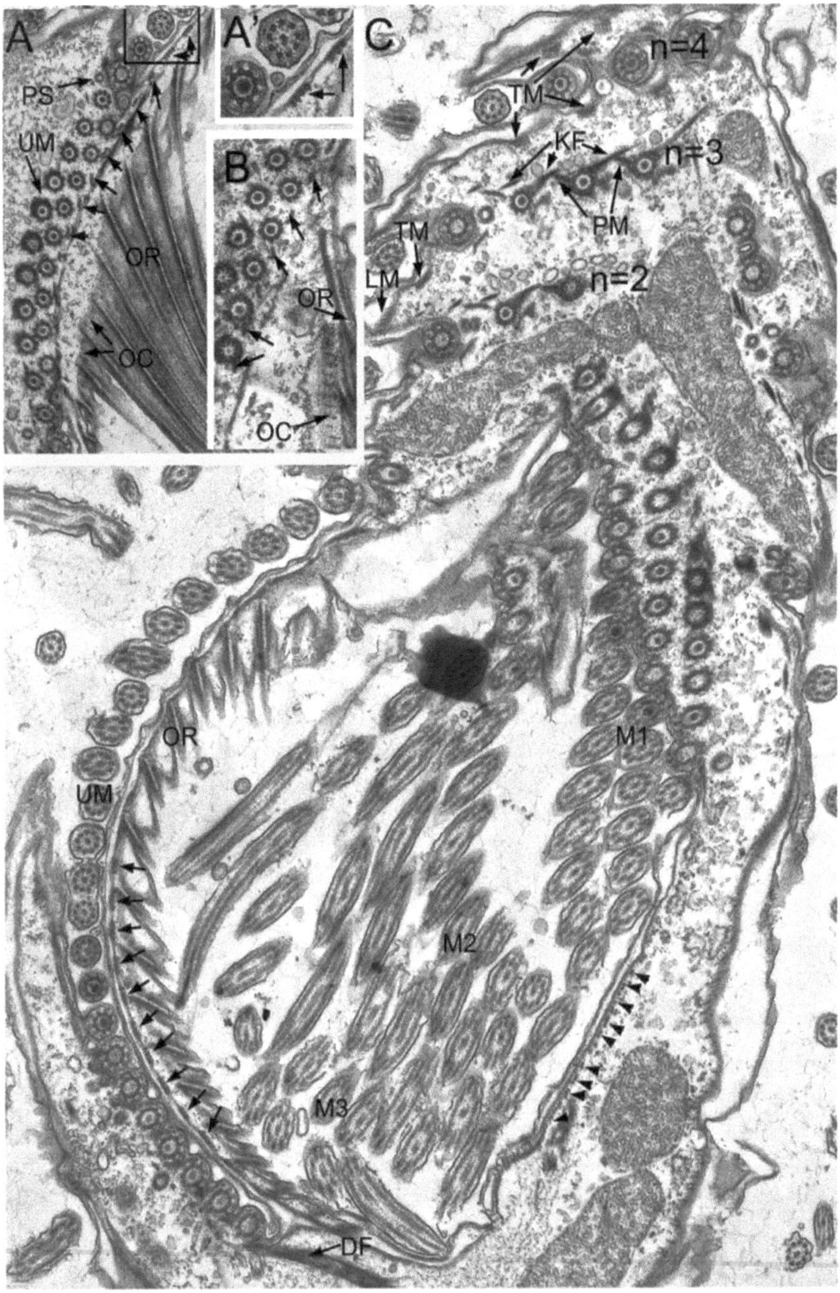

Fig. 4 Ultrastructure of the oral apparatus of *T. thermophila.* (A) A section through the right side of the oral apparatus showing the undulating membrane (UM) composed of two BB rows that border the right edge of oral apparatus; only the outer row is ciliated. Small arrows show postciliary microtubules of the inner BB row of the UM (see a magnified fragment of the UM shown

membranelle (Fig. 4C), is the largest, and M3, the most posterior one, is the smallest and has the most irregular BB organization. The right edge of the OA is bordered by the arched UM composed of two parallel rows of BBs (Fig. 4A, B).

In the mature OA, only the oral BBs in the third, most posterior row of each of the adoral membranelles and BBs in the inner row of the UM are associated with a band of postciliary MTs (Fig. 4A, A′, B and C, arrows) while "…in young OA also the first BBs of all membranellar rows from the right side…" are accompanied by postciliary MTs (Jerka-Dziadosz, 1981a, p. 344). No transverse MTs are observed to accompany BBs in the OA (Jerka-Dziadosz, 1981a).

The BBs of the OA are connected by microtubular and filamentous networks. The former connects membranelles with each other and with the undulating membrane (Forer *et al.*, 1970; Williams and Luft, 1968) while the latter contains abundant tetrins (A, B, and C type), the ciliate-specific filament-forming proteins (Brimmer and Weber, 2000; Honts and Williams, 1990), which are co-localized throughout the oral apparatus (Dress *et al.*, 1992; Frankel, 2000b). The oral ribs (OR, Fig. 4A, B, C), which extend from the inner BB row of the UM, are supported by organized bundles of MTs (Jerka-Dziadosz, 1981a; Sattler and Staehelin, 1979; Williams and Luft, 1968) that extend down along the OA funnel wall as a part of deep fiber bundle (Smith and Buhse, 1983; Williams and Bakowska, 1982). The oral ribs are underlain by a fine filamentous reticulum (Sattler and Staehelin, 1979; Williams and Bakowska, 1982; Williams and Luft, 1968), which appears crescent shaped in immunofluorescence studies and hence is called the "oral crescent" (Jerka-Dziadosz *et al.*, 1995) (OC, Fig. 4A, B).

Immunogold labeling with affinity-purified anti-*Tetrahymena* actin and myosin antibodies indicates the presence of actin (Hoey and Gavin, 1992) and myosins-related epitopes (Garces *et al.*, 1995) in association with BBs in the oral apparatus of *T. thermophila*. Light microscopic observations using a different anti-*Tetrahymena*-actin antibody detected actin in the deep fiber bundle of *T. thermophila* (Gonda *et al.*, 2000), while calmodulin, the Ca^{2+}/calmodulin-binding protein EF-1α, and "p85" (see below) all appear to be present throughout the oral apparatus of *T. thermophila* (Gonda *et al.*, 2000). Immunofluorescent localizations in the oral apparatus need to be viewed with some caution since, with regard to the apparent localization of

as an inset in A'). The oral ribs (OR) underlain by the filaments of oral crescent (OC) extend from the inner row of the UM. Parasomal sacs (PS) accompany the basal bodies of the UM. (B) Another section through a portion of the undulating membrane, showing cross-sections through the basal bodies at different levels, and rows of postciliary MTs (arrows) that accompany the inner basal bodies. Some oral ribs (OR) underlain by banded filaments of the oral crescent (OC) are clearly visible. (C) A section through the oral apparatus showing a row of basal bodies of the undulating membrane (UM) with postciliary MTs (small arrows) and oral ribs (OR), as well as adoral membranelles 1 (M1), 2 (M2) and 3 (M3). The adoral membranelles are obliquely sectioned, and cilia of these membranelles are visible in various orientations within the buccal cavity. The arrowheads point to MTs at the left edge of the oral apparatus. Note a portion of the deep fiber (DF) posterior to the OA. The upper part of this image shows three rows of somatic basal bodies (n = 2, n = 3 and n = 4) with associated structures: kinetodesmal fiber (KF), transverse MTs (TM) and postciliary MTs (PM) as well as longitudinal MTs (LM) extending parallel to the basal-body rows. Unpublished electron micrographs were kindly supplied by Dr. Maria Jerka-Dziadosz.

profilin in the oral apparatus of *T. pyriformis*, Edamatsu *et al.* (1992) state "The fluorescence of oral apparatus and ciliary meridian appear to be nonspecific, since preimmune serum. . .and antigen-absorbed serum. . . stained these structures" (p. 639). Nonetheless, lack of an apparently functional OA in nearly 70% of cells with a profilin knockdown (by the antisense ribosome method) suggests involvement of this protein in OA assembly or function (Wilkes and Otto, 2003).

There is one specific oral localization that is especially striking: a prominent staining of the oral crescent by the well-known 20H5 anti-centrin antibody (Jerka-Dziadosz *et al.*, 1995) and by GFPTtCen3 (Fig. 3C) and GFPTtCen4 (Stemm-Wolf *et al.*, 2005), also seen with antibodies to p85 (Gonda *et al.*, 1999a), calmodulin (Gonda *et al.*, 1999a, 2000), and EF-1α (Gonda *et al.*, 2000; Numata and Gonda, 2001; Numata *et al.*, 2000; Suzuki *et al.*, 1982). Indeed, there appears to be an association between the staining of the apical filament band and the oral crescent, most evident with the 20H5 anti-centrin antibody (Jerka-Dziadosz *et al.*, 1995) and with GFP-centrin 3 (Stemm-Wolf *et al.*, 2005) (Fig. 3C). This association may be related to the similarity in ultrastructure between the apical filament band (Jerka-Dziadosz, 1981b) and the fine filamentous reticulum that makes up the oral crescent (Jerka-Dziadosz, 1981b; Sattler and Staehelin, 1979; Williams and Bakowska, 1982; Williams and Luft, 1968); both possess banded microfilaments of a diameter of approximately 6 nm and periodicity of close to 100 nm in the oral crescent (Sattler and Staehelin, 1979; Williams and Bakowska, 1982) (Fig. 4B) and 100 to 200 nm in the apical band (Jerka-Dziadosz, 1981b). This suggests some underlying functional similarity of these anatomically dissimilar structures, probably as calcium-regulated contractile systems, as has been suggested for the . . . fine filamentous reticulum of *T. thermophila* (Williams and Bakowska, 1982) and for the probably homologous infraciliary lattice of *P. tetraurelia* (Garreau de Loubresse *et al.*, 1988, 1991).

The synchronous beating of cilia of the OA membranelles directs food particles into the OA funnel, and then into the cytostome, the aperture at the base of the buccal cavity that is situated below the membranelle region. This opens into a short passage, called the cytopharynx, at the inner end of which the food vacuoles are formed (Nilsson and Williams, 1966). Mutants with defective food vacuole formation or cells lacking oral cilia or having immotile oral cilia (and thus unable to phagocytose) can be maintained in a rich medium (Brown *et al.*, 1999b; Orias and Rasmussen, 1976; Williams *et al.*, 2006) that is believed to stimulate clathrin-dependent endocytosis (pinocytosis), involving the formation of microvesicles at coated pits (Allen, 1967; Elde *et al.*, 2005; Nilsson and van Deurs, 1983), possibly supplemented and under some circumstances replaced by carrier-mediated transport of amino acids (Orias and Rasmussen, 1979).

Prolonged cell starvation leads to a reduction of the size of the OA, including a reduction in the number of BBs and even the loss of M3; however, changes in the size of the OA are not proportional to the changes in cell size (Bakowska *et al.*, 1982a). Besides environmental factors, the genetic background (mutations) also can affect the structure and size of OA. Specific mutations bring about (1) an increase in the size of the OA due to an increase in the length of the membranelles caused by increased BB proliferation, (2) a reduction or increase of the number of OA adoral membranelles, or (3) fragmentation of the undulating membrane (Frankel *et al.*, 1984a, 1984b; Kaczanowski, 1975, 1976; Lansing *et al.*, 1985; reviewed by Frankel, 2008).

There are two distinct developmental processes leading to the assembly of the OA: divisional morphogenesis in dividing cells and oral replacement in non-dividing cells. Both processes occur in three phases: (1) BB proliferation, (2) BB spatial organization, and (3) oral cavity formation. During preparation for cell division, the formation of the new OA is initiated by random BB proliferation (the "anarchic field" stage of the oral primordium) in the equatorial region of the cell to the left of BB row number 1. Formation of the oral primordium is invariably synchronized with the entry of the micronucleus into mitosis (Frankel *et al.*, 1976a). The new BBs assemble next to already existing ones, even if they are not fully differentiated (Williams and Frankel, 1973). As development progresses, the BBs are gradually organized into the membranelles observed in the mature OA, and BBs become ciliated (Bakowska *et al.*, 1982b; Lansing *et al.*, 1985; Nelsen, 1981). The cleavage furrow is formed anterior to the newly assembling OA, and after cell division the new OA is inherited by the posterior daughter cell. Thus formation of the new OA is synchronized with cell division. This synchronization is physiologically coordinated (Frankel, 1962; Gavin, 1965) and may be presumed to be genetically controlled (Frankel *et al.*, 1976b; Kirk *et al.*, 2008).

Oral replacement is rarely observed in cells from logarithmically growing cultures, but takes place in cells from cultures in stationary phase and especially in starved cells (Frankel, 1969, 1970; Nelsen, 1978; Williams and Frankel, 1973) and during conjugation (Cole and Frankel, 1991; Kiersnowska *et al.,* 1993). Unlike new OA formation during divisional morphogenesis, where the oral primordium is spatially separate from the existing OA and all BBs are assembled anew, during oral replacement the BBs start to proliferate in close association with the existing OA (Frankel, 1969). Moreover, the oral-replacement primordium arises from two portions: (1) pre-existing BBs of the disaggregating UM of the old OA and (2) newly formed BBs adjacent to the anterior end of BB row no. 1. These two moieties fuse to form a single combined oral primordium (Frankel, 1969; Kaczanowski, 1976). The fusion process can be disrupted by the abnormal *mp* (later renamed *mp1–1)* allele that happens to be resident in the D inbred strain of *T. thermophila* (Kaczanowski, 1976).

As the oral-replacement primordium expands anteriorly, the BBs of the existing oral membranelles are simultaneously resorbed, starting with the M3 membranelle. As during cell division, the newly assembled BBs, some with an apparent array of postciliary MTs, gradually align into rows typical for the OA ciliature, first the adoral membranelles and then the undulating membrane (Frankel, 1969; Williams and Frankel, 1973). During the final stages of BB organization, the buccal cavity forms (Williams and Frankel, 1973). The ciliation of OA BBs is a surprisingly early event, as short ciliary shafts can be seen on the most anterior BBs even before membranelle pattern formation (Williams and Frankel, 1973); precocious ciliation within early oral fields has also been described for dividing cells (Bakowska *et al.*, 1982b).

During cell division (after the membranelles have become organized into triple rows and the new UM has started to form), the old OA undergoes extensive remodeling: first, the buccal cavity becomes very shallow, the deep fiber bundle and many of the oral filaments disassemble, the membranellar cilia shorten, and the outer row of the UM peels off and is resorbed (Bakowska *et al.*, 1982b; Nelsen, 1981; Williams *et al.*, 1986). During the subsequent reconstruction, the BBs of the outer row of the UM in both the old and newly forming OAs become assembled at about the same time (Bakowska *et al.*, 1982b; Nelsen, 1981) and the buccal cavity and the fibrillar structures of the old OA reappear in synchrony, with the same processes occurring in the newly developing OA (Williams *et al.*, 1986). Furthermore, oral cilia elongate at the same time in the new and old OA based on their similar size, enrichment in monoglycylated tubulin, and strong presence of kinesin-2 IFT motors (Brown *et al.*, 1999b).

Both remodeling of the existing OA and the assembly of the new one require strict temporal and spatial control in order to coordinate the formation of these structures with the cell developmental program. Thus *Tetrahymena* stomatogenesis is an interesting model in which one can study the molecular basis of the mechanisms that regulate spatial and temporal pattern formation within the confines of a single cell. Moreover, assembly of the new OA is an example of an inducible system for analysis of BB assembly and MT positioning and dynamics (Williams and Frankel, 1973) as well as for the dynamics and function of membranes and the cytoskeleton during phagocytosis.

Methods of OA extraction and isolation can be found, for example, in Williams and Honts (1995) and Wolfe (1970).

VIII. Cytoproct

Undigested particles are expelled from the cell through a single opening, called the cytoproct (sometimes called the cytopyge), located near the posterior end of the cell between the same BB rows as the OA (1 and n) (Corliss, 1973) (Fig. 1A). When closed, the cytoproct appears as a slit-like structure with a width of 0.3–0.8 μm, which increases to about 2 μm when open (Allen and Wolf, 1979). The cytoproct is surrounded by plasma membrane that lacks the subcortical membrane skeleton (Williams *et al.*, 1987) and typical cortical structures such as alveoli; it is bordered by a rim of electron-opaque material located at the termini of alveoli that surround the cytoproct (Allen and Wolf, 1979). The cytoproct is accompanied by individual MTs that seem to originate in the region of the electron-opaque rim and extend into the cytoplasm, where they drape over the membrane of the food vacuoles located near the cytoproct.

The expulsion of the food vacuole contents is a very fast process (Blum and Greenside, 1976) initiated by a fusion between the food vacuole and cytoproct membranes (Allen and Wolf, 1979). After defecation, the fused membrane is retrieved by endocytosis (Allen and Wolf, 1979; Sugita *et al.*, 2009). Food vacuole membrane recycling is an actin-dependent process. The food vacuoles localized near the cytoproct are characterized by membrane invaginations and a coat of microfilaments (Allen and Wolf, 1979). The localization studies with anti-actin antibodies revealed staining around old food vacuoles located near the cytoproct (Hirono *et al.*, 1987; Sugita *et al.*, 2009). Interestingly, Latrunculin B, an inhibitor that prevents assembly of actin filaments, does not prevent defecation of the contents of the food vacuole, but it suppresses the recovery of the food vacuole membrane after expulsion (Sugita *et al.*, 2009) suggesting that membrane retrieval is an actin-dependent process. On the other hand, defecation was affected in cells treated with the MT-destabilizing drug, nocodazole (Sugita *et al.*, 2009), consistent with the suggestion that MTs direct food vacuoles to the cytoproct (Allen and Wolf, 1979). Alternatively, MTs may serve as tracks in the recycling of food-vacuolar membrane, and an inability to transport actin-coated recycling vesicles leads to the formation of the barrier that prevents food vacuoles from accessing the cytoproct (Sugita *et al.*, 2009).

The formation of food vacuoles and their removal at the cytoproct may serve as a splendid model to investigate membrane fusion as well as evolutionary conservation and divergence of the mechanisms that control such processes (see Chapter 6 by Nusblat, Bright, and Turkewitz).

IX. Contractile Vacuole

The *Tetrahymena* cell generally has a single contractile vacuole (CV, Elliott and Kennedy, 1973) that is located near the posterior pole of the cell. This osmoregulatory organelle (Rifkin, 1973) cyclically accumulates (diastole phase) and discharges (systole phase) collected fluid (Organ *et al.*, 1972; Patterson and Sleigh, 1976) through the one to three (2 on average) (Loefer *et al.*, 1966; Nanney, 1966b) contractile vacuole pores (CVPs). These are visible on the cell surface as circular 0.5–1 μm wide openings (Allen and Wolf, 1979), positioned mostly near BB rows number 5 and 6 slightly anterior to the position of the cytoproct (Elliott and Bak, 1964b; Loefer *et al.*, 1966; Nanney, 1966b, 1972; Ng, 1977). The number and the position of the CVPs is correlated with the total number of BB rows in the cell (Nanney, 1966b, 1972), more precisely with cell geometry as cells measure "the *relative* distance between successive . . .right-postoral" BB rows (Frankel, 2000a, p. 91; compare to Nanney, 1966b, p. 316). This description explains the positioning of CVPs in both normal *Tetrahymena* cells (with a single right-postoral BB row, row #1) and in cells with duplicated cortical structures (with two OAs and thus two right-postoral BB rows).

CVPs are closed at their proximal end by two membranes, an outer membrane that is continuous with the CVP wall and an inner membrane that is continuous with the vacuolar membrane (Elliott and Bak, 1964b). The mechanism that regulates the membrane rupture during systole is not clear (Elliott and Bak, 1964b; Organ *et al.*,

1972). The CV forms numerous branches and tubular extensions ("spongiome", Patterson, 1980) that penetrate into the cytoplasm of the posterior cell region (Elliott and Bak, 1964b).

The CVPs are strongly decorated with anti-γ-tubulin antibodies, and the level of accumulated γ-tubulin remains constant during the cell cycle (Shang *et al.*, 2002). The wall of the CVP is surrounded by MTs. Moreover numerous MTs extend from the CVP wall in the direction of the vacuole membrane (Fig. 2G) (Elliott and Bak, 1964b; Organ *et al.*, 1972). Tubulin of both types of MTs is post-translationally modified (Gaertig *et al.*, 1995; Thazhath *et al.*, 2002; Wloga *et al.*, 2008a).

Localization analyses using specific antibodies or protein tagging suggest association of the following proteins with CV structures: calmodulin (Numata and Gonda, 2001; Suzuki *et al.*, 1982), centrin Cen4p (Stemm-Wolf *et al.*, 2005), Nima-related kinases Nrk1p and Nrk2p (Wloga *et al.*, 2006), Rab GTPases TtRabD2, TtRabD10, and TtRabD14 (Bright *et al.*, 2010), and Apm2, a paralog of AP-2 (Elde *et al.*, 2005), the Adaptor Protein-2 that participates in clathrin mediated membrane traffic in eukaryotes (Kirchhausen, 1999).

Recent analysis of the CVs in *Paramecium* cells (and other protozoa) brought about major progress in understanding of the molecular mechanisms that regulate the function of these organelles (for review, see Allen, 2000; Allen and Naitoh, 2002). The interested reader may refer to the following publications focusing on the CV in *Paramecium*: (Grønlien *et al.*, 2002; Iwamoto *et al.*, 2003, 2005; Ladenburger *et al.*, 2006, 2009; Schilde *et al.*, 2006; Stock *et al.*, 2001, 2002; Sugino *et al.*, 2005; Wassmer *et al.*, 2005, 2006).

X. Mitochondria

A. Morphology and Organization

Tetrahymena contains about 1000 of these typical eukaryotic cell organelles (Kay *et al.*, 1974; Poole, 1983). Numerous mitochondria are located in the subcortical region, where they are arranged in rows along the BBs meridians, while the remaining mitochondria are randomly distributed within the cytoplasm (Aufderheide, 1979). It seems that mitochondria can be exchanged between these two (subcortical and cytoplasmic) compartments (Aufderheide, 1979). The number of cytoplasmic mitochondria increases as cells enter into stationary phase. Some of the cytoplasmic mitochondria become incorporated into autophagic vacuoles and are degraded (Elliott and Bak, 1964a).

The subcortical mitochondria are organized either in single rows positioned to the left of the cell's BBs under TMs (the single mitochondrial pattern) or in two rows, on both sides of BBs with the mitochondria to the right of the cell's BBs located under LMs (the double mitochondrial pattern) (Aufderheide, 1979). The pattern of the subcortical mitochondria depends upon nutrient conditions (Aufderheide, 1979). The correlation between mitochondrial pattern and the pattern of the cortical MTs led to the hypothesis that the cortical MTs provide spatial signals (marks) for subcortical localization of mitochondria (Aufderheide, 1980). Indeed, induced changes in the organization of cortical MTs in cells of several *Tetrahymena* mutants that result in inverted (cdaA2), disorganized (disA1), or mirror-image (janus) microtubular patterns led to the re-localization of mitochondria to parallel the new local arrangement of MT arrays (Aufderheide, 1980). Recently, using COS7 fibroblast-like cells, it was shown that cytoplasmic mitochondria preferentially co-localized with acetylated (on K40 of α-tubulin) MTs (Friedman *et al.*, 2010). Interestingly, in *Tetrahymena,* both longitudinal and transverse MTs are acetylated (Gaertig *et al.*, 1995). While acetylation of α-tubulin is not required for normal growth of *Tetrahymena* (Akella *et al.*, 2010), it remains to be determined whether the pattern of mitochondria depends on this tubulin post-translational modification.

Mitochondria are surrounded by two membranes. The outer mitochondrial membrane encloses the mitochondrion while the inner mitochondrial membrane forms tubular cristae that extend into the mitochondrial matrix. Recent analysis of the *Tetrahymena* mitochondrial ATP synthase complex (known also as the F_0F_1 complex or complex V) that is located in tubular cristae revealed major differences in the protein composition and spatial arrangement of subunits within the F_0F_1 complex, when compared to the F_0F_1 complex in other organisms (Balabaskaran Nina *et al.*, 2010). These authors suggest that the formation of tubular cristae, which is unique to alveolates, could be caused by a parallel rather than an angular organization of monomers in the dimeric ATP synthase complexes (Balabaskaran Nina *et al.*, 2010). A dependence between the dimerization of the ATPase synthase and mitochondrial morphology was shown earlier in yeasts (Paumard *et al.*, 2002; Rabl *et al.*, 2009; Velours *et al.*, 2009). These very unusual features of ciliate mitochondria will certainly lead to considerable further investigation for their intrinsic and evolutionary interest, especially when exploring the interrelation between cellular structure and function.

At least some mitochondria in *Tetrahymena* include rod-like structures composed of filaments in a parallel arrangement. Such structures become more frequent as cell divisions slow down (Elliott and Bak, 1964a; Numata *et al.*, 1995b). Immunogold labeling analysis showed that these mitochondrial inclusions contain a citrate synthase, CIT1 (Numata *et al.*, 1995b), the first enzyme of the Krebs tricarboxylic acid cycle. However, it is not known if this enzyme is the inclusion's sole component. The citrate synthase (known also as 49 K protein), when purified from *Tetrahymena* cells or mitochondria, can polymerize *in vitro* into 14-nm filaments (Numata and Watanabe, 1982). *in vitro* filament formation reduces enzyme activity (Takeda *et al.*, 1995). Both activity and polymerization seem to depend upon the level of protein phosphorylation (Kojima and Numata, 2002). The citrate synthase has been suggested to be a bi-functional protein that performs an enzymatic function in mitochondria and also acts as a cytoskeletal protein. Antibodies directed against 49K/citrate synthase protein decorate some structural elements within the cell: the posterior region of the OA in non-dividing cells, filaments in the junction area of conjugating cells, and filaments that surround gametic nuclei during their formation, exchange and fusion in conjugating cells (Numata *et al.*, 1983, 1985, 1991; Takagi *et al.*, 1991). The function of the filaments decorated by anti-49K/citrate synthase antibodies is unknown. It would be informative to express a tagged version of citrate synthase/49K protein to further confirm its dual mitochondrial and cytoskeletal localization.

B. Division of Mitochondria

Analysis of the mitochondria in a wide range of organisms indicates that these organelles frequently undergo fission and fusion (reviewed by Otera and Mihara (2011)). In synchronized *T. pyriformis* cells, mitochondria divide during late macronuclear S phase (Kolb-Bachofen and Vogell, 1975). There are no data concerning fusion of mitochondria in *Tetrahymena* cells; however, the analysis of mitochondrial genetic recombination in *Paramecium* cells suggests the absence of such a process (Adoutte *et al.*, 1979). Other than possessing a dynamin-related protein, *Tetrahymena* lacks apparent orthologs of the proteins involved in mitochondrial fission and fusion in other organisms. One of the *Tetrahymena* dynamin-related proteins, Drp7, tagged with GFP, seems to localize near mitochondria (Elde *et al.*, 2005, supplementary data); however, the function of Drp7 in *Tetrahymena* mitochondrial division is not yet clear.

Surprisingly, septins, evolutionarily conserved GTP-binding proteins that in yeasts and metazoans play an important role, among other processes, in cell division (Cao *et al.*, 2009a), in *Tetrahymena* seem to be involved in the regulation of mitochondrial dynamics (Wloga *et al.*, 2008b). Three GFP tagged septins localize to mitochondria, although their localization patterns are not identical, and vary from the outer mitochondrial membrane (septin 1 and 3) to septa-like structures (septin 2). Knockouts of all three septin genes result in formation of some extremely long mitochondria, pointing to a possible role of septins in mitochondrial fission. However, cells lacking septins have normal growth rates, implying that there is an additional mitochondrial fission mechanism that is not septin dependent (Wloga *et al.*, 2008b). In other organisms, only alternatively spliced forms of septin 4 (human ARTS and murine M-septin) localize to mitochondria (Larisch *et al.*, 2000; Takahashi *et al.*, 2003). The lack of obvious homology outside the central core domain between septins 4 and ciliate septins as well as differences in their function in mitochondria lead to the presumption that the presence of septins in mitochondria of ciliates and mammals could have arisen as evolutionarily independent events (Wloga *et al.*, 2008b).

C. Mitochondrial Apoptosis Inducing Factor (AIF)

Nearly 15 years ago, it was shown that in mammalian cells mitochondria are involved in programmed cell death (apoptosis, Petit *et al.*, 1996). Moreover, mitochondria are also involved in cell death in some invertebrates (Abdelwahid *et al.*,

2011). Upon permeabilization of the mitochondrial outer membrane, mitochondria release (1). cytochrome C that leads to the activation of the caspase-dependent apoptotic pathway (Saelens *et al.*, 2004) and (2). apoptosis-inducing factor, AIF (Daugas *et al.*, 2000) that is translocated to nuclei and activates the caspase-independent apoptotic pathway (Saelens *et al.*, 2004). To answer the question of whether mitochondria are involved in programmed cell death in *Tetrahymena* as in multicellular organisms, one has first to establish if apoptosis or any elements of this process can be detected in this ciliate. The data concerning a potential apoptotic-like process in *Tetrahymena* are equivocal. During conjugation of *Tetrahymena* cells, the old (parental, preconjugational) macronuclei undergo a degradation that results in the formation of oligonucleosome-sized DNA fragments (Davis *et al.*, 1992; Mpoke and Wolfe, 1996). Such DNA degradation may suggest that degradation of old macronuclei is an apoptotic-like process (programmed nuclear death, PND). On the other hand, the phosphorylated histone H2A.X that accumulates at the sites of the DNA double-strand breaks and thus is also present in apoptotic nuclei with double-strand breaks caused by DNA fragmentation (Lu *et al.*, 2006; Mukherjee *et al.*, 2006) was not detected in degenerating old macronuclei of *Tetrahymena* (Song *et al.*, 2007). Moreover, although cysteine protease, caspase-1, -8, -9 -like activity (Ejercito and Wolfe, 2003; Kobayashi and Endoh, 2003) and G-endonuclease-like activity (Kobayashi and Endoh, 2005) were reported in *Tetrahymena* cells, the genes encoding caspases, or G-endonuclease are not present in the *Tetrahymena* macronuclear genome database (Song *et al.*, 2007; Wloga *et al.*, 2008b, discussion). It was shown in plants that other proteases have activity toward caspase substrates (Chichkova *et al.*, 2010, for review Bonneau *et al.*, 2008). Thus, the existence of a caspase-like activity in *Tetrahymena* cells can be explained by the presence of yet to be identified proteases.

Interestingly, two genes encoding proteins orthologous to AIF, the protein that is sufficient to trigger caspase-independent apoptosis of isolated nuclei (Susin *et al.*, 1999), were identified in a search of the *Tetrahymena* macronuclear genome database (Akematsu and Endoh, 2010). Both AIF homologs are predicted with very high probability to be targeted to mitochondria [TTHERM_00622710 = 0.93 and TTHERM_01104910 = 0.99, MitoProt II (http://ihg.gsf.de/ihg/mitoprot.html, (Akematsu and Endoh, 2010 and Wloga, unpublished data)]. One of the GFP-tagged AIF homologs localizes to the cortical and cytoplasmic mitochondria in vegetative cells. In conjugating cells, it is detected in the proximity of preconjugational macronuclei, starting from the stage when pronuclei first appear. A knockout of a single AIF-homolog results in a moderate delay in parental macronuclear condensation and DNA degradation (Akematsu and Endoh, 2010). However, to fully evaluate the role of AIF in the programmed removal of preconjugational macronuclei, analysis of cells with a double knockout of both AIF genes would be necessary.

The mechanism of programmed removal of old macronuclei during conjugation seems to be distinct from apoptotic degradation described in metazoans, and further analysis is required. Interestingly, autophagy seems to play an important role in elimination of the old Mac (Akematsu *et al.*, 2010).

D. Mitochondrial Genome and Proteome

The mitochondrial genome of *Tetrahymena thermophila* is one of the eight fully sequenced and analyzed mitochondrial genomes of ciliates: *P. aurelia* (Pritchard *et al.*, 1990), *Tetrahymena* species: *T. pyriformis* (Burger *et al.*, 2000), *T. thermophila* (Brunk *et al.*, 2003), *T. pigmentosa, T. paravorax,* and *T. malaccensis* (Moradian *et al.*, 2007), *Euplotes minuta* and *Euplotes crassus* (de Graaf *et al.*, 2009). In contrast to the generally circular mitochondrial DNA in most eukaryotes (Bullerwell and Gray, 2004; Gray *et al.*, 1998), the ciliates analyzed, including *T. thermophila* and *T. pyriformis,* have linear mitochondrial chromosomes (Brunk and Hanawalt, 1969; Suyama and Miura, 1968). The telomere flanked (Morin and Cech, 1986) mitochondrial chromosome of *T. thermophila* consists of 47,577 bp (http://www.ncbi.nlm.nih.gov/genome?Db=genome&Cmd=ShowDetailView&TermToSearch=15738). The mitochondrial telomere repeats are different and much longer than the nuclear telomere repeats and are species-specific (Morin and Cech, 1986, 1988). With one exception (duplication of nad9) the mitochondrial genome of *T. thermophila* has a similar organization (the same genes and their arrangement) as the previously sequenced mitochondrial genome of *T. pyriformis* (Brunk *et al.*, 2003).The mitochondrial genome of *T. thermophila* has two transcriptional units. One unit with a centrally located bi-directional promoter encompasses nearly the entire chromosome. The second unit, which has a unidirectional promoter and covers three genes, is inserted in the reverse orientation into the larger transcriptional unit (Brunk *et al.*, 2003). The mitochondrial genome of *T. thermophila* has 59 putative coding regions, including regions encoding rRNA (6 such regions, because large and small subunit ribosomal RNA genes are split), tRNA (8 genes) and 45 proteins (Brunk *et al.*, 2003; Smith *et al.*, 2007 and the above web-site). Out of these 45 protein-coding genes, 19 are open reading frames that encode putative proteins of unknown function with no homologs in other types of organisms (Brunk *et al.*, 2003; Smith *et al.*, 2007). The intergenic fragments are generally very short, with only three intergenic regions longer than 63 bp, which comprise only 3.8% of the mitochondrial genome (Brunk *et al.*, 2003).

As in *T. pyriformis* (Burger *et al.*, 2000), the *T. thermophila* mitochondrial transcriptional machinery utilizes the unorthodox initiation codons ATA, ATT, TTG, and GTG (Brunk *et al.*, 2003) that are recognized as Met codons, in addition to the standard (conventional) initiation codon (ATG). Thus, mitochondria have to have a special mechanism that allows the initiator tRNA Met to recognize these non-canonical start codons. It was suggested that changes detected in the structurally unusual tRNA Met isolated from mitochondria of *T. pyriformis* enable the recognition of unorthodox initiation codons (Burger *et al.*, 2000; Edqvist *et al.*, 2000; Heinonen *et al.*, 1987; Schnare *et al.*, 1995; Suyama *et al.*, 1987). TAA is the sole termination codon in the mitochondrial genome, while TGA codes for tryptophan and the TAG codon is not utilized (Brunk *et al.*, 2003; Burger *et al.*, 2000).

Interestingly, the *T. thermophila* mitochondrial genome (like the mitochondrial genomes of other analyzed ciliates) lacks several evolutionarily well-conserved genes present in animal mitochondrial DNA (Brunk *et al.*, 2003). Only a small percentage of mitochondrial proteins are encoded by the mitochondrial genome. Mass spectrometry analysis of proteins from purified *T. thermophila* mitochondria led to the identification of 573 proteins [mitochondrial proteome, Smith *et al.*, 2007]. Out of these, 545 proteins are encoded by the nuclear genome and only 28 proteins are encoded by the mitochondrial genome. Based on a nearly 50% recovery of proteins encoded by the mitochondrial genome in the prepared mitochondrial proteome, the authors estimated that the *Tetrahymena* mitochondrial proteome may consist of as many as 900–1000 proteins. The functions of nearly 45% of the 545 known mitochondrial proteins encoded by the nuclear genome of *Tetrahymena* are unknown. Moreover, as many as 29% of the proteins identified in the proteome are ciliate specific, and 12% are *Tetrahymena* specific (Smith *et al.*, 2007). The results of this global proteomic analysis fit in well with the unexpected unique aspects of *Tetrahymena* mitochondria summarized above, such as an unusual ATP-synthase complex and the presence of mitochondrial septins. Taken together, they "reinforce an emerging view of the mitochondrion as an evolutionarily flexible organelle, with novel proteins (and presumably functions) being added in a lineage-specific fashion to an ancient, highly conserved functional core" (Smith *et al.*, 2007, p. 837).

Almost 40 years ago, Orias and Roberts isolated *T.thermophila* mutant cells that were resistant to chloramphenicol, a drug that inhibits mitochondrial but not cytoplasmic protein synthesis (Allen and Suyama, 1972; Mager, 1960). It was suggested that the resistance to chloramphenicol was encoded and transferred via mitochondria (Roberts and Orias, 1973a). Despite the analysis of the mitochondrial genome and recently also of the proteome, the gene whose mutation results in this drug-resistance has not yet been identified.

XI. Nuclear Structure and Nuclear-Cortical Interaction

In addition to striking cortical complexity, nuclear dualism is the most characteristic feature of ciliates. The latter is the topic of Chapter 3 of this volume by Karrer, and here we restrict ourselves only to the structure and ultrastructure of nuclear divisions, and to nuclear-cortical interactions in vegetative cells. The spectacular nuclear maneuvers encountered during conjugal development are covered in Chapter 7 of this volume, by Cole and Sugai.

The *T. thermophila* micronucleus (MIC) is known by genetic mapping to possess a haploid set of five chromosomes (Bruns *et al.*, 1983), which can be cytologically distinguished in meiotic prophase (Ray, 1956; Sugai and Hiwatashi, 1974). The non-dividing MIC is situated in a pocket in the surface of the macronucleus (MAC), and is mostly occupied by a clump of highly condensed chromatin (Gorovsky, 1970; Jaeckel-Williams, 1978). Micronuclear centromeres have recently been identified by the presence of centromeric histone H3 encoded by the *CNA1* gene, which is essential for normal micronuclear division and for clonal viability (Cervantes *et al.*, 2006b; Cui and Gorovsky, 2006). During interphase, ten of these Cna1p-GFP labeled dots are attached to the periphery of the MIC, and they return to that location at telophase (Cervantes *et al.*, 2006b; Cui and Gorovsky, 2006).

During mitosis, the chromatin forms separate masses that probably represent the micronuclear chromosomes (Jaeckel-Williams, 1978). The ten duplicated centromeres line up on a somewhat irregular metaphase plate and then separate at anaphase into two groups (of 10 each or less) that migrate to the poles, while the MIC simultaneously elongates (Cui and Gorovsky, 2006).

The limiting membrane of the MIC remains structurally intact at all stages of mitosis (Jaeckel-Williams, 1978). MIC-specific nucleoporins (MicNup98A and MicNup98B) remain associated with the micronuclear membrane through mitosis (Fig. S4 in Iwamoto *et al.*, 2009) supporting the conclusion that "[micro]nuclear division may indeed be fully closed" (Orias *et al.*, 2011, p. 580). This "closed" mitotic configuration is very widespread among Fungi and Protista, and includes all ciliates in which micronuclear mitosis has been examined (Heath, 1980).

A detailed analysis of micronuclear mitosis, involving the tracing of spindle MTs in serial cross sections, was carried out by LaFountain and Davidson (1979, 1980). They found that about one-half of the spindle MTs are arranged as a peripheral sheath just underneath the limiting membrane of the MIC. Some of the remainder are attached to electron-dense masses, which they interpreted as kinetochores. All of the spindle MTs extend longitudinally along the axis of the mitotic spindle, but none extend from one end of the spindle to the other, nor even from a kinetochore to either pole, a situation that apparently is unusual (Heath, 1980). This leaves the identity of the Microtubule Organizing Centers (MTOCs) and the mechanism(s) of chromosome movement unclear. LaFountain and Davidson (1979) suggest a possible nucleation of peripheral sheath MTs at the micronuclear membrane, which is consistent with the later observation that "small particles [of epitope-tagged γ-tubulin] were observed associated with interphase MIC envelopes, the likely MTOCs for MIC spindle MTs" (Shang *et al.*, 2002, p. 1198). Furthermore, an initial attachment of kinetochores to the periphery of the MIC first reported by LaFountain and Davidson (1979) and later confirmed for centromeric histone (Cervantes *et al.*, 2006b) is consistent with a possible derivation of kinetochore MTs from peripheral sheath MTs. The extraordinarily painstaking ultrastructural analysis by LaFountain and Davidson deserves more extensive experimental and molecular follow-up.

As the MIC divides, it departs from its interphase position in a pocket of the MAC and migrates in the direction of the cortex (Gavin, 1965; Jaeckel-Williams, 1978). During late stages of its division, including the separation of the daughter nuclei, the MIC is situated immediately underneath the cell cortex and is aligned along ciliary rows. These events have been intensively investigated by Dr. T. Sugai and his students, and are summarized in Chapter 7 of this volume.

The structure of the macronucleus (MAC) is entirely different from that of the MIC, with approximately 180 different chromosomes bounded by telomeres but lacking centromeres, represented at a multiplicity of about 9000 copies for the rDNA chromosome and on average 45 copies for each of the others (reviewed in Chapter 3 of this volume). The ultrastructure of the interphase MAC, both in the amicronucleate *T. pyriformis*, (Engberg *et al.*, 1974; Nilsson and Leick, 1970) and in *T. thermophila* (Gorovsky, 1973), is characterized by numerous nucleoli situated directly underneath the nuclear envelope and dispersed regions of condensed chromatin, probably heterochromatin (Huang *et al.*, 1998), in the interior. The distribution of macronuclear chromosomes (co-assortment groups) during division of the MAC appears to be random (Wong *et al.*, 2000), allowing allelic assortment (Nanney, 1964) to take place (see Chapter 3), but with the addition of an unknown mechanism that maintains a minimum copy number for each chromosome (Preer and Preer, 1979).

Some recent cytological observations appear to challenge the consensus view of random segregation of the non-centromeric macronuclear chromosomes at macronuclear divisions. In *T. pyriformis,* the macronuclear chromatin granules appear to coalesce into larger "chromatin aggregates" during synchronous division (Nilsson, 1970). This observation was recently repeated at the light-microscopical level in *T. thermophila* by Endo and Sugai (2011), who reported that under certain conditions, including the presence of microtubule inhibitors benomyl and oryzalin, they could detect the presence of about 45 "globular chromatin" aggregates in MACs of dividing cells, which were approximately equally distributed to daughter macronuclei. Furthermore, Endo and Sugai induced unequal division of MACs by centrifuging cells in the presence of benomyl and thereby generated cells with very small MACs, which "subsequently multiplied" upon isolation, a result that they argued is unexpected on the basis of a fully random distribution of macronuclear chromosomes to daughter cells (Endo and Sugai, 2011). Reconciling these observations with the well-established independent assortment of different macronuclear chromosomes creates some severe conceptual difficulties: such a reconciliation would seem to entail the breakup of haploid subnuclei early in the cell cycle for independent assortment of the MAC chromosomes, followed by the rejoining prior to the next division of complete chromosome sets into new haploid subnuclei. Endo and Sugai's provocative observations and analyses therefore should provoke further analytical investigation.

Intramacronuclear MTs have been repeatedly detected by transmission electron microscopy of synchronously dividing *T. thermophila* (see Fujiu and Numata, 2000; Williams and Williams, 1976); however, no comprehensive view of MT distribution in the MAC analogous to that achieved by LaFountain and Davidson (1979, 1980) for MICs has emerged from these studies. This was accomplished to some degree by immunofluorescence at the light microscopical level for *T. pyriformis* (Fujiu and Numata, 2000) and for *T. thermophila* (Kushida *et al.*, 2011). As the MAC begins to elongate, MTs appear in the center of the MAC forming an "aster-like structure". The "astral MTs" resolve themselves into parallel bundles of MTs extending the length of the MAC, followed by a complex rearrangement into two sets of MTs on either side of the prospective fission line, together with prominent MTs along the macronuclear periphery (Kushida *et al.*, 2011). Epitope-tagged γ-tubulin showed a punctate distribution, at first in a central cluster and later spread throughout the

MAC, offering no hint of any systematic orientation of plus and minus ends of intramacronuclear MTs (Kushida *et al.*, 2011).

Exposure of dividing *T. thermophila* cells to the MT inhibitor nocodazole resulted in disappearance of intra-MAC MTs and reduced MAC elongation. It also brought about unequal MAC division, generating some tiny MACs and other unusually large ones (Kushida *et al.*, 2011). Similar results had been obtained earlier in *T. pyriformis* with colchicine (Tamura *et al.*, 1969; Williams and Williams, 1976). Depletion of the *T. thermophila* homolog of "structural maintenance of chromosomes" protein Smc4p, a core protein of the condensin complex that is "necessary for proper chromosome segregation in meiosis and mitosis" of numerous eukaryotes, resulted in extremely unequal macronuclear division, with the smaller of the two daughter macronuclei containing mainly nucleoli and very little chromatin (Cervantes *et al.*, 2006a). Intra-macronuclear microtubules fail to form in cells severely depleted of Smc4p, leading the authors to make the interesting suggestion that "segregation of nucleoli of *Tetrahymena* may occur primarily through their association with the nuclear envelope" (Cervantes *et al*, 2006a, p. 4697).

Depletion of γ-tubulin also affected macronuclear elongation and brought about passive subdivision of the centrally-situated MACs by the cell cleavage furrow (Kushida *et al.*, 2011), a phenomenon that had also been observed earlier in division-synchronized *T. pyriformis* exposed to colchicine (Williams and Williams, 1976).

While the classical MT antagonists, colchicine and nocodazole, bring about depolymerization of MTs, another drug, paclitaxel (taxol), stabilizes MTs. Gaertig and colleagues (Gaertig *et al.*, 1994, 1999) reported a dominant base-substitution mutation [*btu1-1* (K350 M)] in one of the two conventional β-tubulin genes of *T. thermophila* (which encode identical β-tubulin proteins) that confers increased resistance to several microtubule-depolymerizing drugs (including colchicine) while simultaneously bringing about increased sensitivity to the MT-stabilizing drug, paclitaxel, a phenotype consistent with stabilization of MTs (Smith *et al.*, 2004). This mutation had strikingly different effects on the cytoplasm and the nuclei: cortical development and cytokinesis were entirely unaffected, while macronuclei divided unequally, sometimes resulting in amacronucleate cells, an effect similar to that brought about by the microtubule-depolymerizing drugs nocodazole and colchicine (Smith *et al.*, 2004).

A similar macronuclear phenotype, involving a failure of macronuclear elongation and subsequent unequal macronuclear division, was observed following a somatic disruption of the *MYO1* gene, which encodes a class XIV myosin in *T. thermophila* (S.A. Williams *et al.*, 2000). Subsequently, Hosein *et al.* (2003) reported this same abnormal macronuclear phenotype as well as arrest of cytokinesis following transformation of *T. thermophila* with GFP-actin, a phenotype that they interpreted as indicating "a requirement for actin in nuclear elongation and cytokinesis". As pointed out above (Section VI) with regard to cytokinesis, this interpretation requires confirmation with knockouts of actin genes other than, or in addition to, the already accomplished *ACT1* knockout, which does not generate either of the phenotypic abnormalities that were reported for GFP-actin (N.E. Williams *et al.*, 2006).

It is well known that the *"Tetrahymena pyriformis"* sibling-species swarm includes both micronucleate and amicronucleate species (Nanney and Simon, 2000). Amicronucleate species, such as the eponymous *T. pyriformis sensu stricto,* grow perfectly well and are clonally immortal, but also are irretrievably asexual. We can assume that mechanisms exist for occasional successful conversion of micronucleate to amicronucleate species, involving a loss of micronuclei.

Such loss, however, is not well tolerated by *Tetrahymena thermophila*. Haremaki and co-workers (Haremaki *et al.*, 1995) observed that loss of the MIC, resulting from micronuclear misdivision caused by the MT inhibitor nocodazole, brought about regression of oral structures and a failure to maintain the integrity of the cortical cytoskeleton, with eventual death. However, cells that lost their MAC or lost both MIC and MAC ("empty cells") retained their OA and their structural integrity (Haremaki *et al.*, 1996). It was further demonstrated that loss of the OA in amicronucleate cells is an active process dependent upon both transcription and translation (Haremaki *et al.*, 1996). Thus, it appears that the presence of the MIC restrains the MAC from transcribing genes that bring about a disruption of the cell's cortical organization; when the MIC is lost, this restraint is lifted.

How is this restraint exerted? It seems unlikely that it involves any genic action on the part of the MIC, for two reasons. First, while the MIC undergoes a burst of bidirectional transcription during meiotic prophase (Chalker and Yao, 2001; Sugai and Hiwatashi, 1974), micronuclear transcription at other phases of the life cycle has not been detected (Mayo and Orias, 1981). Second, the survival of cell lines containing a great variety of modified micronuclei with nullisomic chromosome deficiencies (Ward *et al.*, 1995) as well as "star" lines containing variously defective micronuclei (Allen and Weremiuk, 1971; Pitts and Doerder, 1988) implies that there is no unique gene locus in the MIC whose activity is essential for maintenance of cellular integrity.

This conclusion was confirmed in a novel manner by Kaczanowski and Kiersnowska (2011). They induced "pulverization" of micronuclear chromosomes by a specific drug regimen (aphidicolin plus caffeine) that is believed to act by overriding an intra-S-phase checkpoint (see Chapter 7). These fragmented chromosomes segregated in an irregular manner, such that some daughter cells retained many tiny chromosome fragments while other daughter cells received none. The cells that possessed multiple micronuclear fragments survived and generally behaved as "star" cells in matings (incapable of normal meiosis or transfer of genetic material – see Chapter 3), whereas the cells that lost all of their MICs consequently resorbed their OAs and suffered a drastic disruption of their cortical organization as described earlier (Haremaki *et al.*, 1995). This disrupted phenotype could be recognized in living cells by a "crinkled" appearance, which had already been recognized much earlier as a hallmark of the amicronucleate condition (Allen and Weremiuk, 1971; Nanney, 1957). The fact that the "amicronucleate" phenotype can be detected in living cells allows the trajectory of micronuclear retention or loss to be followed in clones (Kaczanowski and Kiersnowska, 2011).

The oft-repeated observation that cell integrity depends upon the presence of micronuclear DNA but not on any specific micronuclear gene locus led to a bold suggestion: that the cause of the lethality of cells in which the *CNA1* centromeric histone gene is knocked out is not the loss of specific micronuclear chromosome arms whose segregation depends upon the centromeres, but rather the loss of the centromeres themselves, "…possibly because complete loss of centromeres triggers a checkpoint mechanism in growing cells, which can be bypassed as long as centromeres are present and can segregate" (Cui and Gorovsky, 2006, p. 4508). Another possible structural signal that might prevent a checkpoint-type response could be an interaction of the micronuclear envelope with either the macronuclear envelope or the plasma membrane. These unconventional interpretations, however, need to be confronted by "the exception that proves (i.e. tests) the rule", namely the one strain in which the MIC of *T. thermophila* was lost (so that cells presumably lack micronuclear centromeres) but cells survived. Strain BI3840 was originally isolated by Kaney and Knox (1980) as a melanin-excreting mutant, which was later discovered to be amicronucleate (Kaney and Speare, 1983). It was unaffected by the aphidicolin plus caffeine treatment (Kaczanowski and Kiersnowska, 2011). On the "centromere-mediated checkpoint" hypothesis, one could presume that the BI3840 strain had undergone one or more mutations that severed the regulatory link between the loss of centromeres by the MIC and transcription of the "demolition genes" by the MAC, thereby enabling the MIC to be lost without any negative consequences for the cell. However, it then is surprising that some DNA sequences that normally are restricted to MICs are present in the MAC of BI3840 cells (Karrer *et al.*, 1984) as if they were still needed to protect the cell. Kaczanowski and Kiersnowska therefore suggest that "the maintenance of normal cell morphology depends upon certain, so far unidentified micronuclear transcript(s)" (Kaczanowski and Kiersnowska, 2011, p. 632). A deeper investigation into the function of the micronuclear sequences that reside in the BI3840 MAC, including a possible search for centromeric sequences, might begin to resolve this difficult and fascinating question.

XII. Identification of Genes Involved in Cortical Organization

Forward mutagenesis has been a starting point for studying many different processes in *T. thermophila*. These include the identification of numerous mutant genes that generate drug resistance (Ares and Bruns, 1978; Bleyman and Bruns, 1977; Byrne *et al.*, 1978; Roberts and Morse, 1980; Roberts and Orias, 1973a, 1973b; Roberts *et al.*, 1982). In addition, genes have been found that affect a great variety of cellular processes, including secretory granule formation and exocytosis (reviewed by Turkewitz, 2004), lysosomal secretion (Hunseler *et al.*, 1987; Hunseler and Tiedtke, 1992), pigment secretion (Kaney and Knox, 1980), food vacuole formation (Suhr-Jessen and Orias, 1979a,1979b; Tiedtke *et al.*, 1988), ciliary motility and regeneration (reviewed by Pennock, 2000), membrane excitability (Takahashi, 1992), cytokinesis (Frankel *et al.*, 1980, 1977; Tamura *et al.*, 1984; Yasuda *et al.*,

1984), cell size and shape (Cleffmann, 1989; Doerder *et al.*, 1975; Orias, 1960; Palissa *et al.*, 1985; Schafer and Cleffmann, 1982), cortical development (reviewed by Frankel, 2008), and conjugal development (Cole *et al.*, 1997; Cole and Soelter, 1997).

The vast majority of these mutations were induced by a highly efficient mutagen, *N*-methyl-*N*′-*N*-nitrosoguanidine (MNNG). This mutagen is an alkylating agent that has been extensively studied in *Escherichia coli* and in mammalian cells, where it induces conversion of guanine to O^6-methylguanine (Schendel and Michaeli, 1984), which brings about G-C → A-T transitions (Coulondre and Miller, 1977; Kat *et al.*, 1993) at replication forks (Cerda-Olmedo *et al.*, 1968; Hince and Neale, 1977). A few mutations were induced by another well-known mutagen, ethyl-methane-sulfonate (EMS), which has similar properties (Coulondre and Miller, 1977; Hince and Neale, 1977).

In what follows, we will focus on the mutations affecting cell division and cortical patterning, several of which were already referred to in the preceding sections of this chapter. The great majority of these mutations are MNNG- or EMS-induced recessive micronuclear mutations that were brought to expression either by allelic assortment (Carlson, 1971) in mutagenesis runs up to 1979, or by induced self-fertilization (cytogamy) (Orias *et al.*, 1979) in 1979 and thereafter. Subsequent outcrosses and inbreeding generated stable homozygous lines, which have been maintained in liquid nitrogen from 1978 onward.

Most of these mutations, plus a few genetic variants that do not precisely fit the above description, are described in a recent comprehensive review (Frankel, 2008), and are now available in stock centers, primarily the *Tetrahymena* Stock Center at Cornell University (http://tetrahymena.vet.cornell.edu/); some are also in the American Type Culture Collection (http://www.atcc.org/). The mutant stocks available from the *Tetrahymena* Stock Center include representatives of:

(1) homozygotes of virtually all of the mutated and variant cortical-patterning genes listed in Table 1[1] of Frankel (2008) (exceptions are *disB, disC, disD,* and *doa1*), some available as two or more alleles;
(2) homozygotes of all of the mutated cell division genes, except for *cdaB* (which is lethal as a homozygote), see Frankel *et al.* (1976b, 1977) and Frankel (2008);
(3) selected double homozygotes of mutated genes in the above two categories, including most of those described in Frankel *et al.* (1977), as well as double homozygotes of *janus*, *hypoangular* (*hpo1*), and *broadened cortical domains (bcd1)* mutations;
(4) the amicronucleate BI3840 stock;
(5) the wild-type stocks that were used as parent strains in the mutagenesis runs that generated the series of mutations that affected the cell cortex.

Some of these genes have been mapped to chromosomes or chromosome arms by crossing mutant cells to a battery of nullisomic lines (Bruns *et al.*, 1983). The *PSMA*

[1] The "twisted" mutations *twi1* and *twi2* listed in Table 1 of Frankel (2008) have been renamed "screwy", *scr1* and *scr2*.

(pseudomacrostome A) gene maps on the left arm of micronuclear chromosome 5 while the *CDAA* (cell division arrest A) gene is more finely mapped to a specific macronuclear chromosome from the right arm of micronuclear chromosome 4 (Hamilton and Orias, personal communication). However, none of these genes has as yet been cloned and sequenced.

The identification of the molecular nature of these genes and their products has turned out to be difficult. The strategy of cloning by complementation has thus far not succeeded in *Tetrahymena* for reasons related to the AT-richness of the *Tetrahymena* genome (see Chapter 10, by Orias). With the dramatic and continuous reduction in the cost of sequencing of whole genomes, current interest has shifted toward the use of second (or third)-generation genome sequencing to identify mutated genes. This strategy would be uncomplicated if base-pairing errors introduced by the mutagen were infrequent enough that any difference in DNA base sequence between the mutant and the wild-type stock from which it was derived were located only within the mutated gene of interest and were solely responsible for the altered phenotype of the mutant. However, the very efficiency of this mutagenic procedure with both MNNG and EMS renders this ideal outcome unlikely; these mutagens probably have peppered the genome with G-C → A-T base-pair changes. This difficulty can be mitigated by isolating the mutant to a particular region of the genome via deletion-mapping and/or by using techniques based on out-crossing of the mutant to the wild type followed by inbreeding (see Chapter 10, and Chapter 4 by Coyne, Stover, and Miao).

"Purification" of a mutant allele by the traditional mode of successive outcrossing and inbreeding has already been done for a few of the morphogenetic mutants, notably *cdaA-1* (Jenkins and Frankel, unpublished data) and *janA-1* (Frankel and Jenkins, 1979); however, for most of these genes, the available mutant clones were generated by first outcrossing the original mutated clones to generate F_1 progeny of different mating types, and then crossing these with each other or to a "star" line to generate homozygous progeny (checked by further progeny of testcrosses and F_3s, which, however, were generally not retained). These preserved "F_2" stocks, therefore, are not far from their mutagenic origins, and need to be genetically "cleaned up" before useful base-sequence comparisons can be made.

Utilization of a DNA-sequence-based approach for gene characterization focuses attention on the genealogy of these mutations. For each of the mutations that have been generated at The University of Iowa, this information is provided in individual strain descriptions within the searchable database of the *Tetrahymena* Stock Center at Cornell University, (http://tetrahymena.vet.cornell.edu/strain_search.php). It should, however, briefly be noted that the stock that was mutagenized from 1979 onward was IA264, a *gal1 1/gal1 1* (gal s, II) functional heterokaryon similar to SB210 [the stock that was used for sequencing the macronuclear genome of *T. thermophila* (Eisen *et al.*, 2006)], but with an admixture of genes from an inbred B-strain stock that is homozygous for an *enhancer of janA (eja1-1/eja1-1)*. Crosses subsequent to the initial cytogamy crosses of mutagenized IA264 × non-mutagenized IA267 [a *chx1-1/chx1-1* (cy-s, III) functional heterokaryon of unknown provenance] were carried out either to IA264 itself (to obtain progeny homozygous for *eja1-1*) or, more frequently, to B-2079 or B-2086, the grandchildren and great-grandchildren, respectively, of B-1868, the original foundation stock obtained in 1972 from David Nanney (Jenkins and Frankel, unpublished) (see the *Tetrahymena* Stock Center website for their detailed derivation).

More recently, two genes affecting cell division have been identified by an entirely different method, the "antisense ribosomes" strategy of Sweeney *et al.* (1996), as subsequently implemented by Chilcoat *et al.* (2001) for the creation of an antisense-ribosome cDNA library. This method was utilized by (Zweifel *et al.*, 2009) to discover mutant phenotypes involved in cell division, by screening transformants from the antisense cDNA library of Chilcoat *et al.* (2001) for morphological phenotypes associated with failure of cell division. Two such phenotypic mutants were found, and the responsible silenced genes were designated *CDA12*[2] and *CDA13*. Remarkably, the CDA12 ORF sequence is completely nested within the complementary CDA13 ORF sequence in an antisense orientation (Zweifel *et al.*, 2009). A detailed analysis suggested that both genes "encode proteins that are involved in membrane trafficking events required for cytokinesis and karyokinesis" (Zweifel *et al.*, 2009).

Despite the stunning progress over the past decade in analyzing the *Tetrahymena* genome and proteome, we can be confident that many of the crucial cellular actors have yet to be identified and connected to their functional or developmental roles. For this reason, forward mutagenesis, both conventional and unconventional, still has an important place in the continuing analysis of cellular development in *Tetrahymena*.

Acknowledgments

The authors wish to thank Drs. Jacek Gaertig, Jerry Honts, Maria Jerka-Dziadosz, Eduardo Orias, and Mark Winey for access to unpublished data from their laboratories, Dr. Jerka-Dziadosz for sharing TEM images and Alex Stemm-Wolf, and Mark Winey for providing unpublished light micrographs. We are also grateful to Drs. Gaertig, Jerka-Dziadosz, and Anne W. Frankel for critical reading of the entire manuscript and suggesting numerous improvements. In addition, Drs. Clifford Brunk, Eric Cole, Robert Coyne, Jerry Honts, Kathy Karrer, Eduardo Orias, Chad Pearson, Aaron Turkewitz, and Mark Winey have read sections of the MS and provided helpful comments and suggestions. Any remaining errors are our own responsibility, and we apologize to investigators whose results we may have overlooked. This work was supported by grant N N301 706640 to D.W. from the Ministry of Science and Higher Education and by a Marie Curie International Reintegration Grant within the 7th European Community Framework Programme, also to D.W.

[2] This designation is in accord with the current nomenclatural system for ciliate genes, established by Allen *et al.* (1998). *CDA12* and *CDA13* are, respectively, the 12th and 13th "cell division arrest" (CDA) genes to be discovered in *T. thermophila*. The first eleven such genes, however, were named according to the older system, in which the locus is designated by a letter rather than a number. Thus, *CDA12* follows immediately after *CDAK* (Frankel, 2008) in the previously used nomenclature.

References

Abdelwahid, E., Rolland, S., Teng, X., Conradt, B., Hardwick, J. M., and White, K. (2011). Mitochondrial involvement in cell death of non-mammalian eukaryotes. *Biochim Biophys Acta.* **1813**, 597–607.

Adoutte, A., Knowles, J. K., and Sainsard-Chanet, A. (1979). Absence of detectable mitochondrial recombination in *Paramecium. Genetics* **93**, 797–831.

Akella, J. S., Wloga, D., Kim, J., Starostina, N. G., Lyons-Abbott, S., Morrissette, N. S., Dougan, S. T., Kipreos, E. T., and Gaertig, J. (2010). MEC-17 is an alpha-tubulin acetyltransferase. *Nature* **467**, 218–222.

Akematsu, T., and Endoh, H. (2010). Role of apoptosis-inducing factor (AIF) in programmed nuclear death during conjugation in *Tetrahymena thermophila. BMC Cell. Biol.* **11**, 13.

Akematsu, T., Pearlman, R. E., and Endoh, H. (2010). Gigantic macroautophagy in programed nuclear death of *Tetrahymena thermophila. Autophagy* **6**, 901–911.

Allen, N. E., and Suyama, Y. (1972). Protein synthesis in vitro with *Tetrahymena* mitochondrial ribosomes. *Biochim. Biophys. Acta.* **259**, 369–377.

Allen, R. D. (1967). Fine structure, reconstruction and possible functions of components of the cortex of *Tetrahymena pyriformis. J. Protozool.* **14**, 553–565.

Allen, R. D. (1968). A reinvestigation of cross-sections of cilia. *J. Cell. Biol.* **37**, 825–831.

Allen, R. D. (1969). The morphogenesis of basal bodies and accessory structures of the cortex of the ciliated protozoan *Tetrahymena pyriformis. J. Cell. Biol.* **40**, 716–733.

Allen, R. D. (2000). The contractile vacuole and its membrane dynamics. *Bioessays* **22**, 1035–1042.

Allen, R. D., and Naitoh, Y. (2002). Osmoregulation and contractile vacuoles of protozoa. *Int. Rev. Cytol.* **215**, 351–394.

Allen, R. D., and Wolf, R. W. (1979). Membrane recycling at the cytoproct of *Tetrahymena. J Cell. Sci.* **35**, 217–227.

Allen, S. L., and Weremiuk, S. L. (1971). Defective micronuclei and genomic exclusion in selected C subclones of *Tetrahymena. J. Protozool.* **18**, 509–515.

Allen, S. L., Altschuler, M. I., Bruns, P. J., Cohen, J., Doerder, F. P., Gaertig, J., Gorovsky, M., Orias, E., and Turkewitz, A. (1998). Proposed genetic nomenclature rules for *Tetrahymena thermophila, Paramecium primaurelia* and *Paramecium tetraurelia. Genetics* **149**, 459–462.

Andersen, J. S., Wilkinson, C. J., Mayor, T., Mortensen, P., Nigg, E. A., and Mann, M. (2003). Proteomic characterization of the human centrosome by protein correlation profiling. *Nature* **426**, 570–574.

Angus, S. P., Edelmann, R. E., and Pennock, D. G. (2001). Targeted gene knockout of inner arm 1 in *Tetrahymena thermophila. Eur. J. Cell. Biol.* **80**, 486–497.

Ares Jr, M., and Bruns, P. J. (1978). Isolation and genetic characterization of a mutation affecting ribosomal resistance to cycloheximide in *Tetrahymena. Genetics* **90**, 463–474.

Arnaiz, O., Malinowska, A., Klotz, C., Sperling, L., Dadlez, M., Koll, F., and Cohen, J. (2009). *Cildb: a knowledgebase for centrosomes and cilia.* Database, Oxford bap022.

Asai, D. J., Rajagopalan, V., and Wilkes, D. E. (2009). Dynein-2 and ciliogenesis in *Tetrahymena. Cell. Motil. Cytoskeleton.* **66**, 673–677.

Asai, D. J., and Wilkes, D. E. (2004). The dynein heavy chain family. *J. Eukaryot. Microbiol.* **51**, 23–29.

Aufderheide, K. (1979). Mitochondrial associations with specific microtubular components of the cortex of *Tetrahymena thermophila*: I. Cortical patterning of mitochondria. *J. Cell. Sci.* **39**, 299–312.

Aufderheide, K. J. (1980). Mitochondrial associations with specific microtubular components of the cortex of *Tetrahymena thermophila*: II. Response of the mitochondrial pattern to changes in the microtubule pattern. *J. Cell. Sci.* **42**, 247–260.

Awan, A., Bell, A. J., and Satir, P. (2009). Kin5 knockdown in *Tetrahymena thermophila* using RNAi blocks cargo transport of Gef1. *PLoS One* **4**, e4873.

Awan, A., Bernstein, M., Hamasaki, T., and Satir, P. (2004). Cloning and characterization of Kin5, a novel *Tetrahymena* ciliary kinesin II. *Cell. Motil. Cytoskel.* **58**, 1–9.

Bakowska, J., Frankel, J., and Nelsen, E. M. (1982a). Regulation of the pattern of basal bodies within the oral apparatus of *Tetrahymena thermophila. J. Embryol. Exp. Morphol.* **69**, 83–105.

Bakowska, J., Nelsen, E. M., and Frankel, J. (1982b). Development of the ciliary pattern of the oral apparatus of *Tetrahymena thermophila*. *J. Protozool.* **29**, 366–382.

Balabaskaran Nina, P., Dudkina, N. V., Kane, L. A., van Eyk, J. E., Boekema, E. J., Mather, M. W., and Vaidya, A. B. (2010). Highly divergent mitochondrial ATP synthase complexes in *Tetrahymena thermophila*. *PLoS Biol.* **8**, e1000418.

Beales, P. L., Bland, E., Tobin, J. L., Bacchelli, C., Tuysuz, B., Hill, J., Rix, S., Pearson, C. G., Kai, M., Hartley, J., Johnson, C., Irving, M., Elcioglu, N., Winey, M., Tada, M., and Scambler, P. J. (2007). IFT80, which encodes a conserved intraflagellar transport protein, is mutated in Jeune asphyxiating thoracic dystrophy. *Nat. Genet.* **39**, 727–729.

Beisson, J., Clerot, J. C., Fleury-Aubusson, A., Garreau de Loubresse, N., Ruiz, F., and Klotz, C. (2001). Basal body-associated nucleation center for the centrin-based cortical cytoskeletal network in *Paramecium*. *Protist* **152**, 339–354.

Beisson, J., and Jerka-Dziadosz, M. (1999). Polarities of the centriolar structure: morphogenetic consequences. *Biol. Cell.* **91**, 367–378.

Bleyman, L. K., and Bruns, P. J. (1977). Genetics of cycloheximide resistance in *Tetrahymena*. *Genetics* **87**, 275–284.

Blum, J. J., and Greenside, H. (1976). Particle ejection from the cytoproct of *Tetrahymena*. *J Protozool.* **23**, 500–502.

Bonneau, L., Ge, Y., Drury, G. E., and Gallois, P. (2008). What happened to plant caspases? *J. Exp. Bot.* **59**, 491–499.

Bradley, B. A., and Quarmby, L. M. (2005). A NIMA-related kinase, Cnk2p, regulates both flagellar length and cell size in *Chlamydomonas*. *J. Cell. Sci.* **118**, 3317–3326.

Bright, L. J., Kambesis, N., Nelson, S. B., Jeong, B., and Turkewitz, A. P. (2010). Comprehensive analysis reveals dynamic and evolutionary plasticity of Rab GTPases and membrane traffic in *Tetrahymena thermophila*. *PLoS Genet.* **6**, e1001155.

Brimmer, A., and Weber, K. (2000). The cDNA sequences of three tetrins, the structural proteins of the *Tetrahymena* oral filaments, show that they are novel cytoskeletal proteins. *Protist* **151**, 171–180.

Brown, J. M., Fine, N. A., Pandiyan, G., Thazhath, R., and Gaertig, J. (2003). Hypoxia regulates assembly of cilia in suppressors of *Tetrahymena* lacking an intraflagellar transport subunit gene. *Mol. Biol. Cell.* **14**, 3192–3207.

Brown, J. M., Hardin, C., and Gaertig, J. (1999a). Rotokinesis, a novel phenomenon of cell locomotion-assisted cytokinesis in the ciliate *Tetrahymena thermophila*. *Cell. Biol. Int.* **23**, 841–848.

Brown, J. M., Marsala, C., Kosoy, R., and Gaertig, J. (1999b). Kinesin-II is preferentially targeted to assembling cilia and is required for ciliogenesis and normal cytokinesis in *Tetrahymena*. *Mol. Biol. Cell.* **10**, 3081–3096.

Brunk, C. F., and Hanawalt, P. C. (1969). Mitochondrial DNA in *Tetrahymena pyriformis*. *Exp. Cell Res.* **554**, 143–149.

Brunk, C. F., Lee, L. C., Tran, A. B., and Li, J. (2003). Complete sequence of the mitochondrial genome of *Tetrahymena thermophila* and comparative methods for identifying highly divergent genes. *Nucleic Acids Res.* **31**, 1673–1682.

Bruns, P. J., Brussard, T. B., and Merriam, E. V. (1983). Nullisomic *Tetrahymena*: II. A set of nullisomics define the germinal chromosomes. *Genetics* **104**, 257–270.

Bullerwell, C. E., and Gray, M. W. (2004). Evolution of the mitochondrial genome: protist connections to animals, fungi and plants. *Curr. Opin. Microbiol.* **7**, 528–534.

Burger, G., Zhu, Y., Littlejohn, T. G., Greenwood, S. J., Schnare, M. N., Lang, B. F., and Gray, M. W. (2000). Complete sequence of the mitochondrial genome of *Tetrahymena pyriformis* and comparison with *Paramecium aurelia* mitochondrial DNA. *J. Mol. Biol.* **297**, 365–380.

Byrne, B. C., Brussard, T. B., and Bruns, P. J. (1978). Induced resistance to 6-methylpurine and cycloheximide in tetrahymena: I. Germ line mutants of *T. thermophila*. *Genetics* **89**, 695–702.

Cao, L., Yu, W., Wu, Y., and Yu, L. (2009a). The evolution, complex structures and function of septin proteins. *Cell Mol. Life Sci.* **66**, 3309–3323.

Cao, M., Li, G., and Pan, J. (2009b). Regulation of cilia assembly, disassembly, and length by protein phosphorylation. *Methods Cell. Biol.* **94**, 333–346.

Carlson, P. S. (1971). Mutant selection in *Tetrahymena pyriformis*. *Genetics* **69**, 261–265.

Cerda-Olmedo, E., Hanawalt, P. C., and Guerola, N. (1968). Mutagenesis of the replication point by nitrosoguanidine: map and pattern of replication of the *Escherichia coli* chromosome. *J. Mol. Biol.* **33**, 705–719.

Cervantes, M. D., Coyne, R. S., Xi, X., and Yao, M. C. (2006a). The condensin complex is essential for amitotic segregation of bulk chromosomes but not nucleoli in the ciliate *Tetrahymena thermophila*. *Mol. Cell Biol.* **26**, 4690–4700.

Cervantes, M. D., Xi, X., Vermaak, D., Yao, M. C., and Malik, H. S. (2006b). The CNA1 histone of the ciliate *Tetrahymena thermophila* is essential for chromosome segregation in the germline micronucleus. *Mol. Biol. Cell.* **17**, 485–497.

Chalker, D. L., and Yao, M. C. (2001). Nongenic, bidirectional transcription precedes and may promote developmental DNA deletion in *Tetrahymena thermophila*. *Genes Dev.* **15**, 1287–1298.

Chichkova, N. V., Shaw, J., Galiullina, R. A., Drury, G. E., Tuzhikov, A. I., Kim, S. H., Kalkum, M., Hong, T. B., Gorshkova, E. N., Torrance, L., Vartapetian, A. B., and Taliansky, M. (2010). Phytaspase, a relocalisable cell death promoting plant protease with caspase specificity. *EMBO J.* **29**, 1149–1161.

Chilcoat, N. D., Elde, N. C., and Turkewitz, A. P. (2001). An antisense approach to phenotype-based gene cloning in *Tetrahymena*. *Proc. Natl Acad. Sci. U.S.A.* **98**, 8709–8713.

Cleffmann, G. (1989). The transient division block in the D9ts mutant of *Tetrahymena* is inducible at every cell cycle stage. *J. Cell. Sci.* **92**, 349–352.

Cole, E. S., Anderson, P. C., Fulton, R. B., Majerus, M. E., Rooney, M. G., Savage, J. M., Chalker, D., Honts, J., Welch, M. E., Wentland, A. L., Zweifel, E., and Beussman, D. J. (2008). A proteomics approach to cloning fenestrin from the nuclear exchange junction of *Tetrahymena*. *J. Eukaryot. Microbiol.* **55**, 245–256.

Cole, E. S., Cassidy-Hanley, D., Hemish, J., Tuan, J., and Bruns, P. J. (1997). A mutational analysis of conjugation in *Tetrahymena thermophila*. 1. Phenotypes affecting early development: meiosis to nuclear selection. *Dev. Biol.* **189**, 215–232.

Cole, E. S., and Frankel, J. (1991). Conjugal blocks in *Tetrahymena* pattern mutants and their cytoplasmic rescue: II. janus A. *Dev. Biol.* **148**, 420–428.

Cole, E. S., and Soelter, T. A. (1997). A mutational analysis of conjugation in *Tetrahymena thermophila*. 2. Phenotypes affecting middle and late development: third prezygotic nuclear division, pronuclear exchange, pronuclear fusion, and postzygotic development. *Dev. Biol.* **189**, 233–245.

Corliss, J. O. (1973). History, taxonomy, ecology and evolution of species of Tetrahymena. *In* "Biology of Tetrahymena," (A. M. Elliott, ed.), pp. 1–55. Dowden, Hutchinson & Ross Inc, Stroudsburg, PA.

Coulondre, C., and Miller, J. H. (1977). Genetic studies of the lac repressor. IV. Mutagenic specificity in the lacI gene of *Escherichia coli*. *J. Mol. Biol.* **117**, 577–606.

Cui, B., and Gorovsky, M. A. (2006). Centromeric histone H3 is essential for vegetative cell division and for DNA elimination during conjugation in *Tetrahymena thermophila*. *Mol. Cell. Biol.* **26**, 4499–4510.

Culver, B. P., Meehl, J. B., Giddings Jr., T. H., and Winey, M. (2009). The two SAS-6 homologs in *Tetrahymena thermophila* have distinct functions in basal body assembly. *Mol. Biol. Cell.* **20**, 1865–1877.

Damaj, R., Pomel, S., Bricheux, G., Coffe, G., Vigues, B., Ravet, V., and Bouchard, P. (2009). Cross-study analysis of genomic data defines the ciliate multigenic epiplasmin family: strategies for functional analysis in *Paramecium tetraurelia*. *BMC Evol. Biol.* **9**, 125.

Daugas, E., Nochy, D., Ravagnan, L., Loeffler, M., Susin, S. A., Zamzami, N., and Kroemer, G. (2000). Apoptosis-inducing factor (AIF): a ubiquitous mitochondrial oxidoreductase involved in apoptosis. *FEBS Lett.* **476**, 118–123.

Dave, D., Wloga, D., Sharma, N., and Gaertig, J. (2009). DYF-1 is required for assembly of the axoneme in *Tetrahymena thermophila*. *Eukaryot. Cell* **8**, 1397–1406.

Davis, M. C., Ward, J. G., Herrick, G., and Allis, C. D. (1992). Programmed nuclear death: apoptotic-like degradation of specific nuclei in conjugating *Tetrahymena*. *Dev. Biol.* **154**, 419–432.

de Graaf, R. M., van Alen, T. A., Dutilh, B. E., Kuiper, J. W., van Zoggel, H. J., Huynh, M. B., Gortz, H. D., Huynen, M. A., and Hackstein, J. H. (2009). The mitochondrial genomes of the ciliates *Euplotes minuta* and *Euplotes crassus*. *BMC Genomics*. **10**, 514.

Dentler, W. L. (1984). Attachment of the cap to the central microtubules of *Tetrahymena* cilia. *J. Cell Sci.* **66**, 167–173.

Doerder, F. P., Frankel, J., Jenkins, L. M., and DeBault, L. E. (1975). Form and pattern in ciliated protozoa: analysis of a genic mutant with altered cell shape in *Tetrahymena pyriformis*, Syngen 1. *J. Exp. Zool.* **192**, 237–258.

Dress, V. M., Yi, H., Musal, M. R., and Williams, N. E. (1992). Tetrin polypeptides are colocalized in the cortex of *Tetrahymena*. *J. Struct. Biol.* **108**, 187–194.

Dymek, E. E., Lefebvre, P. A., and Smith, E. F. (2004). PF15p is the *Chlamydomonas* homologue of the Katanin p80 subunit and is required for assembly of flagellar central microtubules. *Eukaryot. Cell* **3**, 870–879.

Edamatsu, M., Hirono, M., and Watanabe, Y. (1992). *Tetrahymena* profilin is localized in the division furrow. *J. Biochem.* **112**, 637–642.

Edqvist, J., Burger, G., and Gray, M. W. (2000). Expression of mitochondrial protein-coding genes in *Tetrahymena pyriformis*. *J. Mol. Biol.* **297**, 381–393.

Eisen, J. A., Coyne, R. S., Wu, M., Wu, D., Thiagarajan, M., Wortman, J. R., Badger, J. H., Ren, Q., Amedeo, P., Jones, K. M., Tallon, L. J., Delcher, A. L., Salzberg, S. L., Silva, J. C., Haas, B. J., Majoros, W. H., Farzad, M., Carlton, J. M., Smith Jr, R. K., Garg, J., Pearlman, R. E., Karrer, K. M., Sun, L., Manning, G., Elde, N. C., Turkewitz, A. P., Asai, D. J., Wilkes, D. E., Wang, Y., Cai, H., Collins, K., Stewart, B. A., Lee, S. R., Wilamowska, K., Weinberg, Z., Ruzzo, W. L., Wloga, D., Gaertig, J., Frankel, J., Tsao, C. C., Gorovsky, M. A., Keeling, P. J., Waller, R. F., Patron, N. J., Cherry, J. M., Stover, N. A., Krieger, C. J., del Toro, C., Ryder, H. F., Williamson, S. C., Barbeau, R. A., Hamilton, E. P., and Orias, E. (2006). Macronuclear genome sequence of the ciliate *Tetrahymena* thermophila, a model eukaryote. *PLoS Biol.* **4**, e286.

Ejercito, M., and Wolfe, J. (2003). Caspase-like activity is required for programmed nuclear elimination during conjugation in *Tetrahymena*. *J. Eukaryot. Microbiol.* **50**, 427–429.

Elde, N. C., Morgan, G., Winey, M., Sperling, L., and Turkewitz, A. P. (2005). Elucidation of clathrin-mediated endocytosis in *Tetrahymena* reveals an evolutionarily convergent recruitment of dynamin. *PLoS Genet.* **1**, e52.

Elliott, A. M., and Bak, I. J. (1964a). The fate of mitochondria during aging in *Tetrahymena pyriformis*. *J. Cell Biol.* **20**, 113–129.

Elliott, A. M., and Bak, I. J. (1964b). The contractile vacuole and related structures in *Tetrahymena pyriformis*. *J. Protozool.* **11**, 250–261.

Elliott, A. M., and Kennedy, J. R. (1973). Morphology of Tetrahymena. *In* "The Biology of Tetrahymena," (A. M. Elliott, ed.), pp. 57–87. Dowden, Hutchinson & Ross Inc, Stroudsburg, PA.

Endo, M., and Sugai, T. (2011). Amitotic division of the macronucleus in *Tetrahymena thermophila*: DNA distribution by genomic unit. *Zool. Sci. (Tokyo)* **28**, 482–490.

Engberg, J., Nilsson, J. R., Pearlman, R. E., and Leick, V. (1974). Induction of nucleolar and mitochondrial DNA replication in *Tetrahymena pyriformis*. *Proc. Natl Acad. Sci. U.S.A.* **71**, 894–898.

Errabolu, R., Sanders, M. A., and Salisbury, J. L. (1994). Cloning of a cDNA encoding human centrin, an EF-hand protein of centrosomes and mitotic spindle poles. *J. Cell. Sci.* **107**, 9–16.

Forer, A., Nilsson, J. R., and Zeuthen, E. (1970). Studies on the oral apparatus of *Tetrahymena pyriformis* Gl. *C. R. Trav. Lab. Carlsberg* **38**, 67–86.

Foth, B. J., Goedecke, M. C., and Soldati, D. (2006). New insights into myosin evolution and classification. *Proc. Natl Acad. Sci. U.S.A.* **103**, 3681–3686.

Frankel, J. (1962). The effects of heat, cold, and p-fluorophenylalanine on the morphogenesis in synchronized *Tetrahymena pyriformis* GL. *C. R. Trav. Lab. Carlsberg*. **33**, 1–52.

Frankel, J. (1969). Participation of the undulating membrane in the formation of oral replacement primordia in *Tetrahymena pyriformis*. *J. Protozool.* **16**, 26–35.
Frankel, J. (1970). The synchronization of oral development without cell division in *Tetrahymena pyriformis* GL-C. *J. Exp. Zool.* **173**, 79–99.
Frankel, J. (1980). Propagation of cortical differences in *Tetrahymena*. *Genetics* **94**, 607–623.
Frankel, J. (2000a). Cell polarity in ciliates. *In* "Cell Polarity," (D. Drubin, ed.), pp. 78–105. Oxford University Press, Oxford.
Frankel, J. (2000b). Cell biology of *Tetrahymena thermophila*. *In* "Methods in Cell Biology," (D. J. Asai, and J. D. Forney, eds.), Vol. 62, pp. 27–125. Academic Press, San Diego.
Frankel, J. (2008). What do genic mutations tell us about the structural patterning of a complex single-celled organism? *Eukaryot. Cell.* **7**, 1617–1639.
Frankel, J., and Jenkins, L. M. (1979). A mutant of *Tetrahymena thermophila* with a partial mirror-image duplication of cell surface pattern: II. Nature of genic control. *J. Embryol. Exp. Morphol.* **49**, 203–227.
Frankel, J., Jenkins, L. M., Bakowska, J., and Nelsen, E. M. (1984a). Mutational analysis of patterning of oral structures in *Tetrahymena*: I. Effects of increased size on organization. *J. Embryol. Exp. Morphol.* **82**, 41–66.
Frankel, J., Jenkins, L. M., and DeBault, L. E. (1976a). Causal relations among cell cycle processes in *Tetrahymena pyriformis*. An analysis employing temperature-sensitive mutants. *J. Cell Biol.* **71**, 242–260.
Frankel, J., Jenkins, L. M., Doerder, F. P., and Nelsen, E. M. (1976b). Mutations affecting cell division in *Tetrahymena pyriformis*: I. Selection and genetic analysis. *Genetics* **83**, 489–506.
Frankel, J., Mohler, J., and Frankel, A. K. (1980). Temperature-sensitive periods of mutations affecting cell division in *Tetrahymena thermophila*. *J. Cell Sci.* **43**, 59–74.
Frankel, J., Nelsen, E. M., Bakowska, J., and Jenkins, L. M. (1984b). Mutational analysis of patterning of oral structures in *Tetrahymena:* II. A graded basis for the individuality of intracellular structural arrays. *J. Embryol. Exp. Morphol.* **82**, 67–95.
Frankel, J., Nelsen, E. M., and Jenkins, L. M. (1977). Mutations affecting cell division in *Tetrahymena pyriformis*, syngen 1: II. Phenotypes of single and double homozygotes. *Dev. Biol.* **58**, 255–275.
Frankel, J., Nelsen, E. M., and Martel, E. (1981). Development of the ciliature of *Tetrahymena thermophila*: II. Spatial subdivision prior to cytokinesis. *Dev. Biol.* **88**, 39–54.
Friedman, J. R., Webster, B. M., Mastronarde, D. N., Verhey, K. J., and Voeltz, G. K. (2010). ER sliding dynamics and ER-mitochondrial contacts occur on acetylated microtubules. *J. Cell Biol.* **190**, 363–375.
Fujiu, K., and Numata, O. (2000). Reorganization of microtubules in the amitotically dividing macronucleus of *Tetrahymena*. *Cell Motil. Cytoskeleton* **46**, 17–27.
Gaertig, J., Cruz, M. A., Bowen, J., Gu, L., Pennock, D. G., and Gorovsky, M. A. (1995). Acetylation of lysine 40 in alpha-tubulin is not essential in *Tetrahymena thermophila*. *J. Cell Biol.* **129**, 1301–1310.
Gaertig, J., Gao, Y., Tishgarten, T., Clark, T. G., and Dickerson, H. W. (1999). Surface display of a parasite antigen in the ciliate *Tetrahymena thermophila*. *Nat. Biotechnol.* **17**, 462–465.
Gaertig, J., Thatcher, T. H., Gu, L., and Gorovsky, M. A. (1994). Electroporation-mediated replacement of a positively and negatively selectable β-tubulin gene in *Tetrahymena thermophila*. *Proc. Natl.Acad. Sci. USA* **91**, 4549–4553.
Garces, J. A., Hoey, J. G., and Gavin, R. H. (1995). Putative myosin heavy and light chains in *Tetrahymena*: co-localization to the basal body-cage complex and association of the heavy chain with skeletal muscle actin filaments *in vitro*. *J. Cell Sci.* **108**, 869–881.
Garreau de Loubresse, N., Keryer, G., Viguès, B., and Beisson, J. (1988). A contractile cytoskeletal network in *Paramecium*: the infraciliary lattice. *J. Cell. Sci.* **90**, 351–364.
Garreau de Loubresse, N., Klotz, C., Viguès, B., Rutin, J., and Beisson, J. (1991). Ca^{2+}-binding proteins and contractility of the infraciliary lattice in *Paramecium*. *Biol. Cell.* **71**, 217–225.
Gavin, R. H. (1965). The effects of heat and cold on cellular development in synchronized *Tetrahymena pyriformis* WH-6. *J. Protozool.* **12**, 307–318.
Giddings Jr., T. H., Meehl, J. B., Pearson, C. G., and Winey, M. (2010). Electron tomography and immunolabeling of *Tetrahymena thermophila* basal bodies. *Methods Cell. Biol.* **96**, 117–141.

Gogendeau, D., Klotz, C., Arnaiz, O., Malinowska, A., Dadlez, M., de Loubresse, N. G., Ruiz, F., Koll, F., and Beisson, J. (2008). Functional diversification of centrins and cell morphological complexity. *J. Cell. Sci.* **121**, 65–74.

Gonda, K., Katoh, M., Hanyu, K., Watanabe, Y., and Numata, O. (1999a). $Ca(^{2+})$/calmodulin and p85 cooperatively regulate an initiation of cytokinesis in *Tetrahymena*. *J. Cell. Sci.* **112**, 3619–3626.

Gonda, K., Komatsu, M., and Numata, O. (2000). Calmodulin and Ca^{2+}/calmodulin-binding proteins are involved in *Tetrahymena thermophila* phagocytosis. *Cell. Struct. Funct.* **25**, 243–251.

Gonda, K., Nishibori, K., Ohba, H., Watanabe, A., and Numata, O. (1999b). Molecular cloning of the gene for p85 that regulates the initiation of cytokinesis in *Tetrahymena*. *Biochem. Biophys. Res. Commun.* **264**, 112–118.

Gonda, K., and Numata, O. (2002). p85 binds to G-actin in a $Ca(^{2+})$/calmodulin-dependent manner, thus regulating the initiation of cytokinesis in *Tetrahymena*. *Biochem. Biophys. Res. Commun.* **292**, 1098–1103.

Gorovsky, M. A. (1970). Studies on nuclear structure and function in *Tetrahymena pyriformis:* II. Isolation of macro- and micronuclei. *J. Cell. Biol.* **47**, 619–630.

Gorovsky, M. A. (1973). Macro- and micronuclei of *Tetrahymena pyriformis*: a model system for studying the structure and function of eukaryotic nuclei. *J. Protozool.* **20**, 19–25.

Gould, S. B., Kraft, L. G., van Dooren, G. G., Goodman, C. D., Ford, K. L., Cassin, A. M., Bacic, A., McFadden, G. I., and Waller, R. F. (2010). Ciliate pellicular proteome identifies novel protein families with characteristic repeat motifs that are common to alveolates. *Mol. Biol. Evol.* **28**, 1319–1331.

Gould, S. B., Tham, W. H., Cowman, A. F., McFadden, G. I., and Waller, R. F. (2008). Alveolins, a new family of cortical proteins that define the protist infrakingdom Alveolata. *Mol. Biol. Evol.* **25**, 1219–1230.

Gray, M. W., Lang, B. F., Cedergren, R., Golding, G. B., Lemieux, C., Sankoff, D., Turmel, M., Brossard, N., Delage, E., Littlejohn, T. G., Plante, I., Rioux, P., Saint-Louis, D., Zhu, Y., and Burger, G. (1998). Genome structure and gene content in protist mitochondrial DNAs. *Nucleic Acids Res.* **26**, 865–878.

Grønlien, H. K., Stock, C., Aihara, M. S., Allen, R. D., and Naitoh, Y. (2002). Relationship between the membrane potential of the contractile vacuole complex and its osmoregulatory activity in *Paramecium multimicronucleatum*. *J. Exp. Biol.* **20**, 3261–3270.

Guerra, C., Wada, Y., Leick, V., Bell, A., and Satir, P. (2003). Cloning, localization, and axonemal function of *Tetrahymena* centrin. *Mol. Biol. Cell.* **14**, 251–261.

Hanyu, K., Numata, O., Takahashi, M., and Watanabe, Y. (1996). Immunofluorescence localization of a 23-kDa *Tetrahymena* calcium-binding protein, TCBP-23, in the cell cortex. *J. Biochem.* **119**, 914–919.

Hanyu, K., Takemasa, T., Numata, O., Takahashi, M., and Watanabe, Y. (1995). Immunofluorescence localization of a 25-kDa *Tetrahymena* EF-hand $Ca(^{2+})$-binding protein, TCBP-25, in the cell cortex and possible involvement in conjugation. *Exp. Cell. Res.* **219**, 487–493.

Hao, L., and Scholey, J. M. (2009). Intraflagellar transport at a glance. *J. Cell. Sci.* **122**, 889–892.

Haremaki, T., Sugai, T., and Takahashi, M. (1995). The vegetative micronucleus has a critical role in maintenance of cortical structure in *Tetrahymena thermophila*. *Cell. Struct. Funct.* **20**, 239–244.

Haremaki, T., Sugai, T., and Takahashi, M. (1996). Involvement of active cellular mechanisms on the disorganization of oral apparatus in amicronucleate cells in *Tetrahymena thermophila*. *Cell. Struct. Funct.* **21**, 73–80.

Harper, J. D. I., Thuet, J., Lechtreck, K., and Hardham, A. R. (2009). Proteins related to green algal striated fiber assemblin are present in stramenopiles and alveolates. *Protoplasma* **236**, 97–101.

Heath, I. B. (1980). Variant mitoses in lower eukaryotes: indicators of the evolution of mitosis. *Int. Rev. Cytol.* **64**, 1–80.

Heinonen, T. Y., Schnare, M. N., Young, P. G., and Gray, M. W. (1987). Rearranged coding segments, separated by a transfer RNA gene, specify the two parts of a discontinuous large subunit ribosomal RNA in *Tetrahymena pyriformis* mitochondria. *J. Biol. Chem.* **262**, 2879–2887.

Hennessey, T. M., Kim, D. Y., Oberski, D. J., Hard, R., Rankin, S. A., and Pennock, D. G. (2002). Inner arm dynein 1 is essential for Ca^{++}-dependent ciliary reversals in *Tetrahymena* thermophila. *Cell. Motil. Cytoskeleton.* **53**, 281–288.

Hince, T. A., and Neale, S. (1977). Physiological modification of alkylating agent induced mutagenesis: II. Influence of the numbers of chromosome replicating forks and gene copies on the frequency of mutations induced in *Escherichia coli*. *Mutat. Res.* **43**, 11–24.
Hirono, M., Nakamura, M., Tsunemoto, M., Yasuda, T., Ohba, H., Numata, O., and Watanabe, Y. (1987). *Tetrahymena* actin: localization and possible biological roles of actin in *Tetrahymena* cells. *J. Biochem.* **102**, 537–545.
Hoey, J. G., and Gavin, R. H. (1992). Localization of actin in the *Tetrahymena* basal body-cage complex. *J. Cell Sci.* **103**, 629–641.
Honts, J. E., and Williams, N. E. (1990). Tetrins: polypeptides that form bundled filaments in *Tetrahymena*. *J. Cell Sci.* **96**, 293–302.
Hosein, R. E., Williams, S. A., Haye, K., and Gavin, R. H. (2003). Expression of GFP-actin leads to failure of nuclear elongation and cytokinesis in *Tetrahymena thermophila*. *J. Eukaryot. Microbiol.* **50**, 403–408.
Hou, Y., Pazour, G. J., and Witman, G. B. (2004). A dynein light intermediate chain, D1bLIC, is required for retrograde intraflagellar transport. *Mol. Biol. Cell* **15**, 4382–4394.
Huang, H., Wiley, E. A., Lending, C. R., and Allis, C. D. (1998). An HP-1-like protein is missing from transcriptionally silent micronuclei of *Tetrahymena*. *Proc. Natl. Acad. Sci. U.S.A* **95**, 13624–13629.
Hunseler, P., Scheidgen-Kleyboldt, G., and Tiedtke, A. (1987). Isolation and characterization of a mutant of *Tetrahymena thermophila* blocked in secretion of lysosomal enzymes. *J. Cell Sci.* **88**, 47–55.
Hunseler, P., and Tiedtke, A. (1992). Genetic characterization of the secretory mutant MS-1 of *Tetrahymena thermophila*: vacuolarization and block in secretion of lysosomal hydrolases are caused by a single gene mutation. *Dev. Genet.* **13**, 167–173.
Iftode, F., Cohen, J., Ruiz, F., Torres-Rueda, A., Chen-Shan, L., Adoutte, A., and Beisson, J. (1989). Development of surface pattern during division in *Paramecium*: I. mapping of duplication and reorganization of cortical cytoskeletal structures in the wild-type. *Development* **105**, 191–211.
Iwamoto, M., Allen, R. D., and Naitoh, Y. (2003). Hypo-osmotic or Ca^{2+}-rich external conditions trigger extra contractile vacuole complex generation in *Paramecium multimicronucleatum*. *J. Exp. Biol.* **206**, 4467–4473.
Iwamoto, M., Mori, C., Kojidani, T., Bunai, F., Hori, T., Fukagawa, T., Hiraoka, Y., and Haraguchi, T. (2009). Two distinct repeat sequences of Nup98 nucleoporins characterize dual nuclei in the binucleated ciliate *Tetrahymena*. *Curr. Biol.* **19**, 843–847.
Iwamoto, M., Sugino, K., Allen, R. D., and Naitoh, Y. (2005). Cell volume control in *Paramecium*: factors that activate the control mechanisms. *J. Exp. Biol.* **208**, 523–537.
Jaeckel-Williams, R. (1978). Nuclear divisions with reduced numbers of microtubules in *Tetrahymena*. *J. Cell Sci.* **34**, 303–319.
Janke, C., Rogowski, K., Wloga, D., Regnard, C., Kajava, A. V., Strub, J. M., Temurak, N., van Dijk, J., Boucher, D., van Dorsselaer, A., Suryavanshi, S., Gaertig, J., and Edde, B. (2005). Tubulin polyglutamylase enzymes are members of the TTL domain protein family. *Science* **308**, 1758–1762.
Jerka-Dziadosz, M. (1981a). Patterning of ciliary structures in janus mutant of *Tetrahymena* with mirror-image cortical duplications. An ultrastructural study. *Acta Protozool.* **20**, 337–356.
Jerka-Dziadosz, M. (1981b). Cytoskeleton-related structures in *Tetrahymena thermophila*: microfilaments at the apical and division-furrow rings. *J. Cell Sci.* **51**, 241–253.
Jerka-Dziadosz, M., Jenkins, L. M., Nelsen, E. M., Williams, N. E., Jaeckel-Williams, R., and Frankel, J. (1995). Cellular polarity in ciliates: persistence of global polarity in a disorganized mutant of *Tetrahymena thermophila* that disrupts cytoskeletal organization. *Dev. Biol.* **169**, 644–661.
Jerka-Dziadosz, M., Muszynska, K., and Krawczynska, W. (1999). A novel healing filament in ciliate regeneration. *J. Eukaryot. Microbiol.* **46**, 507–515.
Jerka-Dziadosz, M., Strzyzewska-Jowko, I., Wojsa-Lugowska, U., Krawczynska, W., and Krzywicka, A. (2001). The dynamics of filamentous structures in the apical band, oral crescent, fission line and the postoral meridional filament in *Tetrahymena thermophila* revealed by monoclonal antibody 12G9. *Protist* **152**, 53–67.
Kaczanowski, A. (1975). A single-gene-dependent abnormality of adoral membranelles in *Tetrahymena pyriformis*, species 1. *Genetics* **81**, 631–639.

Kaczanowski, A. (1976). An analysis of MP gene affected morphogenesis in *Tetrahymena pyriformis*, syngen 1 (Species 1) ciliates. *J. Exp. Zool.* **196**, 215–230.

Kaczanowski, A. (1978). Gradients of proliferation of ciliary basal bodies and the determination of the position of the oral primordium in *Tetrahymena*. *J. Exp. Zool.* **204**, 417–430.

Kaczanowski, A., and Kiersnowska, M. (2011). Inactivation of a macronuclear intra-S-phase checkpoint in *Tetrahymena thermophila* with caffeine affects the integrity of the micronuclear genome. *Protist* **162**, 616–636.

Kaczanowska, J., Buzanska, L., and Ostrowski, M. (1993). Relationship between spatial pattern of basal bodies and membrane skeleton (epiplasm) during the cell cycle of *Tetrahymena*: cdaA mutant and anti-membrane skeleton immunostaining. *J. Eukaryot Microbiol.* **40**, 747–754.

Kaczanowska, J., Joachimiak, E., Kiersnowska, M., Krzywicka, A., Golinska, K., and Kaczanowski, A. (2003). The fenestrin antigen in submembrane skeleton of the ciliate *Tetrahymena thermophila* is proposed as a marker of cell polarity during cell division and in oral replacement. *Protist* **154**, 251–264.

Kaney, A. R., and Knox, G. W. (1980). Production of melanin precursors by a mutant of *Tetrahymena thermophila*. *J. Protozool* **27**, 339–341.

Kaney, A. R., and Speare, V. J. (1983). An amicronucleate mutant of *Tetrahymena thermophila*. *Exp. Cell Res.* **143**, 461–467.

Karrer, K., Stein-Gavens, S., and Allitto, B. A. (1984). Micronucleus-specific DNA sequences in an amicronucleate mutant of *Tetrahymena*. *Dev. Biol.* **105**, 121–129.

Kat, A., Thilly, W. G., Fang, W. H., Longley, M. J., Li, G. M., and Modrich, P. (1993). An alkylation-tolerant, mutator human cell line is deficient in strand-specific mismatch repair. *Proc. Natl. Acad. Sci. U.S.A.* **90**, 6424–6428.

Kay, E., Rohatgi, K., and Krawiec, S. (1974). Morphometric studies of mitochondria in *Tetrahymena pyriformis* exposed to chloramphenicol or ethidium bromide. *J. Protozool.* **21**, 608–612.

Keller, L. C., Geimer, S., Romijn, E., Yates 3rd, J., Zamora, I., and Marshall, W. F. (2009). Molecular architecture of the centriole proteome: the conserved WD40 domain protein POC1 is required for centriole duplication and length control. *Mol. Biol. Cell.* **20**, 1150–1166.

Keller, L. C., Romijn, E. P., Zamora, I., Yates 3rd, J. R., and Marshall, W. F. (2005). Proteomic analysis of isolated *Chlamydomonas* centrioles reveals orthologs of ciliary-disease genes. *Curr. Biol.* **15**, 1090–1098.

Kiersnowska, M., and Golinska, K. (1996). Pattern of phosphorylated structures in the morphostatic ciliate *Tetrahymena thermophila*: MPM-2 immunogold labeling. *Acta Protozool.* **35**, 297–308.

Kiersnowska, M., Kaczanowski, A., and deHaller, G. (1993). Inhibition of oral morphogenesis during conjugation of *Tetrahymena thermophila* and its resumption after cell separation. *Eur. J. Protistol.* **29**, 359–369.

Kilburn, C. L., Pearson, C. G., Romijn, E. P., Meehl, J. B., Giddings Jr, T. H., Culver, B. P., Yates 3rd, J. R., and Winey, M. (2007). New *Tetrahymena* basal body protein components identify basal body domain structure. *J. Cell Biol.* **178**, 905–912.

Kirchhausen, T. (1999). Adaptors for clathrin-mediated traffic. *Annu. Rev. Cell Dev. Biol.* **15**, 705–732.

Kirk, K. E., Christ, C., McGuire, J. M., Paul, A. G., Vahedi, M., Stuart, K. R., and Cole, E. S. (2008). Abnormal micronuclear telomeres lead to an unusual cell cycle checkpoint and defects in *Tetrahymena* oral morphogenesis. *Eukaryot. Cell.* **7**, 1712–1723.

Kobayashi, T., and Endoh, H. (2003). Caspase-like activity in programmed nuclear death during conjugation of *Tetrahymena thermophila*. *Cell Death Differ.* **10**, 634–640.

Kobayashi, T., and Endoh, H. (2005). A possible role of mitochondria in the apoptotic-like programmed nuclear death of *Tetrahymena thermophila*. *FEBS J.* **272**, 5378–5387.

Kojima, H., and Numata, O. (2002). Enzymatic form and cytoskeletal form of bifunctional *Tetrahymena* 49 kDa protein is regulated by phosphorylation. *Zoolog. Sci.* **19**, 37–42.

Kolb-Bachofen, V., and Vogell, W. (1975). Mitochondrial proliferation in synchronized cells of *Tetrahymena pyriformis*. A morphometric study by electron microscopy on the biogenesis of mitochondria during the cell cycle. *Exp. Cell Res.* **94**, 95–105.

Kubo, T., Yanagisawa, H. A., Yagi, T., Hirono, M., and Kamiya, R. (2010). Tubulin polyglutamylation regulates axonemal motility by modulating activities of inner-arm dyneins. *Curr. Biol.* **20**, 441–445.

Kuribara, S., Kato, M., Kato-Minoura, T., and Numata, O. (2006). Identification of a novel actin-related protein in *Tetrahymena* cilia. *Cell Motil. Cytoskeleton.* **63**, 437–446.

Kurz, S., and Tiedtke, A. (1993). The Golgi apparatus of *Tetrahymena thermophila. J. Eukaryot. Microbiol.* **40**, 10–13.

Kushida, Y., Nakano, K., and Numata, O. (2011). Amitosis requires gamma-tubulin-mediated microtubule assembly in *Tetrahymena thermophila. Cytoskeleton (Hoboken)* **68**, 89–96.

Lacroix, B., van Dijk, J., Gold, N. D., Guizetti, J., Aldrian-Herrada, G., Rogowski, K., Gerlich, D. W., and Janke, C. (2010). Tubulin polyglutamylation stimulates spastin-mediated microtubule severing. *J. Cell Biol.* **189**, 945–954.

Ladenburger, E. M., Korn, I., Kasielke, N., Wassmer, T., and Plattner, H. (2006). An Ins(1,4,5)P3 receptor in *Paramecium* is associated with the osmoregulatory system. *J. Cell Sci.* **119**, 3705–3717.

Ladenburger, E. M., Sehring, I. M., Korn, I., and Plattner, H. (2009). Novel types of Ca^{2+} release channels participate in the secretory cycle of *Paramecium* cells. *Mol. Cell Biol.* **29**, 3605–3622.

LaFountain Jr., J. R., and Davidson, L. A. (1979). An analysis of spindle ultrastructure during prometaphase and metaphase of micronuclear division in *Tetrahymena. Chromosoma* **75**, 293–308.

LaFountain Jr., J. R., and Davidson, L. A. (1980). An analysis of spindle ultrastructure during anaphase of micronuclear division in *Tetrahymena. Cell Motil.* **1**, 41–61.

Lansing, T. J., Frankel, J., and Jenkins, L. M. (1985). Oral ultrastructure and oral development of the misaligned undulating membrane mutant of *Tetrahymena thermophila. J. Protozool.* **32**, 126–139.

Larisch, S., Yi, Y., Lotan, R., Kerner, H., Eimerl, S., Tony Parks, W., Gottfried, Y., Birkey Reffey, S., de Caestecker, M. P., Danielpour, D., Book-Melamed, N., Timberg, R., Duckett, C. S., Lechleider, R. J., Steller, H., Orly, J., Kim, S. J., and Roberts, A. B. (2000). A novel mitochondrial septin-like protein, ARTS, mediates apoptosis dependent on its P-loop motif. *Nat. Cell Biol.* **2**, 915–921.

Lechtreck, K. F., and Melkonian, M. (1991). Striated microtubule-associated fibers: identification of assemblin, a novel 34-kD protein that forms paracrystals of 2-nm filaments *in vitro. J. Cell Biol.* **115**, 705–716.

Liang, A., Ruiz, F., Heckmann, K., Klotz, C., Tollon, Y., Beisson, J., and Wright, M. (1996). Gamma-tubulin is permanently associated with basal bodies in ciliates. *Eur. J. Cell Biol.* **70**, 331–338.

Liu, S., Hard, R., Rankin, S., Hennessey, T., and Pennock, D. G. (2004). Disruption of genes encoding predicted inner arm dynein heavy chains causes motility phenotypes in *Tetrahymena. Cell Motil. Cytoskeleton.* **59**, 201–214.

Liu, Q., Tan, G., Levenkova, N., Li, T., Pugh Jr, E. N., Rux, J. J., Speicher, D. W., and Pierce, E. A. (2007). The proteome of the mouse photoreceptor sensory cilium complex. *Mol. Cell Proteomics.* **6**, 1299–1317.

Loefer, J. B., Small, E. B., and Furgason, W. H. (1966). Range of variation in the somatic infraciliature and contractile vacuole pores of *Tetrahymena pyriformis. J. Protozool.* **13**, 90–102.

Lu, C., Zhu, F., Cho, Y. Y., Tang, F., Zykova, T., Ma, W. Y., Bode, A. M., and Dong, Z. (2006). Cell apoptosis: requirement of H2AX in DNA ladder formation, but not for the activation of caspase-3. *Mol. Cell.* **23**, 121–132.

Lynn, D. H. (1981). The organization and evolution of microtubular organelles in ciliated protozoa. *Biol. Rev. Cambr.* **56**, 243–292.

Lynn, D. H., and Small, E. B. (1981). Protist kinetids: structural conservatism, kinetid structure, and ancestral states. *Biosystems* **14**, 377–385.

Mager, J. (1960). Chloramphenicol and chlortetracycline inhibition of amino acid incorporation into proteins in a cell-free system from *Tetrahymena pyriformis. Biochim. Biophys. Acta.* **38**, 150–152.

Mayo, K. A., and Orias, E. (1981). Further evidence for lack of gene expression in the *Tetrahymena* micronucleus. *Genetics* **98**, 747–762.

McCoy, J. W. (1974). New features of the Tetrahymenid cortex revealed by protargol stainning. *Acta Protozool.* **13**, 155–160.

Meehl, J. B., Giddings Jr., T. H., and Winey, M. (2009). High pressure freezing, electron microscopy, and immuno-electron microscopy of *Tetrahymena thermophila* basal bodies. *Methods Mol. Biol.* **586**, 227–241.

Moradian, M. M., Beglaryan, D., Skozylas, J. M., and Kerikorian, V. (2007). Complete mitochondrial genome sequence of three *Tetrahymena* species reveals mutation hot spots and accelerated nonsynonymous substitutions in *Ymf* genes. *PLoS One* **2**, e650.

Morin, G. B., and Cech, T. R. (1986). The telomeres of the linear mitochondrial DNA of *Tetrahymena thermophila* consist of 53 bp tandem repeats. *Cell* **46**, 873–883.

Morin, G. B., and Cech, T. R. (1988). Mitochondrial telomeres: surprising diversity of repeated telomeric DNA sequences among six species of *Tetrahymena*. *Cell* **52**, 367–374.

Mpoke, S., and Wolfe, J. (1996). DNA digestion and chromatin condensation during nuclear death in *Tetrahymena*. *Exp. Cell Res.* **225**, 357–365.

Mukherjee, B., Kessinger, C., Kobayashi, J., Chen, B. P., Chen, D. J., Chatterjee, A., and Burma, S. (2006). DNA-PK phosphorylates histone H2AX during apoptotic DNA fragmentation in mammalian cells. *DNA Repair (Amst)* **5**, 575–590.

Muto, E., Edamatsu, M., Hirono, M., and Kamiya, R. (1994). Immunological detection of actin in the 14S ciliary dynein of *Tetrahymena*. *FEBS Lett.* **343**, 173–177.

Nahon, P., Coffe, G., Le Guyader, H., Darmanaden-Delorme, J., Jeanmaire-Wolf, R., Clérot, J. -C., and Adoutte, A. (1993). Identification of the epiplasmins, a new set of cortical proteins of the membrane cytoskeleton in *Paramecium*. *J. Cell Sci.* **104**, 975–990.

Nakagawa, T., Fujiu, K., Cole, E. S., and Numata, O. (2008). Involvement of a 25 kDa *Tetrahymena* Ca^{2+}-binding protein in pronuclear exchange. *Cell Struct. Funct.* **33**, 151–162.

Nanney, D. L. (1957). Inbreeding degeneration in *Tetrahymena*. *Genetics* **42**, 137–146.

Nanney, D. L. (1964). Macronuclear differentiation and subnuclear assortment in ciliates. *In* "The Role of Chromosomes in Development," (M. M. Locke, ed.), pp. 253–273. Academic Press, New York.

Nanney, D. L. (1966a). Corticotype transmission in *Tetrahymena*. *Genetics* **54**, 955–968.

Nanney, D. L. (1966b). Cortical integration in *Tetrahymena:* an exercise in cytogeometry. *J. Exp. Zool* **161**, 307–317.

Nanney, D. L. (1971). The constancy of the cortical units in *Tetrahymena* with varying numbers of ciliary rows. *J. Exp. Zool* **178**, 177–182.

Nanney, D. L. (1972). Cytogeometric integration in the ciliate cortex. *Ann. N. Y. Acad. Sci.* **193**, 14–28.

Nanney, D. L. (1975). Patterns of basal body addition in ciliary rows in *Tetrahymena*. *J. Cell Biol.* **65**, 503–512.

Nanney, D. L., Chen, S. S., and Meyer, E. B. (1978). Scalar constraints in *Tetrahymena* evolution. Quantitative basal body variations within and between species. *J. Cell Biol.* **79**, 727–736.

Nanney, D. L., and Simon, E. M. (2000). Laboratory and evolutionary history of Tetrahymena thermophila. *In* "Methods in Cell Biology," (D. J. Asai, and J. D. Forney, eds.), Vol. 62, pp. 3–25. Academic Press, San Diego.

Nelsen, E. M. (1975). Regulation of tubulin during ciliary regeneration in non-growing *Tetrahymena*. *Exp. Cell Res.* **94**, 152–158.

Nelsen, E. M. (1978). Transformation in *Tetrahymena thermophila*. Development of an inducible phenotype. *Dev. Biol.* **66**, 17–31.

Nelsen, E. M. (1981). The undulating membrane of *Tetrahymena*: formation and reconstruction. *Trans. Am. Microsc. Soc.* **100**, 285–295.

Nelsen, E. M., and Debault, L. E. (1978). Transformation in *Tetrahymena pyriformis:* description of an inducible phenotype. *J. Protozool.* **25**, 113–119.

Nelsen, E. M., Williams, N. E., Yi, H., Knaak, J., and Frankel, J. (1994). "Fenestrin" and conjugation in *Tetrahymena thermophila*. *J. Eukaryot. Microbiol.* **41**, 483–495.

Ng, S. F. (1977). Analysis of contractile vacuole pore morphogenesis in *Tetrahymena pyriformis* by 180 degree rotation of ciliary meridians. *J. Cell Sci.* **25**, 233–246.

Ng, S. F. (1978). Directionality of microtubule assembly: an *in vivo* study with the ciliate *Tetrahymena*. *J. Cell Sci.* **33**, 227–234.

Ng, S. F. (1979). Unidirectional regeneration is an intrinsic property of longitudinal microtubules in *Tetrahymena* – an *in vivo* study. *J. Cell Sci.* **36**, 109–119.

Nilsson, J. R. (1970). Suggestive structural evidence for macronuclear "subnuclei" in *Tetrahymena pyriformis* GL. *J. Protozool.* **17**, 539–548.
Nilsson, J. R. (1979). Phagotrophy in Tetrahymena. *In* "Biochemistry & Physiology of Protozoa," (M. Levandowsky, and S. H. Hutner, eds.), Vol. 2, pp. 329–379. Academic Press, New York.
Nilsson, J. R., and Leick, V. (1970). Nucleolar organization and ribosome formation in *Tetrahymena pyriformis* GL. *Exp. Cell Res.* **60**, 361–372.
Nilsson, J. R., and van Deurs, B. (1983). Coated pits with pinocytosis in *Tetrahymena. J. Cell Sci.* **63**, 209–222.
Nilsson, J. R., and Williams, N. E. (1966). An electron microscope study of the oral apparatus of *Tetrahymena pyriformis. C. R. Trav. Lab. Carlsberg* **35**, 119–141.
Numata, O., and Gonda, K. (2001). Determination of division plane and organization of contractile ring in *Tetrahymena. Cell Struct. Funct.* **26**, 593–601.
Numata, O., Hirono, M., and Watanabe, Y. (1983). Involvement of *Tetrahymena* intermediate filament protein, a 49K protein, in the oral morphogenesis. *Exp. Cell Res.* **148**, 207–220.
Numata, O., Kurasawa, Y., Gonda, K., and Watanabe, Y. (2000). *Tetrahymena* elongation factor-1 alpha is localized with calmodulin in the division furrow. *J. Biochem.* **127**, 51–56.
Numata, O., Sugai, T., and Watanabe, Y. (1985). Control of germ cell nuclear behaviour at fertilization by *Tetrahymena* intermediate filament protein. *Nature* **314**, 192–194.
Numata, O., Suzuki, H., Ohba, H., and Watanabe, Y. (1995a). The mutant gene product of a *Tetrahymena* cell-division-arrest mutant cdaA is localized in the accessory structure of specialized basal body close to the division furrow. *Zoolog. Sci.* **12**, 133–135.
Numata, O., Takemasa, T., Takagi, I., Hirono, M., Hirano, H., Chiba, J., and Watanabe, Y. (1991). *Tetrahymena* 14-nm filament-forming protein has citrate synthase activity. *Biochem. Biophys. Res. Commun.* **174**, 1028–1034.
Numata, O., and Watanabe, Y. (1982). *in vitro* assembly and disassembly of 14-nm filament from *Tetrahymena pyriformis*. The protein component of 14-nm filament is a 49,000-Dalton protein. *J. Biochem.* **91**, 1563–1573.
Numata, O., Yasuda, T., and Watanabe, Y. (1995b). *Tetrahymena* intramitochondrial filamentous inclusions contain 14-nm filament protein/citrate synthase. *Exp. Cell Res.* **218**, 123–131.
Ohba, H., Ohmori, I., Numata, O., and Watanabe, Y. (1986). Purification and immunofluorescence localization of the mutant gene product of a *Tetrahymena* cdaA1 mutant affecting cell division. *J. Biochem.* **100**, 797–808.
Organ, A. E., Bovee, E. C., and Jahn, T. L. (1972). The mechanism of the water expulsion vesicle of the ciliate *Tetrahymena pyriformis. J. Cell Biol.* **55**, 644–652.
Orias, E. (1960). The genetic control of two lethal traits in variety 1, *Tetrahymena pyriformis. J. Protozool.* **7**, 64–68.
Orias, E., Cervantes, M. D., and Hamilton, E. P. (2011). *Tetrahymena thermophila*, a unicellular eukaryote with separate germline and somatic genomes. *Res. Microbiol.* **162**, 578–586.
Orias, E., Hamilton, E. P., and Flacks, M. (1979). Osmotic shock prevents nuclear exchange and produces whole-genome homozygotes in conjugating *Tetrahymena. Science* **203**, 660–663.
Orias, E., and Rasmussen, L. (1976). Dual capacity for nutrient uptake in *Tetrahymena*: IV. Growth without food vacuoles and its implications. *Exp. Cell Res.* **102**, 127–137.
Orias, E., and Rasmussen, L. (1979). Dual capacity for nutrient uptake in *Tetrahymena*: V. Utilization of amino acids and proteins. *J. Cell Sci.* **36**, 343–353.
Otera, H., and Mihara, K. (2011). Molecular mechanisms and physiologic functions of mitochondrial dynamics. *J. Biochem.* **149**, 241–251.
Palissa, H., Seyfert, H. M., and Cleffmann, G. (1985). Analysis of a mutant expressing temperature-sensitive changes of cell size in *Tetrahymena. J. Cell Physiol.* **123**, 361–368.
Patterson, D. J. (1980). Contractile vacuoles and associated structures: their organization and function. *Biol. Rev. Cambr.* **55**, 1–46.
Patterson, D. J., and Sleigh, M. A. (1976). Behavior of the contractile vacuole of *Tetrahymena pyriformis* W: a redescription with comments on the terminology. *J. Protozool.* **23**, 410–417.

Paumard, P., Vaillier, J., Coulary, B., Schaeffer, J., Soubannier, V., Mueller, D. M., Brethes, D., di Rago, J. P., and Velours, J. (2002). The ATP synthase is involved in generating mitochondrial cristae morphology. *EMBO J.* **21**, 221–230.

Pazour, G. J., Dickert, B. L., and Witman, G. B. (1999). The DHC1b (DHC2) isoform of cytoplasmic dynein is required for flagellar assembly. *J. Cell Biol.* **144**, 473–481.

Pearson, C. G., Giddings Jr, T. H., and Winey, M. (2009a). Basal body components exhibit differential protein dynamics during nascent basal body assembly. *Mol. Biol. Cell.* **20**, 904–914.

Pearson, C. G., Osborn, D. P., Giddings Jr, T. H., Beales, P. L., and Winey, M. (2009b). Basal body stability and ciliogenesis requires the conserved component Poc1. *J. Cell Biol.* **187**, 905–920.

Pearson, C. G., and Winey, M. (2009). Basal body assembly in ciliates: the power of numbers. *Traffic* **10**, 461–471.

Pennock, D. G. (2000). Selection of motility mutants. *In* "Methods in Cell Biology," (D. J. Asai, and J. D. Forney, eds.), Vol. 62, pp. 281–290. Academic Press, San Diego.

Petit, P. X., Susin, S. A., Zamzami, N., Mignotte, B., and Kroemer, G. (1996). Mitochondria and programmed cell death: back to the future. *FEBS Lett.* **396**, 7–13.

Pitelka, D. R. (1961). Fine structure of the silverline and fibrillar systems of three Tetrahymenid ciliates. *J Protozool.* **8**, 75–89.

Pitts, R. A., and Doerder, F. P. (1988). Genomic exclusion and other micronuclear anomalies are common in genetically defective clones of *Tetrahymena thermophila*. *Genetics* **120**, 135–143.

Pollard, T. D. (2010). Mechanics of cytokinesis in eukaryotes. *Curr. Opin. Cell Biol.* **22**, 50–56.

Pomel, S., Diogon, M., Bouchard, P., Pradel, L., Ravet, V., Coffe, G., and Vigues, B. (2006). The membrane skeleton in *Paramecium*: molecular characterization of a novel epiplasmin family and preliminary GFP expression results. *Protist* **157**, 61–75.

Poole, R. K. (1983). Mitochondria of *Tetrahymena pyriformis*: enumeration and sizing of isolated organelles using a Coulter Counter and pulse-height analyser. *J. Cell Sci.* **61**, 437–451.

Preer, J. R. J., and Preer, L. B. (1979). The size of macronuclear DNA and its relationship to models for maintaining nuclear balance. *J. Protozool.* **26**, 14–18.

Pritchard, A. E., Seilhamer, J. J., Mahalingam, R., Sable, C. L., Venuti, S. E., and Cummings, D. J. (1990). Nucleotide sequence of the mitochondrial genome of *Paramecium*. *Nucleic Acids Res.* **18**, 173–180.

Rabl, R., Soubannier, V., Scholz, R., Vogel, F., Mendl, N., Vasiljev-Neumeyer, A., Korner, C., Jagasia, R., Keil, T., Baumeister, W., Cyrklaff, M., Neupert, W., and Reichert, A. S. (2009). Formation of cristae and crista junctions in mitochondria depends on antagonism between Fcj1 and Su e/g. *J. Cell Biol.* **185**, 1047–1063.

Rajagopalan, V., Corpuz, E. O., Hubenschmidt, M. J., Townsend, C. R., Asai, D. J., and Wilkes, D. E. (2009a). Analysis of properties of cilia using *Tetrahymena thermophila*. *In* "Methods in Cell Biology," (D. J. Asai, and J. D. Forney, eds.), Vol. 62, pp. 283–299. Academic Press, San Diego.

Rajagopalan, V., Subramanian, A., Wilkes, D. E., Pennock, D. G., and Asai, D. J. (2009b). Dynein-2 affects the regulation of ciliary length but is not required for ciliogenesis in Tetrahymena thermophila. *Mol. Biol. Cell.* **20**, 708–720.

Ray Jr., C. (1956). Meiosis and nuclear behavior in. *Tetrahymena pyriformis*. *J. Protozool.* **3**, 88–96.

Richards, . A., and Cavalier-Smith, T. (2005). Myosin domain evolution and the primary divergence of eukaryotes. *Nature* **436**, 1113–1118.

Rifkin, J. L. (1973). The role of the contractile vacuole in the osmoregulation of *Tetrahymena pyriformis*. *J. Protozool.* **20**, 108–114.

Roberts Jr., C. T., and Morse, D. E. (1980). Galactokinase-deficient mutants of *Tetrahymena thermophila*: selection and characterization. *Mol. Gen. Genet.* **180**, 129–134.

Roberts Jr, C. T., and Orias, E. (1973a). Cytoplasmic inheritance of chloramphenicol resistance in *Tetrahymena*. *Genetics* **73**, 259–272.

Roberts Jr, C. T., and Orias, E. (1973b). A cycloheximide-resistant mutant of *Tetrahymena pyriformis*. *Exp. Cell Res.* **81**, 312–316.

Roberts Jr., C. T., Lavine, J. E., and Morse, D. E. (1982). Selection and characterization of a glucokinase-deficient mutant of *Tetrahymena thermophila*. *Mol. Cell Biol.* **2**, 378–385.

Rohatgi, R., and Snell, W. J. (2010). The ciliary membrane. *Curr. Opin. Cell Biol.* **22**, 541–546.
Ruiz, F., Garreau de Loubresse, N., Klotz, C., Beisson, J., and Koll, F. (2005). Centrin deficiency in Paramecium affects the geometry of basal-body duplication. *Curr. Biol.* **15**, 2097–2106.
Saelens, X., Festjens, N., Vande Walle, L., van Gurp, M., van Loo, G., and Vandenabeele, P. (2004). Toxic proteins released from mitochondria in cell death. *Oncogene* **23**, 2861–2874.
Satir, P. (1984). The generation of ciliary motion. *J. Protozool.* **31**, 8–12.
Sattler, C. A., and Staehelin, L. A. (1974). Ciliary membrane differentiations in *Tetrahymena pyriformis. Tetrahymena* has four types of cilia. *J. Cell Biol.* **62**, 473–490.
Sattler, C. A., and Staehelin, L. A. (1976). Reconstruction of oral cavity of *Tetrahymena pyriformis* utilizing high voltage electron microscopy. *Tissue Cell.* **8**, 1–18.
Sattler, C. A., and Staehelin, L. A. (1979). Oral cavity of *Tetrahymena pyriformis.* A freeze-fracture and high-voltage electron microscopy study of the oral ribs, cytostome, and forming food vacuole. *J. Ultrastruct Res.* **66**, 132–150.
Schafer, E., and Cleffmann, G. (1982). Division and growth kinetics of the division mutant conical of *Tetrahymena*. A contribution to regulation of generation time. *Exp. Cell Res.* **137**, 277–284.
Schendel, P. F., and Michaeli, I. (1984). A model for the mechanism of alkylation mutagenesis. *Mutat. Res.* **125**, 1–14.
Schilde, C., Wassmer, T., Mansfeld, J., Plattner, H., and Kissmehl, R. (2006). A multigene family encoding R-SNAREs in the ciliate *Paramecium tetraurelia. Traffic* **7**, 440–455.
Schnare, M. N., Greenwood, S. J., and Gray, M. W. (1995). Primary sequence and post-transcriptional modification pattern of an unusual mitochondrial tRNA(Met) from *Tetrahymena pyriformis. FEBS Lett.* **362**, 24–28.
Schuldt, A. (2011). Cytoskeleton: SAS-6 turns a cartwheel trick. *Nat. Rev. Mol. Cell Biol.* **12**, 137.
Sehring, I. M., Reiner, C., Mansfeld, J., Plattner, H., and Kissmehl, R. (2007). A broad spectrum of actin paralogs in *Paramecium tetraurelia* cells display differential localization and function. *J Cell Sci.* **120**, 177–190.
Sehring, I. M., Reiner, C., and Plattner, H. (2010). The actin subfamily PtAct4, out of many subfamilies, is differentially localized for specific local functions in *Paramecium tetraurelia* cells. *Eur. J. Cell Biol.* **89**, 509–524.
Shang, Y., Li, B., and Gorovsky, M. A. (2002). *Tetrahymena thermophila* contains a conventional gamma-tubulin that is differentially required for the maintenance of different microtubule-organizing centers. *J. Cell Biol.* **158**, 1195–1206.
Shang, Y., Tsao, C. C., and Gorovsky, M. A. (2005). Mutational analyses reveal a novel function of the nucleotide-binding domain of gamma-tubulin in the regulation of basal body biogenesis. *J. Cell Biol.* **171**, 1035–1044.
Sharma, N., Bryant, J., Wloga, D., Donaldson, R., Davis, R. C., Jerka-Dziadosz, M., and Gaertig, J. (2007). Katanin regulates dynamics of microtubules and biogenesis of motile cilia. *J. Cell Biol.* **178**, 1065–1079.
Shirayama, S., and Numata, O. (2003). *Tetrahymena* fimbrin localized in the division furrow bundles actin filaments in a calcium-independent manner. *J. Biochem.* **134**, 591–598.
Smith, D. G., Gawryluk, R. M., Spencer, D. F., Pearlman, R. E., Siu, K. W., and Gray, M. W. (2007). Exploring the mitochondrial proteome of the ciliate protozoon *Tetrahymena* thermophila: direct analysis by tandem mass spectrometry. *J. Mol. Biol.* **374**, 837–863.
Smith, H. E., and Buhse Jr., H. E. (1983). On the origin of the deep fiber in *Tetrahymena*. *Trans. Am. Microsc. Soc.* **102**, 264–272.
Smith, J. C., Northey, J. G., Garg, J., Pearlman, R. E., and Siu, K. W. (2005). Robust method for proteome analysis by MS/MS using an entire translated genome: demonstration on the ciliome of *Tetrahymena thermophila. J. Proteome Res.* **4**, 909–919.
Smith, J. J., Yakisich, J. S., Kapler, G. M., Cole, E. S., and Romero, D. P. (2004). A beta-tubulin mutation selectively uncouples nuclear division and cytokinesis in *Tetrahymena thermophila. Eukaryot Cell.* **3**, 1217–1226.

Song, X., Gjoneska, E., Ren, Q., Taverna, S. D., Allis, C. D., and Gorovsky, M. A. (2007). Phosphorylation of the SQ H2A.X motif is required for proper meiosis and mitosis in *Tetrahymena thermophila*. *Mol. Cell Biol.* **27**, 2648–2660.

Sperling, L., Keryer, G., Ruiz, F., and Beisson, J. (1991). Cortical morphogenesis in *Paramecium*: a transcellular wave of protein phosphorylation in ciliary rootlet disassembly. *Dev. Biol.* **148**, 205–218.

Stargell, L. A., Heruth, D. P., Gaertig, J., and Gorovsky, M. A. (1992). Drugs affecting microtubule dynamics increase alpha-tubulin mRNA accumulation via transcription in *Tetrahymena thermophila*. *Mol. Cell Biol.* **12**, 1443–1450.

Stemm-Wolf, A. J., Morgan, G., Giddings Jr, T. H., White, E. A., Marchione, R., McDonald, H. B., and Winey, M. (2005). Basal body duplication and maintenance require one member of the *Tetrahymena thermophila* centrin gene family. *Mol. Biol. Cell.* **16**, 3606–3619.

Stock, C., Allen, R. D., and Naitoh, Y. (2001). How external osmolarity affects the activity of the contractile vacuole complex, the cytosolic osmolarity and the water permeability of the plasma membrane in *Paramecium multimicronucleatum*. *J. Exp. Biol.* **204**, 291–304.

Stock, C., Gronlien, H. K., Allen, R. D., and Naitoh, Y. (2002). Osmoregulation in *Paramecium*: *in situ* ion gradients permit water to cascade through the cytosol to the contractile vacuole. *J. Cell Sci.* **115**, 2339–2348.

Strzyzewska-Jowko, I., Jerka-Dziadosz, M., and Frankel, J. (2003). Effect of alteration in the global body plan on the deployment of morphogenesis-related protein epitopes labeled by the monoclonal antibody 12G9 in *Tetrahymena thermophila*. *Protist* **154**, 71–90.

Sugai, T., and Hiwatashi, K. (1974). Cytologic and autoradiographic studies of the micronucleus at meiotic prophase in *Tetrahymena pyriformis*. *J. Protozool.* **21**, 542–548.

Sugino, K., Tominaga, T., Allen, R. D., and Naitoh, Y. (2005). Electrical properties and fusion dynamics of in vitro membrane vesicles derived from separate parts of the contractile vacuole complex of Paramecium multimicronucleatum. *J. Exp. Biol.* **208**, 3957–3969.

Sugita, M., Iwataki, Y., Nakano, K., and Numata, O. (2011). Unique sequences and predicted functions of myosins in *Tetrahymena thermophila*. *Gene* **480**, 10–20.

Sugita, M., Nakano, K., Sato, M., Toyooka, K., and Numata, O. (2009). The roles of actin cytoskeleton and microtubules for membrane recycling of a food vacuole in *Tetrahymena thermophila*. *Cell Motil. Cytoskeleton* **66**, 371–377.

Suhr-Jessen, P. B., and Orias, E. (1979a). Mutants of *Tetrahymena thermophila* with temperature-sensitive food vacuole formation: I. Isolation and genetic characterization. *Genetics* **92**, 1061–1077.

Suhr-Jessen, P. B., and Orias, E. (1979b). Mutants of *Tetrahymena thermophila* with temperature sensitive food vacuole formation: II. Physiological and morphological studies. *Exp. Cell Res.* **124**, 317–327.

Suprenant, K. A., and Dentler, W. L. (1988). Release of intact microtubule-capping structures from *Tetrahymena* cilia. *J. Cell Biol.* **107**, 2259–2269.

Suryavanshi, S., Edde, B., Fox, L. A., Guerrero, S., Hard, R., Hennessey, T., Kabi, A., Malison, D., Pennock, D., Sale, W. S., Wloga, D., and Gaertig, J. (2010). Tubulin glutamylation regulates ciliary motility by altering inner dynein arm activity. *Curr. Biol.* **20**, 435–440.

Susin, S. A., Lorenzo, H. K., Zamzami, N., Marzo, I., Snow, B. E., Brothers, G. M., Mangion, J., Jacotot, E., Costantini, P., Loeffler, M., Larochette, N., Goodlett, D. R., Aebersold, R., Siderovski, D. P., Penninger, J. M., and Kroemer, G. (1999). Molecular characterization of mitochondrial apoptosis-inducing factor. *Nature* **397**, 441–446.

Suyama, Y., Jenney, F., and Okawa, N. (1987). Two transfer RNA sequences abut the large ribosomal RNA gene in *Tetrahymena* mitochondrial DNA: tRNA(leu) (anticodon UAA) and tRNA(met) (anticodon CAU). *Curr. Genet.* **11**, 327–330.

Suyama, Y., and Miura, K. (1968). Size and structural variations of mitochondrial DNA. *Proc. Natl Acad. Sci. U.S.A.* **60**, 235–242.

Suzuki, Y., Ohnishi, K., Hirabayashi, T., and Watanabe, Y. (1982). Tetrahymena calmodulin. Characterization of an anti-*Tetrahymena* calmodulin and the immunofluorescent localization in Tetrahymena. *Exp. Cell Res.* **137**, 1–14.

Sweeney, R., Fan, Q., and Yao, M. C. (1996). Antisense ribosomes: rRNA as a vehicle for antisense RNAs. *Proc. Natl Acad. Sci. U.S.A.* **93**, 8518–8523.

Takagi, I., Numata, O., and Watanabe, Y. (1991). Involvement of 14-nm filament-forming protein and tubulin in gametic pronuclear behavior during conjugation in *Tetrahymena*. *J. Protozool.* **38**, 345–351.

Takahashi, M. (1992). Locus-dependent profiles of the rescue of nonexcitable behavioral mutants during conjugation in *Tetrahymena thermophila*. *Dev. Genet.* **13**, 174–179.

Takahashi, S., Inatome, R., Yamamura, H., and Yanagi, S. (2003). Isolation and expression of a novel mitochondrial septin that interacts with CRMP/CRAM in the developing neurones. *Genes Cells.* **8**, 81–93.

Takeda, T., Kurasawa, Y., Watanabe, Y., and Numata, O. (1995). Polymerization of highly purified *Tetrahymena* 14-nm filament protein/citrate synthase into filaments and its possible role in regulation of enzymatic activity. *J. Biochem.* **117**, 869–874.

Tamura, R., Takahashi, M., and Watanabe, Y. (1984). Molecular mechanisms of cell division in *Tetrahymena thermophila:* I. Analysis of the execution period of a division-arrest ts-mutant. *Zool. Sci. (Tokyo)* **1**, 50–61.

Tamura, S., Tsuruhara, T., and Watanabe, Y. (1969). Function of nuclear microtubules in macronuclear division of *Tetrahymena pyriformis*. *Exp. Cell Res.* **55**, 351–358.

Thazhath, R., Jerka-Dziadosz, M., Duan, J., Wloga, D., Gorovsky, M. A., Frankel, J., and Gaertig, J. (2004). Cell context-specific effects of the beta-tubulin glycylation domain on assembly and size of microtubular organelles. *Mol. Biol. Cell.* **15**, 4136–4147.

Thazhath, R., Liu, C., and Gaertig, J. (2002). Polyglycylation domain of beta-tubulin maintains axonemal architecture and affects cytokinesis in *Tetrahymena*. *Nat. Cell Biol.* **4**, 256–259.

Tiedtke, A., Hünselar, P., and Rasmussen, L. (1988). Growth requirement of a new food-vacuole-less mutant of *Tetrahymena*. *Eur. J. Protistol.* **23**, 350–353.

Tsao, C. C., and Gorovsky, M. A. (2008a). Different effects of *Tetrahymena* IFT172 domains on anterograde and retrograde intraflagellar transport. *Mol. Biol. Cell.* **19**, 1450–1461.

Tsao, C. C., and Gorovsky, M. A. (2008b). *Tetrahymena* IFT122A is not essential for cilia assembly but plays a role in returning IFT proteins from the ciliary tip to the cell body. *J. Cell Sci.* **121**, 428–436.

Turkewitz, A. P. (2004). Out with a bang! *Tetrahymena* as a model system to study secretory granule biogenesis. *Traffic* **5**, 63–68.

Turkewitz, A. P., and Kelly, R. B. (1992). Immunocytochemical analysis of secretion mutants of *Tetrahymena* using a mucocyst-specific monoclonal antibody. *Dev. Genet.* **13**, 151–159.

Ueno, H., Gonda, K., Takeda, T., and Numata, O. (2003). Identification of elongation factor-1alpha as a Ca^{2+}/calmodulin-binding protein in *Tetrahymena* cilia. *Cell Motil. Cytoskeleton.* **55**, 51–60.

Ueno, H., Iwataki, Y., and Numata, O. (2006). Homologues of radial spoke head proteins interact with Ca^{2} $^{+}$/calmodulin in *Tetrahymena* cilia. *J. Biochem.* **140**, 525–533.

Velours, J., Dautant, A., Salin, B., Sagot, I., and Brethes, D. (2009). Mitochondrial F1F0-ATP synthase and organellar internal architecture. *Int. J. Biochem. Cell Biol.* **41**, 1783–1789.

Verhey, K. J., and Gaertig, J. (2007). The tubulin code. *Cell Cycle* **6**, 2152–2160.

Vonderfecht, T., Stemm-Wolf, A. J., Hendershott, M., Giddings Jr, T. H., Meehl, J. B., and Winey, M. (2011). The two domains of centrin have distinct basal body functions in *Tetrahymena*. *Mol. Biol. Cell.* **22**, 2221–2234.

Ward, J. G., Davis, M. C., Allis, C. D., and Herrick, G. (1995). Effects of nullisomic chromosome deficiencies on conjugation events in *Tetrahymena thermophila*: insufficiency of the parental macronucleus to direct postzygotic development. *Genetics* **140**, 989–1005.

Wassmer, T., Froissard, M., Plattner, H., Kissmehl, R., and Cohen, J. (2005). The vacuolar proton-ATPase plays a major role in several membrane-bounded organelles in *Paramecium*. *J. Cell Sci.* **118**, 2813–2825.

Wassmer, T., Kissmehl, R., Cohen, J., and Plattner, H. (2006). Seventeen a-subunit isoforms of *Paramecium* V-ATPase provide high specialization in localization and function. *Mol. Biol. Cell.* **17**, 917–930.

Wickstead, B., and Gull, K. (2006). A "holistic" kinesin phylogeny reveals new kinesin families and predicts protein functions. *Mol. Biol. Cell.* **17**, 1734–1743.

Wilkes, D. E., and Otto, J. J. (2003). Profilin functions in cytokinesis, nuclear positioning, and stomatogenesis in *Tetrahymena thermophila*. *J. Eukaryot. Microbiol.* **50**, 252–262.

Williams, N. E. (2004). The epiplasm gene EPC1 influences cell shape and cortical pattern in *Tetrahymena thermophila*. *J. Eukaryot. Microbiol.* **51**, 201–206.

Williams, N. E., and Bakowska, J. (1982). Scanning electron microscopy of cytoskeletal elements in the oral apparatus of *Tetrahymena*. *J. Protozool.* **29**, 382–389.

Williams, N. E., and Frankel, J. (1973). Regulation of microtubules in *Tetrahymena*: I. Electron microscopy of oral replacement. *J. Cell Biol.* **56**, 441–457.

Williams, N. E., and Honts, J. E. (1995). Isolation and fractionation of the *Tetrahymena* cytoskeleton and oral apparatus. *Methods Cell Biol.* **47**, 301–306.

Williams, N. E., Honts, J. E., Dress, V. M., Nelsen, E. M., and Frankel, J. (1995). Monoclonal antibodies reveal complex structure in the membrane skeleton of *Tetrahymena*. *J. Eukaryot. Microbiol.* **42**, 422–427.

Williams, N. E., Honts, J. E., and Graeff, R. W. (1986). Oral filament proteins and their regulation in *Tetrahymena pyriformis*. *Exp. Cell Res.* **164**, 295–310.

Williams, N. E., Honts, J. E., and Jaeckel-Williams, R. F. (1987). Regional differentiation of the membrane skeleton in *Tetrahymena*. *J. Cell Sci.* **87**, 457–463.

Williams, N. E., Honts, J. E., and Kaczanowska, J. (1990). The formation of basal body domains in the membrane skeleton of *Tetrahymena*. *Development* **109**, 935–942.

Williams, N. E., Honts, J. E., Lu, Q., Olson, C. L., and Moore, K. C. (1989a). Identification and localization of major cortical proteins in the ciliated protozoan, *Euplotes eurystomus*. *J. Cell Sci.* **92**, 433–439.

Williams, N. E., Honts, J. E., and Stuart, K. R. (1989b). Properties of microtubule-free cortical residues isolated from *Paramecium tetraurelia*. *J. Cell Sci.* **92**, 427–432.

Williams, S. A., Hosein, R. E., Garces, J. A., and Gavin, R. H. (2000). MYO1, a novel, unconventional myosin gene affects endocytosis and macronuclear elongation in *Tetrahymena thermophila*. *J. Eukaryot. Microbiol.* **47**, 561–568.

Williams, N. E., and Luft, J. H. (1968). Use of a nitrogen mustard derivative in fixation for electron microscopy and observations on the ultrastructure of *Tetrahymena*. *J. Ultrastructure Res.* **25**, 271–292.

Williams, N. E., and Nelsen, E. M. (1997). HSP70 and HSP90 homologs are associated with tubulin in hetero-oligomeric complexes, cilia and the cortex of *Tetrahymena*. *J. Cell Sci.* **110**, 1665–1672.

Williams, N. E., Tsao, C. C., Bowen, J., Hehman, G. L., Williams, R. J., and Frankel, J. (2006). The actin gene ACT1 is required for phagocytosis, motility, and cell separation of *Tetrahymena thermophila*. *Eukaryot. Cell* **5**, 555–567.

Williams, N. E., and Williams, R. J. (1976). Macronuclear division with and without microtubules in *Tetrahymena*. *J. Cell Sci.* **20**, 61–77.

Wloga, D., Camba, A., Rogowski, K., Manning, G., Jerka-Dziadosz, M., and Gaertig, J. (2006). Members of the NIMA-related kinase family promote disassembly of cilia by multiple mechanisms. *Mol. Biol. Cell.* **17**, 2799–2810.

Wloga, D., Dave, D., Meagley, J., Rogowski, K., Jerka-Dziadosz, M., and Gaertig, J. (2010). Hyperglutamylation of tubulin can either stabilize or destabilize microtubules in the same cell. *Eukaryot. Cell* **9**, 184–193.

Wloga, D., and Gaertig, J. (2010). Post-translational modifications of microtubules. *J. Cell Sci.* **123**, 3447–3455.

Wloga, D., Rogowski, K., Sharma, N., Van Dijk, J., Janke, C., Edde, B., Bre, M. H., Levilliers, N., Redeker, V., Duan, J., Gorovsky, M. A., Jerka-Dziadosz, M., and Gaertig, J. (2008a). Glutamylation on alpha-tubulin is not essential but affects the assembly and functions of a subset of microtubules in *Tetrahymena thermophila*. *Eukaryot. Cell* **7**, 1362–1372.

Wloga, D., Strzyzewska-Jowko, I., Gaertig, J., and Jerka-Dziadosz, M. (2008b). Septins stabilize mitochondria in *Tetrahymena thermophila*. *Eukaryot. Cell.* **7**, 1373–1386.

Wloga, D., Webster, D. M., Rogowski, K., Bre, M. H., Levilliers, N., Jerka-Dziadosz, M., Janke, C., Dougan, S. T., and Gaertig, J. (2009). TTLL3 is a tubulin glycine ligase that regulates the assembly of cilia. *Dev. Cell.* **16**, 867–876.

Wolfe, J. (1970). Structural analysis of basal bodies of the isolated oral apparatus of *Tetrahymena pyriformis*. *J. Cell Sci.* **6**, 679–700.

Wong, L., Klionsky, L., Wickert, S., Merriam, V., Orias, E., and Hamilton, E. P. (2000). Autonomously replicating macronuclear DNA pieces are the physical basis of co-assortment groups in *Tetrahymena thermophila*. *Genetics* **155**, 1119–1125.

Wood, C. R., Hard, R., and Hennessey, T. M. (2007). Targeted gene disruption of dynein heavy chain 7 of *Tetrahymena thermophila* results in altered ciliary waveform and reduced swim speed. *J. Cell Sci.* **120**, 3075–3085.

Xia, L., Hai, B., Gao, Y., Burnette, D., Thazhath, R., Duan, J., Bre, M. H., Levilliers, N., Gorovsky, M. A., and Gaertig, J. (2000). Polyglycylation of tubulin is essential and affects cell motility and division in *Tetrahymena thermophila*. *J. Cell Biol.* **149**, 1097–1106.

Yasuda, T., Tamura, R., and Watanabe, Y. (1984). Molecular mechanisms of cell division in *Tetrahymena*. Ultrastructural changes found in division-arrest ts mutant. *Zool. Sci. (Tokyo)* **1**, 62–73.

Zweifel, E., Smith, J., Romero, D., Giddings Jr, T. H., Winey, M., Honts, J., Dahlseid, J., Schneider, B., and Cole, E. S. (2009). Nested genes CDA12 and CDA13 encode proteins associated with membrane trafficking in the ciliate *Tetrahymena thermophila*. *Eukaryot. Cell.* **8**, 899–912.

CHAPTER 6

Conservation and Innovation in *Tetrahymena* Membrane Traffic: Proteins, Lipids, and Compartments

Alejandro D. Nusblat[*], **Lydia J. Bright**[†] **and Aaron P. Turkewitz**[‡,§]

[*]Cátedra de Biotecnología y Microbiología Industrial, Facultad de Farmacia y Bioquímica, Universidad de Buenos Aires, Junín 956, (C1113AAD) Buenos Aires, Argentina

[†]Department of Biology, University of Indiana, Bloomington, 327 Jordan Hall, 1001 East 3rd Street Bloomington, IN 47405-7005, USA

[‡]Department of Molecular Genetics and Cell Biology, The University of Chicago, 920 E. 58th Street, Chicago IL 60637, USA

Abstract

The past decade has seen a significant expansion in our understanding of membrane traffic in *Tetrahymena thermophila*, facilitated by the development of new experimental tools and by the availability of the macronuclear genome sequence.

[§]Corresponding author

0091-679X/10 $35.00
DOI 10.1016/B978-0-12-385967-9.00006-2

Here we review studies on multiple pathways of uptake and secretion, as well as work on metabolism of membrane lipids. We discuss evidence for conservation *versus* innovation in the mechanisms used in ciliates compared with those in other eukaryotic lineages, and raise the possibility that existing gene expression databases can be exploited to analyze specific pathways of membrane traffic in these cells.

I. Introduction

The partitioning of the cytoplasm into functionally distinct compartments called organelles, delimited by lipidic membranes, is a hallmark of eukaryotic cells. By concentrating enzymes as well as substrates, each organelle is specialized for a set of reactions, for which compartmental conditions can be optimized. In addition to increasing the efficiency of individual and coupled reactions, compartmental organization also enables eukaryotic cell to simultaneously execute reactions that might be incompatible if pursued in a continuous cytoplasm. Membrane-bound compartments can also serve for dynamic storage, for example of calcium that can be rapidly transported from the lumen of an organelle via trans-membrane channels to the cytoplasm in response to signaling pathways. Additionally, molecules within the lumen of organelles can be exported to the cell exterior, which occurs upon fusion of the organellar and plasma membranes. Secretion via membrane fusion is the major mechanism of protein secretion from eukaryotic cells, a phenomenon of great physiological significance that allows cells to influence their environments in a multitude of ways, for both unicellular and multicellular organisms. The secretory apparatus consists of an array of morphologically and biochemically complex organelles, whose concerted activity is responsible for first translocating proteins out of the cytoplasm into the lumen of the secretory pathway, sorting those proteins to accommodate different modes of secretion, and subsequently releasing them into the environment. In many cases, the proteins are covalently modified during this process. A second complex network of compartments, partially overlapping with the first, exists to receive and sort molecules that are internalized from the cell surface.

A major thrust of eukaryotic cell biology in the last 50 years has been to understand organelle activities and organization. Broadly speaking, a first aim has been to understand the basic mechanisms underlying organelle biogenesis and function, which are thought to be widely conserved among eukaryotes. Numerous insights have come from exploiting powerful approaches in budding yeast, in many cases providing initial identification of key components, or detailed mechanistic information, for steps that are conserved in mammalian cells. Generally, however, the pathways of membrane traffic in mammalian cells are more complex than in yeast, so a second broad aim has been to understand the additional features of specific mammalian cell types. For example, some polarized mammalian cells can maintain two or more parallel pathways of protein secretion, with each pathway directed toward a distinct cell surface (Weisz and Rodriguez-Boulan, 2009). These two lines of inquiry have also been explored in evolutionary terms. In particular, mechanisms that are conserved among eukaryotes are inferred to have been present in a shared eukaryotic ancestor, and therefore are very ancient (Dacks and Field, 2007). In contrast, pathways and mechanisms that are specifically required for mammalian complexity can be inferred to have arisen as more recent adaptations that are consequently restricted to a subset of modern lineages.

When compared to pathways of membrane traffic in mammalian cells or fungi, the pathways in *Tetrahymena* are still relatively unexplored. Many historical studies, as reviewed by Frankel (Frankel, 2000), rely chiefly on morphological analysis. Published electron micrographs provide an excellent overview of many aspects of subcellular organization and offer a strong starting point for molecular mechanistic studies. However, for many compartments there were and remain few if any identified molecular components. Such molecular markers are powerful tools for direct observations of compartments and their dynamics in living cells (e.g., by green fluorescent protein tagging). Secondly, in some cases, molecular markers are essential for interpreting the compartments that are visualized. While some membrane structures are unmistakeable, such as the nuclear envelope, many others are highly pleiomorphic in other cells where they have been studied. Structures such as the Golgi, trans-Golgi network, and distinct classes of endosomes adopt different appearances depending on the cell type and cell activity and can therefore be difficult to identify without molecular markers. Fortunately, the resources available from the sequenced Tetrahymena genome should accelerate the development of such markers, in part by facilitating the identification of Tetrahymena homologs for markers established in other systems. The genomic data also facilitate proteomic approaches to identify the components of isolated organelles.

Although many details are lacking, we know that *Tetrahymena thermophila* maintains a highly complex network of membrane trafficking pathways. For example, there is evidence for at least four distinct pathways of endocytic uptake. Cells form small endocytic vesicles at cortical invaginations called parasomal sacs (Nilsson and Van Deurs, 1983), while much larger vesicles (phagosomes) arise at the base of the oral apparatus (Nilsson, 1979). While it is clear that different mechanisms are involved in endocytosis from parasomal sacs versus the oral apparatus, neither of these pathways has been dissected in detail. A third pathway of endocytosis, which can be inferred from work in Paramecium, is coupled with the exocytosis of dense-core secretory vesicles and facilitates the rapid recovery of vesicle membranes (Hausmann and Allen, 1976). Fourth, endocytic membrane recovery also occurs upon phagosome fusion, at a cortical site called the cytoproct (Allen and Wolf, 1979). Similarly, Tetrahymena secrete proteins by at least three different routes: a pathway of rapid constitutive release of newly synthesized proteins (for which the vesicular carriers have not been identified) (Bowman and Turkewitz, 2001; Madinger *et al.*, 2010); regulated exocytosis from docked mucocysts (Turkewitz, 2004); and release of hydrolytic enzymes via lysosome exocytosis (Kiy *et al.*, 1993). There is also indirect evidence for cytoplasmic protein release via an exosome-like mechanism (Madinger *et al.*, 2010). This list understates the complexity of the pathways of cell surface delivery, since for example there is also vesicle trafficking from the cytoproct (the site of phagosome exocytosis) to the oral apparatus (Allen and Fok, 1980; Bright *et al.*, 2010). It is also not clear whether there is a single pathway of endocytosis from parasomal sacs or whether parallel pathways exist, as will be discussed below.

In our view, more detailed molecular studies of trafficking in *T. thermophila* could be significant for several reasons. While it may not be possible or even desirable to obtain a comprehensive understanding of the entire network of trafficking steps, there are individual pathways whose analysis could make major contributions to both of the aims outlined above. Unlike budding yeast whose successful evolutionary strategy was to become small and relatively simple, the ciliates, like animal cells, have undergone large expansions in gene families encoding key determinants of membrane trafficking, as the membrane trafficking pathways themselves grew increasingly complex (Bright *et al.*, 2010; Eisen *et al.*, 2006; Saito-Nakano *et al.*, 2010). Because the adaptations to membrane traffic occurred independently in ciliates and animals, a comparison of these lineages offers one the chance to ask whether specific pathways were prone to expansion and adaptation. This question is also being asked more broadly, taking advantage of the wealth of sequenced genomes now available to ask whether the determinants of specific pathways have tended to expand in multiple lineages, to generate large gene subfamilies. For example, phylogenetic analysis of SNARE proteins in many lineages suggests that SNARE subfamilies associated with endocytosis have undergone more expansion than other subfamilies, suggesting that endocytosis has been a particularly rich substrate for innovations in membrane traffic (Kienle *et al.*, 2009). These metagenomic studies can be complemented by more in-depth studies of individual organisms. Studying such questions in a specific complex nonanimal lineage, such as higher plants or ciliates, allows one to confirm the phylogenetic predictions, that is, test the underlying assumption that sequence comparison are reliable for assigning function. Secondly, such single-species studies are critical to understanding both how, and to what purpose, the genetic innovations have modified conserved pathways or generated new ones. In other words, how has selection acted on the organization and function of membrane trafficking pathways? Critically, the experimental tools available in *T. thermophila* already facilitate asking complex cell biological questions, and new approaches continue to be developed (Turkewitz *et al.*, 2002). One such novel approach to studying membrane traffic in particular is discussed at the end of this chapter, based on the availability of extensive whole genome expression data.

Effective use of *T. thermophila* may come from exploiting unique features of its complex and unusual organization. For example, many sites of specific membrane trafficking steps at the plasma membrane are organized as precise arrays, allowing a microscopist to analyze multiple sites, simultaneously, at predictable locations. This aspect of ciliate organization has recently been brilliantly exploited to analyze basal bodies (Pearson and Winey, 2009). For membrane trafficking, such organized domains include sites of clathrin-mediated endocytosis and of regulated exocytosis (Allen, 1967; Elde *et al.*, 2005; Satir *et al.*, 1973). A second striking aspect of Tetrahymena is that there are structurally and functionally distinct variants of several organelles, maintained in the same cytoplasm. This is best known for the nucleus, where studies exploiting the differences between the macro- and micronucleus have made pivotal contributions to molecular biology (Pederson, 2010). Nuclear dimorphism in Tetrahymena has recently been exploited to analyze the role of nuclear pore components. Nuclear pores are selective gates that regulate traffic of cytosolic and membrane proteins into the nucleoplasm, and a major question in the field is how the components of nuclear pores act as gatekeepers, with much attention focusing on iterative motifs consisting of glycine-leucine-phenylalanine-glycine (GLFG) that are abundant in many proteins lining the pores (nucleoporins, or nups). In ciliates, the two functionally distinct nuclei contain different sets of nucleoplasmic proteins, implying that nuclear pores in Mics and Macs are also distinct. Haraguchi and colleagues recently identified micronuclear- and macronuclear-specific versions of *NUP98* (Iwamoto *et al.*, 2009). The repeats in the micronuclear (but not macronuclear) Nup98p were NIFN, rather than the canonical GLFG, and domain-swapping experiments provided evidence that the change in the Nup repeat motif has functional consequences for gatekeeping. This line of work therefore holds promise both to reveal mechanisms underlying nuclear dimorphism in Tetrahymena as well as providing a unique model system for dissecting features of nuclear pore selectivity.

Another example of organellar differentiation in Tetrahymena is that each cell contains both a "standard" endoplasmic reticulum (ER), including the nuclear envelope, but also distinct flattened cisternae called alveoli that tightly underlayer the plasma membraneto. While alveoli have been only glancingly studied in Tetrahymena, data from Paramecium make a strong case that alveoli function as a major store for mobilizable calcium, a classical activity of the ER, and also contain some ER proteins (Plattner *et al.*, 1999; Stelly *et al.*, 1995). ER subdomains in animal cells are recognized as an important aspect of the secretory pathway and of cellular signaling, and understanding the biogenesis and maintenance of the ER and alveoli in Tetrahymena may offer exceptional opportunities for illuminating mechanisms of protein and lipid sublocalization in this organelle.

Lastly, Tetrahymena, because of the strong experimental tools that have been developed, may be an excellent organism to appreciate "cell biodiversity," namely the range of adaptations that have evolved in eukaryotes that are deeply divergent from animals. For example, the contractile vacuole is a multipart organelle that collects water from the cytoplasm to pump it out of the cell and is essential for osmotic homeostasis in fresh water organisms lacking cell walls. The remarkable properties of the ciliate contractile vacuole have been investigated in Paramecium, but virtually nothing is known about assembly or mechanism of action at the molecular level (Allen, 2000). Contractile vacuoles are also present in Amoebozoa and other distantly related lineages, but whether these are homologous organelles to those in Ciliates (i.e., inherited from a common ancestor), or whether organelles as complex as contractile vacuoles have arisen multiple times, independently, is an open question. Pursuing such organelles in Tetrahymena, as they are also being studied in Dictyostelium, could help to provide a new perspective on the relative importance of inheritance versus innovation in the structures that animate modern eukaryotes.

II. Recent Work on Membrane Traffic in *Tetrahymena*

Studies of membrane traffic in Tetrahymena up until the last decade have been authoritatively reviewed by Frankel, and we will therefore focus on work reported since that review (Frankel, 2000). At the end of each section, we list additional papers reporting observations or reagents that may be of interest to those investigating membrane traffic.

A. Protein Secretion

1. Constitutive Secretion

Constitutive secretion refers to secretion of newly synthesized proteins in the absence of specific extracellular stimulation. Rapid secretion of newly synthesized proteins in *T. thermophila* was visualized using pulse-chase biosynthetic labeling, where it was demonstrated that the proteins released via this pathway were different from those released via regulated exocytosis from mucocysts (Bowman and Turkewitz, 2001). It is likely that these proteins, after transport through the ER and Golgi, are transported from the trans-Golgi to the cell surface in vesicles or membrane tubules, but this process has not been directly visualized in Tetrahymena. Moreover, the site(s) of such protein release represents an interesting problem since most of the plasma membrane, with which vesicles must fuse to release their contents, is inaccessible from the cytoplasm because of the intervening alveoli (Allen, 1978).

A proteomic analysis of *T. thermophila* culture supernatants has recently supplied the first relatively comprehensive view of what this unicellular protest is releasing into its environment (Madinger *et al.*, 2010). That list includes 207 proteins including many hydrolytic enzymes as well as novel proteins of unknown function, with the cohort of secreted proteins changing significantly depending on whether cells were incubated under nutritive versus starvation conditions. The authors also characterize protein secretion from a previously isolated Mendelian mutant, SB281, whose most striking defect is the failure to synthesize mucocysts. They showed that the constitutive proteins released in this mutant were different from wildtype cells, consistent with previous descriptive evidence (Bowman and Turkewitz, 2001). Interestingly, the authors find evidence that some proteins known to be released via mucocysts may also be released via constitutive exocytosis, and that some specific mucocyst proteins may be released differentially under growth versus starvation conditions. Following up on these observations may be interesting both from a mechanistic perspective and also to illuminate the role of mucocysts for Tetrahymena, which is not yet known.

Part of the interest in characterizing secretion comes from the question of whether Tetrahymena can be usefully engineered to produce and secrete heterologous proteins (Aldag *et al.*, 2011; Hartmann *et al.*, 2000).

2. Regulated Secretion

In animal cells, proteins can be released in response to extracellular stimuli both from small vesicles such as synaptic vesicles, or from larger dense-core secretory vesicles, also called secretory granules. These two types of vesicles arise via distinct biosynthetic pathways and serve a wide range of physiological roles in different tissues (Gumbiner and Kelly, 1982). Ciliates synthesize vesicles containing dense cores, which because of their distinct appearance are prominent features in many cytological studies (Rosati and Modeo, 2003). In *T. thermophila* these vesicles are called mucocysts, and they have been explored as a system for understanding biosynthetic mechanisms that are still rather poorly understood in animals, but which appear to differ from canonical mechanisms involved in vesicle formation (Tooze *et al.*, 2001; Turkewitz, 2004). Mucocysts contain two major families of soluble (i.e., not membrane-bound) proteins, which are coordinately expressed (Rahaman *et al.*, 2009). The first, whose corresponding genes were called *GRL* for granule lattice, are required to form a protein crystal that comprises the bulk of the dense core (Cowan *et al.*, 2005). A subset of the *GRL*-encoded proteins may however play roles that are less structural than regulatory, since gene knockout did not affect the appearance of the dense core but reduced mucocyst accumulation (Cowan *et al.*, 2005). The *GRL*s have been identified both biochemically but also by forward genetics, using an unbiased screen based on antisense ribosomes (Cowan *et al.*, 2005). The Grl proteins are synthesized as pro-proteins, and endoproteolytic cleavage is closely connected with assembly of the dense core (Verbsky and Turkewitz, 1998). This assembly appears to take place in a post-Golgi vesicular compartment and intermediates can be visualized by EM (Bowman *et al.*, 2005a). A link between proprotein processing and dense core assembly is established in mammalian systems, and the similarity in ciliates is intriguing (Creemers *et al.*, 1998). In addition, biochemical and genetic experiments demonstrated that granule assembly intermediates in *Tetrahymena* form in the endoplasmic reticulum (Cowan *et al.*, 2005).

A second family of granule proteins in *T. thermophila* is defined by a common C-terminal β/γ crystalline domain, the remainder of the proteins consisting of a variable number of repeats of several different domains (Bowman *et al.*, 2005b). Two members of this family have been investigated and neither is proteolytically processed (Bowman *et al.*, 2005a; Haddad *et al.*, 2002). In addition, none of the genes in this family that have been disrupted is essential for core assembly, but the double disruption of two related genes subtly changed the properties of the mucocyst core following its exocytic release, so apparently the proteins in this family are playing distinct roles from the Grls (Rahaman *et al.*, 2009). Only one protein in the non-Grl family has been localized and was found, remarkably, to be highly concentrated at the end of the secretory granule where it docks at the plasma membrane, prior to exocytosis (Bowman *et al.*, 2005a). The protein, called Grt1p for Granule tip, fails to polarize in two Mendelian mutants that are defective in a late stage of mucocyst assembly as well as mucocyst docking (Bowman *et al.*, 2005a). One possibility is that Grt1p and other proteins in that family can interact with specific proteins in the mucocyst membrane and thereby organize membrane zones with specific activities, such as docking and exocytic fusion. Interestingly, some electron micrographs of mucocysts appear to show that, in addition to the crystalline core, there are other components that are more closely associated with the membrane (Williams and Luft, 1968).

The mechanisms enabling accurate sorting to mucocysts of the Grl and β/γ crystallin-containing protein families are not yet known. One model, based on work in animal cells, is that proteins destined for dense core vesicles have a predisposition to co-aggregate in a late Golgi compartment (Chanat and Huttner, 1991). However, neither genetic nor biochemical experiments has demonstrated any interaction between the two families of proteins in *T. thermophila* dense core vesicles (Rahaman *et al.*, 2009). These are suggestive rather than conclusive results, since the system is sufficiently complex so that physiologically important interactions may have escaped detection, for example, due to functional redundancy. In addition, biochemical interactions may be highly sensitive to the ionic conditions within specific compartments of the secretory pathway, about which very little in known in ciliates.

A third demonstrated pathway of secretion in *T. thermophila* is that of secretory lysosomes, and is reviewed in Frankel (Frankel, 2000). A pathway in Tetrahymena that has not been investigated at the molecular level is secretion from the contractile vacuole upon its cyclic fusion at plasma membrane pores.

Other Studies of Interest in *T. thermophila*

1. Identification of *PGP1*, an HSP70 homolog, of the GRP170 subfamily, whose product localizes to the endoplasmic reticulum and is shown to be a glycyosylated, glycylated protein. *PGP1* is induced on cell stress but also essential for vegetative growth (Xie *et al.*, 2007). The C-terminal peptide KQTDL functions as an ER-retention signal. KDEL and related sequences have previously been shown to act as functional ER retention signals in *T. thermophila* (Cowan *et al.*, 2005).
2. A novel gene, *CDA13*, encodes a predicted transmembrane protein which may reside in a post-Golgi compartment of the secretory pathway (Zweifel *et al.*, 2009).
3. Identification of *DRP6*, a highly divergent dynamin-related protein that localizes to the macronuclear envelope and a vesicular ER-like compartment, and which is essential for nuclear remodeling during conjugation (Rahaman *et al.*, 2008).
4. Analysis, including localization and functional studies, of the nucleoporins and karyopherins involved in nuclear import (Malone *et al.*, 2008).
5. Analysis of the expression of Ser antigens, the best-known cell surface proteins in this system (Doerder and Gerber, 2000). Work on antigenic variation in ciliates has recently been reviewed (Simon and Schmidt, 2007).
6. Analysis of the carbohydrate structure of secretory proteins (Becker and Rusing, 2003).
7. Analysis of extracellular cysteine proteases (Herrmann *et al.*, 2006).

B. Endocytosis

In mammalian cells, endocytosis is a critical pathway for the uptake of extracellular macromolecules, modulation of signaling pathways, and turnover of membrane proteins. A number of different endocytic pathways exist in animals, the best characterized of which involves assembly of the protein clathrin, which interacts with the heterotetrameric adaptor protein AP-2 during endocytic vesicle formation. The scission of the vesicle membrane from the plasma membrane, releasing it into the cytoplasm, involves a GTPase of the dynamin family, and actin is involved at multiple steps in the process (Kirchhausen, 2009).

In Tetrahymena, the alveoli limit contact between the cytoplasm and plasma membrane. One interruption in the alveoli are indentations called parasomal sacs that are found proximal to each ciliary basal body, and these are sites of endocytosis as shown by EM studies of cationized ferritin uptake (Nilsson and Van Deurs, 1983). Endocytosis at parasomal sacs has more recently been confirmed in living *T. thermophila* using a styryl dye, FM1-43, which had been shown in other systems to be a useful marker for endocytic vesicles (Cousin and Robinson, 1999). *Tetrahymena* incubated with FM dyes first show fluorescence in small puncta that form an array near the cell cortex, as would be expected for endocytic vesicles that have just undergone scission at parasomal sacs (Elde *et al.*, 2005). Thereafter the puncta are highly mobile and within minutes appear to collect toward the cell posterior. The appearance of the posterior fluorescent structures, which can on the basis of the FM1-43 labeling be classified as endosomes, suggests that the initial endocytic vesicles have undergone fusion events to create larger, heterogeneous structures. The FM1-43 accumulation in these posterior endosomes persists for at least tens of minutes. A related dye, FM4-64, has also been shown to accumulate after long labeling periods in moderate-sized vesicles at some distance from the cortex, located throughout the cell (Zweifel *et al.*, 2009).

The FM1-43 uptake assay facilitated analysis of the protein requirements for endocytosis. Using GFP-tagging, the authors demonstrated that clathrin was localized to parasomal sacs and involved in endocytosis since induced expression of a truncated clathrin heavy chain, which acts as a dominant negative form, suppressed FM1-43 uptake (Elde *et al.*, 2005). Similarly, GFP-tagging was used to localize four AP complexes, and only AP-2 was found to localize to parasomal sacs. A third similarity with animal cells was a requirement for dynamin in endocytosis. *T. thermophila* was found to encode eight members of the dynamin family, called *DRP1-8* for *dynamin-related proteins*, a remarkably large number for a unicellular organism. Drp1p and Drp2p were found to localize to parasomal sacs (Elde *et al.*, 2005; Rahaman and Turkewitz, unpublished) and the endocytic activity of Drp1p, an essential gene, was demonstrated by several genetic approaches. A 28 amino acid stretch of Drp1p was sufficient, when exchanged with the same region of a different Drp family member, to redirect the chimeric protein to sites of endocytosis. Surprisingly, actin did not appear to be required for endocytosis, judging by results with pharmacological actin inhibitors. If this result is correct, *Tetrahymena* may be unique in having evolved actin-independent endocytic mechanisms. However, the experimental results could not rule out the possibility that divergent actin isoforms, which are insensitive to the drugs used, are involved in endocytosis. Other interesting possibilities are discussed below, as well as additional endocytic markers identified in a screen of Rab GTPases.

It is worth noting that neither the expression of the dominant negative alleles of clathrin or *DRP1*, nor disruption of the endogenous *DRP1* gene, led to a complete block in FM1-43 uptake (Elde *et al.*, 2005). The appearance of the residual FM accumulation in these strains suggested that the signal was due to vesicle formation at parasomal sacs. These results raise the question of whether Tetrahymena also has a clathrin- and *DRP1*-independent pathway of endocytosis. To investigate this, it will be important to analyze the function of Drp2p, which also localizes to parasomal sacs (Rahaman and Turkewitz, unpublished).

Additional Papers of Interest

1. Analysis of a novel predicted membrane protein, Cda12p, which localizes to a putative endocytic compartment. The knockdown (via antisense ribosomes) phenotype includes defects in cytokinesis and some aspect of endosome formation. During cytokinesis as well as in mating cells, the protein localizes in regions where there is likely to be active membrane remodeling. This provides a hint about how remodeling of endosomal compartments might underlie structural changes at specific stages in the life cycle (Zweifel *et al.*, 2009).

C. Phagocytosis and Phagosome Maturation

T. thermophila is magnificently adapted for bactivory, sweeping small particles into the base of the oral apparatus where they are ingested via formation of large food vacuoles called phagosomes (Frankel, 2000). The digestion of phagosome contents takes place via a series of remodeling steps, collectively termed maturation, in which the nascent phagosome fuses with vesicles that deliver acidification machinery as well as hydrolytic enzymes, while other components are selectively removed/recycled via vesicle budding (Stuart and Ezekowitz, 2005). Remodeling of the phagosome membrane by cytosolic factors is also likely to be important, as has been shown for mammalian cells (Huynh *et al.*, 2007). Compartment maturation is an important theme in membrane traffic, and phagosomes are a particularly attractive pathway for detailed analysis because phagosomes are large and easily labeled by loading specific cargo (e.g., fluorescent bacteria or bacteria-sized latex beads).

Phagosome formation and maturation in *Tetrahymena* are actin-dependent processes. First, there is a clear requirement for dynamic actin during phagosome formation, which has been demonstrated using pharmacological inhibition of actin dynamics but also on disruption of genes encoding the major actin gene, and the actin-assembly cofactor, profilin (Wilkes and Otto, 2003; Williams *et al.*, 2006). Actin filaments may also be indirectly involved in phagosome maturation, based on reports that the movement of phagosomes from the oral apparatus in the cell anterior, to the cytoproct at the cell posterior, involves an actin-based myosin motor encoded by *MYO1* (Hosein *et al.*, 2005). In addition, the microtubule-based dynein motor *DYH1* has also been implicated in phagosome formation (Lee *et al.*, 1999).

The dynamin Drp1p, involved in clathrin-mediated endocytosis at parasomal sacs, is also required for phagocytosis. Cells expressing a dominant-negative *DRP1* allele (K51E) or in which the level of wildtype gene expression is reduced by methods described above, showed no discernable phagocytic uptake of particles (e.g., India ink particles) from the medium (N. Elde and A.P. Turkewitz, unpublished) (N.Elde, PhD thesis). This phenotype was distinct from that of cells treated with actin inhibitors. The actin-inhibited cells showed no accumulation of ink particles in cytoplasmic phagosomes, but nearly all cells accumulated ink in a large vesicle that formed at, but failed to detach from, the base of the oral apparatus. *DRP1* mutant cells failed to accumulate even this single frustrated phagosome, suggesting that Drp1p acts upstream of actin during phagosome formation. A potential hint of the function of Drp1p in this pathway is that GFP-tagged Drp1p labels puncta along the so-called deep fiber, a cytoskeletal filament that extends from the base of the oral apparatus that appears to act as a vesicle highway. The deep fiber and nearby structures also appear to be sites of localization of calmodulin and a number of calmodulin-binding proteins, and pharmacological inhibition of calmodulin activity and calcium-based signaling block phagosome formation (Gonda *et al.*, 2000; Moya and Jacobs, 2006).

The sequencing of the *T. thermophila* genome made it possible to easily identify homologs to many proteins previously studied in other systems, facilitating many of the studies cited above (Eisen *et al.*, 2006). A second important consequence was facilitation of proteomic studies since the predicted *T. thermophila* proteome could be used to identify proteins in isolated organelle fractions using mass spectrometry data. This has been very fruitfully applied to phagosomes, which were highly purified by taking advantage of the aforementioned ability to identify, and change the fractionating properties of, phagosomes that had taken up polystyrene beads (Jacobs *et al.*, 2006). Proteins associated with the purified phagosomes were then analyzed by mass spectrometry, resulting in the identification of 73 genes. This extensive list allowed the authors to gauge the similarity of phagosomes between *T. thermophila* and other organisms, since 28 of the proteins had been associated with phagocytosis in other organisms. In addition, the authors choose four candidate genes from the survey and, by expressing these as GFP-tagged copies, demonstrated that three of these were phagosome-associated. Taken together, these results suggest that many mechanisms are conserved between the phagosome pathways in multiple lineages. Since many of the conserved proteins including several associated with human disease have functions that are not well understood, *T. thermophila* may offer an attractive system for addressing questions about the mammalian phagosome pathway.

Additional Papers of Interest

1. A study showing that Pseudopterosin A, a marine natural product, inhibits phagocytosis, with pharmacological evidence arguing for a G protein-coupled

receptor mechanism of action involving a calcium-dependent step (Moya and Jacobs, 2006).

2. A study showing that degradation of the old macronucleus during conjugation has features of an unusual autophagy (Akematsu *et al.*, 2010). Additional insightful studies from the same group focus on the role of mitochondria and mitochondrial signaling factors during macronuclear breakdown (Akematsu and Endoh, 2010).
3. Mass spectrometric analysis of the mitochondrial proteome (Smith *et al.*, 2007b), the conjugation junction (Cole *et al.*, 2008), and the ciliome (Smith *et al.*, 2005).
4. A particularly elegant study identifying a basal body proteome, including extensive ultrastructural analysis (Kilburn *et al.*, 2007).
5. Studies on the effect of passage through the phagosome on bacterial conjugation and infectivity (Klobutcher *et al.*, 2006; Matsuo *et al.*, 2010).
6. The important role of cytoskeletal-based motor proteins in membrane traffic is well established. Most work on motor proteins in *T. thermophila* has focused on ciliary beat, but informatics-based surveys indicate that a large number of cytosolic proteins remain to be explored (Sugita *et al.*, 2011; Wilkes *et al.*, 2008).

D. Rab GTPases as Markers for Membrane Traffic

Rabs are small GTPases that act as key determinants of compartmental specificity by recruiting, when in their GTP-bound, membrane-tethered state, a wide range of effectors (Segev, 2001). Rabs exist as products of large gene families in which each family member associates with one or a small number of cellular compartments. Thus, the number of Rabs expressed in a cell likely reflects the complexity of membrane trafficking pathways in that cell (Stenmark and Olkkonen, 2001). There are 12 Rabs in *Saccharomyces cerevisiae*, compared with 63 in humans (Pereira-Leal and Seabra, 2001).

Two groups have recently characterized the Rabs in *T. thermophila*, which were identified via homology searches in the macronuclear database based on the fact that Rabs can be distinguished from other related small GTPases based on a number of conserved motifs. Bright *et al.* characterized Rabs by combining phylogenetic and expression analysis with localization data, the last based on GFP-tagging the large majority of the Rab family members (Bright *et al.*, 2010). Phylogenetic and expression analysis of the Rab superfamily was also reported by Numata and colleagues (Saito-Nakano *et al.*, 2010). Where they overlap, the results of the two groups are largely similar, the most important difference being that different criteria were used to assign orthology, an issue discussed below (Turkewitz and Bright, in press). In addition, the Numata group took the valuable step of experimentally verifying the sequences predicted by genome annotation.

The *T. thermophila* genome encodes 63 Rabs, a number greater than the 33 in *Drosophila melanogaster* or 29 in *Caenorhabditis elegans*. There are an additional

25 Rab-like proteins, which differ from Rabs in lacking identifiable C terminal prenylation motifs (Saito-Nakano *et al.*, 2010). A first question was how many of these Rabs are expressed concurrently. In animal cells, many Rab isoforms are expressed preferentially in particular tissues, so that the total number of isoforms reflects the range of adaptations of membrane traffic for distinct tissues (Zhang *et al.*, 2007). For unicellular organisms that possess large Rab families, one possibility is that subsets are expressed under different conditions, that is, that the large number reflects adaptations of membrane traffic for different environments or life stages. For *Tetrahymena*, this question was answered by mining a public database in which all transcripts were measured in cultures sampled at a variety of physiologically relevant states (Miao *et al.*, 2009). A small number of Rabs showed strikingly stage-specific expression (e.g., undetectable expression under growth conditions and high expression upon starvation or during mating) (Bright *et al.*, 2010; Saito-Nakano *et al.*, 2010). However, the large majority of Rabs were co-expressed, and many at very high levels, in growing cells. The large number of co-expressed Rabs suggests that *T. thermophila* maintains a membrane trafficking network that is roughly as elaborate as that of mammalian cells.

To investigate the functions of the large set of Rabs, Bright *et al.* expressed each in *T. thermophila* as an N-terminal GFP fusion, and took advantage of a novel thermally controlled gel to immobilize the normally fast-swimming cells to capture time-lapse movies showing the dynamics of the GFP-Rab-labeled structures (Jeong *et al.*, 2007). Given the paucity of molecular markers for many pathways in Ciliates, the individual Rabs may become important tools since they localize to a wide range of cellular structures, many of which could be tentatively identified even at the level of light microscopy. For example, Rabs associated with endocytosis were identified by using FM4-64 as an endocytic tracer, while another set of Rabs could be assigned to phagocytosis-related structures (i.e., the oral apparatus, phagosomes, or the cytoproct) by labeling phagosomes with fluorescent bacteria or India ink. Rabs associated with unique large structures, such as the contractile vacuole, could be assigned in the absence of any other compartmental marker. Movies showing the dynamic behavior of many of the GFP-Rab-labeled structures can be viewed at http://tetrahymenacell.uchicago.edu.

Since the Rabs are GFP-tagged and can be viewed in living cells, they offer the possibility of studying membrane dynamics in these cells. For example, phagocytosis in *Tetrahymena* and in human macrophages have many similarities but one difference is that, in the former, egestion of undigested contents in fully matured phagosomes occurs at a unique site on the plasma membrane called the cytoproct (Allen and Wolf, 1979). Several Rabs were found to associate only with phagosomes that were positioned right at the cytoproct, and time-lapse movies showed that egestion resulted in transient transfer of those Rabs to the PM and subsequent retrieval (Bright *et al.*, 2010). These "terminal Rabs" may be activated by proteins (e.g., Rab-GEFs or GTP-exchange factors) that are present at the cytoproct itself, so that some stages in phagosome maturation are influenced by cortical determinants. Another Rab is uniquely associated with what appear to be elongated vesicles being transported, primarily toward the cell anterior, along cytoplasmic microtubules that extend from the cytoproct region toward the oral apparatus. These vesicles may underlie the recycling of phagosome membrane components following retrieval of the phagosome membrane at the cytoproct, an actin-dependent process (Allen and Fok, 1980; Sugita *et al.*, 2009).

III. Studies on Membrane Lipids in *Tetrahymena*

While the discussion above has focused on the role of proteins in membrane traffic, it is also increasingly appreciated that lipids are not merely passive structural elements in cells but also key determinants. We therefore review recent work in Tetrahymena on three different aspects of lipids.

A. Phosphoinositides

Lipid-anchored phosphoinositols (which are called phosphatidyl inositols or PtdIns) can serve as important determinants of compartmental identity and are therefore key elements of membrane traffic. Their activity depends upon the fact that the inositol ring, when phosphorylated in specific combinations at the 3, 4, and/or 5 positions, can be recognized by large numbers of proteins, which can thereby be activated or recruited. In metazoa, this has important consequences for cytosolic signaling, membrane trafficking, nuclear events, cytoskeleton integrity, permeability, and transport (Di Paolo and De Camilli, 2006).

Because of their ease of culture, including the possibility of precisely defined growth medium, Tetrahymena were used in a large number of classical studies on lipid metabolism. More recently, several groups have focused on phosphoinositides in Tetrahymena species. Work in *Tetahymena vorax*, *Tetahymena pyriformis,* and *thermophila* was reviewed by Ryals in 2009, focusing primarily on biochemical aspects (Ryals, 2009). Our goal here is to view these data from a genetic/molecular perspective, focusing on *T. thermophila* and exploiting information that can be gathered using, for example, the genomic and gene expression databases of *T. thermophila* (TGD and TGED, respectively).

The metabolic pathways that generate inositol derivatives are schematized in Fig. 1. This figure, adapted from Michell (2008), shows the steps for which *T. thermophila* enzymes have been identified or can be inferred, as described below in the text.

To begin, cells can take up inositol from the environment or synthesize it from D-glucose-6-phosphate. Synthesis involves two enzymes: *myo*-inositol-3-phosphate synthase (MIPS), which catalyzes the cyclization of D-glucose-6-phosphate to D-*myo*-inositol-3-phosphate (Ins3P) (Michell, 2008), and inositolmonophosphatase (InsPase), which dephosphorylates Ins3P. While neither activity has been reported in *T. thermophila*, there are clear homologs encoded in the genome. TTHERM_00519810 shows a high homology ($E < 1.0 \times 10^{-120}$) with MIPS from

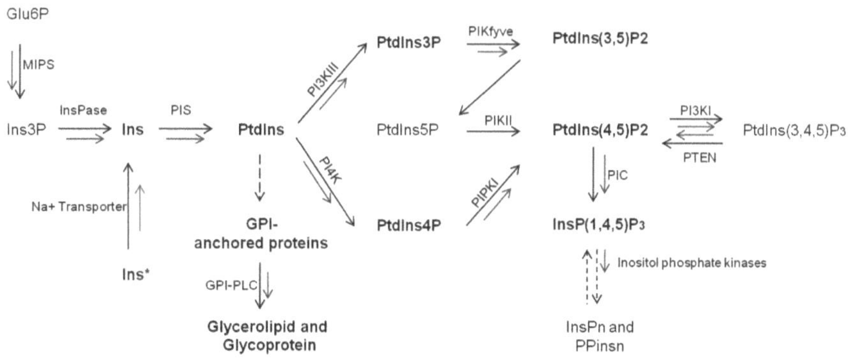

Fig. 1 Putative metabolic pathways of inositol derivatives in *Tetrahymena thermophila.* Blue arrows show reactions for which the corresponding enzymes appear present in the *T. thermophila* genome database. Compounds shown in bold indicate were identified in Tetrahymena. Ins* denotes environmental inositol. MIPS, *myo*-inositol-3-phosphate syntase; InsPase, inositolmonophosphatase; PIS, phosphatidylinositol synthase; PI3KI, class I phosphoinositide-3-kinase; PI3KIII, class III phosphoinositide-3-kinase; PI4K, phosphatidylinositol-4-kinase, PIKII, phosphatidylinositol 5 phosphate-4-kinase; PIPKI, phosphatidylinositol 4 phosphate-5-kinase; PIKfyve, phosphatidylinositol 3 phosphate-5-kinase; PTEN, phosphatidylinositol 3,4,5 triphosphate-3-phosphatase. The figure is modified from (Michell, 2008). (For interpretation of the references to color in this figure legend, the reader is referred to the web version of this book.)

plants (*Arabidopsis thaliana*), animals (*Homo sapiens*), fungi (*S. cerevisiae*), amobozoa (*Dictyostelium discoideum*), and kinetoplastids (Trypanosoma *cruzi*). The second enzyme, TTHERM_00318610, contains a highly conserved inositol monophosphate domain (PFAM PF00459). Both enzymes are expressed in growing, starved, and conjugating cell cultures (TGED) (Miao *et al*., 2009).

Environmental inositol import in *Tetahymena vorax* involves a sodium-dependent mechanism with features similar to those in other organisms (Ryals and Kersting, 1999). The genes are also likely to be similar, and homology searches identify a strong set of candidates (i.e., TTHERM_00852790, TTHERM_01080450 and TTHERM_00473200) that are related to other eukaryotic Na^+/*myo*-inositol symporters. Inositol exists as a number of stereoisomers, the most common of which is myo-inositol, but other isomers are also present in cells including *scyllo-*, *neo-*, *epi-D-chiro-*, and muco-inositols. In *T. vorax*, non-*myo*-inositols have also been detected, with evidence that these can be taken up from the medium (Kersting *et al*., 2003; Kersting and Ryals, 2004; Ryals and Kersting, 1999). Conversion between inositol isomers may also occur via inositol epimerases (Sun *et al*., 2002), though no conclusive genetic or biochemical data have been published in Tetrahymena to date. An open question is whether the free inositols in Tetrahymena have a function independent of their role as biosynthetic precursors. In other organisms, the presence of these organic solutes, together with other polyalcohols, has been linked to a cytoprotective response against environmental stress. For example, in mammalian kidney cells, inositols play a role in osmoregulation, while in plants and archaea they play a role in the stabilization of cellular proteins (Yancey, 2005).

Inositols are a substrate for PtdIns synthesis, carried out by phosphatidylinositol synthase (PIS) starting with inositol and cytidine diphosphate-diacylglycerol (CDP-DAG). This reaction has been characterized in *T. vorax* microsomes, where the substrates included both *myo*- and non-*myo*-inositol isomers (Riggs *et al.*, 2007). *T. thermophila* has a single PIS gene (TTHERM_00678350) whose expression profile is similar to the enzymes involved in the synthesis of inositol (MIPS and InsPase). Its high homology (E value between 1.0×10^{-20} and 1.0×10^{-35}) with PIS of species belonging to different eukaryotic supergroups supports its broad conservation, as previously reported (Michell, 2008).

Starting from PtdIns, a set of kinases and phosphatases are involved in the synthesis of phosphatidylinositol phosphates (PtdInsPs, also called phosphoinositides), generating seven possible species (see Fig. 1). The species actually present have been determined by HNMR (see below) (Leondaritis and Galanopoulou, 2000). PtdIns in *T. pyriformis* strain W, which makes up approximately 4% of total phospholipids, is found exclusively as diacylphospholipid (Pieringer and Conner, 1979), whereas other phospholipids (phosphatidylcholine (PC) and aminoethylphosphonoglyceride (AEPL)) are predominantly found as alkylacyl lipids (Leondaritis and Galanopoulou, 2000). Other differences between PtdIns and PC/PE in *T. pyriformis* include high myristic acid content, fully saturated acyl chains, and the absence of C18 fatty acid. PC and PE may have similar fatty acid content because they derive from a shared biosynthetic pathway, whereas PtdIns synthesis depends on a distinct phosphatidic acid pool for the generation of CDP- DAG (Leondaritis and Galanopoulou, 2000).

Work by Galanopoulou and colleagues on phosphoinositides in *T. pyriformis* and *thermophila* identified PtdIns(3)P, PtdIns(4)P, PtdIns(3,5)P2, and PtdIns(4,5)P2, but not PtdIns(5)P, PtdIns(3,4)P2, or PtdIns(3,4,5)P3 (Deli *et al.*, 2008; Leondaritis *et al.*, 2005). One major source of interest in phosphoinositides in mammalian cells comes from their role as compartment-specific determinants, based on the ability of individual phosphoinositide species to recruit proteins with the corresponding phosphoinositide-binding domains. As described earlier in this chapter, the pathways of membrane trafficking in ciliates depend on a network of Rab GTPase determinants as extensive as that in animals, so an interesting question is whether phosphoinositide isomers provide a second set of determinants in ciliates as they do in animals. However, we still lack information about localization of the identified phosphoinositide species in Tetrahymena. One important question is whether the individual phosphoinositides are concentrated in specific compartments or act in specific pathways. In many eukaryotes, PtdIns(3)P and PtdIns(3,5)P2 are determinants for endocytic traffic, and PtdIns(4)P has been implicated in maintaining Golgi structure and function (Di Paolo and De Camilli, 2006). A hint regarding phosphoinositide function in *T. thermophila* comes from studies using wortmannin, a specific inhibitor of phosphoinositide 3-kinases (PI3Ks), which reduces the levels of PtdIns(3)P and PtdIns(3,5)P2 but not D4-phosphoinositides. In *T. thermophila*, wortmannin treatment led to increased secretion of lysosomal enzymes (Kovacs and Pallinger, 2003; Leondaritis *et al.*, 2005). Interestingly, this enhancement was absent or reduced in two mutant strains that are deficient, respectively, in a late stage of secretion from lysosomes (MS-1), and in phagocytosis (A2). Unfortunately, neither the precise cell biological nor genetic deficiencies are known in these mutants, but the fact that the mutations attenuate the effect of wortmannin may indicate, as suggested by the authors, that an InsPtd phosphorylated at the 3-position primarily functions in phagosomes/phagolysosomes A complicating factor is that the MS-1 mutant shows elevated levels of PtdIns(4)P, which could be a direct or indirect effect of the mutation (Deli *et al.*, 2008). Wortmannin treatment also significantly inhibited phagocytic activity in *T. pyriformis,* possibly due to inhibition of actin polymerization (Kovacs and Pallinger, 2003). Further studies, particularly if more fully characterized mutant strains become available should reveal more details of pathway regulation by phosphoinositides.

The enzymes involved in phosphoinositide synthesis have also been investigated, at least by informatics approaches. Based on work in other eukaryotes, phosphoinositide-3-kinase activity can be encoded by enzymes belonging to three classes, all of which have a PI3K enzymatic core inhibited by wortmannin but which diverge in other structural features. Importantly, because the three classes use overlapping but not-identical substrates, they can generate different 3-phosphoinositides that consequently recruit or activate distinct effectors (Vanhaesebroeck *et al.*, 2010). In the genome of *T. thermophila*, four putative PI3Ks have been identified, three of which bear structural features of group I enzymes (TtPI3K 1-3: TTHERM_00655270, TTHERM_00323020, TTHERM_00951960) and one of group III (TtPI3K III (TTHERM_00649380) (Leondaritis *et al.*, 2005). Interestingly, the Class I enzymes appear to be absent in plants, fungi, and many other protozoa (Michell, 2008). All four *T. thermophila* genes are expressed in growing, starved, and conjugating cell cultures, but with different expression patterns suggesting nonredundant functions. The only published information on the roles of these putative PI3K genes comes from pharmacologic studies. Inhibition of PI3K activity by wortmannin, LY294002 and 3-methyladenine (which do not distinguish between the three groups of PI3Ks (Vanhaesebroeck *et al.*, 2001; Wu *et al.*, 2010)) blocked programmed nuclear degradation (PND) in conjugating *T. thermophila* leading to accumulation of additional micronuclei and macronuclei (Yakisich and Kapler, 2004). The authors proposed that a product of PI3K is associated with PND activation and the degradation of nonexchanged pronuclei and later the macronucleus, and suggested that the active species may be PtdIns(3,4,5)P3. Understanding the precise role of PI3K in PND is complicated by the uncertain specificity of the available inhibitors, but future work may be able to illuminate noted similarities between PND in *Tetrahymena* and autophagy in other organisms, a pathway in which different classes of kinases are known to act (Wu *et al.*, 2010). It may also be useful to consider enzymes that potentially degrade or convert PI3K. We detected four genes encoding putative homologs of PTEN, a phosphatase responsible for degrading PtdIns(3,4,5)P3 to PtdIns(4,5)P2 (TTHERM_00538980, TTHERM_00313160, TTHERM_00421160, and TTHERM_00467300). Interestingly, the expression of three of these is maximal during the stages of conjugation (C6 and C8) where PND occurs.

Another way to interrogate the roles of phosphoinositides is to identify proteins with putative phosphoinositide-binding domains. Many such proteins appear to be encoded in the *T. thermophila* genome. We used a domain-based search, with the SMART database in Genomic Mode (Letunic *et al.*, 2006), to identify 49 genes possessing a pleckstrin homology (PH) domain, 39 possessing a phox homology (PX) domain, and 7 possessing an FYVE domain. In other searches, we uncovered additional elements of the *T. thermophila* phosphoinositide genetic toolkit. For example, PIKfyve, which produces PtdIns(3,5)P2 from PtdIns3P, is present in the genome of *T. thermophila* in a single copy (TTHERM_01005090). In addition, several putative homologs can be found for PI4K, which converts PtdIns to PtdIns (4)P, and PIPKI, which converts PtdIns(4)P to PtdIns(4,5)P2, (Fig. 1). These suggestive hits, which will need to be confirmed using genetic and biochemical approaches, suggest that *T. thermophila* could be an excellent model system for understanding how phosphoinositides contribute to the organization of complex cells and could be particularly useful for investigating the Class I PI3K, as noted above.

In addition to their role as membrane-linked determinants, PtdIns are used as substrates to generate soluble molecules acting as secondary messengers. In particular, metazoans activate phospholipase C (PLC) via G protein-coupled receptors (GPCR) to hydrolyze PtdIns(4,5)P2, forming the second messengers Ins(1,4,5)P3 and DAG, with downstream effects on Ca^{2+} mobilization and protein phosphorylation. Outside of the metazoa, there is little convincing evidence linking PtdIns(4,5) P2 to these functions (Michell, 2008). However, the picture is becoming clearer in ciliates, through the study of PLC in *Tetrahymena* and the Ca^{2+} release channels (CRC) in *Paramecium*. Though beyond the scope of this chapter focused on membrane traffic, both bacterial-like and eukaryotic PI-PLCs have been identified in *T. thermophila* and *pyriformis* and putative homologs in both classes are encoded in the *T. thermophila* genome (Leondaritis *et al.*, 2011). The bacterial PI-PLCs, which are likely to have been acquired via lateral gene transfer, may function in hydrolysis of the ciliate GPI anchors and the degradation of phospholipids in the extracellular space, along with other phospholipases (Florin-Christensen *et al.*, 1986).

B. Sterol Metabolism

Sterols affect membrane fluidity and permeability (Ohvo-Rekila *et al.*, 2002). In addition, they are essential components of the "lipid rafts" that have been characterized principally in animal cells, which are currently understood as membrane microdomains whose formation depends upon the affinity of sterols for sphingolipids. The partitioning of proteins in lipid rafts may be important for regulation of signal transduction pathways (Simons and Toomre, 2000). Sterols also serve as precursors of bile salts and steroid hormones in mammals, brassinosteroids in plants, and fungi and ecdysteroids in arthropods.

Eukaryotic organisms can satisfy their sterol requirement by *de novo* synthesis in vertebrates (cholesterol), plants (stigmasterol, sitosterol, and campesterol), and fungi (ergosterol), or by obtaining them from food. Sterol auxotrophs include invertebrates (nematodes and arthropods), some ciliates (*Paramecium tetraurelia*), apicomplexans (*Plasmodium falciparum*), and some flagellated parasites (*Giardia intestinalis* and *Trichomonas vaginalis*). *T. thermophila* is unusual in this regard, having no detectable sterols in its membranes and, accordingly, no sterol requirement. Instead, it synthesizes tetrahymanol, a compound similar to hopanoids found in bacteria, which acts as a surrogate sterol. However, when sterols are added to the growth medium, tetrahymanol synthesis is suppressed and *T. thermophila* incorporates the exogenous sterol, either with or without modifications(Conner *et al.*, 1968). In particular, the ciliate desaturates sterols at positions C5(6), C7(8), and C22(23) and removes the C24 ethyl group in C29 sterols (phytosterols) (Mallory and Conner, 1971). By the activity of these three sterol desaturases (C-5, C-7, and C-22 sterol desaturases) and C-24 sterol deethylation, the ciliate modifies exogenous sterols and accumulates the tri-unsaturated products in its membrane.

C-22 sterol desaturases have been characterized in other eukaryotes. In *T. thermophila*, the C-7 and C-22 sterol-desaturating activities, found mainly in a microsomal fraction, require cytochrome b_5 as shown by their inhibition with azide and cyanide (Nusblat *et al.*, 2005; Valcarce *et al.*, 2000). This cytochrome b_5 dependence is not characteristic of the C22 desaturases of plants and fungi, which require cytochrome P450. The difference is underscored by the insensitivity of the ciliate C22 desaturase to azole, a compound that strongly inhibits the corresponding plant and fungal activities. Moreover, no clear orthologs can be found in the *T. thermophila* genome for known C-22 sterol desaturases (Morikawa *et al.*, 2006). These observations suggest that the *T. thermophila* enzyme represents a new class of C-22 sterol desaturases.

The C-5 sterol desaturase present in most eukaryotic cells belongs to the fatty acid hydroxylase (FAH) superfamily of integral membrane proteins that bind an iron cofactor via a 3-histidine motif. The C-5 sterol desaturase in *T. thermophila*, *DES5A*, was identified by characterizing the phenotype resulting from deletion of a putative FAH gene (Nusblat *et al.*, 2009). The deletion mutant, which was fully viable, showed strongly diminished C-5 sterol desaturase activity, while C-7(8) and C-22 (23) desaturase activities were unaffected.

The gene involved in C-24 sterol deethylation, *DES24*, was similarly confirmed by the disruption of putative FAH genes (Tomazic *et al.*, 2011), resulting in a strain unable to eliminate the C-24 ethyl group from different phytosterols, and probably defective at the first step in dealkylation. Interestingly, the mutant strain was highly sensitive to phytosterols in the culture media, showing defects in growth and morphology and altered tetrahymanol biosynthesis. This observation suggests that C29 sterols can impair the normal growth of *Tetrahymena*. While C-24 sterol deethylation activity has been characterized in other organisms including nematodes, arthropods, and green algae, the *Tetrahymena* enzyme represents the first molecular characterization. However, *DES24* clusters phylogenetically with bacterial FAH sequences of unknown function, with no obvious orthologs in other eukaryotes, and may therefore have been acquired by lateral transfer. A variety of other observations, including substrate specificity and inhibitor studies, are also consistent with the hypothesis that the mechanism of *T. thermophila* C-24 deethylation differs from that in other eukaryotes.

T. thermophila, which is exposed in its environment to phytoplankton, higher plants and algae, may have developed the ability to metabolize otherwise-harmful phytosterols upon acquisition of *DES24* from bacteria. Interestingly, however, *Paramecium tetraurelia* does not have C-24 dealkylation activity (Conner *et al.*, 1971) and requires phytosterols (Whitaker and Nelson, 1987). Overall, sterol metabolism in *T. thermophila* seems to be the evolutionary product of a fascinating combination of gene losses (e.g., typical eukaryotic genes involved in sterol biosynthesis) combined with acquisition of bacterial genes to allow for synthesis of unusual compounds, with potentially novel mechanisms of sterol modification. This evolutionary history may be illuminated by interrogating the genomes of other Tetrahymena species as these are sequenced. In addition, further studies of the sterol pathways in *T. thermophila* may yield more information about lipid diversity and function.

C. Role of Lipids in Membrane Curvature

Membrane fusion occurs when two separate lipid membranes merge into a single continuous bilayer. It underlies all membrane traffic as well as other important intra- and inter-cellular phenomena. The propensity of lipid bilayers to fuse *in vitro* is sensitive to lipid composition. One potential factor is that different lipids prefer, from an energetic perspective, to form surfaces with specific curvatures, and curvature affects fusogenicity in experimental pure lipid systems. Cone-shaped lipid like phosphatidylethanolamine (PE) and diacylglycerol (DAG) induce negative spontaneous curvature whereas inverted cone-shaped lipids like lysophosphatidylcholine (LPC) can induce positive spontaneous curvature. On the other hand, cylindrical phosphatidylcholine (PC) forms an almost flat monolayer. However, real biological membranes are also rich in proteins, and assessing the relative contributions of proteins and lipids to membrane fusion is a long-standing challenge. One issue has been whether lipids can drive changes in membrane curvature or simply accommodate changes that are driven by proteins. An approach that has been pioneered in *Tetrahymena* is to examine the distribution of lipids with fine resolution in subcellular membranes of defined curvature.

T. thermophila cultures can be induced to undergo synchronous mating, during which they form conjugation junctions containing hundreds of fusion pores in a small, well-defined zone, though which micronuclei are exchanged between paired cells (Wolfe, 1982, 1985). During formation of this zone, local lipid composition could change due to either *de novo* lipid synthesis, known to be required during conjugation, or to lipid exchange between cellular membranes. Ewing and colleagues exploited synchronous conjugation in *Tetrahymena*, combined with secondary ion *mass spectrometry (SIMS),* to ask whether the lipids in the fusion zone were enriched in species predicted to favor positively curved membranes, and also depleted in lipids whose shapes would resist such curvature (Ostrowski *et al.*, 2004). The SIMs technique allows for visualization of the spatial distribution of molecular species according to the mass/charge ion ratio (Murphy *et al.*, 2009). (For a detailed description of the technique, see Heien *et al.* (2010).) The results indicated that the mating junction has a lower concentration of phosphatidylcholine, relative to the cell body, as expected since PC tends to favor flat membranes. In contrast, the mating junction contained a higher concentration of 2-aminoethylphosphonolipid, a phosphonolipid analog of phosphatidylethanolamine whose cone-shape would favor highly curved membranes.

These results, important in showing a set of predicted deviations within the lipid composition of an *in vivo* fusogenic zone, could not address the question of whether such lipid domains were present prior to or following the formation of the fusion pores. If the former, the lipids could be acting as fusion determinants; if the latter, the change in lipid composition could be accommodating the curvature imposed by other mechanisms, for example, proteins. In a second paper, the same group addressed this by studying pairs during a time course of *Tetrahymena* mating, once again exploiting the synchronicity that can be achieved in the laboratory, and using the stability of formed pairs as a proxy for whether a zone of fusion pores had formed, based on earlier EM studies (Kurczy *et al.*, 2010). The results suggested that change in lipid composition of the mating cell junction occur after fusion pores have formed, supporting a model in which changes in lipid composition arise subsequent to structural changes that are imposed by proteins. These studies, although still based on correlations, represent a beautiful example of using advanced technology to build upon the wealth of classical studies in *T. thermophila*, and using unique features to address fundamental questions in cell biology. If fusion pore formation is driven by proteins that are selectively expressed during conjugation, the identification of such proteins and disruption of the corresponding genes could facilitate future studies on lipid composition in which the correlations established in the studies described above could be tested by direct manipulation.

Space limitations prevent us from discussing other interesting work in the lipid field, including studies on sphingolipids, phospholipase D, and endocannabinoids (Wang *et al.*, 2001, 2002) (Anagnostopoulos *et al.*, 2010).

IV. Conservation Versus Innovation

With the data outlined above, one can begin to ask questions about the extent of evolutionary innovation to generate complex pathways of membrane traffic. A view ensconced in many textbooks is that modern cells are overwhelmingly similar, their shared features and pathways reflecting the shared inheritance from a common ancestor. A potential problem with such a blanket conclusion is that it stems from cell biology studies that have historically been conducted on a narrow swath of eukaryotic diversity, with all animal and fungal "model organisms" belonging to a single lineage, the Opisthokonts (Parfrey *et al.*, 2006). In this regard, cell biological studies in higher plants are of great value since these constitute a more divergent evolutionary branch. The greater divergence means that gain-of-function mutations in membrane trafficking determinants may have arisen and been positively selected following the split from Opisthokonts. Similarly, Ciliates represent another deeply divergent eukaryotic branch. The question in such divergent lineages is not whether innovations occurred (since they must have), but how they may have shaped specific pathways. An exceptionally interesting example of radical innovation was illustrated in recent work showing that *T. thermophila*, and probably all Alveolates, has invented a novel means to operate the mitochondrial ATP synthase complex that was previously believed to be conserved in structure and mechanism throughout eukaryotes (Balabaskaran Nina *et al.*, 2010).

With regard to membrane traffic, one potential example of innovation generating a largely novel pathway in Tetrahymena, rather than simply tweaking a pre-existing pathway, is regulated secretion, that is, the synthesis of the dense-core secretory vesicles called mucocysts. A large number of protein components of mucocysts have been deduced using biochemical and genetic approaches, as well as by identifying *T. thermophila* homologs of proteins required for trichocyst exocytosis in Paramecium (reviewed in (Bowman *et al.*, 2005b; Turkewitz, 2004)) (A. Turkewitz, unpublished). While some of these proteins contain identifiable domains, for example, β/γ crystallin domains, none of the proteins has an identifiable homolog outside of ciliates, with some possible weak exceptions in the related Apicomplexans. Thus all of the identified components of dense core vesicles in ciliates appear to reflect mutations that occurred after ciliates (or perhaps Alveolates) had branched from other organisms. Furthermore, it appears that the endoproteases responsible for processing of the dense core vesicle proproteins (*GRL*-encoded in Tetrahymena) are not related to the endoproteases that serve the homologous function in mammalian endocrine dense core granules (P. Romei and A. Turkewitz, unpublished). A tentative conclusion, based on these data, is that the striking functional similarities between the regulated secretory pathways in animals and ciliates primarily reflect independent innovation in the two lineages, shaped by similar selective pressures (Elde *et al.*, 2007). However, such conclusions should be considered tentative. First, the genes that have been identified to date may reflect biases in the genetic and biochemical methods used. Secondly, there is no information yet available on the cellular machinery involved in mucocyst synthesis, so one possibility is that animals and ciliates have both adapted the same conserved biosynthetic machinery to create dense core vesicles, albeit from different ingredients.

The analysis of endocytosis from parasomal sacs revealed significant conservation between ciliates and animals in a clathrin-dependent pathway that also involved AP-2 (Elde *et al.*, 2005). Another apparent similarity is the endocytic involvement of dynamin in both lineages, but phylogenetic analysis revealed that this conservation has an innovative twist. In particular, the phylogenetic reconstruction argued that dynamin existed in a common ancestor of ciliates and animals, but that ancestral dynamin was unlikely to be involved in endocytosis (Elde *et al.*, 2005). Instead, independent mutations in animals and ciliates subsequently led to the targeting of a dynamin paralog (i.e., a dynamin gene arising from a gene duplication within each lineage) to the endocytic pathway. An inference of this analysis was that the mechanism of targeting of the animal and ciliate dynamins could be different. While the ciliate mechanism is not yet known, the targeting motif identified in Drp1p does not resemble the known motif in the mammalian dynamins, consistent with the hypothesis (Elde *et al.*, 2005). An interesting question is why dynamin was recruited for endocytosis in *Tetrahymena*, since this does not appear to have occurred in many other protist lineages. As discussed above, *Tetrahymena* may be very unusual in not using dynamic actin during endocytosis, and dynamin may be providing a function contributed by actin in animal cells.

A broader but shallower dataset to assess the relative contribution of innovation to membrane traffic exists in the Rab GTPase survey discussed above. Rabs are particularly well suited to addressing such questions because they determine compartmental identity, and must have co-evolved with their associated compartments (Pereira-Leal and Seabra, 2001). Consistent with this idea, the phylogenetic and functional comparison of yeast and human Rabs confirms that sequence-relatedness correlates with functional relatedness (Pereira-Leal, 2008). This indicates that many Rabs existed in the common ancestor of yeast and humans, associated with compartments that were retained in both the fungal and animal lineages.

Roughly one-fourth of the *T. thermophila* Rabs, falling into six major branches, are conserved with homologs in distant lineages (Bright *et al.*, 2010; Turkewitz and Bright, in press). Five of these six branches have previously been argued to represent "core Rabs" (Dacks and Field, 2007). The human Rabs in these branches are Rabs 4, 5, 7, 11, and 21 (all associated with stages of endocytosis), Rab 1 (associated with ER-to-Golgi traffic), and Rab 6 (associated with retrograde Golgi traffic). *Tetrahymena* appears to be missing Rabs in the three other identified core groups, corresponding to Golgi-related (two clades), and regulated exocytic pathways. This may reflect lineage-restricted loss, but could be an artifact of failing to detect true homologs due to excessive sequence divergence. (Note that the absence of a *Tetrahymena* Rab in the regulated exocytic branch would be consistent with independent evolution of mucocysts and secretory vesicles in animals.) Two *Tetrahymena* Rabs fall into a robust branch with human Rab32 as well as RabE in *D. discoideum*, suggesting that this group may be a ninth highly conserved Rab clade. Rab32 has been associated with several different organelles in mammals, including mitochondria and lysosome-related organelles (Tamura *et al.*, 2009).

The fact that most conserved *Tetrahymena* Rabs belong to endocytic clades suggests that much of the endocytic machinery in *Tetrahymena* was inherited from an ancient eukaryotic ancestor, but this does not tell the whole story. Of the Rabs that were experimentally determined to be associated with endocytic compartments based on co-localization with FM4-64, roughly half belonged to conserved endocytic clades, while the rest were highly divergent (RabsD4, D5, D24, D27, D28, and D35) (Bright *et al.*, 2010). This suggests that a substantial part of the expansion within Rabs during *Tetrahymena* evolution was devoted to lineage-specific adaptations to endocytic pathways. A surprising observation was that four Rabs (Rabs4A, 4B, 11B, and 31) that were assigned to the endocytic clade based on sequence did not co-localize with FM4-64. This suggests that some Rabs may have retained the sequence signatures of conserved clades but have changed their compartmental localization. If such role-switching has indeed occurred, such unexpected plasticity in sequence-conserved Rabs would mean that inferring Rab function in divergent lineages simply based on sequence may sometimes be misleading. However, it is important to note that robust phylogenetic clustering of *Tetrahymena* Rabs required that the hypervariable C-terminal domains be excluded during the tree building. Since these domains may contain targeting information, the role-switching may have been driven by specific C-terminal mutations that would be invisible to the phylogenetic methods used by both groups that have analyzed this important gene family to date.

Almost one-third of the *T. thermophila* Rabs localized to phagosomes or structures associated with the phagocytic pathway, that is, the oral apparatus and cytoproct (Bright *et al.*, 2010). A similarly large number of Rabs have been associated with phagosomes in mammalian cells (Smith *et al.*, 2007a). Given the results of the phagosome proteome project cited above (i.e., 28/73 *Tetrahymena* proteins were homologous to putative phagosome proteins in other systems), the expectation was that a large set of *Tetrahymena* and mammalian phagosomal Rabs would be mutually orthologous. However, only two of the phagosomal Rabs appeared orthologous (Bright *et al.*, 2010). Moreover, these two Rabs are also associated with late endosomal compartments, so the orthology may reflect conservation in endocytic pathways that also intersect with the phagocytic pathway. In this regard, it will be important to learn what fraction of the phagosome proteome constituents are restricted to phagosomes. The Rab data, taken by itself, do not strongly support a common origin for the phagocytic pathways in ciliates and animals. However, it is also possible that the failure to detect orthology is due to the limitations of phylogenetic analysis for highly divergent lineages such as ciliates.

As implied by the discussion above, it is not always straightforward to generate robust phylogenetic trees using *Tetrahymena* gene sequences. An example of this can be seen in a comparison between two analyses of the myosin family in *T. thermophila*, in which the earlier survey underestimated the similarity of *Tetrahymena* myosins to those in other eukayotes (Sugita *et al.*, 2011; Williams and Gavin, 2005). A very important issue in analyzing members of gene families is being able to distinguish orthologs (homologs that diverged in sequence following a speciation event, and which often retain the same function) versus paralogs (homologs that diverged within a species and often diverged in function). While the two papers surveying *Tetrahymena* Rabs reach identical conclusions for many of the family members, they also differ in some cases. This is both because the datasets are slightly different and because different criteria were used to assign orthology. Some of the differences have been resolved, but researchers using phylogenetic tools to analyze *Tetrahymena* genes should be aware that some standard approaches may yield ambiguous or spurious results when applied to such divergent sequences, so collaboration with experts may be useful. In the relatively near future, some aspects of phylogenetic analysis should be simplified when additional Tetrahymena species genomes are sequenced.

Additional Paper of Interest

1. Genetic and functional analysis of the septins, a family of three genes. The data indicate that the Tetrahymena septins are primarily involved in

mitochondrial functions. Though septins are found throughout eukaryotes, such mitochondrial roles have previously been found only in mammals, suggesting independent recruitment for similar functions in mammals and ciliates (Wloga *et al.*, 2008).

V. Using Expression Data to Elucidate Pathways of Membrane Traffic

To better understand both mechanistic and evolutionary aspects of membrane traffic in *Tetrahymena*, it will be important to assemble a more complete parts list of the proteins associated with specific compartments and pathways. While significant progress has been made by simply pursuing *Tetrahymena* homologs of relevant animal and fungal proteins, as outlined above, this approach is clearly limited, in part because distinguishing orthologs from paralogs is not always possible. Secondly, the elucidation of membrane traffic is far from complete even in the best-studied systems like budding yeast, so limiting oneself to a homology-based approach would eliminate any contribution that *Tetrahymena* could make to identifying new factors in membrane traffic. One less biased approach would be to use biochemical approaches to identify Rab-interacting proteins starting with Rabs associated with specific compartments, which has been a powerful approach in mammalian cells (Christoforidis and Zerial, 2000).

A relatively novel approach that appears promising as a tool in *Tetrahymena* is based on a systems biology approach in mammalian cells, namely correlating the transcriptional profiles of genes involved in membrane trafficking factors. Balch and colleagues compiled microarray expression data from a large number of different tissues and cell lines, collected under a wide range of conditions, and calculated the degree of co-regulation between genes known to be involved in membrane trafficking (Gurkan *et al.*, 2005). Based on the observed patterns of co-regulation, the authors proposed that membrane trafficking events are orchestrated by Rab-regulated protein “hubs” that are transcriptionally linked to the machinery involved in processes including coat formation, tethering and membrane fusion at those hubs. To test the significance of the observed co-regulation, the authors turned to the extensive biochemical and genetic data on membrane trafficking available in both mammalian and fungal systems. In some cases, these experimental data strongly supported the idea that co-regulated genes encoded products that were associated with the same hub. While the significance of many of their findings remains to be tested, the suggestion is that expression data, which are relatively simple to collect, could potentially be used to identify novel components of hubs, or to identify new hubs. Moreover, the data could also give hints regarding the conditions under which specific hubs are most physiologically significant.

In *T. thermophila*, whole-genome microarrays have been used for a variety of purposes, including to identify genes upregulated during induced synthesis of mucocysts (L. Bright and A. Turkewitz, unpublished). A particularly rich dataset, referred to above, was collected by Gorovsky and colleagues by sampling *Tetrahymena* cultures under growing and starved conditions as well as during conjugation (Miao *et al.*, 2009). The processed data are publicly available, curated by Miao and colleagues, in a format in which co-expressed genes can be identified based on Pearson correlation coefficients (http://tged.ihb.ac.cn/). More recently, this group has used additional approaches to recognize co-regulation within the dataset (Xiong *et al.*, 2011).

We have begun to ask whether the *Tetrahymena* expression database can provide insights into the organization of membrane traffic in this organism. A hint that this might be true was mentioned earlier, namely that the genes encoding all known protein components of *Tetrahymena* mucocysts are co-regulated (Rahaman *et al.*, 2009). However, this represents a very specific example and is limited to the proteins residing in the mucocyst lumen and membrane, rather than the still-unknown proteins involved in mucocyst biosynthesis. To pursue the larger questions, we have begun by asking whether proteins involved in membrane trafficking are co-regulated according to the Gorovsky expression database. Heterotetrameric adaptor complexes, such as AP2 discussed above, play highly conserved roles including the recruitment of coat proteins to membranes during vesicle budding. *T. thermophila* has five predicted AP μ subunits, including two μ subunits that fall into the AP1 cluster, and one each of AP2, AP3, and AP4, which would be expected to act at different sites and with different effectors. In the terms of Balch and colleagues, each AP complex represents a different hub. In Table I, we show the results of querying the database to identify genes that are coregulated with each of these micron subunits. In each case, we started with the APμ gene (here called *APM*, to conform to *T. thermophila* genetic nomenclature) as a "seed" and asked what genes in the dataset were most highly co-regulated, judging by the Pearson correlation coefficients. We then analyzed the hits using BLAST searches and gene ontology databases. We considered all hits with correlation coefficients $\geq$0.85, based on data presented on co-regulation in the whole genome dataset (Miao *et al.*, 2009).

Many of the results are intriguing. For example, all of the top co-regulated genes with the AP1 μ subunit encode other genes involved in membrane traffic, including two dynamin-family proteins and the ε subunit of the AP4 adaptor. For a second example, the genes co-regulated with APM3 include multiple sortilin homologs, which are receptors known to be sorted in an AP3-dependent mechanism in other organisms, while genes co-regulated with APM4 include multiple coatamer subunits and a collection of actin-binding proteins. Interestingly, while the majority of coregulated genes are nonoverlapping between the five lists, a small number of genes (whose TTHERM IDs are emboldened in the table) appear to be co-regulated with more than one of the AP μ genes. One of these is a SNARE, a protein involved in membrane fusion, while a second is NSF, a complex required for SNARE disassembly. While the significance of the hits has not been tested, these preliminary results suggest that exploiting *Tetrahymena* expression databases may be a rich new vein to explore for cell biologists interested in membrane traffic.

Table I
Genes coregulated with the micron subunit genes of five heterotetrameric adaptors in *T. thermophila*[a]

Coregulated gene name (or putative)	Ttherm ID	PCC
APM1A query (TTHERM 01108620)		
Drp1, dynamin-related protein 1	_00486790	0.9434
Drp2, dynamin-related protein 2	_00188910	0.9433
Adaptor-related protein complex 4, epsilon 1	_00316010	0.9339
SNARE; syntaxin 5-2	**_00558250**	0.9298
Rab6C GTPase	_00079900	0.9042
N-terminal YjeF-domain	**_00149620**	0.9021
RabD4 GTPase	_00825210	0.8965
Ubiquitin carboxyl-terminal hydrolase family	_00077370	0.8861
ARF5 GTPase	_00735240	0.8846
RCC1-domain protein	_00586630	0.8793
Endonuclease/exonuclease/phosphatase family	_00316140	0.8783
Protein kinase domain-containing	_00128930	0.8760
Rab GGTB qeranylgeranyltransferase	_00630460	0.8739
Yip1 domain family (GDI displacement factor)	_00193890	0.8737
HECTdomain and RCC1-like domain	_00586640	0.8721
NSF (N-ethylmaleimide-sensitive factor)	**_00039210**	0.8718
von Willebrand factor type A domain-containing	_00809300	0.8698
TRAF-type family	_00666410	0.8695
DHHC zinc finger domain-containing	_00581830	0.8683
Adaptor-related protein complex 1, gamma 1	_00715750	0.8673
APM1B query (TTHERM 00455300)		
Phosphatidylinositol qlycan, class Q (PIGQ)	_00945240	0.9392
RabD40 GTPase	_01129680	0.9346
Hydrolase, NUDIX family	_01084270	0.9214
TB2/DP1, HVA22 family	_00270350	0.9184
Adaptor-related protein complex 4, beta-1	_00225780	0.9141
NSF (N-ethylmaleimide-sensitive factor)	**_00039210**	0.9078
SNARE; syntaxin 5-2	**_00558250**	0.8983
HIT domain-containing	_00378790	0.8978
N-terminal YjeF-domain	**_00149620**	0.8977
Oxidoreductase	_00334360	0.8941
Eukaryotic phosphomannomutase family	_00059380	0.8860
Protein kinase domain-containing	_00715960	0.8851
Adaptor-related protein complex 4, sigma 1	_01227730	0.8843
Adaptor-related protein complex 1, gamma 1	_00287890	0.8783
RabX23 Rab-like protein; GTPase	_00298510	0.8676
Adaptor-related protein complex 4, mu 1	**_00545860**	0.8604
Hydrolase, NUDIX family	_00113100	0.8536
ADP-ribosylation factor	_00448910	0.8528
EFTu C-terminal domain-containing	_01234330	0.8505

(Continued)

Table I *(Continued)*

Coregulated gene name (or putative)	Ttherm ID	PCC
APM2 query (TTHERM 00577030)		
Fimbrin-like	**_00136510**	0.9148
Succinate dehydroqenase/fumarate reductase	_00241700	0.9073
Proteasome A-type and B-type family	_00487110	0.9072
Mov34/MPN/PAD-1 family	_00049450	0.9066
Proteasome A-type and s-tvoe family	_00147560	0.9044
Phosphoglycerate mutase 1 family	_00641240	0.9021
Ubiquinol-cytochrome c reductase	_00295080	0.8969
Actin family	**_01020680**	0.8962
Adaptor-related protein complex 1, beta 1	_00219370	0.8959
Protein phosphatase 2C-containing	_00446510	0.8955
Coatomer protein complex, subunit beta 1	_00488350	0.8937
V-type ATPase, D subunit	_00821870	0.8936
EF hand-containing	_00059190	0.8917
Proteasome A-type and B-type family	_00316530	0.8912
Cytochrome C1 family	_00918500	0.8908
DnaJ domain-containing	_00195890	0.8888
ATP synthase F1, delta	_00684790	0.8886
Ribosomal protein L13	_01207660	0.8884
Proteasome A-type and B-type family	_00043880	0.8834
IQ calmodulin-binding motif family	_01194640	0.8822
APM3 query (TTHERM 00572100)		
Adaptor-related protein complex 3, beta 1	_00703490	0.9581
Protein kinase domain-containing	_01015890	0.9467
VPS10 domain (sortilin 2)	_00410210	0.9428
Protein kinase domain-containing	_01052880	0.9378
VPS10 domain (sortilin 4)	_00313130	0.9350
MAC/Perforin domain-containing	_01141420	0.9231
Rab4B GTPase	_01097960	0.9190
Thioredoxin (TRX) family	_00664030	0.9175
Golgin; GRIP domain-containing	_01143870	0.9168
Eukaryotic aspartyl protease family	_00440510	0.9156
SNARE; synaptobrevin 3	_00152090	0.9116
Amidase family	_00361390	0.9086
Taurine catabolism dioxygenase TauD, TfdA family	_00823760	0.8995
Rab21A GTPase	_00540110	0.8758
HAD-superfamily hydrolase, subfamily IIA	_01093670	0.8722
Histidine acid phosphatase family	_00649220	0.8666
Protein kinase domain-containing	_01164090	0.8638
Calpain family cysteine protease	_00196650	0.8618
Mitochondrial carrier protein	_00637710	0.8609

(Continued)

Table I *(Continued)*

Coregulated gene name (or putative)	Ttherm ID	PCC
APM4 query (TTHERM 00545860)		
Oxidoreductase family	_00334360	0.9517
ARP2/3 complex 20 kDa subunit (ARPC4)	_00857810	0.9397
Amidase family	_00361450	0.9383
Coatomer zeta coat	_00052600	0.9378
EF hand-containing	_01055480	0.9364
F-actin capping protein, beta subunit-containing	_00784430	0.9339
Coatomer epsilon subunit family	_00459320	0.9304
Actin family	**_01020680**	0.9299
Fimbrin-like 71 K	**_00136510**	0.9291
ADP-ribosylation factor	_00340150	0.9286
Signal peptidase I family	_01093560	0.9283
Peptidyl-prolyl cis-trans isomerase	_00548130	0.9279
Peptidyl-prolyl cis-trans isomerase	_00532620	0.9239
Phosphatidylserine decarboxylase family	_00859340	0.9221
HCaRG	_00059120	0.9205
Coatomer protein complex, gamma 2	_00444320	0.9199
Coatomer alpha	_01345820	0.9196
Endomembrane protein 70-containing	_00723110	0.9187
Roadblock/LC7 domain-containing	_00348650	0.9180
EF hand-containing	_00497600	0.9177

[a] We list, in order of decreasing correlation coefficients, all coregulated genes with Pearson's correlation coefficients $\geq$0.85 for each micron subunit (to a maximum of 20, for reasons of space). PCC = Pearson's Correlation Coefficient. Not included in these lists are coregulated genes for which no functions or conserved domains can be specified (denoted "hypothetical protein" in NCBI). The number of such hypothetical genes with correlation coefficients equal or larger than those of the genes shown is as follows: for *AMP1A*, 14; for *APM1B*, 35; for *APM2*, 21; for *APM4*, 24; for *APM5*, 55.

The TTHERM IDs of genes that appear more than once in the table, that is, that appear to be coregulated with more than one adaptor, are emboldened.

Acknowledgments

We thank Joe Briguglio for insightful comments on the manuscript and, and Mimi Pixque for support. Work in APT's laboratory is supported by NSF MCB0422011 and NIH GM077607. ADN was supported by the International funding program of the National Scientific and Technical Research Council (CONICET) of Argentina. L.J.B. was supported by Genetics and Regulation Training Grant T32 GM007197. APT is a member of the scientific advisory board of TetraGenetics, Inc.

References

Akematsu, T., and Endoh, H. (2010). Role of apoptosis-inducing factor (AIF) in programmed nuclear death during conjugation in *Tetrahymena thermophila*. *BMC Cell Biol.* **11**, 13.

Akematsu, T., Pearlman, R. E., and Endoh, H. (2010). Gigantic macroautophagy in programmed nuclear death of *Tetrahymena thermophila*. *Autophagy* **6**, 901–911.

Aldag, I., Bockau, U., Rossdorf, J., Laarmann, S., Raaben, W., Herrmann, L., Weide, T., and Hartmann, M. W. (2011). Expression, secretion and surface display of a human alkaline phosphatase by the ciliate *Tetrahymena thermophila*. *BMC Biotechnol.* **11**, 11.

Allen, R. D. (1967). Fine structure, reconstruction and possible functions of components of the cortex of *Tetrahymena pyriformis*. *J. Protozool.* **14**, 553–565.

Allen, R. D. (1978). Membranes of ciliates: ultrastructure, biochemistry and fusion. pp. 657–763. Elsevier/North-Holland Biomed. Press, Amsterdam.

Allen, R. D. (2000). The contractile vacuole and its membrane dynamics. *Bioessays* **22**, 1035–1042.

Allen, R. D., and Fok, A. K. (1980). Membrane recycling and endocytosis in Paramecium confirmed by horseradish peroxidase pulse-chase studies. *J. Cell Sci.* **45**, 131–145.

Allen, R. D., and Wolf, R. W. (1979). Membrane recycling at the cytoproct of Tetrahymena. *J. Cell Sci.* **35**, 217–227.

Anagnostopoulos, D., Rakiec, C., Wood, J., Pandarinathan, L., Zvonok, N., Makriyannis, A., and Siafaka-Kapadai, A. (2010). Identification of endocannabinoids and related *N*-acylethanolamines in tetrahymena. A new class of compounds for Tetrahymena. *Protist* **161**, 452–465.

Balabaskaran Nina, P., Dudkina, N. V., Kane, L. A., van Eyk, J. E., Boekema, E. J., Mather, M. W., and Vaidya, A. B. (2010). Highly divergent mitochondrial ATP synthase complexes in *Tetrahymena thermophila*. *PLoS Biol.* **8**, e1000418.

Becker, B., and Rusing, M. (2003). Structure of *N*-glycosidic carbohydrates of secretory proteins of *Tetrahymena thermophila*. *J. Eukaryot. Microbiol.* **50**, 235–239.

Bowman, G. R., Elde, N. C., Morgan, G., Winey, M., and Turkewitz, A. P. (2005a). Core formation and the acquisition of fusion competence are linked during secretory granule maturation in Tetrahymena. *Traffic* **6**, 303–323.

Bowman, G. R., Smith, D. G., Michael Siu, K. W., Pearlman, R. E., and Turkewitz, A. P. (2005b). Genomic and proteomic evidence for a second family of dense core granule cargo proteins in Tetrahymena thermophila. *J. Eukaryot. Microbiol.* **52**, 291–297.

Bowman, G. R., and Turkewitz, A. P. (2001). Analysis of a mutant exhibiting conditional sorting to dense core secretory granules in *Tetrahymena thermophila*. *Genetics* **159**, 1605–1616.

Bright, L. J., Kambesis, N., Nelson, S. B., Jeong, B., and Turkewitz, A. P. (2010). Comprehensive analysis reveals dynamic and evolutionary plasticity of Rab GTPases and membrane traffic in *Tetrahymena thermophila*. *PLoS Genetics* **6**, e1001155.

Chanat, E., and Huttner, W. B. (1991). Milieu-induced, selective aggregation of regulated secretory proteins in the trans-Golgi network. *J. Cell Biol.* **115**, 1505–1519.

Christoforidis, S., and Zerial, M. (2000). Purification and identification of novel Rab effectors using affinity chromatography. *Methods* **20**, 403–410.

Cole, E. S., Anderson, P. C., Fulton, R. B., Majerus, M. E., Rooney, M. G., Savage, J. M., Chalker, D., Honts, J., Welch, M. E., Wentland, A. L., Zweifel, E., and Beussman, D. J. (2008). A proteomics approach to cloning fenestrin from the nuclear exchange junction of Tetrahymena. *J. Eukaryot. Microbiol.* **55**, 245–256.

Conner, R. L., Landrey, J. R., Burns, C. H., and Mallory, F. B. (1968). Cholesterol inhibition of pentacyclic triterpenoid biosynthesis in *Tetrahymena pyriformis*. *J. Protozool.* **15**, 600–605.

Conner, R. L., Landrey, J. R., Kaneshiro, E. S., and Van Wagtendonk, W. J. (1971). The metabolism of stigmasterol and cholesterol by *Paramecium aurelia*. *Biochim. Biophys. Acta* **239**, 312–319.

Cousin, M. A., and Robinson, P. J. (1999). Mechanisms of synaptic vesicle recycling illuminated by fluorescent dyes. *J. Neurochem.* **73**, 2227–2239.

Cowan, A. T., Bowman, G. R., Edwards, K. F., Emerson, J. J., and Turkewitz, A. P. (2005). Genetic, genomic, and functional analysis of the granule lattice proteins in Tetrahymena secretory granules. *Mol. Biol. Cell* **16**, 4046–4060.

Creemers, J. W., Jackson, R. S., and Hutton, J. C. (1998). Molecular and cellular regulation of prohormone processing. *Semin. Cell Dev. Biol.* **9**, 3–10.

Dacks, J. B., and Field, M. C. (2007). Evolution of the eukaryotic membrane-trafficking system: origin, tempo and mode. *J. Cell Sci.* **120**, 2977–2985.

Deli, D., Leondaritis, G., Tiedtke, A., and Galanopoulou, D. (2008). Deficiency in lysosomal enzyme secretion is associated with upregulation of phosphatidylinositol 4-phosphate in tetrahymena. *J. Eukaryot. Microbiol.* **55**, 343–350.

Di Paolo, G., and De Camilli, P. (2006). Phosphoinositides in cell regulation and membrane dynamics. *Nature* **443**, 651–657.

Doerder, F. P., and Gerber, C. A. (2000). Molecular characterization of the SerL paralogs of *Tetrahymena thermophila*. *Biochem. Biophys. Res. Commun.* **278**, 621–626.

Eisen, J. A., Coyne, R. S., Wu, M., Wu, D., Thiagarajan, M., Wortman, J. R., Badger, J. H., Ren, Q., Amedeo, P., Jones, K. M., Tallon, L. J., Delcher, A. L., Salzberg, S. L., Silva, J. C., Haas, B. J., Majoros, W. H., Farzad, M., Carlton, J. M., Smith Jr., R. K., Garg, J., Pearlman, R. E., Karrer, K. M., Sun, L., Manning, G., Elde, N. C., Turkewitz, A. P., Asai, D. J., Wilkes, D. E., Wang, Y., Cai, H., Collins, K., Stewart, B. A., Lee, S. R., Wilamowska, K., Weinberg, Z., Ruzzo, W. L., Wloga, D., Gaertig, J., Frankel, J., Tsao, C. C., Gorovsky, M. A., Keeling, P. J., Waller, R. F., Patron, N. J., Cherry, J. M., Stover, N. A., Krieger, C. J., del Toro, C., Ryder, H. F., Williamson, S. C., Barbeau, R. A., Hamilton, E. P., and Orias, E. (2006). Macronuclear genome sequence of the ciliate *Tetrahymena thermophila*, a model eukaryote. *PLoS Biol.* **4**, e286.

Elde, N. C., Long, M., and Turkewitz, A. P. (2007). A role for convergent evolution in the secretory life of cells. *Trends Cell Biol.* **17**, 157–164.

Elde, N. C., Morgan, G., Winey, M., Sperling, L., and Turkewitz, A. P. (2005). Elucidation of clathrin-mediated endocytosis in Tetrahymena reveals an evolutionarily convergent recruitment of dynamin. *PLoS Genet.* **1**, e52.

Florin-Christensen, J., Florin-Christensen, M., Knudsen, J., and Rasmussen, L. (1986). Phospholipases and phosphonolipids in a ciliate: an attack and defense system? *Trends Biochem. Sci.* **11**, 354–355.

Frankel, J. (2000). Cell biology of *Tetrahymena thermophila*. *Methods Cell. Biol.* **62**, 27–125.

Gonda, K., Komatsu, M., and Numata, O. (2000). Calmodulin and Ca^{2+}/calmodulin-binding proteins are involved in *Tetrahymena thermophila* phagocytosis. *Cell Struct. Funct.* **25**, 243–251.

Gumbiner, B., and Kelly, R. B. (1982). Two distinct intracellular pathways transport secretory and membrane glycoproteins to the surface of pituitary tumor cells. *Cell* **28**, 51–59.

Gurkan, C., Lapp, H., Alory, C., Su, A. I., Hogenesch, J. B., and Balch, W. E. (2005). Large-scale profiling of Rab GTPase trafficking networks: the membrome. *Mol. Biol. Cell* **16**, 3847–3864.

Haddad, A., Bowman, G. R., and Turkewitz, A. P. (2002). A new class of cargo protein in *Tetrahymena thermophila* dense core secretory granules. *Eukaryot. Cell* **1**, 583–593.

Hartmann, M., Guberman, A., Florin-Christensen, M., and Tiedtke, A. (2000). Screening for and characterization of phospholipase A1 hypersecretory mutants of *Tetrahymena thermophila*. *Appl. Microbiol. Biotechnol.* **54**, 390–396.

Hausmann, K., and Allen, R. D. (1976). Membrane behavior of exocytic vesicles: II. Fate of the trichocyst membranes in Paramecium after induced trichocyst discharge. *J. Cell Biol.* **69**, 313–326.

Heien, M. L., Piehowski, P. D., Winograd, N., and Ewing, A. G. (2010). Lipid detection, identification, and imaging single cells with SIMS. *Methods Mol. Biol.* **656**, 85–97.

Herrmann, L., Erkelenz, M., Aldag, I., Tiedtke, A., and Hartmann, M. W. (2006). Biochemical and molecular characterisation of *Tetrahymena thermophila* extracellular cysteine proteases. *BMC Microbiol.* **6**, 19.

Hosein, R. E., Williams, S. A., and Gavin, R. H. (2005). Directed motility of phagosomes in *Tetrahymena thermophila* requires actin and Myo1p, a novel unconventional myosin. *Cell Motil. Cytoskeleton* **61**, 49–60.

Huynh, K. K., Kay, J. G., Stow, J. L., and Grinstein, S. (2007). Fusion, fission, and secretion during phagocytosis. *Physiology (Bethesda)* **22**, 366–372.

Iwamoto, M., Mori, C., Kojidani, T., Bunai, F., Hori, T., Fukagawa, T., Hiraoka, Y., and Haraguchi, T. (2009). Two distinct repeat sequences of Nup98 nucleoporins characterize dual nuclei in the binucleated ciliate tetrahymena. *Curr. Biol.* **19**, 843–847.

Jacobs, M. E., DeSouza, L. V., Samaranayake, H., Pearlman, R. E., Siu, K. W., and Klobutcher, L. A. (2006). The *Tetrahymena thermophila* phagosome proteome. *Eukaryot Cell* **5**, 1990–2000.

Jeong, Y., Joo, M. K., Sohn, Y. S., and Jeong, B. (2007). Reverse thermal organogels. *Adv. Mater.* **19**, 3947–3950.

Kersting, M. C., Boyette, M., Massey, J. H., and Ryals, P. E. (2003). Identification of the inositol isomers present in Tetrahymena. *J. Eukaryot. Microbiol.* **50**, 164–168.
Kersting, M. C., and Ryals, P. E. (2004). Sodium-dependent transport of [3H](1D)chiro-inositol by Tetrahymena. *J. Eukaryot. Microbiol.* **51**, 307–311.
Kienle, N., Kloepper, T. H., and Fasshauer, D. (2009). Differences in the SNARE evolution of fungi and metazoa. *Biochem. Soc. Trans.* **37**, 787–791.
Kilburn, C. L., Pearson, C. G., Romijn, E. P., Meehl, J. B., Giddings Jr., T. H., Culver, B. P., Yates 3rd, J. R., and Winey, M. (2007). New Tetrahymena basal body protein components identify basal body domain structure. *J. Cell Biol.* **178**, 905–912.
Kirchhausen, T. (2009). Imaging endocytic clathrin structures in living cells. *Trends Cell Biol.* **19**, 596–605.
Kiy, T., Vosskuhler, C., Rasmussen, L., and Tiedtke, A. (1993). Three pools of lysosomal enzymes in Tetrahymena thermophila. *Exp. Cell Res.* **205**, 286–292.
Klobutcher, L. A., Ragkousi, K., and Setlow, P. (2006). The *Bacillus subtilis* spore coat provides "eat resistance" during phagocytic predation by the protozoan *Tetrahymena thermophila*. *Proc. Natl Acad. Sci. U.S.A.* **103**, 165–170.
Kovacs, P., and Pallinger, E. (2003). Phosphatidylinositol 3-kinase-like activity in Tetrahymena. Effects of wortmannin and LY294002. *Acta Protozool.* **42**, 277–285.
Kurczy, M. E., Piehowski, P. D., Van Bell, C. T., Heien, M. L., Winograd, N., and Ewing, A. G. (2010). Mass spectrometry imaging of mating Tetrahymena show that changes in cell morphology regulate lipid domain formation. *Proc. Natl Acad. Sci. U.S.A.* **107**, 2751–2756.
Lee, S., Wisniewski, J. C., Dentler, W. L., and Asai, D. J. (1999). Gene knockouts reveal separate functions for two cytoplasmic dyneins in *Tetrahymena thermophila*. *Mol. Biol. Cell* **10**, 771–784.
Leondaritis, G., and Galanopoulou, D. (2000). Characterization of inositol phospholipids and identification of a mastoparan-induced polyphosphoinositide response in *Tetrahymena pyriformis*. *Lipids* **35**, 525–532.
Leondaritis, G., Sarri, T., Dafnis, I., Efstathiou, A., and Galanopoulou, D. (2011). Biochemical and genetic evidence for the presence of multiple phosphatidylinositol- and phosphatidylinositol 4,5-bisphosphate-specific phospholipases C in Tetrahymena. *Eukaryot. Cell* **10**, 412–422.
Leondaritis, G., Tiedtke, A., and Galanopoulou, D. (2005). D-3 phosphoinositides of the ciliate Tetrahymena: characterization and study of their regulatory role in lysosomal enzyme secretion. *Biochim. Biophys. Acta* **1745**, 330–341.
Letunic, I., Copley, R. R., Pils, B., Pinkert, S., Schultz, J., and Bork, P. (2006). SMART 5: domains in the context of genomes and networks. *Nucleic Acids Res.* **34**, D257–D260.
Madinger, C. L., Collins, K., Fields, L. G., Taron, C. H., and Benner, J. S. (2010). Constitutive secretion in *Tetrahymena thermophila*. *Eukaryot. Cell* **9**, 674–681.
Mallory, F. B., and Conner, R. L. (1971). Dehydrogenation and dealkylation of various sterols by *Tetrahymena pyriformis*. *Lipids* **6**, 149–153.
Malone, C. D., Falkowska, K. A., Li, A. Y., Galanti, S. E., Kanuru, R. C., LaMont, E. G., Mazzarella, K. C., Micev, A. J., Osman, M. M., Piotrowski, N. K., Suszko, J. W., Timm, A. C., Xu, M. M., Liu, L., and Chalker, D. L. (2008). Nucleus-specific importin alpha proteins and nucleoporins regulate protein import and nuclear division in the binucleate *Tetrahymena thermophila*. *Eukaryot. Cell* **7**, 1487–1499.
Matsuo, J., Oguri, S., Nakamura, S., Hanawa, T., Fukumoto, T., Hayashi, Y., Kawaguchi, K., Mizutani, Y., Yao, T., Akizawa, K., Suzuki, H., Simizu, C., Matsuno, K., Kamiya, S., and Yamaguchi, H. (2010). Ciliates rapidly enhance the frequency of conjugation between *Escherichia coli* strains through bacterial accumulation in vesicles. *Res. Microbiol.* **161**, 711–719.
Miao, W., Xiong, J., Bowen, J., Wang, W., Liu, Y., Braguinets, O., Grigull, J., Pearlman, R. E., Orias, E., and Gorovsky, M. A. (2009). Microarray analyses of gene expression during the *Tetrahymena thermophila* life cycle. *PLoS One* **4**, e4429.
Michell, R. H. (2008). Inositol derivatives: evolution and functions. *Nat. Rev. Mol. Cell Biol.* **9**, 151–161.

Morikawa, T., Mizutani, M., Aoki, N., Watanabe, B., Saga, H., Saito, S., Oikawa, A., Suzuki, H., Sakurai, N., Shibata, D., Wadano, A., Sakata, K., and Ohta, D. (2006). Cytochrome P450 CYP710A encodes the sterol C-22 desaturase in Arabidopsis and tomato. *Plant Cell* **18**, 1008–1022.

Moya, C. E., and Jacobs, R. S. (2006). Pseudopterosin A inhibits phagocytosis and alters intracellular calcium turnover in a pertussis toxin sensitive site in *Tetrahymena thermophila*. *Comp. Biochem. Physiol. C Toxicol. Pharmacol.* **143**, 436–443.

Murphy, R. C., Hankin, J. A., and Barkley, R. M. (2009). Imaging of lipid species by MALDI mass spectrometry. *J. Lipid. Res.* **50**(Suppl), S317–S322.

Nilsson, J. R. (1979). Phagotrophy in Tetrahymena. *In* "Biochemistry and Physiology of Protozoa," (M. Levandowsky, and S. H. Hutner, eds.), pp. 339–379. Academic Press, New York.

Nilsson, J. R., and Van Deurs, B. (1983). Coated pits and pinocytosis in Tetrahymena. *J. Cell Sci.* **63**, 209–222.

Nusblat, A. D., Munoz, L., Valcarce, G. A., and Nudel, C. B. (2005). Characterization and properties of cholesterol desaturases from the ciliate *Tetrahymena thermophila*. *J. Eukaryot. Microbiol.* **52**, 61–67.

Nusblat, A. D., Najle, S. R., Tomazic, M. L., Uttaro, A. D., and Nudel, C. B. (2009). C-5(6) sterol desaturase from *Tetrahymena thermophila*: gene identification and knockout, sequence analysis, and comparison to other C-5(6) sterol desaturases. *Eukaryot. Cell* **8**, 1287–1297.

Ohvo-Rekila, H., Ramstedt, B., Leppimaki, P., and Slotte, J. P. (2002). Cholesterol interactions with phospholipids in membranes. *Prog. Lipid Res.* **41**, 66–97.

Ostrowski, S. G., Van Bell, C. T., Winograd, N., and Ewing, A. G. (2004). Mass spectrometric imaging of highly curved membranes during Tetrahymena mating. *Science* **305**, 71–73.

Parfrey, L. W., Barbero, E., Lasser, E., Dunthorn, M., Bhattacharya, D., Patterson, D. J., and Katz, L. A. (2006). Evaluating support for the current classification of eukaryotic diversity. *PLoS Genet* **2**, e220.

Pearson, C. G., and Winey, M. (2009). Basal body assembly in ciliates: the power of numbers. *Traffic* **10**, 461–471.

Pederson, T. (2010). An olympian protozoan. *Nucleus* **1**, 2–3.

Pereira-Leal, J. B. (2008). The Ypt/Rab family and the evolution of trafficking in fungi. *Traffic* **9**, 27–38.

Pereira-Leal, J. B., and Seabra, M. C. (2001). Evolution of the Rab family of small GTP-binding proteins. *J. Mol. Biol.* **313**, 889–901.

Pieringer, J., and Conner, R. L. (1979). Positional distribution of fatty acids in the glycerophospholipids of *Tetrahymena pyriformis*. *J. Lipid Res.* **20**, 363–370.

Plattner, H., Flotenmeyer, M., Kissmehl, R., Pavlovic, N., Hauser, K., Momayezi, M., Braun, N., Tack, J., and Bachmann, L. (1999). Microdomain arrangement of the SERCA-type Ca^{2+} pump (Ca^{2+}-ATPase) in subplasmalemmal calcium stores of paramecium cells. *J. Histochem. Cytochem.* **47**, 841–854.

Rahaman, A., Elde, N. C., and Turkewitz, A. P. (2008). A dynamin-related protein required for nuclear remodeling in Tetrahymena. *Curr. Biol.* **18**, 1227–1233.

Rahaman, A., Miao, W., and Turkewitz, A. P. (2009). Independent transport and sorting of functionally distinct protein families in *Tetrahymena thermophila* dense core secretory granules. *Eukaryot. Cell* **8**, 1575–1583.

Riggs, B. M., Lansley, T. A., and Ryals, P. E. (2007). Phosphatidylinositol synthase of Tetrahymena: inositol isomers as substrates in phosphatidylinositol biosynthesis and headgroup exchange reactions. *J. Eukaryot. Microbiol.* **54**, 119–124.

Rosati, G., and Modeo, L. (2003). Extrusomes in ciliates: diversification, distribution, and phylogenetic implications. *J. Eukaryot. Microbiol.* **50**, 383–402.

Ryals, P. E. (2009). Inositols and phosphoinositides in Tetrahymena. *Acta Protozool.* **48**, 191–202.

Ryals, P. E., and Kersting, M. C. (1999). Sodium-dependent uptake of [3H]scyllo-inositol by Tetrahymena: incorporation into phosphatidylinositol, phosphatidylinositol-linked glycans, and polyphosphoinositols. *Arch. Biochem. Biophys.* **366**, 261–266.

Saito-Nakano, Y., Nakahara, T., Nakano, K., Nozaki, T., and Numata, O. (2010). Marked amplification and diversification of products of ras genes from rat brain, Rab GTPases, in the ciliates Tetrahymena thermophila and Paramecium tetraurelia. *J. Eukaryot. Microbiol.* **57**, 389–399.

Satir, B. H., Schooley, C., and Satir, P. (1973). Membrane fusion in a model system. Mucocyst secretion in Tetrahymena. *J. Cell Biol.* **56**, 153–176.

Segev, N. (2001). Ypt/rab gtpases: regulators of protein trafficking. *Sci STKE* **2001**, RE11.

Simon, M. C., and Schmidt, H. J. (2007). Antigenic variation in ciliates: antigen structure, function, expression. *J. Eukaryot. Microbiol.* **54**, 1–7.

Simons, K., and Toomre, D. (2000). Lipid rafts and signal transduction. *Nat. Rev. Mol. Cell. Biol.* **1**, 31–39.

Smith, D. G., Gawryluk, R. M., Spencer, D. F., Pearlman, R. E., Siu, K. W., and Gray, M. W. (2007b). Exploring the mitochondrial proteome of the ciliate protozoon Tetrahymena thermophila: direct analysis by tandem mass spectrometry. *J. Mol. Biol.* **374**, 837–863.

Smith, A. C., Heo, W. D., Braun, V., Jiang, X., Macrae, C., Casanova, J. E., Scidmore, M. A., Grinstein, S., Meyer, T., and Brumell, J. H. (2007a). A network of Rab GTPases controls phagosome maturation and is modulated by Salmonella enterica serovar Typhimurium. *J. Cell Biol.* **176**, 263–268.

Smith, J. C., Northey, J. G., Garg, J., Pearlman, R. E., and Siu, K. W. (2005). Robust method for proteome analysis by MS/MS using an entire translated genome: demonstration on the ciliome of *Tetrahymena thermophila*. *J. Proteome. Res.* **4**, 909–919.

Stelly, N., Halpern, S., Nicolas, G., Fragu, P., and Adoutte, A. (1995). Direct visualization of a vast cortical calcium compartment in Paramecium by secondary ion mass spectrometry (SIMS) microscopy: possible involvement in exocytosis. *J. Cell Sci.* **108**, 1895–1909.

Stenmark, H., and Olkkonen, V. M. (2001). The Rab GTPase family. *Genome Biol.* **2**, REVIEWS3007.

Stuart, L. M., and Ezekowitz, R. A. (2005). Phagocytosis: elegant complexity. *Immunity* **22**, 539–550.

Sugita, M., Iwataki, Y., Nakano, K., and Numata, O. (2011). Unique sequences and predicted functions of myosins in *Tetrahymena thermophila*. *Gene* **480**, 10–20.

Sugita, M., Nakano, K., Sato, M., Toyooka, K., and Numata, O. (2009). The roles of actin cytoskeleton and microtubules for membrane recycling of a food vacuole in *Tetrahymena thermophila*. *Cell Motil. Cytoskeleton* **66**, 371–377.

Sun, T. H., Heimark, D. B., Nguygen, T., Nadler, J. L., and Larner, J. (2002). Both myo-inositol to chiro-inositol epimerase activities and chiro-inositol to myo-inositol ratios are decreased in tissues of GK type 2 diabetic rats compared to Wistar controls. *Biochem. Biophys. Res. Commun.* **293**, 1092–1098.

Tamura, K., Ohbayashi, N., Maruta, Y., Kanno, E., Itoh, T., and Fukuda, M. (2009). Varp is a novel Rab32/38-binding protein that regulates Tyrp1 trafficking in melanocytes. *Mol. Biol. Cell* **20**, 2900–2908.

Tomazic, M. L., Najle, S. R., Nusblat, A. D., Uttaro, A. D., and Nudel, C. B. (2011). A novel sterol desaturase-like protein promoting dealkylation of phytosterols in *Tetrahymena thermophila*. *Eukaryot. Cell* **10**, 423–434.

Tooze, S. A., Martens, G. J., and Huttner, W. B. (2001). Secretory granule biogenesis: rafting to the SNARE. *Trends Cell Biol.* **11**, 116–122.

Turkewitz, A. P. (2004). Out with a bang! Tetrahymena as a model system to study secretory granule biogenesis. *Traffic* **5**, 63–68.

Turkewitz, A. P., and Bright, L. J. (2011). A Rab-based view of membrane traffic in the ciliate *Tetrahymena thermophila*. *Small GTPases* **2**, 222–226.

Turkewitz, A. P., Orias, E., and Kapler, G. (2002). Functional genomics: the coming of age for *Tetrahymena thermophila*. *Trends Genet.* **18**, 35–40.

Valcarce, G., Florin-Christensen, J., and Nudel, C. (2000). Isolation of a delta7-cholesterol desaturase from *Tetrahymena thermophila*. *Appl. Microbiol. Biotechnol.* **53**, 591–595.

Vanhaesebroeck, B., Guillermet-Guibert, J., Graupera, M., and Bilanges, B. (2010). The emerging mechanisms of isoform-specific PI3K signalling. *Nat. Rev. Mol. Cell. Biol.* **11**, 329–341.

Vanhaesebroeck, B., Leevers, S. J., Ahmadi, K., Timms, J., Katso, R., Driscoll, P. C., Woscholski, R., Parker, P. J., and Waterfield, M. D. (2001). Synthesis and function of 3-phosphorylated inositol lipids. *Annu. Rev. Biochem.* **70**, 535–602.

Verbsky, J. W., and Turkewitz, A. P. (1998). Proteolytic processing and Ca^{2+}-binding activity of dense-core vesicle polypeptides in Tetrahymena. *Mol. Biol. Cell* **9**, 497–511.

Wang, S., Banno, Y., Nakashima, S., and Nozawa, Y. (2001). Enzymatic characterization of phospholipase D of protozoan Tetrahymena cells. *J. Eukaryot. Microbiol.* **48**, 194–201.

Wang, S., Banno, Y., and Nozawa, Y. (2002). Two forms of membrane-bound sphingosine kinase in Tetrahymena and activity changes during growth and the cell cycle. *J. Eukaryot. Microbiol.* **49**, 305–311.

Weisz, O. A., and Rodriguez-Boulan, E. (2009). Apical trafficking in epithelial cells: signals, clusters and motors. *J. Cell Sci.* **122**, 4253–4266.

Whitaker, B. D., and Nelson, D. L. (1987). Growth support and metabolism of phytosterols in *Paramecium tetraurelia*. *Lipids* **22**, 386–396.

Wilkes, D. E., and Otto, J. J. (2003). Profilin functions in cytokinesis, nuclear positioning, and stomatogenesis in *Tetrahymena thermophila*. *J. Eukaryot. Microbiol.* **50**, 252–262.

Wilkes, D. E., Watson, H. E., Mitchell, D. R., and Asai, D. J. (2008). Twenty-five dyneins in Tetrahymena: a re-examination of the multidynein hypothesis. *Cell Motil. Cytoskeleton.* **65**, 342–351.

Williams, S. A., and Gavin, R. H. (2005). Myosin genes in Tetrahymena. *Cell Motil. Cytoskeleton.* **61**, 237–243.

Williams, N. E., and Luft, J. H. (1968). Use of a nitrogen mustard derivative in fixation for electron microscopy and observations of the ultrastructure of Tetrahymena. *J. Ultrastruct. Res.* **25**, 271–292.

Williams, N. E., Tsao, C. C., Bowen, J., Hehman, G. L., Williams, R. J., and Frankel, J. (2006). The actin gene ACT1 is required for phagocytosis, motility, and cell separation of *Tetrahymena thermophila*. *Eukaryot. Cell* **5**, 555–567.

Wloga, D., Strzyzewska-Jowko, I., Gaertig, J., and Jerka-Dziadosz, M. (2008). Septins stabilize mitochondria in *Tetrahymena thermophila*. *Eukaryot. Cell* **7**, 1373–1386.

Wolfe, J. (1982). The conjugation junction of Tetrahymena: its structure and development. *J. Morphol.* **172**, 159–178.

Wolfe, J. (1985). Cytoskeletal reorganization and plasma membrane fusion in conjugating Tetrahymena. *J. Cell Sci.* **73**, 69–85.

Wu, Y. T., Tan, H. L., Shui, G., Bauvy, C., Huang, Q., Wenk, M. R., Ong, C. N., Codogno, P., and Shen, H. M. (2010). Dual role of 3-methyladenine in modulation of autophagy via different temporal patterns of inhibition on class I and III phosphoinositide 3-kinase. *J. Biol. Chem.* **285**, 10850–10861.

Xie, R., Clark, K. M., and Gorovsky, M. A. (2007). Endoplasmic reticulum retention signal-dependent glycylation of the Hsp70/Grp170-related Pgp1p in Tetrahymena. *Eukaryot. Cell* **6**, 388–397.

Xiong, J., Yuan, D., Fillingham, J. S., Garg, J., Lu, X., Chang, Y., Liu, Y., Fu, C., Pearlman, R. E., and Miao, W. (2011). Gene network landscape of the ciliate *Tetrahymena thermophila*. *PLoS One* **6**, e20124.

Yakisich, J. S., and Kapler, G. M. (2004). The effect of phosphoinositide 3-kinase inhibitors on programmed nuclear degradation in Tetrahymena and fate of surviving nuclei. *Cell Death Differ* **11**, 1146–1149.

Yancey, P. H. (2005). Organic osmolytes as compatible, metabolic and counteracting cytoprotectants in high osmolarity and other stresses. *J. Exp. Biol.* **208**, 2819–2830.

Zhang, J., Schulze, K. L., Hiesinger, P. R., Suyama, K., Wang, S., Fish, M., Acar, M., Hoskins, R. A., Bellen, H. J., and Scott, M. P. (2007). Thirty-one flavors of Drosophila rab proteins. *Genetics* **176**, 1307–1322.

Zweifel, E., Smith, J., Romero, D., Giddings Jr., T. H., Winey, M., Honts, J., Dahlseid, J., Schneider, B., and Cole, E. S. (2009). Nested genes CDA12 and CDA13 encode proteins associated with membrane trafficking in the ciliate *Tetrahymena thermophila*. *Eukaryot. Cell* **8**, 899–912.

CHAPTER 7

Developmental progression of *Tetrahymena* through the cell cycle and conjugation

Eric Cole* and Toshiro Sugai†

*Biology Department St. Olaf College, Northfield, MN 55057, USA

†Department of Biology, College of Science, Ibaraki University, Mito, Ibaraki 310, Japan

0091-679X/10 $35.00
DOI 10.1016/B978-0-12-385967-9.00007-4

Abstract

The ciliate *Tetrahymena thermophila* can be said to undergo a variety of developmental programs. During vegetative growth, cells coordinate a variety of cell-cycle operations including macronuclear DNA synthesis and a-mitotic fission, micronuclear DNA synthesis and mitosis, cytokinesis and an elaborate program of cortical morphogenesis that replicates the cortical organelles. When starved, cells undergo oral replacement, transformation into fast-swimming dispersal forms or, when encountering cells of a complementary mating type, conjugation. Conjugation involves a 12 hour program of meiosis, mitosis, nuclear exchange and karyogamy, and two postzygotic divisions of the fertilization nucleus. This chapter reviews experimental data exploring the developmental dependencies associated with both vegetative and conjugal development.

I. Introduction

Tetrahymena provides a remarkable theater for exploring both sexual and asexual programs of unicellular reproduction. Vegetative development (the program of events leading to asexual fission) is a decidedly complex phenomenon. On the one hand, *Tetrahymena* runs two seemingly independent nuclear division pathways involving replication of the somatic macronucleus by amitotic fission, and replication of the germline micronucleus by a more conventional mitosis involving centromeres and an intra-nuclear mitotic spindle. On the other hand, the ciliate cortex represents a unique problem in that each cell division requires duplication of a host of cortical organelles including the basal-body rich oral apparatus, the cytoproct, and the contractile vacuole pores with their associated vacuole system (Fig. 1). As one can imagine, driving these very different morphogenetic programs and coordinating them with rounds of macronuclear and micronuclear DNA synthesis throughout the vegetative cell cycle provides significant logistical challenges.

Conjugation, or sexual reproduction, is even more elaborate. Mating pairs of *Tetrahymena* undergo a 12 h developmental program involving six nuclear divisions: meiosis I and II, a third gametogenic division of a single "selected" meiotic product, nuclear exchange and fusion (karyogamy), two post-zygotic nuclear divisions, and a final micronuclear division as cells re-enter the cell division pathway after mating. Following meiosis, three of the four meiotic products are triggered to undergo programmed nuclear degeneration (PND). Following the second post-zygotic division, the parental macronuclei also undergo a form of PND resembling apoptosis, while anterior division products undergo extensive genome remodeling and amplification.

The focus of this chapter will be to provide an overview of the cytological events associated with vegetative division and conjugation and a survey of what is known regarding developmental contingencies and checkpoints controlling progress through both vegetative and conjugal developmental pathways as well as the transition from conjugal to vegetative development, and to map these events onto a detailed description of the cytogenetic program. Other chapters will take up more structural aspects of

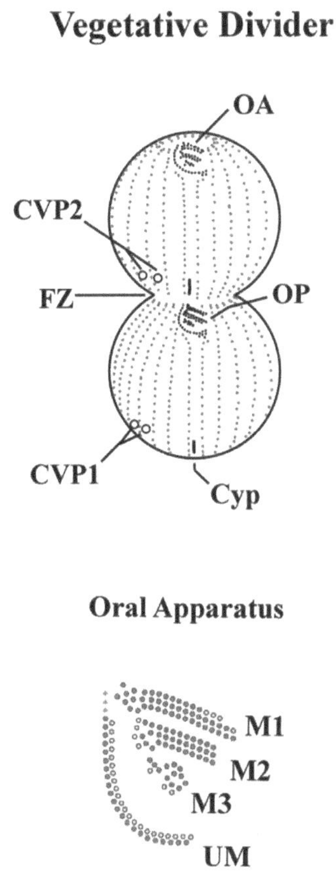

Fig. 1 A diagram of the dividing *Tetrahymena* cell cortex showing relevant organelles. OA, oral apparatus of the mature cell; OP, the developing oral primordium that will serve the posterior daughter cell; Cyp, cytoproct; CVP, contractile vacuole pores (typically two). CVP1 represents the CVPs of the mature cell that will be inherited by the posterior daughter cell. CVP2 represents the newly developing CVPs that will serve the anterior daughter cell. FZ, fission zone. The lower diagram shows a detailed drawing of the mature oral apparatus. M1–M3 are the three oral "membranelles": triple-rows of ciliated basal bodies. The UM is the fourth undulating membrane. Black dots represent ciliated basal bodies. Grey dots represent unciliated basal bodies.

the cell, and processes associated with macronuclear differentiation and programmed nuclear elimination (See Chapters by Wloga and Frankel, and Karrer).

II. Checkpoints and Developmental Contingencies

Progress through a developmental program (a temporal sequence of biological events), can be mediated by either independent or dependent processes. In the case of independent pathways, one process can be disrupted or blocked without perturbing progress in another. These can be viewed as parallel programs or pathways. Dependent pathways appear as developmental contingencies in which one event requires completion of an earlier event. This can be relatively straightforward as within a metabolic pathway in which the product of one reaction serves as the substrate for a subsequent reaction. A more complex form of developmental contingency arises when an organism has in place a surveillance mechanism that monitors completion of one step in a pathway and inhibits further progress until that step is completed. Such developmental contingencies can be uncoupled by mutating single genes within the surveillance/feedback mechanism, and these types of surveillance mechanisms are referred to as checkpoints (Hartwell and Weinert, 1988; Hartwell and Weinert, 1989).

III. Vegetative Development (Asexual Reproduction by Cellular Fission)

A. A Note About Cytology

The micronucleus and macronucleus (MIC and MAC) undergo dynamic changes during both cell division and conjugation. Throughout the cell's life cycle, the nuclear envelope never breaks down. Nuclear cytology has been described mainly in fixed cells. For light microscopy, the nucleus is usually stained with orcein, Giemsa, or DAPI. These "nuclear stains" do not actually stain the nucleus but rather the chromatin or DNA contents within it. Consequently, when the nuclear shape is different from the chromatin shape such as within the mitotic nucleus and many stages of conjugation, staining can give a misleading interpretation of nuclear size, position, and behavior. By electron microscopy the nuclear envelope may be clearly visible, but the shape and position of the dynamic, sometimes elongate, MIC can be difficult to observe. In various immunofluorescence studies in which microtubules have been imaged the nucleus is often counter-stained with DAPI or PI. In these cases, whether microtubule location is intranuclear or extranuclear cannot be easily determined.

The nuclear envelope can be stained with vital dyes, such as DiOC6, in living and fixed cells, and DNA can be stained with Hoechst 33342. Continuous observation of nuclear events in living cells by phase contrast or differential interference contrast (DIC) microscopy, or by using nuclear envelope and DNA stains shows complex and often very rapid nuclear changes during cell division and conjugation. Although it is known that nuclear positioning is important, nuclear position of many stages is easily disturbed by pipetting, centrifuging, and flattening with a cover glass. The following is a summary of nuclear events observed employing light microscopy of living cells with the least invasive practices and is based on unpublished observations by Sugai. An illustration of the events described appears in Fig. 2.

B. Nuclear Events During Cell Division

During most of the cell division cycle the MIC is attached to the MAC surface, half buried in a pocket of the MAC. The MIC's position on the MAC is random relative to

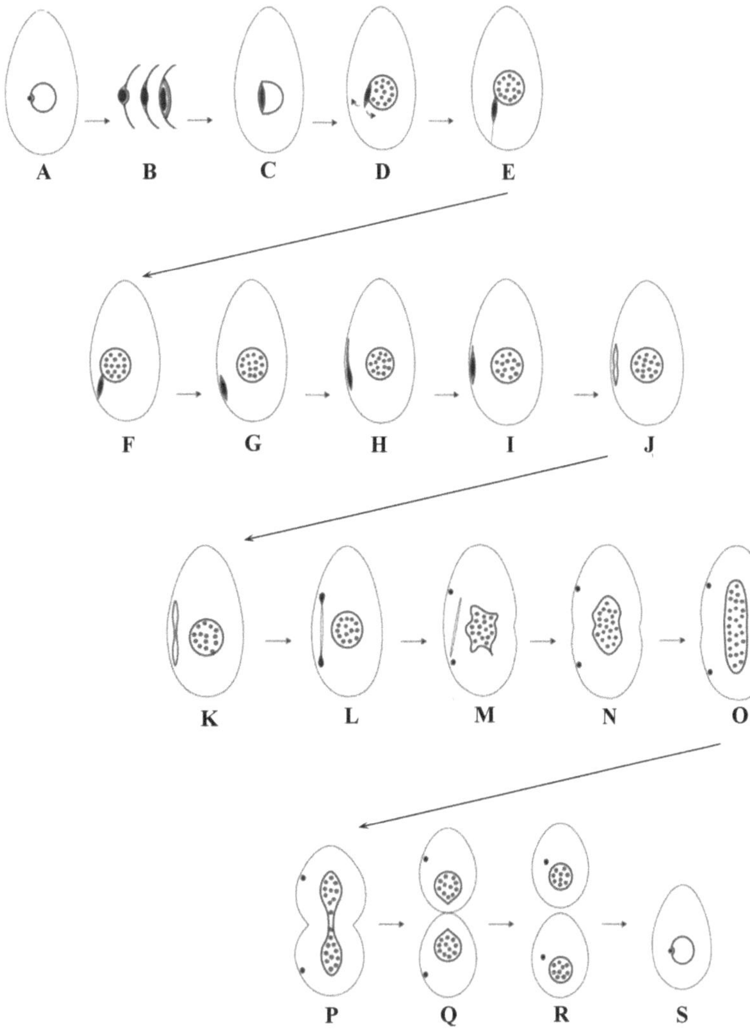

Fig. 2 Various stages of nuclear division during the vegetative cell cycle of *Tetrahymena* (details in text). In Fig. 2A, the macronucleus (MAC) is represented by the large circle. The micronucleus (MIC) is the small black dot nested in a pocket of the MAC. Red dots in the MAC represent division-specific chromatin granules (Endoh and Sugai, 2005). (For interpretation of the references to color in this figure legend, the reader is referred to the web version of this book.)

cell polarity during G1, S, and G2, but then becomes fixed at the equator of the MAC just before mitosis (Fig. 2A).

When mitosis begins, the MIC becomes spindle shaped, lying parallel to the cell's long axis, but still half buried on MAC's surface (Fig. 2B,C). Both spindle poles are attached to MAC. Next, the posterior end of the MIC detaches from the MAC and appears to be pulled by a thin thread to the cell cortex near the posterior end of the cell, often near the contractile vacuole, while the anterior end remains firmly attached to the MAC (Fig. 2D,E). At this point, the MAC rotates slightly.

The point of attachment of this thread to the cell cortex then moves anteriorly. The thread then shortens and disappears, leaving the MIC directly attached to the cortex at one end and attached to the MAC at other end (Fig. 2F). This is accompanied by movement of the MAC from its central position to a place near the cortical position adjacent to the MIC. The MIC then releases the MAC, with the former remaining attached to the cortex (Fig. 2G). The MIC is never free in the cytoplasm, and the MAC returns to its central position. The spindle shaped MIC appears to attach to the cell cortex oriented parallel to the ciliary rows and in a posterior position within the cell. It then moves anteriorly along a ciliary row toward the cell's equator where MIC division occurs (Fig. 2H, I).

At the onset of MIC anaphase, the MIC elongates and becomes dumbbell shaped with two round portions that contain the chromosomes and a middle portion (the separation spindle) that contains no chromosomes or DNA (Fig. 2J–L). The middle portion is cut off at the end of anaphase, discarded into the cytoplasm, and eventually absorbed (Fig. 2M). The daughter MICs remain attached to the cell cortex during cytokinesis and return to the MAC following cytokinesis (Fig. 2N and later).

During macronuclear amitosis, the MAC moves slightly from its interphase position to a mid-position just before amitosis begins. Just as MIC anaphase is ending, the MAC shape becomes irregular for a short interval and then becomes rod shaped prior to its fission that coincides with the division of the cell.

C. Nested Developmental Programs

In the vegetative cell-division pathway, there are multiple developmental "programs" operating concurrently with varying degrees of interdependence (Fig. 3). These include MIC DNA synthesis and division, MAC DNA synthesis and division, cytokinesis, and "cortical development" loosely defined as the process by which the cell assembles a complete set of cortical organelles including an oral primordium (OP) located just posterior to the fission zone and along the right post-oral stomatogenic ciliary row (Figs. 1 and 4), contractile vacuole pores (CVPs) located just anterior to the fission zone, and to the cell's right of the stomatogenic ciliary row, and cytoproct (CY) located just anterior to the fission zone and along the stomatogenic ciliary row (Fig. 1). Coupled with assembly of these prominent cortical organelles is the process of somatic basal body proliferation and ciliation. Assays for completion of these various programs include simple light microscopy for cytokinesis, the protein-silver or "protargol" technique (Ng and Nelsen, 1977; Aufderheide, 1982), and the Chatton-Lwoff silver impregnation method (Frankel and Heckmann, 1968; Nelsen and DeBault, 1978), for determining various stages of somatic ciliature or oral development (stomatogenesis), and a variety of nuclear dyes for MIC and MAC division (Feulgen staining: Doerder and DeBault, 1975; DAPI staining: Cole *et al.*, 1987).

Studies into the cell cycle have relied on establishing a reliable method of synchronizing the cell division pathway. It is a little known fact that *Tetrahymena* was the

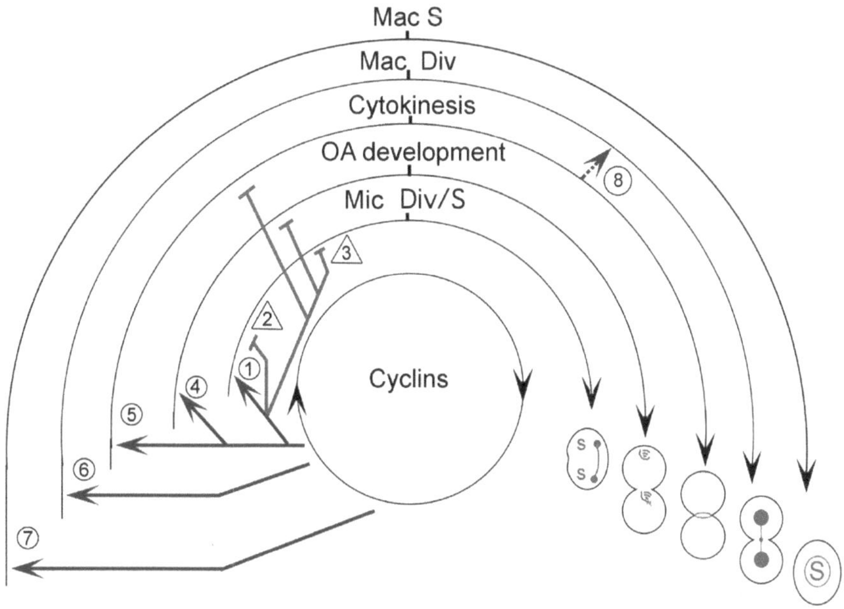

Fig. 3 This diagram represents the various nuclear and cortical pathways that are coordinated during the cell division cycle in *Tetrahymena* and their hypothetical relationship to a variety of cyclin-driven timing mechanisms. Mic Div/S refers to the micronuclear division pathway that includes a round of DNA synthesis that appears to be imbedded in anaphase. "OA development" refers to the sequence of events resulting in localized basal body proliferation and patterning that assemble into the new oral primordium which matures into the daughter cell's oral apparatus. This process is sometimes referred to as "stomatogenesis" (see Fig. 4 for details). Cytokinesis refers to the cortical events leading to formation of a fission zone and cytokinetic furrow, ultimately cleaving the cell in two. "MAC div" (macronuclear division) and "MAC S" (macronuclear DNA synthesis) occur at very different times in the cell cycle (unlike MIC division and synthesis). Blue arrows suggest that each cytological pathway is driven by independent timing mechanisms (where arrows are separate) or that they may be driven in a coordinate fashion (where arrows branch from a common stem). Grouping of arrows 1, 4, and 5 reflect the rather tight coupling of stomatogenesis with MIC division and cytokinesis. Agents that block or reset MIC division appear to also block and reset oral development and cytokinesis. The centripetal arrangement of the pathways also reflects the fact that disrupting outer pathways rarely disrupts inner pathways, though the opposite is not true. One can induce blocks in cytokinesis, for example, while stomatogenesis and MIC division continue to proceed. Blocking cytokinesis, on the other hand, often leads to MAC replication arrest. Dashed line "8" represents that cytokinesis (while not essential for MAC division) can assist with macronuclear fission. Red lines represent specific, demonstrated inhibitory checkpoints that regulate particular pathways. Triangle 2 represents the early anaphase arrest best revealed by a RAD51 knockout. MICs in RAD51 cell lines enter early anaphase and arrest, without affecting progress in stomatogenesis, cytokinesis, or MAC division. Triangle 3 represents a late anaphase arrest seen in Kirk *et al.* (2008), where chromosome-separation failure leads to a "dynamic pause" in stomatogenesis and a block in cytokinesis and MAC division. (See color plate.)

first eukaryotic cell in which induced cell division synchrony was achieved through a program of heat shocks (Scherbaum and Zeuthen, 1954; Zeuthen, 1964). Unfortunately the earliest experiments utilizing heat-shock-mediated synchronization involved *Tetrahymena pyriformis* GL, a genetically sterile, amicronucleate sibling species to *T. thermophila* with which most of the modern genetic work has been possible. The method was later applied successfully to a fertile strain *T. thermophila,* referred to in the literature of the time as *T. pyriformis* mating type 1, variety 1 (Holz *et al.*, 1957). Heat shock synchronization of *T. thermophila* appears to be more difficult than that of *T. pyriformis* (J. Frankel, pers. comm.). Since then, researchers have developed other synchronization methods including vinblastine treatment (Stone, 1968), starvation and re-feeding (Cameron and Jeter, 1970; Mohammad *et al.*, 2003), and centrifugal elutriation: a method of centrifugal size-selection of G1 cells for subsequent cell culture (Tang *et al.*, 1994; Naduparambil *et al.*, 2001; Cole *et al.*, 2001). The experiments we shall describe are a compilation of observations from the early studies using *T. pyriformis* and more recent work with *T. thermophila.* Some reasonable caution should be exercised in assuming common mechanisms and timelines for these two species.

Various types of developmental disruption have been delivered throughout the cell division cycle in order to discern which events are necessary for subsequent development to proceed and to what extent the various programs depend on one another. Such developmental disruptions have included temperature shock, both heat (Scherbaum and Zeuthen, 1954; Frankel, 1962, 1964, 1967a; Frankel *et al.*, 1980) and cold (Gavin, 1965; Nachtwey, 1967a; Frankel, 1967b), pharmacological inhibitors including those that block synthesis of DNA (Yakisich *et al.*, 2006; Kaczanowski and Kiersnowska, 2011), RNA (Nachtwey and Dickinson, 1967b; Haremaki *et al.*, 1996) and protein (Frankel, 1962, 1969b; Rasmussen and Zeuthen, 1962). Cell-cycle disruption has also been performed using reagents that interfere with cytoskeletal assembly such as vinblastine (Stone, 1968a,b) and colchicine (Tamura *et al.*, 1969; Nelsen, 1970; Williams and Williams, 1976). Insights into cell-cycle dependencies have also been gleaned by examining the phenotypes from a large collection of TS-cell division arrest mutants (Frankel *et al.*, 1976a,b, 1980a,b) as well as a variety of targeted gene mutations affecting cytoskeleton, membrane trafficking, DNA replication and repair, and telomere synthesis (among others). The latter will be referenced as they are discussed.

D. Temperature Shock

The oldest studies into ciliate cell-cycle control involved subjecting *Tetrahymena* cells to heat shock and characterizing the developmental delays that ensued. The development of the oral primordium (stomatogenesis) provides a richly detailed timer of developmental events late in the cell cycle (Frankel and Nelsen, 2001; Kirk *et al.*, 2008). We have recreated the events of oral development in Fig. 4, correlating them with nuclear events.

Following cell division in enriched medium, individual *Tetrahymena* cells typically grow and divide again within about 150 min at 30 °C. When cells are subjected

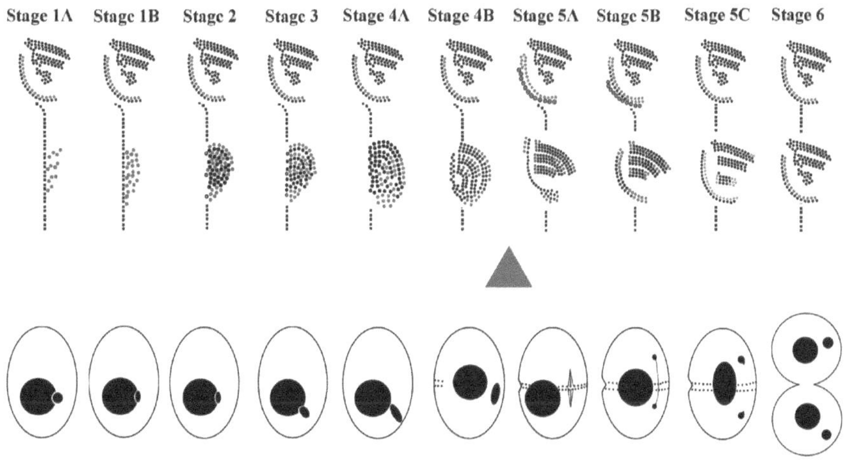

Fig. 4 This figure attempts to correlate cortical events associated with stomatogenesis with nuclear events. It draws heavily on published accounts (Kirk *et al.*, 2008; Lansing *et al.*, 1985; Nelsen, 1981). In the upper panels, basal body proliferation is detailed leading to stomatogenesis. (The stages are modified from those described by Lansing *et al.*, 1985 and Nelsen, 1981.) Black dots represent ciliated basal bodies. Grey dots represent unciliated basal bodies. Grey dots with black circles represent basal bodies from the mature OA that are destined for resorption. During stage 1, basal bodies proliferate from ciliated basal bodies in the first post-oral ciliary row where they assemble into an "anarchic field," and the MIC begins to elongate "in pocket." In stage 2, ciliation begins (black dots) defining the future oral membranelles. These ciliated basal bodies become coupled with unciliated basal bodies (it is likely they nucleate synthesis of their doublet partners). By stage 3, the basal body couplets have aligned into curving "pro-membranelles," and the MIC has angled out of the pocket. By stage 4A, a second wave of ciliation has begun, and the MIC has attached to the cortex. In stage 4B, a third row of basal bodies begins to form along each membranelle, a chaotic string of basal bodies marks the forming UM, and the MIC has left the MAC pocket to align at the cortex in an anterior-posterior orientation. The end of this stage marks the beginning of the "Physiological Transition Point" (red triangle) at which heat shock and other treatments can no longer prevent subsequent developmental events. In stage 5A, the triple-row membranelles swing into a transverse orientation, one row (the inner row) of UM basal bodies is now ciliated, and a second (outer) row is being formed. Stage 5A also marks the start of MIC anaphase, and the mature OA begins to disassemble the outer row of UM basal bodies, replacing it with a new "outer row" that forms in between. The fission zone is now prominent at midbody, and an asymmetric furrow-cleft has formed. In stage 5B, the OP UM has two rows, the inner is being de-ciliated and the outer is being ciliated. In the mature OA, the new outer UM row is being ciliated in synchrony with the OP outer UM row. MICs enter late anaphase. In stage 5C, the MIC division is complete, the MAC begins to elongate and OP ciliation is nearly complete. In stage 6, a depressed, buccal cavity forms in the OP (and reforms in the OA), the MAC divides, and cytokinetic furrow is prominent. (For interpretation of the references to color in this figure legend, the reader is referred to the web version of this book.)

to a heat shock followed by a return to normal temperatures, they exhibit a delay in the onset of their next cell division that is longer than the duration of the heat shock itself (Thormar, 1959). This "excess delay" phenomenon becomes more pronounced the later in the cell cycle that the shock is delivered (Zeuthen, 1958). This holds true up to a "transition point," after which a heat shock no longer

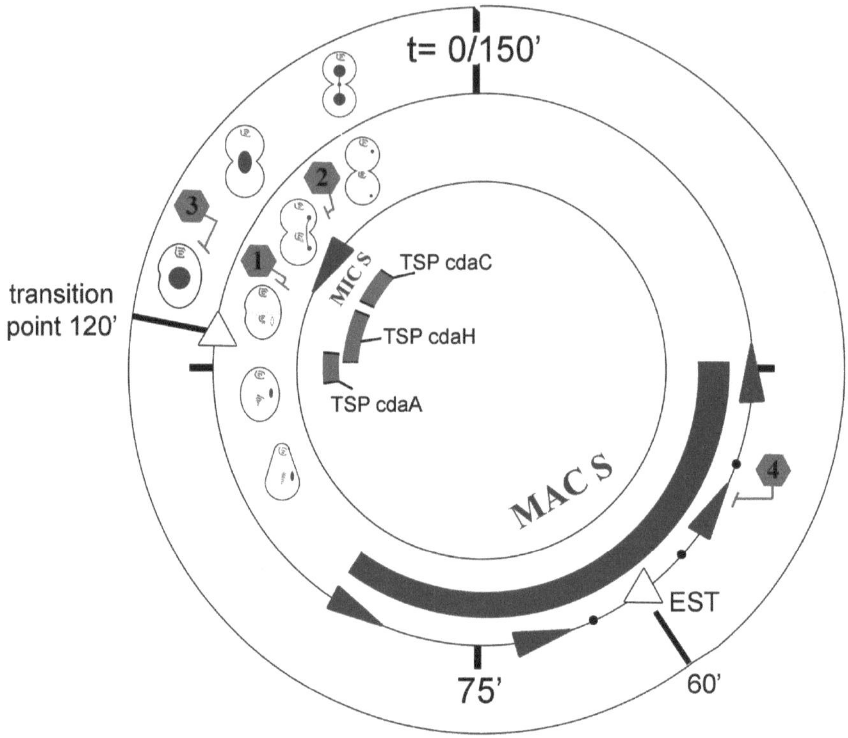

Fig. 5 This figure attempts to summarize a number of significant cytological events and observations. The outer circle represents the roughly 150 min cell cycle one observes at 30 °C. The transition point is marked (at 120 min), as is the "end of the synchronization treatment" at approximately 60 min (90 min before division) (EST). In the inner circle, the heat-sensitive periods for three of the most thoroughly studied TS-mutants are noted (CdaA, CdaC, and CdaH). In the next circle outside, the period of MAC DNA synthesis in the lower right, is delineated in blue. MIC DNA S is shown in the upper left (also in blue, triangle). Rough sketches of cell division stages are illustrated around the transition point attempting to show the stage4B/5A transition point, and stage sin MAC division are represented in the outer-most circle (upper left). Four red hexagons represent "checkpoints" demonstrated by a number of investigations (see text). They are (1) The MIC early-anaphase arrest seen when DNA damage is left unrepaired [yet the cell is not blocked by the MAC intra-S checkpoint (#4)]. MICs arrest in early anaphase and are segregated to one or the other daughter cell. (MIC division is arrested without impact on other pathways.) (2) Late anaphase arrest is seen when chromosomes fail to separate. This checkpoint arrests all pathways (cortical and nuclear) and appears to be triggered by a mechanical failure in chromosome separation. (3) MAC fission block. This appears when MAC fission is delayed, or when excess MAC DNA is shed via chromatin elimination bodies. It is not an arrest, but a delay, and other cycles appear un-involved. (4) Intra-S-checkpoint: This is a cell-cycle arrest in interphase during MAC DNA synthesis. It is triggered by DNA-damaging agents (in cells with an intact ATR-checkpoint mechanism). (See color plate.)

interferes with the completion of cell division. Curiously, the excess delay is typically about 15–20 min following a heat shock delivered early in the cell cycle, but increases dramatically as the time of onset of the heat shock approaches the "transition point" (Fig. 5).

When cells are subjected to a schedule of repetitive heat shocks (alternating from 28 to 34 °C for *T. pyriformis,* and 35 to 42.3 °C precisely for *T. thermophila),* with 30-min intervals between temperature shifts, cells collect at a characteristic stage early in stomatogenesis (late stage 1B, Fig. 4) with their micronuclei slightly elongate (Holz *et al.*, 1957; Gavin, 1965). Cells then resume oral development and undergo a synchronous division burst at 55–60 min (in *T. thermophila)* after the end of the last synchronizing shock (Gavin, 1965). It is the pooling of cells at a common developmental stage following repetitive heat shock treatment that lent such utility to this technique in driving cell synchrony (Scherbaum and Zeuthen, 1954). Furthermore, these studies led to the postulation of a heat-labile protein whose accumulation might trigger cell division (Zeuthen, 1964). It is worth noting that Zeuthen's description of a "division protein" occurred nearly 30 years before the discovery of cyclin.

Williams and Macey (1991) commemorated Zeuthen's insights by demonstrating that a 60-kDa *Tetrahymena* protein exhibited a cyclin-like pattern of accumulation and degradation following progress through the cell division cycle, and that heat shock led to its premature degradation. These observations supported the notion that Zeuthen's heat-labile "division protein" was one of the first published references to a "cyclin" protein in the literature. A *Tetrahymena* cyclin-dependent protein kinase TtCDK1 was cloned and characterized in 2002 (Zhang *et al.*, 2002), and recently, with the publication of the *Tetrahymena* genome (Eisen *et al.*, 2006), a family of 26 *Tetrahymena* genes exhibiting cyclin domains was identified and their individual expression profiles through conjugation were determined (Stover and Rice, 2011). Nine CDKs have also been identified (Stover, personal communication).

When cells are analyzed morphologically, it can be seen that if additional heat shocks are delivered early after the end of the synchronizing heat shocks (Stage 1 of oral development, Fig. 4), the cells pause briefly and then resume development following termination of the shock (Gavin, 1965). Cells subjected to heat shocks delivered at stage 2 will revert to stage 1, and then proceed normally. Following shocks delivered later, especially after basal bodies have begun to organize into promembranelles (Fig. 4; stages 3, 4), cells resorb the entire oral primordium, and are set back to an earlier stage of the cell cycle (Frankel, 1964). It is possible that the extra time required to disassemble the developing oral primordium before initiating a new round of stomatogenesis accounts for the accentuated delay in cells subjected to late stage heat shock treatments (Williams, 1964).

It is worth noting that there may be strain-specific differences to the precise staging of oral development that correlates with the "transition point." In *Tetrahymena pyriformis,* the transition point appears to vary between early stage 4 (Williams, 1964) and the stage 4–5 boundary (Frankel, 1962) depending upon the specific strain used, whereas for *T. pyriformis* WH-6 (later renamed *T. thermophila),* the transition point appears in late stage 4 to early stage 5 (Gavin, 1965). Development typically continues to completion following heat shocks delivered after the transition point; hence, the "transition point" for cell division corresponds to a "stabilization point" for the developing oral apparatus.

There are indications from studies using a high concentration of the protein synthesis inhibitor cycloheximide (Gavin and Frankel, 1969) that the extended excess delay of cell division, and the resorption of stage 3–4 oral primordia, are not consequences of a direct dependence of these processes upon protein synthesis, but rather reflect the operation of an active "checkpoint" mechanism that can be triggered by heat shocks or other environmental insults (Frankel, 1964). It is uncertain how the difference between the amicronucleate condition of *T. pyriformis* and the presence of a micronucleus in *T. thermophila* might play into this issue, but it seems that the specific events during oral development *per se* are unlikely to be the direct drivers of this "checkpoint" mechanism, and the micronucleus may play an active role in those *Tetrahymena* species that have one (see below).

It is tempting to hypothesize about the role of cyclins in these experimental phenomena. First, heat-labile cyclins may well be driving cortical development in *Tetrahymena.* If so, then when they are destroyed by heat shock, the cells must rebuild their pool of active protein before advancing. Pursuing this thinking, it is apparent that the early stages of stomatogenesis can survive a heat-shock-induced ebb in the cyclin titers (or whatever is destroyed by heat shock). The late stage 2 primordium may simply wait for the cell to rebuild its cyclin reservoir and pick up where it left off. What is curious is that as soon as the anarchic field of basal bodies begins to form couplets, and these couplets align into pro-membranelles, the whole oral primordium becomes sensitive to complete demolition triggered by heat shock (again possibly by eliminating the cell's pool of cyclin). Another transition takes place in late stage 4B or early stage 5A primordia, in which the primordium stabilizes and becomes insensitive to heat shock, and the process of its continued maturation is similarly robust. It is difficult to be absolutely sure, but it appears that this stage corresponds with the following suite of characteristics: the undulating membrane [a curving "C"-shaped double row of basal bodies that cups around the three principal membranelles (M1,2,3) forming the oral apparatus, Figs. 1 and 4] is half-way organized, a triplet row of basal bodies is assembled anterior to the existing doublet row in each pro-membranelle, a gap appears in the somatic ciliary rows marking the future cleavage furrow, and the micronucleus enters anaphase. It is worth noting that micronuclear anaphase also marks the beginning of DNA synthesis or MIC S phase (Doerder and Shabatura, 1980). It is tempting to propose that early events in stomatogenesis (basal body proliferation, doublet formation, and alignment of doublets into pro-membranelles) are driven progressively by cyclin-dependent kinase activity. We might also propose (although the careful cytological analysis has not yet demonstrated this definitively) that this corresponds with events leading up to micronuclear metaphase (Stage 4B, Fig. 4). The transition from stage 4B to 5A is a critical period. Temperature shock delivered here triggers demolition rather than a simple developmental pause. This stage may represent some type of temperature-sensitive checkpoint. If micronuclear anaphase initiates well, a suite of cortical events proceeds including completion of organization of the undulating membrane, completion of the third row of basal bodies anterior to each of the promembranelles, and a gap forming in the equatorial region of the somatic ciliary rows foreshadowing assembly of the fission zone. If the metaphase to anaphase transition encounters heat-shock-induced problems, a separate program of oral degeneration may be triggered. Beyond this heat-sensitive checkpoint (red triangle in Fig. 4), development becomes relatively insensitive to agents that block development. We might imagine that cell-cycle proteins necessary for initiation and maintenance of early- and mid-stage events initiate a late-stage assembly program that no longer requires those proteins.

Early reports held that micronuclear "S" phase occurred late in micronuclear division (McDonald, 1962; Prescott and Stone, 1967). A more accurate measurement was reported by Woodard *et al.* (1972), pinpointing late anaphase as the time of micronuclear DNA replication. If we are allowed to extrapolate from cytofluorimetric analyses of conjugal mitoses, it appears that MIC DNA synthesis may be triggered even earlier during anaphase, even as chromosomes are separating (Doerder and DeBault, 1975). This might make onset of DNA synthesis concurrent with the cell's physiological "transition point." It is interesting to note the differences and greater complexity of cell-cycle events in *Tetrahymena* as compared to other eukaryotic cells. Most cells are characterized by a G1-S-G2-M cell division cycle. This roughly coincides with the MAC amitosis-G1-S-G2 pattern seen in *Tetrahymena.* Curiously, the MIC, which undergoes a slightly more conventional mitosis, is dramatically out of phase with the MAC cycle. M and S are compressed into a common interval for the MIC, giving us a pattern that looks like this: M/S-G2-M/S, with no apparent G1 interval.

On a final note, ciliates have evolved an elaborate cortical architecture compared to other cells, and the cyclical remodeling of that architecture has had to integrate with the machinery driving the cell division cycle. It is quite likely that we will find cyclins and their kinases co-opted to coordinate cortical events with nuclear events. On this note, Zhang *et al.* (2002) demonstrated that a cdk protein (Ttcdk1) localizes to a ring around each basal body in the cell cortex changing localization from a circumferential to punctate pattern during cell division. This unusual pattern of cdk localization suggests that cyclins and their associated kinases may be intimately associated with patterns of basal body duplication. Clearly, the time is right for a return to these interesting phenomenological discoveries armed with molecular tools.

E. Cell Division Mutants

In the 1970s a concerted effort was made to perform saturation mutagenesis on *T. thermophila* using a visual screen of TS-fission arrest phenotypes (See Frankel, 2008 for review). Of the gallery of 11 mutant loci identified, the two most thoroughly studied, and most relevant to an analysis of checkpoints and developmental dependencies, were cdaA and cdaC (also known as mo1 and mo3, respectively).

The *cdaA-1 (mo1^{a})* phenotype shows a 10 min temperature-sensitive period just prior to appearance of the fission zone (and just preceding the physiological transition point, Fig. 4, stages 4b–5a; see also Fig. 5) (Frankel *et al.*, 1980a). Heat-induced phenotypes of temperature-sensitive mutants, including *cdaA-1*, can be triggered at 36 °C, a temperature below that which induces a heat-shock delay (Frankel *et al.*, 1980b). *cdaA-1* mutants exposed to high temperature at this stage complete oral development, yet fail to form fission zones, and fail to undergo cytokinesis. Curiously, when maintained at the restrictive temperature and fed, these cells underwent multiple rounds of micronuclear division and stomatogenesis, suggesting that cytokinesis could be blocked without disrupting either of these other two pathways. Macronuclei generally failed to divide, and then arrested (Frankel *et al.*, 1976b). It is unclear whether this was because fission arrest is coupled to mac division, or whether the cell is monitoring total macronuclear DNA content and inhibiting macronuclear S as it becomes multi-nucleate. The simplest interpretation of this phenotype is that there is a cdaA-dependent step in assembly of the future fission zone, and that blocking this process does not interfere with progression through mic, mac, or cortical development.

The cdaC phenotype is another fission-zone phenotype whose temperature-sensitive period is considerably later than that for cdaA, falls after the transition point, and directly coincides with the process of cytokinesis (Fig. 5). These cells assemble cortical features that mark the future fission zone, and even initiate cytokinesis, yet fail to separate individual daughter cells (Frankel *et al.*, 1980a). In this mutant, fission block does not appear to interfere with other cell-cycle programs, revealing that their progression is independent of successful completion of cytokinesis.

F. Pseudomacrostome Mutants

Another group of conditional mutants result in cell-division failures. These are the four "pseudomacrostome" mutants *(psmA-D,* Frankel *et al.*, 1984). These mutants appear to bypass cell division altogether by rerouting the cell through an alternative (albeit aberrant) developmental pathway. Oral replacement is a developmental pathway that *Tetrahymena* undergo in response to starvation (Frankel, 1969a). It can occur on its own (Kaczanowska *et al.*, 2008), in conjunction with post-conjugal events (Cole and Frankel, 1991), or in association with a transformation of starved cells into a fast-swimming dispersal form (Nelsen and DeBault, 1978). The pseudomacrostome mutants appear to undergo oral replacement despite the presence of a complete set of nutrients, and the process appears to be exaggerated resulting in cells with expanded oral replacement fields leading to enlarged oral apparatuses.

Unfortunately, all of the *cda and psm* mutants were developed through chemical mutagenesis, and we do not yet have precise insight into their molecular nature.

G. The *cda12* Mutant

The CDA12 gene encodes a putative membrane trafficking protein that appears to be associated with the recycling endosome (Zweifel *et al.*, 2009). An antisense-ribosome targeted to silence the CDA12 mRNA resulted in a prominent cell division arrest phenotype. This analysis suggested that membrane trafficking in *Tetrahymena* plays a role during cytokinesis, presumably through membrane deposition during furrow ingression. Subsequent to the onset of this phenotype, it was apparent (once again) that blocking cytokinesis does not interfere with either micronuclear fission or stomatogenesis. Again, a limited number of macronuclear fissions occurred raising the question of whether DNA-content assessment might not be playing a role in ending macronuclear DNA synthesis at some point following multiple rounds of mac replication.

The take-home message from fission arrest phenotypes seems to be that cytokinesis, although triggered in a cell-cycle-dependent fashion, is not necessary for either nuclear division cycles or periodic rounds of stomatogenesis and cortical development to proceed.

H. DNA Damage Checkpoints During Vegetative Development

Biochemical and mutational analyses have revealed a set of DNA damage checkpoints in *Tetrahymena* regulating both macronuclear and micronuclear divisions. In most eukaryotic cells, DNA damage resulting in double-stranded breaks (DSBs) activates a cell-cycle checkpoint during S phase (intra-S-phase checkpoint). As the DNA synthetic machinery encounters double-stranded breaks, the replication fork stalls. This results in recruitment and activation of an ATM or ATR master kinase (ATR: ataxia telangiectasia and Rad3-related protein, ATM: ataxia telangiectasia mutated, reviewed by Abraham, 2001; Lambert and Carr, 2005). ATR subsequently activates downstream effector kinases that in turn activate the DNA damage-repair machinery involving Rad51. Several lines of evidence from *Tetrahymena* point to mitotic checkpoints similar to those described in other eukaryotes. There are unique challenges, however, to interpreting data from cells having two nuclei with out-of-phase periods of DNA replication, especially since a MAC S-phase arrest may prevent a cell's entry into mitosis, and therefore mask expression of a MIC S-phase arrest that would occur during mitotic anaphase.

1. MIC and MAC Fission Delays Following DNA Damage and an Intra-S-Checkpoint

When DNA damage is induced and left unrepaired during the cell cycle, either because the DNA-damage checkpoint fails to trigger, or because the DNA-repair mechanism has been compromised, we see characteristic MIC and MAC fission delays during mitosis. MICs appear to enter anaphase, producing an elongate mitotic figure, but fail to complete nuclear fission before cytokinesis. In mild phenotypes, we see cleavage furrows disrupting an incomplete MIC spindle. In more extreme forms, the entire MIC spindle gets distributed to one daughter cell creating an amicronucleate division product of the sister cell. (MICs normally complete fission well before a cleavage furrow has even formed; Fig. 4). In similarly compromised cells, MACs undergo nuclear elongation, but large amounts of chromatin appear trapped at the fission zone. This results in abnormal, extranuclear chromatin extrusion bodies (CEBs) that accumulate in the cytoplasm. A number of treatments and mutations produce a similar set of nuclear misbehaviors.

2. DNA Damage in *Tetrahymena* (a Checkpoint-Mediated Cell Cycle Arrest)

Agents that induce DNA damage [UV radiation, hydroxyurea (HU), methyl methane sulfonate (MMS), and aphidicolin], produce an interphase (MAC-S) cell-cycle arrest (Nachtwey and Giese, 1968; Yakisich *et al.*, 2006, Kaczanowski and Kiersnowska, 2011). Such arrests have been correlated with elevated Rad51p expression in *Tetrahymena* (Campbell and Romero, 1998; Smith *et al.*, 2004a) and are distinct from the heat-shock arrests described above (Nachtwey and Giese, 1968) in that they fall early during macronuclear S phase (Yakisich *et al.*, 2006). Oddly, in the case of aphidicolin treatment (blocking DNA synthesis) multiple rounds of MAC DNA synthesis proceed despite failure of cell-cycle progression in low concentrations of APD. These cells also exhibit arrested development of their somatic ciliature (a subtlety not examined in other treatments). These results implicate a conventional checkpoint mechanism acting during MAC S-phase that prevents cell-cycle progression until DNA damage is repaired.

3. DNA Damage Without ATR-Checkpoint (Bypassing the Cell Cycle Arrest)

Hydroxyurea treatment (resulting in DNA damage) produces a cell-cycle arrest that can be bypassed by exposing cells to caffeine, a reagent that blocks ATR-mediated checkpoints in other systems (Yakisich *et al.*, 2006). HU/caffeine-treated cells that bypass the interphase arrest undergo aberrant nuclear fissions during the subsequent mitosis. Macronuclear divisions appear delayed with respect to cytokinesis producing large chromatin elimination bodies (CEBs) associated with excess Mac DNA (Morrison *et al.*, 2005). This phenotype reminded the authors of the "cut" phenotypes seen in fission yeast mutants (Yanagida, 1998), although there are questions regarding this parallel. Such cells also exhibit delays in early MIC anaphase, with spindles persisting well into cytokinesis, and long after MIC division should be complete.

Aphidicolin-treated cells exhibit a similar intra-S-phase arrest that can be bypassed by treatment with caffeine (Kaczanowski and Kiersnowska, 2011). These cells have excessive DNA damage resulting in pulverized MIC chromosomes. They also show generation of macronuclear CEBs and delayed, elongate MIC anaphase figures.

Finally, we see evidence of an intra-S-phase DNA damage checkpoint in a mutant effecting telomere synthesis. The *POT1a* mutant (a conditional knock-down allele), (Naduparambil *et al.*, 2007), also produces a macronuclear-S-phase arrest that can be relieved by caffeine. It is proposed that Pot1 Ap protects telomeric G-overhangs in *Tetrahymena* from over-active telomerase and also from DNA-damage detection. It is also proposed that without *POT1a* activity,

DNA damage detection is activated in an ATR-dependent way and this arrests the cell in interphase (MAC S phase).

The bypass of a DNA-damage-induced cell-cycle arrest through caffeine exposure suggests that *Tetrahymena* has a robust ATR-dependent DNA damage repair pathway involving a macronuclear, intra-S checkpoint.

4. Mutations that Bypass the MAC Intra-S-Checkpoint

Four different mutations have been studied that may block the DNA-damage (ATR-dependent) checkpoint, or the DNA-repair machinery activated by theATR-kinase. These are mutations in the genes: *TIF1, RAD51*, Histone *H2A*, or the *Tetrahymena ATR* homolog itself (Morrison *et al.*, 2005; Yaksisich *et al.*, 2006; Marsh *et al.*, 2000; Song *et al.*, 2007; Loidl and Mochizuki, 2009).

tif1

The *TIF1* gene product binds to rDNA mini chromosome origins of replication and inhibits "firing" of the origin, thereby regulating the timing of DNA replication (Morrison *et al.*, 2005). *tif1* mutants show premature "firing" of the rDNA replication origin, and are hypersensitive to DNA-damaging agents (Yakisich *et al.*, 2006). Hydroxyurea-challenged *tif1* mutants fail to arrest in S phase or exhibit caffeine-sensitive *RAD51* overexpression, indicating the involvement of *TIF1* in *ATR-like,* checkpoint activation. It is hypothesized that Tif1p may be involved in triggering the ATR-dependent, DNA damage response checkpoint. Of interest to our survey of developmental dependencies, *tif1* mutants (defective in triggering theATR-checkpoint), which are subjected to DNA-damaging agents (HU, MMS), fail to arrest at MAC-S, enter mitosis, and exhibit both the MAC CEB phenotype and a delayed MIC anaphase (Yakisich *et al.*, 2006).

rad51

RAD51 is a gene whose product is intimately associated with repair of double-stranded DNA breaks (Campbell and Romero, 1998). Rad51p is activated via the ATR-kinase pathway in response to DNA damage. *rad51* (knockout) mutants fail to mediate repair of induced DNA damage. This can result in accumulation of DNA damage through successive rounds of DNA replication, even in the absence of DNA-damaging agents. *rad51* mutants exhibit vegetative nuclear division defects during mitosis (Marsh *et al.*, 2000). Cells with no *RAD51* activity enter mitosis, yet exhibit the MAC CEB phenotype generating substantial chromatin extrusion bodies (as described in *tif1* mutants, Morrison *et al.*, 2005) and micronuclei become severely arrested in early anaphase (red hexagon #3 in Fig. 5). Despite this MIC anaphase arrest, cytokinesis proceeds as does MAC division and cortical development. This results in one daughter cell failing to inherit a micronucleus, while the other inherits an early-anaphase-arrested MIC. This null phenotype appears to be a more severe version of the *tif1* (hypomorph) phenotype and the HU/Caffeine phenocopy, yet even this severe response does not interfere with stomatogenesis and subsequent macronuclear division and cytokinesis.

Histone H2A

A site-directed histone mutation results in failure to repair DNA damage by preventing phosphorylation of the histone H2Ap protein (Song *et al.*, 2007). Such cells are compromised in their ability to repair normal DNA damage incurred during MIC mitosis and MAC amitosis. The resulting vegetative phenotype includes cells with multiple chromatin elimination bodies (resembling the CEB phenotype described above), and again, MICs are delayed in anaphase.

atr

Recently a *Tetrahymena ATR* homolog has been identified and knocked out. The resulting cells undergo (unspecified) vegetative defects and MIC instability. No homolog to ATM has been found (Loidl and Mochizuki, 2009).

There is remarkable consensus in the cell-cycle responses described here. When cells acquire DNA damage (exposure to UV, hydroxyurea, MMS, APD) or when cell's "perceive" DNA damage (exposed telomeres), they arrest in macronuclear "S" phase. In some situations, residual MAC DNA synthesis seems to proceed causing a gradual buildup of the MAC DNA content in the otherwise stalled cells. This consistent DNA-damage response argues for a robust DNA-repair checkpoint. When caffeine is applied, cells break their cell-cycle arrest and initiate mitosis. This suggests an ATR-dependent checkpoint mechanism.

When events downstream from the DNA-damage event are compromised: *tif1* (knockdown) mutants fail to activate *ATR*; *rad51* (knockout) mutants fail to mediate repair; mutation of a histone *H2A* results in a failure to repair DNA damage; or when one suppresses the damage-induced ATR-responses (with caffeine), there is no MAC intra-S phase arrest. Instead, cells enter mitosis and exhibit two distinct nuclear phenotypes: a delay in MIC anaphase resulting in segregation of the entire anaphase spindle to one daughter cell (in the most extreme cases), and a MAC that appears "lodged" in the cytokinetic furrow (the severed-MAC or "Cut" phenotype) resulting in accumulation of large, extranuclear chromatin bodies (CEBs) in the cytoplasm.

These latter phenotypes can be interpreted in one of two ways. DNA damage might trigger two checkpoint responses: an ATR-dependent response that arrests cells in macronuclear "S" phase; and an ATR-independent set of "M" phase delays that are only revealed if one suppresses the epistatic "S"-phase response. Alternatively, the MIC anaphase phenotype and MAC-CEB phenotype could be viewed as "mechanical" downstream consequences to sustained chromatin damage. In case of the MIC, either broken chromosomes might lead to anaphase delays through the mechanical difficulties associated with trying to segregate fragments that may have lost centromeres, or there may be a second checkpoint (unrelated to the ATR-triggered checkpoint, perhaps a "spindle assembly checkpoint") triggered when chromosomes have difficulties assembling at the metaphase configuration. The MAC "cut-like" (CEB) phenotype may also be a mechanical consequence rather than an actual checkpoint phenomenon. In many of the S-phase arrests described above (especially those involving complete loss-of-function mutants), MAC DNA continued to increase despite division arrest. When released from checkpoint arrest, these MACs may simply be shedding excess macronuclear DNA through the production of chromatin elimination bodies (Scherbaum *et al.*, 1958; Worthington *et al.*, 1976; Cleffmann, 1980). This matter remains to be resolved.

With regard to developmental contingencies, micronuclear arrest early in anaphase does not affect stomatogenesis, cytokinesis, or macronuclear division. In other words, progress in these other pathways is not contingent on passage out of early MIC anaphase.

I. A Late-Anaphase Checkpoint in Micronuclear Division

The DNA-damage-induced, early-anaphase MIC arrest stands in stark contrast with observations from another kind of telomere defect. Kirk *et al.* (2008) altered the telomerase template in *Tetrahymena*, resulting in ever-shortening micronuclear telomeres. Such phenotypes have been shown in other model organisms to result in aberrant end-joining of chromosomes that can result in chromosome mis-segregation. In *Tetrahymena,* this mutation resulted in a remarkable cell-cycle arrest. The micronucleus entered mitosis and became arrested in a profoundly elongate late-stage anaphase configuration (red polygon 2 in Fig. 5). Unlike the early-anaphase arrest described above whose effects were confined to the micronucleus, this late-anaphase arrest had profound consequences across various cell-cycle theaters. Cytokinesis was blocked, macronuclear fission was blocked, and stomatogenesis arrested in what was described as a "dynamic pause" in which basal bodies continued to proliferate producing grossly elongated oral primordia. Furthermore, basal bodies within the oral primordium nucleated doublets, and these doublets aligned into pro-membranelles. What did not happen was there was no nucleation of the third row of basal bodies, there was no undulating membrane assembly, and no fission zone appeared. This appears to be a different kind of spindle checkpoint operating downstream from the earlier DNA-damage-induced anaphase delay. It is remarkable in that it reveals cross-talk across the MIC-division/stomatogenesis/and MAC division pathways (see red lines, and triangle 3, Fig. 3).

Stomatogenesis, cytokinesis, and macronuclear fission all depend on a cell traversing a late MIC-anaphase checkpoint.

J. The Independence of Macronucleus Fission

There is ample evidence that rounds of MAC DNA synthesis (when untroubled by DNA damage) are relatively uncoupled from other aspects of cell-cycle control. This can be seen from reports that *Tetrahymena* with unusually large amounts of MAC DNA can skip a period of DNA synthesis, while others with low DNA content can undergo multiple rounds of MAC DNA synthesis within a single cell cycle (Cleffman, 1968; Doerder and DeBault, 1978; Doerder, 1979). There is also evidence showing that disrupting MAC division does not greatly impact other aspects of the cell division cycle. In a paper by Smith *et al.* (2004), a dominant beta-tubulin mutation resulted in failure of macronuclear fission. Despite failure of the poly-copy macronucleus to divide, cytokinesis, MIC division, and stomatogenesis appeared relatively unaffected. A myosin mutation resulted in a similar phenotype in which the MAC failed to elongate or divide without affecting other aspects of the cell division cycle (Williams *et al.*, 2000).

Completion of the amitotic macronuclear fission does not significantly affect the other cell-cycle pathways.

K. Micronuclear Persistence and Cell Division

The presence of some minimal MIC chromatin content has been shown to be crucial for cell-cycle persistence (reviewed by Wloga and Frankel in this volume). In particular, loss of a micronucleus typically results in a failure to maintain the oral apparatus, cell crinkling, and ultimately cell death (Kaczanowski and Kiersnowsa, 2011; Haremaki *et al.*, 1995, 1996). These observations suggest a positive, and necessary (most likely structural) role for the MIC in maintaining various aspects of cortical architecture.

Several situations can lead to progressive MIC chromosome fragmentation and loss. Cells treated with aphidicolin and caffeine sustain massive MIC DNA damage. This can lead to progressive aneuploidy resulting in MIC elimination. This usually leads to oral resorption, crinkling, and death (Kaczanowski and Kiersnowsa, 2011). Cells that retain some MIC chromatin fragments survive and replicate successfully. When cells are subject to genetic knockdown of the *CNA1* gene (a centromeric histone *H3*), they, too, undergo a progressive MIC deterioration leading to aneuploidy (Cui and Gorovsky, 2006). Complete *cna1* knockouts arrest and die after only 10 cell divisions (Cervantes *et al.*, 2006). The detailed cytology of these progressive aneuploids was not reported.

Cells that become completely amicronucleate die (Haremaki *et al.*, 1995; Ng, 1986). Exceptions to this rule are the large number of amicronucleate cell lines recovered from natural populations (Nanney and McCOy, 1976), and a single laboratory strain (Kaney and Speare, 1983), which reportedly has incorporated some MIC-specific DNA within its MAC genome (Karrer *et al.*, 1984). Cui and Gorovsky (2006) make the interesting suggestion that: "It may be the specific loss of centromeres that is responsible for the lethality in *cna1* knockout cells, possibly because complete loss of centromeres triggers a checkpoint mechanism in growing cells, which can be bypassed as long as centromeres are present and can segregate." It would be interesting to learn whether MIC centromeres themselves provide some positive signal necessary for maintaining the oral apparatus and cortical integrity.

L. Development of the Somatic Ciliature and Fission Zone

Not only *Tetrahymena* must assemble their oral primordium, CVPs, and cytoproct prior to each cell division, but new somatic basal bodies must be generated as well. (Kaczanowska *et al.*, 1993; Kaczanowska *et al.*, 1999; Kaczanowski, 1978, Nanney, 1975). Very few studies have explored the interdependence of somatic ciliature with other aspects of the *Tetrahymena* cell-divison cycle. These phenomena are reviewed by Wloga and Frankel (this volume).

M. Overview of the Developmental Logic of Cell Division

In summary, we find evidence that the MAC division cycle is largely independent of most of the other cell-cycle-related programs (see Fig. 3). Blocking MAC division has little effect on other aspects of the cell division pathway, and cells can vary the number of rounds of DNA synthesis associated with MAC replication during a single cell division cycle. For this reason, in Fig. 3, we have drawn independent lines from the presumed cyclin archive toward both MAC DNA synthesis and MAC division. Evidence does support a positive role for cytokinesis in helping macronuclei divide, yet this is clearly not a strong dependency. Cytokinesis failures do not inevitably lead to MAC fission failure. In cases where MAC divisions progress up to a point and then cease following a cytokinesis arrest, one could argue that the sheer amount of MAC DNA present in a multi-MAC syncytium results in down-regulation of further MAC synthesis and division. We have found that MIC synthesis and division, stomatogenesis (cortical development), and cytokinesis exhibit some developmental interdependencies. Early anaphase arrests (induced by DNA damage) can result in the segregation of the entire mitotic MIC apparatus to one daughter cell, and yet cytokinesis and cortical development proceed unaffected. Later anaphase arrests, on the other hand (induced, presumably, by chromosome end-joining), can cause an arrest that extends from the MIC division pathway to stomatogenesis and cytokinesis and even MAC division. While cytokinesis may depend on successful events occurring during MIC mitosis, the opposite is not true. Blocking cytokinesis has little effect on subsequent rounds of nuclear division or cortical development. Regarding the interdependence of stomatogenesis and MIC division, it is difficult to tell whether events occurring at the cell cortex impact MIC division dynamics. Environmental-shock and temperature-sensitive mutant phenotypes that affect stomatogenesis also seem to arrest MIC division, and vice versa. It would not be surprising to learn that specifically blocking oral apparatus assembly resulted in feedback that arrested MIC division, but we have not seen evidence to support this. With the discovery of the *Tetrahymena* cyclin and CDK repertoire, it seems the time is ripe to return to these many cell-cycle-related phenomena and unravel their molecular mechanisms.

IV. Alternatives to Cell Division

Nutrient starvation has been shown to induce three different developmental pathways, depending on conditions: oral replacement, metamorphosis into a dispersal form, and conjugation (or sexual reproduction). Oral replacement is commonly triggered in starved cells within 2 h of starvation (Frankel, 1969a; 1970; Nelsen, 1981). In this pathway, the mature OA is disassembled and resorbed while a new primordium (the oral replacement primordium, ORP) assembles just posterior to the degenerating OA. Another alternative to the cell division pathway (or to conjugation) is the metamorphosis of *Tetrahymena* into a fast-swimming dispersal form, which occurs in conjunction with oral replacement (Nelsen and DeBault, 1978; Nelsen, 1978, 1981). These pathways are described in more detail by Wloga and Frankel (this volume). Sexual reproduction is described in detail below.

V. Conjugal Development

A. Pre-pairing events

Events leading to formation of a mating junction between *Tetrahymena* cells have been reviewed extensively elsewhere (Cole, 2006). In brief, cells must first be starved in a process referred to as "initiation" (Bruns and Brussard, 1974). Following starvation-induced initiation, cells of complementary mating type must be in physical contact with one another for 1–2 h in order to undergo "co-stimulation" (Bruns and Palestine, 1975; Finley and Bruns, 1980; Wellnitz and Bruns, 1982). During this stage, there is a dramatic remodeling of the anterior cell cortex involving novel membrane synthesis, and glycoprotein "capping" otherwise known as tip transformation (Wolfe and Grimes, 1979; Watanabe *et al.*, 1981; Wolfe, 1982; Wolfe *et al.*, 1986).

Mating pairs form first loose contacts, and then a region of tight membrane–membrane adhesion separated by a 50-nm gap (Wolfe, 1985). This mating junction or fusion plate becomes perforated by hundreds of evenly spaced 0.1- to 0.2-μmn pores, the result of limited, focal membrane fusions. It appears that these pores expand, ultimately transforming the perforated fusion plate into a curtain of branching membrane tubules with a uniform diameter (roughly 90 nm, Orias 1986; Orias *et al.*, 1983). By 2 h after mixing, stable pairs form, and cytoplasmic materials begin to be exchanged (McDonald, 1966), transferred through the expanding pores of the exchange junction. Later, the expanded pores actually permit exchange of entire nuclei. More recently, a method of purifying exchange junctions has allowed mass spectrometric analysis of some of its protein constituents including the epiplasm constituent, fenestrin (Cole *et al.*, 2008).

B. Nuclear Events During Conjugation

Nuclear events during conjugation consist of meiosis, gametic pronucleus formation, nuclear exchange, fertilization, nuclear differentiation, and apotosis-like nuclear absorption (Figs. 6 and 7). There are six MIC divisions, three prezygotic

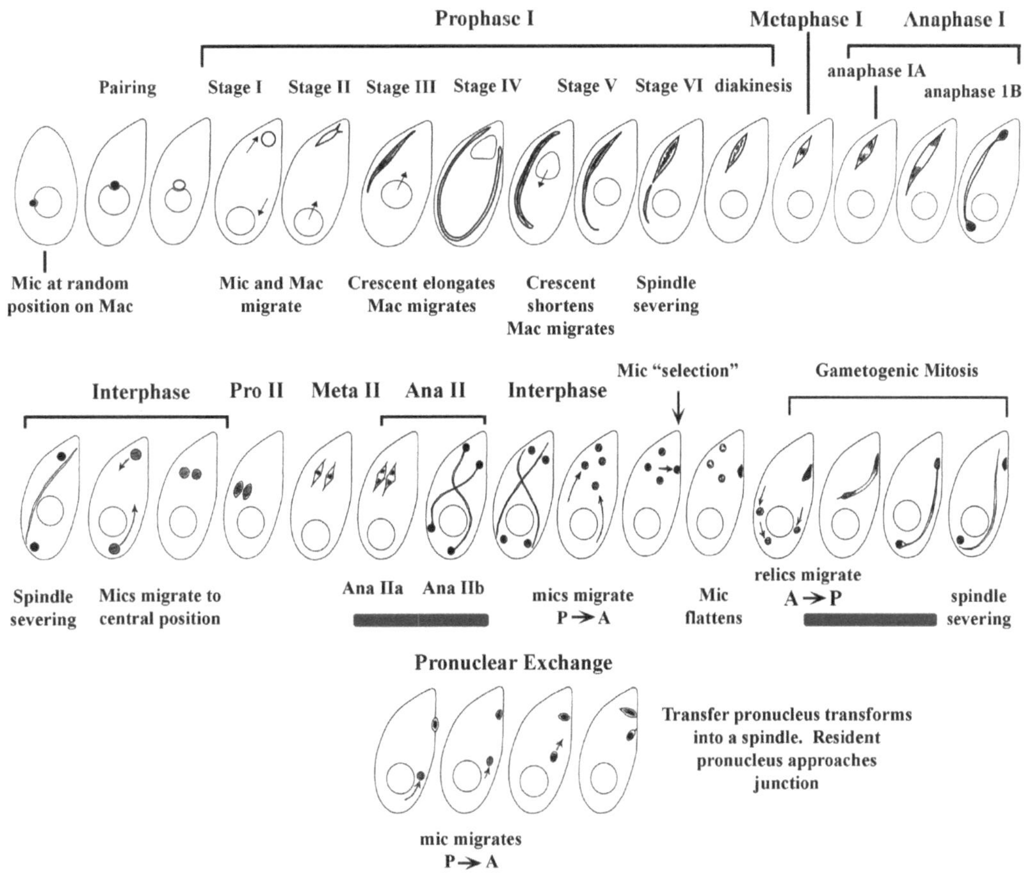

Fig. 6 These illustrations show early conjugal development leading up to pronuclear exchange. Each image represents one partner of a mating pair. Blue bars represent periods of DNA synthesis. Details appear in the text. (See color plate.)

divisions, and three postzygotic divisions to resume the original state with one MAC and one MIC in a cell. Each division is unique in the size and morphology of the dividing nucleus. As in vegetative mitosis, when a conjugal MIC divides mitotically or meiotically, it forms a long separation spindle and divides into three portions: two round nuclei that contain the chromosomes and one separation spindle. The position of the MAC also changes depending on the stage. The following represents a compilation of observations made during live-cell imaging that are largely unpublished (Sugai, unpublished).

1. Initiation

Cells must undergo a period of starvation in order to be mating reactive. During this period, cell size is down-regulated. If the cell is large, a pre-conjugal cell division

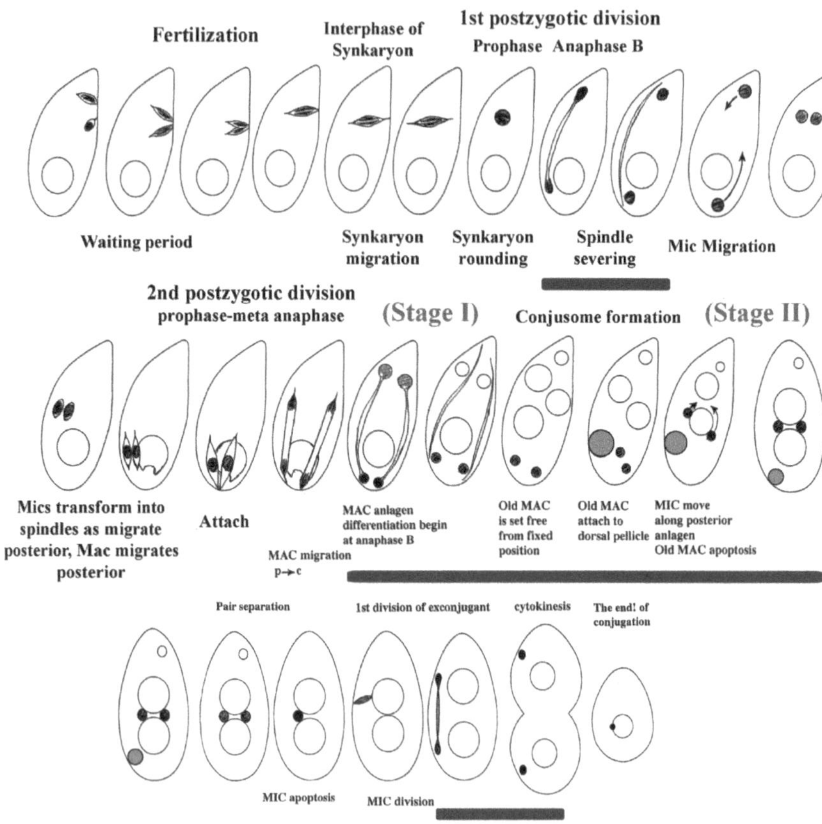

Fig. 7 These illustrations show late conjugal developments leading from pronuclear exchange to pair-separation, nuclear elimination, and re-entry into the cell division cycle. Blue bars represent period of DNA synthesis. Red "Stage I" and "Stage II" represent nomenclature conventionally used to indicate early macronuclear anlagen stage (MA still in the anterior cytoplasm), and late macronuclear anlagen stage pairs (with MAs lining up vertically and MICs nestled into the space between the two macronuclear anlagen). (See color plate.)

will occur. During initiation, the number of food vacuoles is also reduced. Just prior to conjugation, the MIC position with respect to the MAC is random relative to the cell's polarity.

2. Co-stimulation

Co-stimulation involves physical contact between starved cells of complementary mating type. During this period, the MIC position on the MAC moves to the anterior half of the MAC.

3. Meiosis I (first prezygotic division)

Prophase of meiosis I has been divided into six stages (Sugai and Hiwatashi, 1974, Fig. 6).

Stage I

Stage I begins 30 min after pair formation. The MIC swells while attached to the MAC, detaches from it, and then migrates anteriorly. The MAC migrates to the posterior of the cell.

Stage II

The MIC begins to elongate from anterior to posterior along the dorsal side of the cell, becoming spindle shaped. The MAC begins to return to a central position.

Stage III

The MIC continues to elongate along the dorsal cortex toward the posterior end of the cell.

Stage IV

The MIC achieves maximum elongation during this "Crescent" stage. It curves around the posterior end of the cell along the ventral side extending to the anterior end of the cell. Its length becomes twice that of the cell. Both ends seem to attach to the conjugation junction. The MAC also comes to lie near the junction. The MAC's shape is no longer round but conical.

Stage V and VI

The crescent shortens and distribution of chromatin becomes uneven. One end of the MIC becomes slender.

Metaphase I

This is the largest spindle formed during conjugation. It is located anterior to the MAC. A clear metaphase plate is formed. A thick bundle of microtubules attached to bivalent chromosomes can be seen. The centromeric chromatin of the bivalent is pulled long toward pole before anaphase begins.

Anaphase I

Chromosomes segregate within the intact nuclear envelope and migrate toward the anterior and posterior ends of the cell. In the early stage (Anaphase A), the MIC appears spindle shaped. After reaching a critical length, the chromatin rounds up into two spherical masses that are propelled to opposite ends of the cell by a separation spindle (Anaphase B).

Interphase

The posterior MIC migrates into the anterior cytoplasm coming to lie near the anterior MIC product.

4. Meiosis II (second prezygotic division)

Metaphase II

Two spindles lay side by side, anterior to MAC. A clear metaphase plate is formed. Metaphase plates are clearly visible only during meiosis.

Anaphase II

Anaphase A and anaphase B occur similar to anaphase I. The separation spindle is thin and very long, the longest among all spindles throughout life cycle. After division, two meiotic products are located anterior of the cell and the other two are posterior.

5. Interphase

The two posterior meiotic products move toward the anterior end of the cell joining the other two meiotic products. Together the four meiotic products reside near the conjugation junction. One of the four meiotic products appears to be pulled to the lower right side of the junction where it becomes attached. This process has been referred to as "nuclear selection." The round, selected nucleus located at the junction becomes flattened.

6. Programmed nuclear degeneration (PND)

The three unselected nuclei move toward the posterior end of the cell and become pyknotic and are eventually absorbed. These have been referred to as "relics." It is likely that alteration of these three meiotic products begins while they are near the junction as heterochromatin begins to form.

7. The third prezygotic or gametogenic division

The flattened, "selected" nucleus becomes triangular and divides obliquely toward the posterior half of the cell. One end of the spindle remains firmly attached to the junction. The spindle lies slightly ventral. No metaphase plate is visible. The separation spindle is not as long as the second meiotic spindle. After completion of this division, stationary and migratory pronuclei are formed. The migratory (or *transfer*) pronucleus remains at the junction while the stationary (or *resident*) pronucleus becomes located more posteriorly.

8. Migration of resident pronucleus

The resident pronucleus migrates toward the conjugation junction along the ventral side of the cell. As it approaches the junction, the round resident pronucleus gradually becomes spindle shaped and attaches to the junction by one of its pointed ends.

9. Pronuclear exchange

While the resident pronucleus migrates, the transfer pronuclei are exchanged. The transfer pronucleus is flattened on the junction before exchange and transfers without rotation. After exchange, the transfer pronucleus becomes spindle shaped and attaches to the conjugation junction by one of its pointed ends. The resident pronucleus approaches the junction, becomes spindle shaped, and attaches to the junction by one end just posterior to the now anchored exchange pronucleus.

10. Fertilization

The attached ends of the two pronuclei are very close each other at the junction. This arrangement lasts for a while (the "waiting period"); then the two spindle-shaped pronuclei make contact near their attachment with the junction. Fusion of nuclear envelopes starts at this contact region and spreads to the other end within 20 s.

11. Interphase

The fused nucleus, the synkaryon, is still attached to the junction and remains spindle shaped. Then it detaches from the junction and comes to lie anterior of the MAC. The chromatin is thick and thread-like and lies parallel to the long axis of the spindle-shaped nucleus. This might be considered the interphase of the synkaryon.

12. First postzygotic division

The spindle-shaped synkaryon assumes a round shape for a very short interval before dividing. This spindle is relatively large. After division, the posterior nucleus migrates back to the anterior cytoplasm and becomes positioned side by side with the other nucleus.

13. Second postzygotic division and macronuclear differentiation

Both nuclear products of the previous division are round in shape and migrate toward the posterior cytoplasm in side-by-side position. During this migration, the round nuclei change gradually into spindle shapes with a round chromatin mass in their centers (this could be a prophase or metaphase configuration). The spindle-shaped nuclei move to the cell's posterior and attach at one restricted point of the cell cortex at the very posterior end of the cell. The parental MAC and numerous cytoplasmic granules also migrate to the posterior. The rounded chromatin in the center of each nucleus disperses and anaphase begins. The dividing nucleus remains attached to the posterior end and the spindle extends toward the anterior. Before completion of anaphase B, the anterior, round portion that contains a 2N set of chromosomes begins swelling. This represents the first visible change initiating differentiation of the macronuclear anlagen (MA). The swollen anterior portion of the spindle is located near center of the cell, not at anterior end of the cell. Anaphase ends, leaving young MAC anlagen in the center and new MICs at the posterior end of the cell. These spindles are relatively short. During the initial swelling of macronuclear anlagen, the conjusome appears at anterior region of the cell (Janetopoulos *et al.*, 1998). This is a non-membrane-bound organelle that transiently houses a number of proteins destined for the developing macronuclear anlagen during an intense period of genome reorganization (See Karrer, this volume).

14. "Endgame."

The parental MAC loses its central position and moves into a posterior region of the cell where it become pyknotic and is absorbed. Chromatin associated with the parental MAC becomes pyknotic quite rapidly. The two MAC anlages are positioned tandemly in the center of the cell. The two MICs at the posterior end of the cell move anteriorly toward the posterior-most MAC anlagen, and, sliding along the surface, assume a position between the two anlages. As pairs separate, one of the MICs disappears. The timing of MIC resorption and parental MAC resorption with regard to pair separation shows some variability. Following MIC resorption, and if switched to nutrient medium, the cells will undergo a unique division in which the existing pair of MACs are segregated (without fission) to the daughter cells, while the surviving MIC undergoes a sixth, mitotic division, thereby restoring each daughter cell to its vegetative state with one MAC and one MIC.

VI. Developmental Disruptions

A variety of experiments that disrupt conjugation have provided insight into the developmental dependencies and checkpoints that regulate passage through the underlying developmental program. These include physical disruptions, applications of pharmacological inhibitors that target cytoskeleton, biosynthetic pathways, and signaling pathways, and genetic changes ranging from large-scale aneuploidy within the germ-line micronucleus, through randomly targeted mutations and ultimately to specifically targeted gene disruptions. What this literature has uncovered is a richly regulated program with complex gating of cellular passage through meiosis and nuclear selection, a dramatic threshold leading from the pre-zygotic to post-zygotic developmental programs, and even regulation of the exit from conjugation and re-entry into the vegetative cell division cycle once conjugation is complete. The following is an attempt to summarize these studies, highlighting evidence that sheds light on the developmental logic of conjugation.

A. Physical disruption

1. Pre-pairing Events

Tetrahymena require a period of nutrient deprivation for a minimum of 45 min in order to be competent to form pairs. This "initiation" event can be disrupted in a number of ways. A hyperosmotic shock can erase the initiation-driven events if applied after 34 min of starvation. A 15-min exposure to food can also erase whatever program of events are started by the starvation interval (Wellnitz and Bruns, 1979, 1982). Starved cells must also undergo a period of "co-stimulation" involving physical contact with cells of a complementary mating type. This obligatory period of contact lasts between 1 and 2 h (Bruns and Palestine, 1975; Finley and Bruns, 1980) and leads to differentiation of a fusion plaque at the anterior end of the cell (reviewed by Cole, 2006). Pair formation can still be inhibited following initiation and co-stimulation if cultures of mixed cells are kept under mild mechanical agitation.

2. Mechanical Pair-Separation

In two studies, mating pairs have been disrupted by mechanical means, glass-bead vortexing in one case (Virtue and Cole, 1999), and vigorous pipetting in another (Kiersnowska *et al.*, 2000). In both cases, once pairs had formed, and prophase had been initiated, disrupted conjugants were able to complete normal development without attachment to a mating partner. (This result was best produced using the gentler Kiersnowska technique.) The ability of disrupted pairs to complete normal development suggests that once initiated (provided the cells have healthy diploid micronuclei), the conjugation program can proceed in a cell-autonomous fashion. It is likely that this only occurs when there is an intact exchange junction complex at the anterior end of the cell.

3. Centrifugation

David Nanney conducted a series of centrifugation experiments that resulted in redistribution of cytoplasmic contents. Of particular note: when centrifugal treatments dislodged post-zygotic nuclei from their anchorage at the posterior cortex, they invariably differentiated into macronuclear anlagen rather than germinal micronuclei as they would without treatment. Nanney also suspected that delivering a nucleus to anchorage at the posterior cell cortex late in development might provoke the final postzygotic nuclear division (Nanney, 1953). These observations introduce the idea that there may be cortical determinants of nuclear behavior, and that anchorage of nuclei at these sites might lead them down significant developmental pathways.

A more curious result is produced when starved (initiated) cells are subjected to centrifugation (Iwamoto *et al.*, 2004). In this case, mild centrifugation resulted in a fairly synchronous round of cell division 2 h after the mechanical stimulus. This also resembles a synchronous round of cell divisions triggered by co-stimulation of starved cells (Wolfe, 1974, 1976).

4. Electrofusion

Electrofusion was first applied to *Tetrahymena* by Gaertig *et al.* (1988) as a means of creating parabiotic, live-cell fusions. Since then, live-cell fusion experiments have shed light on nuclear–cortical interactions during both vegetative development (Gaertig and Iftode, 1989) and conjugal development (Kaczanowski and Kiersnowska, 1996a; Gaertig and Cole, 2000; Cole *et al.*, 2001). In one set of experiments, vegetative cells were parabiotically fused and these were then allowed to mate with normal partners. Nuclear configurations were examined by Giemsa staining. By examining a variety of cytogenetic configurations some surprising insights into conjugal development were revealed.

1. In early meiosis, during the crescent stage, the crescent spindle becomes straight and not curved in double-long partners, suggesting the crescent spindle curvature is normally due to simple space constraints and not some intrinsic morphogenetic process.
2. In "unipolar" matings (the anterior of one cell is fused to the posterior of an actively pairing second cell), both fusion cell MICs migrate toward the anterior cytoplasm after meiosis, but only the anterior end that is actively fused to a partner will capture a nucleus and bring about "nuclear selection." There may be some residual "selection" activity in the region of the posterior fusion cell corresponding to the para-oral exchange plaque that may serve to shield adjacent nuclei from programmed nuclear degeneration (PND).
3. The meiotic spindles will elongate to span a double-cell length on the fusion side of a mating pair.
4. The second postzygotic division spindle remains short delivering nuclear products to the mid-body region between parabiotic fusions.
5. Nuclei from the anterior cell of a mating unipolar fusion that are delivered to this mid-body region remain determined to form germline micronuclei.

This latter result was reported by both Gaertig and Cole (2000) and Kaczanowski and Kiersnowski (1996a) who fused already mating pairs together. Together, observations 2 and 5 strongly reinforce the idea that there are cortical determinants that influence nuclear fate. More specifically, the para-oral cortex (near the exchange junction) not only anchors "selected" meiotic nuclei, but may shield them from cytoplasmic signals that trigger PND. These "selected" nuclei become available to respond to subsequent mitotic triggers leading to the third gametogenic division. Similarly, there appear to be cortical determinants in the cell's posterior that shield nuclei derived from the second postzygotic division from cytoplasmic signals triggering differentiation of macronuclear anlagen. It is intriguing that in both cases, nuclei anchored to these sites are shielded from cytoplasmic signals that otherwise affect untethered nuclei. (This is a different interpretation from Nanney's in which anchorage was seen to drive a developmental event rather than "shield" nuclei from some cytoplasmic signal.)

In a very different set of electrofusion experiments, mating pairs were allowed to progress into conjugation, and then electrically fused to either vegetative cells or other mating pairs (Cole *et al.*, 2001). These experiments revealed several things about the developmental program of mating *Tetrahymena*.

1. Dividing cells produce some soluble activity that can inhibit conjugal development.
2. This activity peaks during mitosis and drops during interphase.
3. Mating pairs are sensitive to this arrest activity (division factor) up until the second postzygotic division. This is a "terminal commitment point" or point of no return.
4. Pairs undergoing "genomic exclusion" [a form of abortive conjugal development triggered when diploid cells are mated to severely aneuploidy partners, (see below)] also exhibit a "conjugal arrest activity" blocking development when fused to normal diploid partners.
5. The arrest activity expressed by genomic exclusion pairs peaks in strength just after meiosis I.
6. Healthy diploid pairs are sensitive to this arrest activity (abort factor), up until the second postzygotic division (as with the division factor arrest).

5. Osmotic Shock

Hyperosmotic shock has been shown to disrupt mating pairs at two different time points. First, as described earlier, osmotic shock will erase the preparative changes brought about by "initiation" (nutrient deprivation) (Wellnitz and Bruns, 1979). A more dramatic change is brought about by osmotic shock delivered during the period of pronuclear exchange (Orias and Hamilton, 1979; Orias *et al.*, 1979; Cole and Bruns, 1992). In mating pairs exposed to hyperosmotic shock, (1.5% glucose) a form of self-fertilization occurs. The mechanism proposed for this suggests that the migratory (transfer) pronucleus is prevented from crossing the exchange junction, and ends up fusing with its own resident pronucleus in a form of self-fertilization. This seems likely, in that a hyperosmotic shock should result in cell volume shrinkage that could interfere with pronuclear transfer. This process (cytogamy) and a related version (uniparental cytogamy) in which osmotic shock is delivered to genomic exclusion pairs have provided powerful tools for converting cells with heterozygous, germline mutations into whole-genome homozygotes, ready for mutant screening or selection. Uniparental cytogamy, in particular, has proven effective at helping design screens for mutations affecting mucocyst discharge (Melia *et al.*, 1998) and conjugation itself (Cole *et al.*, 1997; Cole and Soelter, 1997). Of relevance to this discussion, it seems that postzygotic development can proceed in the absence of nuclear exchange (although perhaps nuclear fusion is necessary). On a related note, Nanney (1976) demonstrated that when 0.1 M $CaCl_2$ is applied to mating cells at the time of macronuclear anlagen formation (even after pronuclear fusion), pairs retain their parental MACs creating heterozygous heterokaryons.

6. UV irradiation

When mating pairs are subjected to UV-B radiation, they show a dramatic, stage-specific developmental arrest (Kobayashi and Endoh, 1998). Specifically, mating pairs irradiated at late meiotic prophase (stages IV and V in Fig. 6) fail to complete development. These pairs exhibit nuclear selection failure and follow an abortive conjugal pathway similar to that seen with matings that involve aneuploidy (star) cell partners in a process termed "genomic exclusion" (see below). This report suggests that DNA damage incurred during late meiotic prophase can trigger an early developmental arrest, possibly by causing nuclear selection failure.

B. Cytoskeletal Inhibitors

Microtubules are involved throughout conjugation as dynamic components of the meiotic and mitotic intranuclear spindles, as part of a cytoplasmic infrastructure involved in nuclear movements and cortical tethering, and probably as the means of propelling nuclei across the exchange junction during pronuclear transfer and bringing about fertilization. The various roles of microtubules during conjugation have been determined using immunofluorescence microscopy (Gaertig and Fleury, 1992), TEM (Orias *et al.*, 1983), and a range of microtubule disrupting agents such as vinblastine sulfate (Hamilton and Suhr-Jessen, 1980; Hamilton, 1984; Hamilton *et al.*, 1988) and nocodazole (Kaczanowski *et al.*, 1985; Gaertig *et al.*, 1986; Kaczanowski *et al.*, 1991). Microtubule disruption has been particularly effective in helping investigate developmental dependencies that occur during conjugation.

1. Vinblastine (VB)

In a very thorough investigation, Hamilton "dissected" conjugation in *Tetrahymena* using vinblastine sulfate (Hamilton, 1984). Treatments involving continuous exposures at three micromolar concentrations revealed that disrupting microtubules before 5.25 h (at 38 °C) resulted in "macronuclear retention" (pairs failed to initiate postzygotic development). This time-point corresponds with the completion of the third, gametogenic nuclear division, and the onset of the pronuclear exchange configuration. Mating pairs treated successively later than the 5.25 h time-point formed macronuclear anlagen in greater greater proportion (they successfully launched their post-zygotic developmental program).

This result is worth highlighting, as it represents a profoundly significant developmental checkpoint. The postzygotic developmental program includes two postzygotic nuclear divisions, condensation and degeneration of the parental MAC, differentiation of macronuclear anlagen, pair separation, MIC elimination, and transcriptional activation of the newly formed somatic MAC (see Fig. 8). When pairs commit to this, they are committing to the termination of their parental genome and investment of their future survival in their newly recombinant zygotic genome. It is, perhaps, not surprising that this commitment step appears to be carefully gated. These early vinblastine results suggest that, in very broad strokes, disrupting events occurring prior to nuclear exchange will result in cells retaining their parental nuclei, and aborting postzygotic development.

This first set of experiments highlights some event associated with pronuclear exchange or possibly pronuclear fusion as a potential trigger for postzygotic development. (The time course of vinblastine-driven post-zygotic development matches the time course of pairs entering pronuclear exchange configuration. It cannot be rule out, however, that there might be some lag between the time of drug application and the precise time it takes effect.) Vinblastine treatment after pronuclear exchange disrupted a number of postzygotic events, but typically

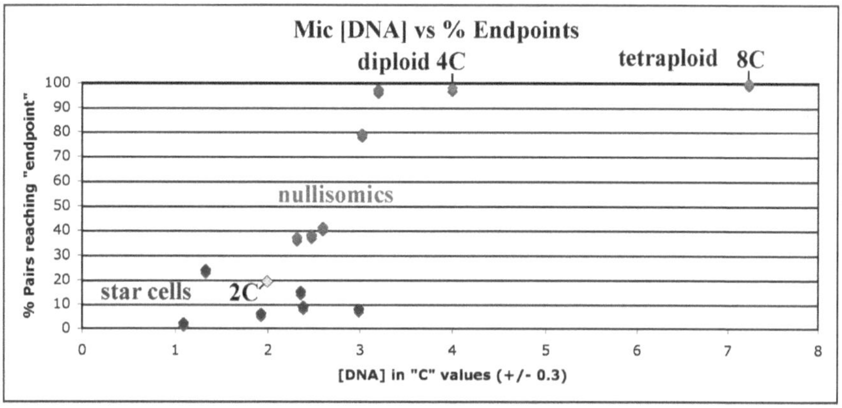

Fig. 8 This figure shows the correlation between conjugal success (measured as the percent of pairs that achieve Stage I or Stage II macronuclear anlagen stage) with MIC DNA content as measured by Feulgen cytophotometry (Cole, unpublished results). Blue dots are a variety of star cells. Red dots represent a gallery of nullisomic cell lines. Haploid, diploid, and tetraploid cells lines are also indicated (2C, 4C, and 8C, respectively). (See color plate.)

failed to prevent either degeneration of the parental MAC or differentiation of the macronuclear anlagen.

Continuous treatment of pairs at or after 5.25 h was successful at blocking pronuclear exchange, pronuclear fusion, and both postzygotic mitoses, highlighting their dependence on microtubule activities. Again, treatments subsequent to pronuclear exchange (though effective at blocking individual steps in the pathway) failed to prevent parental MAC elimination or MAC anlagen differentiation.

Relevant to this discussion were two more observations of vinblastine-induced conjugal anomalies (Hamilton and Suhr-Jessen, 1980; and Hamilton *et al.*, 1988). First, it was shown that vinblastine could result in fertilization failure following nuclear exchange. What is striking about this result is that pairs in which nuclear exchange was successful but fertilization was blocked completed postzygotic development, eliminating their parental MACs and developing homozygous MAC anlagen. This result eliminates fertilization per se as the trigger for postzygotic development. On a related note, vinblastine (and osmotic shock) provokes a type of self-fertilization in which pronuclei are not exchanged but postzygotic development is triggered. In the case of vinblastine-induced "self-fertilization," the question becomes, how does self-fertilization occur if microtubules are required for karyogamy? It seems plausible that what were identified as vinblastine-induced self-fertilizers were actually pairs in which a gametogenic division occurred but transfer was blocked, and the resultant unexchanged pronuclei simply entered the postzygotic pathway. (The author's assays would not be able to distinguish true self-fertilizers from unexchanged, unfused pronuclei that traversed post-zygotic development.) Hamilton's thoughtful conclusion regarding developmental dependencies is that the most likely candidate for a trigger of post-zygotic development was bilateral association of gametic pronuclei at the nuclear exchange junction regardless of whether or not exchange was successful, and regardless of whether or not pronuclear fusion was successful.

2. Nocodazole

In a parallel set of investigations, nocodazole was used to dissect conjugal events in *Tetrahymena* (Kaczanowski *et al.*, 1985; Gaertig *et al.*, 1986; Kaczanowski *et al.*, 1991). In an especially careful study, Kaczanowski *et al.* (1985) explored the consequences of microtubule disruption at various times during early development. What they learned was the following:

1. 10–20 μg/mL nocodazole (in DMSO) blocks pair formation when applied during co-stimulation.
2. Nocadazole added up to 5 h post-mixing (at 30 °C) causes MAC retention (as seen with VB treatments described above).
3. Curiously, 1 μg/mL nocodazole (a concentration that does not block nuclear behavior prior to exchange) is effective at blocking pronuclear fusion.
4. Nocadazole applied prior to crescent elongation prevents crescent elongation.

5. When applied after crescent elongation is initiated, but prior to full clongation, there is evidence that chromosomes fail to synapse, appearing as unpaired monovalents in metaphase I.
6. When nocodazole is applied after complete crescent elongation, metaphase chromosomes appear as bivalents once again.
7. Applied during anaphase or telophase of either meiotic division, nocodazole blocks nuclear division but allows a subsequent round of DNA synthesis (presumably associate with the third prezygotic division) to occur. This can result in polyploid micronuclei.

(Results 4-6 nicely foreshadow recent molecular confirmation that recombination occurs during crescent elongation, and that double-stranded breaks associated with recombination may even trigger crescent elongation: Loidl and Scherthan, 2004; Mochizuki *et al.*, 2008; Loidl and Mochizuki, 2009 to be discussed later).

In a follow-up to this study, Gaertig *et al.* (1986) showed that continuous treatment of mating pairs with 30 μg/mL nocodazole starting 4 h after mixing (toward the end of Meiosis I at 29 °C) resulted in pairs that retained their parental MAC (failed to enter the post-zygotic pathway) and maintained a single enlarged MIC that underwent at least two rounds of DNA replication expanding from 4C to 20C or more. This suggests that, despite a failure to initiate post-zygotic development, and despite a failure to divide, post-meiotic MICs underwent multiple rounds of DNA synthesis.

The published interpretation of this result bears re-examination. The authors noted that the final DNA content achieved by nocodazole-treated pairs matched the DNA content that would have been achieved if normal development had proceeded, and one added up total DNA content for every nucleus (including degenerate relics).

Doerder and DeBault (1975) count DNA reps in the conjugal micronuclei as follows: following meiosis II, the 1C haploid MICs undergo replication (2C). The selected MIC divides and restores its 2C count by a second round of DNA synthesis. Karyogamy results in a 4C fertilization synkaryon. The first post-zygotic mitosis of the synkaryon is accompanied by immediate DNA synthesis (during anaphase), again restoring the two MIC products to 4C. The 21 second post-zygotic division results in two 2C MAC anlagen and two 2C MICs. The MICs do not undergo rapid DNA synthesis (remaining about 2-3C at pair separation), but the two new Mac anlagen replicate immediately to 4C and rise to 8C by pair separation.

One way to re-examine the Gaertig *et al.*'s result is to count not the total DNA content that would be manufactured in a mating cell (28C at pair separation including degenerate relics), but the numbers of rounds of division. There are five rounds of DNA synthesis associated with conjugal events preceding pair separation: just following MII, after the third pre-zygotic division, after each of the two post-zygotic divisions, and a subsequent round in the MAC anlagen. If the single (4C) MIC were prevented from dividing, yet subject to that same schedule of DNA synthesis events, we would anticipate somewhere between 64C and 128C.

It is appealing to hypothesize that conjugants replicate their DNA up to some final whole-cell mass (limited, perhaps by cytosolic levels of nucleotide substrates), and that this same set-point determines DNA content in a MIC that cannot divide. It might be more valuable to consider a different scenario, however. Nocodazole-treated pairs do not enter the post-zygotic developmental pathway, and so, should, perhaps not be expected to execute that schedule of DNA synthesis. Rather, they should execute the DNA synthesis schedule appropriate to pairs that abort development and undergo genomic exclusion. This pathway has also been analyzed with regard to DNA synthesis (Doerder and Shabatura, 1980).

The schedule of DNA synthesis for genomic exclusion pairs (pairs that abort development prior to entering the post-zygotic pathway) include two full rounds of DNA synthesis: the first following Meiosis II (producing a 2C MIC) and the second following the third gametogenic mitosis (again, restoring a 2C condition). Curiously, in pairs that abort normal development at this point, the gametic pronucleus (later termed the hemikaryon after unilateral transfer) appears to undergo a third round of synthesis without division. Two reports place this round of DNA synthesis as beginning prior to nuclear exchange, or shortly thereafter (Allen, 1967; Doerder and Shabatura, 1980, Table 4). A third report suggests that this endomitotic process occurs only after pair separation (Kaczanowski *et al.*, 1989). It may be worth noting that different "star" strains were employed in studies by these various laboratories, and strain differences have been shown to exhibit significantly different sensitivities (Cole and Bruns, 1992). The "Star" strains employed by Doerder and Shabatura as well as by Allen (demonstrating endomitosis during pairing) were A*III and C*. That employed by Kaczanowski *et al.* (showing endomitosis after pair separation) was A*V.

If the nocodazole-treated pairs followed the abortive (genomic exclusion) pathway, with only two to three rounds of DNA synthesis, and we supposed that these rounds occurred reasonably on-schedule despite blocks in nuclear division, we would anticipate a final polyploid MIC with a DNA content between 16C and 32C as was seen by Gaertig *et al.* (1986). (It is again worth noting that Gaertig's experiments utilized A*III, which has shown evidence of a third round of pre-separation DNA synthesis or endomitosis.) Regardless of whether nocodazole-arrested pairs follow a complete conjugation schedule (and monitor final DNA content) or follow the abort pathway (and count "S" periods), it is clear that rounds of DNA synthesis can proceed in mating *Tetrahymena* MICs despite nuclear division failure.

Finally, studies were conducted on the consequences of blocking the postzygotic nuclear divisions (Kaczanowski *et al.*, 1991). Depending on when nocodazole was applied (at synkaryon or after the first postzygotic division), exconjugants were produced with either one large macronuclear anlagen, or two (and occasionally one macronuclear anlagen and one MIC). DNA content for individual macronuclear anlagen measured 12 h into conjugation at 30° were roughly 4C (for controls), 8C for cells with two anlagen (and no MICs), and 16C for cells with one single macronuclear anlagen, suggesting that no rounds of DNA synthesis were skipped despite the failure of nuclear division.

C. Biosynthesis Inhibitors

A number of studies have examined the effect of pharmacological inhibitors on conjugal development. These include inhibitors of DNA synthesis, RNA synthesis, and protein and lipid biosynthesis. As with the cytoskeletal poisons, biosynthesis inhibitors help us in two ways. First, they reveal which steps in a developmental program require a certain type of macromolecular synthesis. Second, by blocking a particular stage in development, they may help reveal which subsequent steps in development are dependent upon an earlier, synthesis-dependent process.

1. Inhibition of DNA Synthesis

Aphidicolin (APD)

DNA polymerase was demonstrated in *Tetrahymena* by Pearlman and Westergaard (1969). Furukawa *et al.* (1979) later isolated two DNA polymerase activities from *Tetrahymena,* and an aphidicolin-sensitive DNA polymerase activity was described by Sakai and Watanabe in 1982. Kaczanowski and Kiersnowski (2011) were the first to explore the consequences of treating vegetative cells with APD (6 μM in DMSO) + caffeine (0.3 mM), thereby inducing chromosomal damage (see above), and then challenging such cells to mate. When (APD + caffeine)-treated cells are released from toxic exposure and challenged to mate with a healthy, diploid partner, they typically abort development following unilateral pronuclear transfer and perform "genomic exclusion" (described below). Curiously, a small fraction of clones derived from an (APD+ caffeine)-treated culture, when mated with a diploid partner, result in a cytoplasmic-dominant metaphase I arrest of the diploid partner. This extraordinary result resembles a similar phenomenon described by Kaczanowski *et al.* in 2004. In that report, cells treated with both nocodazole (to disrupt microtubules) and etoposide (to block DNA ligation and introduce double-stranded DNA breaks) resulted in a similar cytoplasmic-dominant, metaphase I arrest in outcrosses with a diploid partner. These striking results raise two interesting possibilities. First, it is possible that in certain circumstances, double-stranded breaks in a mating cell's MIC can generate a potent meiotic arrest activity that is transferable, and bring about a metaphase I arrest. A second possibility is that certain kinds of DNA damage can be detected by a diploid partner, causing the diploid partner to express a developmental arrest. The latter possibility raises the interesting question of whether or not mating cells exert some kind of genome surveillance over their mating partners.

Etoposide

On a related note, mating *Tetrahymena* has also been exposed to the topoisomerase II inhibitor etoposide, which has been shown to generate double stranded DNA breaks (Kaczanowski *et al.*, 2004; Kaczanowski and Kiersnowska, 2011). Etoposide arrests conjugants at post-meiotic interphase. Of particular interest to this chapter, in arrested pairs none of the four meiotic products was "selected" following etoposide treatment, suggesting that DNA damage induced during meiosis can result in a failure to trigger cortical tethering at the nuclear exchange junction. Even more curious, cells treated with etoposide and nocodazole produced an exconjugant clone that subsequently triggered a Metaphase I arrest whenever it was mated to a normal diploid partner (as described above for APD-treated pairs). Such pairs frequently underwent spontaneous, parabiotic fusion.

2. Inhibition of RNA Synthesis

Actinomycin D (ActD)

Allewell *et al.* (1976) demonstrated that blocking transcription before cells pair would prevent paring, suggesting that novel gene expression is required for this process. Once pairs formed, however, treatment with ActD did not lead to pair separation. Ward and Herrick (1996) explored transcriptional requirements at later stages of conjugation. They discovered that ActD (50 μg/mL) applied prior to fertilization produced a response that resembles abortive (genomic exclusion) development. Pairs separate and retain their parental MACs. ActD applied just after fertilization blocked all postzygotic development in pairs including pair separation. This is significant in that it suggests that the trigger for post-zygotic development (hypothesized to involve bilateral pronucleus association with the exchange junction, Hamilton, 1984) requires novel gene expression. Later treatments with ActD revealed a series of conjugal block configurations, suggesting that transcription is essential for the following events or transitions (notes in parentheses refer to the time at which ActD was first added to produce specific type of arrest):

1. Launching post-zygotic development. Pairs retain parental MAC and fail to separate (appears with addition of ACT just after fertilization).
2. Exiting macronuclear anlagen (MA) "stage 2": the two MA are located in a vertical orientation at mid-body, the two MICs have positioned themselves in the groove between the two MA, and the parental MAC has condensed and moved to the posterior. Pairs fail to separate (ActD added at the beginning of MAC anlagen stage 1).
3. As in #2 except MA show diminished DAPI staining, suggesting under-replicated DNA (ActD added roughly 9 h into mating, at the end of MA stage I).
4. Pair separation occurs but condensed parental MAC and both MICs persist (ActD added at the beginning of stage II MA).
5. As with #4 except MA show diminished DAPI staining, suggesting under-replicated DNA (ActD added at stage II MA).
6. Pair separation has occurred, parental MAC is gone, and two MICs persist. (ActD added at the onset of pair separation).
7. As with # 6 except MA show diminished DAPI staining, suggesting under-replicated DNA (ActD added at the onset of old MAC resorption).
8. Parental MAC and one MIC are gone; MA show diminished DAPI staining, suggesting under-replicated DNA (ActD added at the onset of MIC elimination).

9. As with # 8 except DNA replication has proceeded. Cells fail to enter cell division cycle (ActD added after MIC elimination).

These defects (aberrant anlagen DNA loss, block to pair separation, old MAC resorption, new MIC elimination) are reminiscent of matings involving cells with MICs that are deficient in one or more chromosomes (Davis *et al.*, 1992; Ward *et al.*, 1995). These observations allowed the authors to conclude that pair separation, old MAC resorption, and new MIC elimination occur independently of defects in anlagen DNA replication, and require sequential periods of gene expression. Their work also supported an early (9.5 h) time for initiation of zygotic, macronuclear anlagen gene expression (even before programmed DNA rearrangements).

3. Inhibition of Protein Synthesis

Cycloheximide (Chx)

Dose–response curves for *Tetrahymena* grown in CHX have been established for both wild-type and CHX-resistant mutants (Frankel, 1969b; Roberts and Orias, 1973, 1974). Allewell *et al.* (1976) demonstrated that pair formation is sensitive to cycloheximide (1–2 μg/mL), and pairs can be disrupted by blocking protein synthesis up until 1.5 h (at 28 °C). Kaczanowski *et al.* (1989), made creative use of protein synthesis inhibitors by applying CHX (10 μg/mL) to genomic exclusion pairs between 6 and 8 h of mating (30 °C). CHX apparently prevents triggering of the third, endomitotic round of DNA synthesis in these pairs resulting in production of viable, haploid cell lines. This demonstrated the need for protein synthesis to drive endomitotic replication of MIC DNA in genomic exclusion (abortive) conjugants.

Kaczanowski and Kaczanowska (1996b) performed a systematic investigation into which developmental events occurring during conjugation are dependent on proteins synthesis. They uncovered the following:

1. Pairs exposed to CHX during early prophase (Stages I–III) remained in Prophase I.
2. Pairs exposed during the interval from pachytene (Stages IV–VI) to metaphase became blocked in a unique pro-metaphase arrest with over-condensed swollen bivalents (centromeres remained intact).
3. Pairs exposed to CHX at Metaphase I completed nuclear division and arrested in Interphase I.
4. Pairs exposed during second meiosis arrested after completion of the second nuclear division.
5. Pairs exposed during the third prezygotic division arrested at the pronucleus stage.
6. Pairs arrested during the interval from pronuclear exchange to karyogamy or during the first postzygotic division arrested after completion of the first post-zygotic division.

7. Pairs exposed to CHX during the second postzygotic division had an interesting arrest having initiated macronuclear anlagen formation while retaining their parental MAC.
8. Pairs exposed following the second postzygotic division arrested as pairs with condensed (but not eliminated) parental MACs, two MICs and two MACs (in the form of macronuclear anlagen).

This ensemble of developmental arrests reveal that novel protein synthesis is required in order that mating pairs can move forward at numerous stages throughout conjugation. Given that many of these stages are likely gated by specific cyclin proteins and their associated CDKs (Stover and Rice, 2011; Tang *et al.*, 1997), and that cyclin synthesis and accumulation may be responsible for the timing of many of these events, the result may not be surprising. The results that are surprising include appearance of over-condensed metaphase chromosomes. The authors hypothesize that chromosome condensation in normal pairs requires synthesis of a protein that limits or terminates this process prior to anaphase. This could represent another type of "dynamic pause" in which progress through the developmental program is arrested although a particular process characteristic of the arrested stage remains active. In this case, the process involves progressive chromatin condensation. Another surprising result is that no new protein synthesis is required for passage from Meiosis I to Meiosis II. Finally, it is interesting that macronuclear anlagen differentiation can be initiated without protein synthesis during late stages of development, but degeneration of the parental MAC cannot, and pair separation also seems to require novel synthesis.

4. Lipid and Glycoprotein Synthesis Inhibitors

De novo lipid synthesis has been shown to be required for pair formation in *Tetrahymena* (Frisch *et al.*, 1978). This was done by applying 8–20 μg/mL of the antibiotic cerulenin, a compound shown to block steps in the biosynthesis of lipids. These authors also demonstrated a dramatic shift in the ratio of saturated: unsaturated fatty acids from 0.30 in unpaired cells to 0.45 in conjugants. This is unexpected in that such a shift is associated with loss of membrane fluidity (counter-intuitive in a system that must undergo membrane fusion).

Similarly, tunicamycin has been shown to block pair formation in *Tetrahymena* (Frisch *et al.*, 1976). This drug is an inhibitor of glycoproteins synthesis. This result is of interest in that cell–cell adhesion during co-stimulation is probably mediated by transmembrane proteins with glycosyl side groups. Related to this is the recent discovery of the *Tetrahymena* mating type locus that encodes a transmembrane protein (Cervantes and Orias, personal communication).

Related to these findings is the demonstration that concanavalin A (a plant lectin that binds specific sugar residues on membrane glycoproteins) binds to the developing mating plaque in co-stimulated *Tetrahymena* cells and can prevent pair formation (Frisch and Loyter, 1977, Watanabe *et al.*, 1988 reviewed by Cole, 2006).

These studies demonstrate the importancc of novcl mcmbranc biosynthcsis and membrane remodeling as *Tetrahymena* cells prepare to mate.

5. Cell Signal Inhibitors

cAMP

Several studies suggest that intracellular cAMP levels must drop to permit pair formation in *Tetrahymena*. Ding and Weijie (1987) have shown that intracellular cAMP levels (in a related species) drop prior to pairing. Furthermore, phosphodiesterase inhibitors (theophylline, caffeine) appear to be effective at preventing pair formation (Allewell *et al.*, 1976).

GTP

Iwamoto and Nakaoka (2002) and Iwamoto *et al.* (2004), demonstrated that starved (initiated) cells would launch a single round of cell divisions not only in response to mechanical stimulation and co-stimulation, but in response to GTP (an effective *Tetrahymena* chemo-repellent, Francis and Hennessey, 1995).

Serine/Threonine Kinase Inhibition

6-Dimethylaminopurine (DMAP) is an inhibitor of serine/threonine protein kinases. When applied to mating *Tetrahymena,* DMAP (0.5 mM) appears to arrest pairs at the pronucleus stage, blocking both pronuclear fusion and entry into the postzygotic development pathway (Kaczanowski and Kaczanowska, 1996). Although only brief mention is made of this result by the authors, it is very intriguing in that it suggests that protein phosphorylation may be necessary for triggering postzygotic development.

PI3-Kinase inhibition

Recently, a number of agents that block PI3 kinase activity have been applied to mating *Tetrahymena*. Yakisich and Kapler (2004), in particular, have applied wortmannin (250 nM), 3-methyladenine (10 mM), and LY294002 (100 μM) at various times during conjugal development to assay the role of IP3 kinase in triggering PND. These treatments produced a dramatic inhibition in PND resulting in mating pairs with supernumerary nuclei completing development. The authors conclude that:

1. The PI 3-kinase pathway is involved in PND in *T. thermophila,* and is required for acidification and degradation of both non-exchanged pronuclei and the (old) parental macronucleus.
2. Pronuclei that are normally degraded can be re-programmed to differentiate into micro- and macronuclei when PND is blocked.
3. PI 3-kinase activates PND of three haploid pronuclei early in development and degradation of the parental macronucleus at a later time.

In other work, 10 nM Wortmannin and 50 μM of LY294002 have also been shown to block pair formation, prevent stabilization of pair formation (between 1.5 h and 2.0 h), and prevent early meiosis in mated pairs (Takahiko Akematsu, personal communication). Clearly, PI3 kinase signaling is a promising area for future research into ciliate conjugal development.

Sirtuin Inactivation Leads to Failure of PND

Recently it has been demonstrated that nicotinamide, a general inhibitor of sirtuin-like histone deacetylases, blocks both resorption of the parental MAC during late stages of conjugation and PND of the various meiotic products early in development (Slade *et al.*, 2011). It has been suggested that a specific sirtuin (Thd14) is delivered to those nuclei destined or targeted for programmed degeneration. Its role in PND may be to bring about global chromatin condensation within the targeted nuclei, prior to its degradation.

D. Micronuclear Ploidy Manipulations

When *Tetrahymena* mate, the outcome is profoundly affected by the condition of the two partners' germline nuclei. When both MICs are diploid with no appreciable genetic damage, cells execute a complete program of conjugal development. Variations in micronuclear ploidy or chromosome integrity frequently lead to an abortive form of conjugation termed "genomic exclusion" (Allen, 1967a,b; Allen *et al.*, 1967; Pitts, 1979; Doerder and Shabatura, 1980; Pitts and Doerder, 1988; Karrer, 2000; Orias, this volume). In this alternative program (Fig. 8), development is more or less normal in both mating partners up through meiosis II. In a mating partner that has developed some form of micronuclear aneuploidy *(a "Star" cell),* subsequent events are perturbed. Most commonly, all four meiotic products in the "star" partner undergo PND. None of the star-partner's MICs are "selected," tethered to the exchange junction, protected from PND, and triggered to undergo a third gametogenic division. Development is relatively normal in the diploid partner, although mating events are somewhat delayed (Gaertig and Kaczanowsi, 1987), and a peculiar third round of DNA synthesis brings about restoration of the diploid condition in the progeny. This unilateral normalcy extends so far as to seeing the diploid partner pass its transfer pronucleus across to its sterile "Star" partner while receiving no nuclear material in return. Such pairs dissociate (about 9 h at 30 °C), retaining their parental MACs and performing none of the events associated with postzygotic development. This form of interrupted mating is sometimes referred to as a "backout."

Following pair separation, these exconjugants are competent to re-mate immediately. The consequences of such a re-pairing event are that the pairs complete a full course of conjugal development and emerge as whole-genome homozygotes. These two rounds of mating are referred to as first and second round genomic exclusion, reflecting the fact that one cell line's genome has been effectively eliminated from the F2 generation.

Two questions loom large over this scenario. What is the signal that triggers conjugal arrest in genomic exclusion partners, and what conditions must be met to trigger postzygotic development and completion of the conjugal program in normal, diploid partners? In its simplest form, it appears that some form of micronuclear aneuploidy leads to a failure of nuclear selection within a "Star" partner. Nuclear selection failure, in turn, prohibits formation of the bilateral exchange configuration in which gametic pronuclei come to lie in intimate association and on either side of the nuclear exchange junction, enmeshed in baskets of microtubules. Instead, an aberrant unilateral configuration is observed. It has been argued that it is the bilateral nature of the exchange configuration that triggers postzygotic development (Hamilton, 1984).

This pushes the question back to, "what conditions result in nuclear selection failure?" One possibility is that it is simply DNA content or ploidy. A survey of conjugal success in a variety of cell lines (Fig. 8) at first seems to support this. The most aneuploidy star lines also have the lowest conjugal success, whereas diploids and tetraploids enjoy high conjugal success. Furthermore, nullisomic cell lines (cells that have lost entire chromosomes from their micronuclear genomes, Bruns *et al.*, 1983) also exhibit diminished conjugal success that is at least somewhat proportional to the number of chromosomes missing. There are significant outliers, however: a star cell with a nearly 3C micronuclear content (B* VII) has far lower conjugal success than a nullisomic (CU376) with a comparable micronuclear C-level. The finding that DNA-damaging agents can result in star-like behavior suggests that unrepaired broken chromosome ends, or missing homologous partners, might play a part as well (Kaczanowski *et al.*, 2004).

Even more extreme situations of aneuploidy have been generated through treatment of vegetative cells with aphidicolin and caffeine (Kaczanowski and Kiersnowska, 2011). These, too, result in star behavior when mated to a normal, diploid partner. In short, the genomic exclusion pathway serves to highlight that conjugal events occurring early in development have far-reaching consequences during mid-conjugal development. More specifically, events associated with meiosis determine whether or not a mating pair aborts development after nuclear exchange, or triggers the entire postzygotic developmental program.

E. Mutants

Nitrosoguanidine has been used in a number of schemes to screen for developmental mutants (Orias and Flacks, 1973; Orias and Bruns, 1976; Orias and Hamilton, 1979; Bruns, 1986; Frankel *et al.*, 1976a, 1977, 1984; Cole *et al.*, 1997; Cole and Soelter, 1997; Frankel, 2008). Nitrosoguanidine tends to create loss-of-function point mutations. The result of such searches has been the assembly of a small panel of recessive mutants that affect a variety of stages in conjugal development. The upside of this type of research is that one can fish for genes whose function is critical to particular steps in conjugal development. The resultant loci frequently could not have been deduced to play a role in mating a priori, and can be used in a genetic dissection of developmental events. The down-side to this type of study is that genes identified in this fashion are rarely (if ever) identified. One can do much to deduce the wild-type gene's function, and one can explore which steps in development can be blocked without preventing subsequent developmental events, but it has not yet been possible to identify a specific gene product associated with a particular genetic lesion. With the development of the Illumina Sequencing Platform in *Paramecium* (Arnaiz and Sperling, 2011), it is now finally possible to identify genes in this organism associated with such historically anonymous mutations. Similar developments can be anticipated for *Tetrahymena.*

1. The *cnj* mutant panel

In 1997, a novel, one-step method of expressing nitrosoguanidine mutations as whole-genome homozygotes (Cole and Bruns, 1992) was deployed in a screen for mutants that affect conjugation (Cole *et al.*, 1997; Cole and Soelter, 1997). The gallery of mutants generated included nine distinct conjugal phenotypes. It should be noted that this initial screen for conjugation mutants was by no means as thorough as earlier searches for cytokinesis defects in which multiple alleles for most loci were recovered (Frankel *et al.*, 1976a, 1977). In fact only a single allele for most conjugation mutants was likely recovered based on their distinctive phenotypes (the possible exception being *cnj1* and *cnj2,* which exhibit similar phenotypes; complementation tests and chromosome mapping have not been done with most of these mutants to date). In short, the number of genes whose loss-of-function phenotype would produce cells viable during vegetative growth, but defective in conjugal development, has barely been tapped. Contributing to this gallery of conjugal mutants are two pattern mutants from the Frankel laboratory that exhibit pleiotropic defects during mating *(bcd and janA:* Cole, 1991; Cole and Frankel, 1991). Kaczanowski has also described a conjugal-defect mutant *mra* (Kaczanowski, 1992), and produced the only known, cytoplasmic-dominant phenotype through various means that introduce MIC chromosome damage (see above, Kaczanowski *et al.*, 2004; Kaczanowski and Kiersnowska, 2011).

As the ciliate molecular toolbox has expanded, researchers have been able to target specific genes for knock-out or knock-down loss-of-function analysis. Alternatively, the ribosome-antisense approach to mutational screening produces screenable phenotypes whose genes can subsequently be identified (Chilcoat *et al.*, 2001). Although the list of identified genes whose loss-of-function phenotype involves conjugation is still reasonably small, it promises to grow rather dramatically with recent developments. At the time of publishing, we have seen targeted mutations involving meiotic recombination, a host of chromatin-remodeling proteins, and a potential membrane trafficking gene that all result in conjugation phenotypes.

2. Early Conjugal Phenotypes: Meiotic Defects. *spo11*. Double-Stranded DNA Breaks (SDBs)

spo11 brings about double-stranded DNA breaks during meiosis (Keeney *et al.*, 1997; Bergerat *et al.*, 1997). A *Tetrahymena* homolog was subjected to targeted somatic gene disruption. Pairs without *SPO11* failed to form chiasmata and exhibited chromosome mis-segregation. Failure in meiotic cross-over was demonstrated by the absence of labeling for Rad51p (a DNA-repair protein) and gamma yH2A.X (a phosphorylated histone variant), both markers of meiotic DSBs undergoing repair (Mochizuki *et al.*, 2008; Loidl and Mochizuki, 2009). These authors make a convincing (and visually stunning) case for a model suggesting that double-stranded breaks are necessary in order to trigger crescent elongation in meiosis I. They further propose that the crescent is a useful structure for promoting DSB repair in the absence of a synaptonemal complex.

Other targeted gene mutations that affect meiotic recombination include those encoding *HOP2a* and a phosphorylatable form of *yH2A.X* (Mochizuki *et al.*, 2008; Song *et al.*, 2007). Both of these gene products are involved in RAD51-mediated DSB repair. *hop2a* mutants exhibit what appear to be fragmented chromosomes at the end of meiosis I. Their fertility is destroyed as well, suggesting a "star"-like failure to select meiotic products. Blocking phosphorylation of the histone H2A resulted in cells behaving "star"-like, completing two rounds of meiosis, but failing to undergo nuclear selection.

Finally, the *Tetrahymena ATR* homolog has been identified and subject to somatic disruption (Loidl and Mochizuki, 2009). Mating pairs deficient in *ATR* expression fail to produce the extended crescent configuration (and exhibit vegetative defects in MIC maintenance).

3. Summary of the Meiotic Phenotypes

All these results suggest that crescent formation during meiosis I is dependent upon the formation of double-stranded DNA breaks. More specifically, DSBs must form and trigger the ATR checkpoint kinase in order for crescent formation. Subsequent involvement of the actual repair machinery (RAD51, HOP2A, and a phosphorylated H2A.X) are not needed for crescent formation, but damage left unrepaired by these proteins leads to a failure of nuclear selection following meiosis, and cells abort conjugation without triggering the post-zygotic developmental pathway.

4. Midconjugal Phenotypes

Mutations affecting mid-stages of conjugal development (nuclear selection, pronuclear exchange, and pronuclear fusion) have relevance to our exploration of what triggers the post-zygotic developmental pathway. In particular, they allow us to test the hypothesis that symmetrical or bilateral pronuclear association with the exchange junction is a necessary and sufficient trigger for post-zygotic development. We have already seen one violation of the model in that mechanically disrupted singlets can complete postzygotic development (Kiersnowska *et al.*, 2000; Virtue and Cole, 1999). One might argue that failing to be in cytoplasmic continuity with a partner in which nuclear selection has failed represents a special case. In support of the hypothesis, every mutant that results in nuclear selection failure: *cnj1*, *cnj2, cnj4,* and a variety of mutants with meiotic defects, fail to trigger post-zygotic development. In *cnj5* mutants, both meiotic divisions are typically skipped, yet some partners "select" the unreduced MIC that subsequently undergoes a gametogenic mitosis producing a symmetrical, bilateral exchange configuration involving diploid MICs. Some of these pairs (few) complete postzygotic development. It is unclear whether *cnj5* pairs that reach endpoint are the result of a particularly successful bilateral exchange configuration, or the result of incomplete penetrance of the phenotype (Cole *et al.*, 1997). *cnj7* and *cnj8* mutants complete meiosis, appear to skip the third gametogenic mitosis, and yet assemble a bilateral exchange configuration. (It appears as if two of the meiotic products assume the role of gametic pronuclei without the gametogenic mitosis.) These pairs skip one postzygotic division, yet achieve endpoints with one macronuclear anlagen and one enlarged MIC (Cole and Soelter, 1997). These examples all lend support to the model suggesting that the bilateral exchange configuration is a necessary and sufficient trigger for postzygotic development regardless of how it is achieved. The one dissenting example is the *bcd* mutant (Cole, 1991). These pairs assemble a bilateral exchange configuration, even trade pronuclei, yet fail to trigger postzygotic development. There are several unique features to the *bcd* conjugal arrest. First, multiple meiotic products appear to be selected and undergo a gametogenic mitosis. Second, multiple gametic nuclei assemble at the exchange junction (we see four or even six pronuclei at the junction). Third, despite nuclear exchange, nuclei remain trapped at the junction, decorated with both fenestrin (Cole, unpublished observations) and TCBP25 (Nakagawa *et al.*, 2008). This exception to the rule may hold a key to our understanding of this developmental threshold, yet the mystery remains.

5. Exconjugant Phenotypes

Many mutants have been identified that exhibit phenotypes affecting only the very last stages of development. We can organize these temporally by the terminal phenotype reflecting the latest stages in development that are successfully completed. These would include "*Pair-separation failures*" that develop macronuclear anlagen, yet retain a condensed parental MAC and both MICs *(cnj6, mra,janA);* a "*two-MIC, two-MAC*" phenotype in which pairs separate, the parental MAC is eliminated, but MIC elimination has failed and cells fail to re-enter the cell division pathway; and a "*one-MIC, two-MAC*" phenotype similar to the preceding except that pairs do eliminate one of their two MICs.

Many of these mutations are in genes associated with differentiation of the developing macronuclear anlagen. These events are reviewed elsewhere (Karrer, 2000). In brief, the MIC genome that is destined to become a MAC is modified through a number of carefully regulated steps: (1) the initial MIC chromosomes are broken at precisely defined sites (CBS: chromosome breakage sequences); (2) the resultant mini-chromosomes of the MAC have their ends modified through active telomerase; (3) specific regions of the MIC genome are eliminated by DNA excision with ligation of the boundaries of the eliminated sequence (IES: internal eliminated sequences are targeted for elimination via a small scnRNA-mediated process that targets DNA for histone methylation, followed by heterochromatinization and excision in specialized protein foci); (4) surviving MAC-destined chromatin is amplified. Mutations that block these processes result in a few iconic conjugal arrest phenotypes.

6. The Pair Separation Phenotype

In these mutants, mating pairs seem to complete nuclear events leading up to Stage II endpoints (Fig. 7). At this point, pairs fail to separate. Mutants resulting in this phenotype include *cnj6* (Cole and Soelter, 1997), *janA* (Cole and Frankel, 1991), and *mra* (Kaczanowski, 1992). We have little insight into this phenotype. It has been suggested that it can result from cortical anomalies in *the janA* phenotype (Cole and Frankel, 1991), but this is not apparent in either of the other two mutants. Pair separation failure can be phenocopied by blocking gene expression late in conjugation with ActD (Ward and Herrick, 1994), or by mating cells with any MIC chromosome nullisomy, provided that both partners are missing the same MIC chromosome (Ward *et al.*, 1995).

7. CDA13

When screening the vegetative-antisense ribosome library generated in the laboratory of Aaron Turkewitz, Zweifel *et al.* (2008) identified a cell division arrest phenotype associated with an antisense knockdown of the *CDA12* gene. Nested within this gene, and in antisense orientation to it, was a second open reading frame (labeled *CDA13).* GFP-tagging of this nested sequence gave TEM immunogold localization over cytological elements consistent with a trans-Golgi network assignment. An antisense knock-down directed at this second, nested sequence produced a phenotype that blocked exit from conjugation. Curiously, *cda13* {mutant X mutant} pairs completed the nuclear division program of conjugation, failed to separate, and died with 74% penetrance. When outcrossed to a wild-type partner, lethality was reduced, but survivors frequently produced a monster phenotype: partners failed to separate, yet at least one partner launched into the cell division pathway despite being parabiotically fused to a non-dividing partner. This abnormal, one-sided development included multiple rounds of normal-looking oral development and MIC division, but the MAC and the cytoplasm remained uncleaved. These fission-arrested monsters became syncytial aggregates reaching dimensions of 0.5 mm in some cases. When outcrossed to a "star" cell, most {mutant X A*III} pairs produced viable exconjugant clones. About 5% of these viable clones showed "monster" phenotypes, a proportion equal to the rate at which A*III matings complete conjugal development rather than the genomic exclusion pathway that aborts post-conjugal development. One interpretation of this result is that the *cda13* monster phenotype was triggered by completion of the conjugation pathway. Another possible interpretation is that the *cda13* antisense mutant cell lines carry a dominant, lethal gene mutation buried in their MICs that only comes into expression following successful conjugation.

8. The "two-MIC, two-MAC" exconjugant phenotype

Cells with whole-genome knockouts in genes necessary for programmed DNA rearrangement (IES excision and/or CBS breakage) complete the cytological events associated with conjugation up through pair separation and parental MAC degeneration. Cells then arrest in a "two-MIC, two-MAC" phenotype, unable to eliminate one of their two MICs or re-enter the cell division cycle. These phenotypes also show failure to amplify DNA in the macronuclear anlagen.

LIA1, PDD1, PDD2, DCL1, DIE5 Genes Required for RNA-Mediated IES Excision

All five of these genes have been shown to be necessary for programmed IES excision and DNA rearrangement during macronuclear anlagen differentiation (Rexer and Chalker, 2007; Coyne *et al.*, 1999; Nikiforov *et al.*, 1999; Malone *et al.*, 2005; Mochizuki *et al.*, 2005; Matsuda *et al.*, 2010). Mutations in Dicerlike *(DCL1)* and *DIE5* have also been shown to be defective in CBS breakage (Chalker, personal communication; Matsuda *et al.*, 2010). Knockout of any of these leads to the "two-MIC, two-MAC" phenotype. The *die5* phenotype is unique in that it not only fails to amplify macronuclear anlagen DNA, but appears to lose anlagen DNA over time, although the MAC envelope remains present.

TWI1, a Member of the Argonaute Family Involved in Small scnRNA-Mediated DNA elimination

Cells lacking TWI1, a gene involved in RNA-mediated DNA elimination (see Mochizuki and Gorovski, 2004; Chalker *et al.*, 2005, Chalker, 2008 for review), are defective in both IES excision and CBS cutting. The knockout-conjugation phenotype includes normal pair separation and old MAC resorption, but abnormal retention of both post-zygotic MICs (a two-MIC, two-MAC exconjugant phenotype) and a failure to re-enter division (Mochizuki *et al.*, 2002).

EMA1, a DExH Box RNA Helicase

Cells lacking EMA1, a gene whose product is implicated in mediating interactions between Twi1p and chromatin during DNA elimination by stimulating base-pairing interactions between scnRNAs and noncoding transcripts in both parental and developing new macronuclei (Aronica *et al.*, 2008), also produce this two-MIC, two-MAC exconjugant phenotype.

CNA1: a Gene Encoding a Centromeric Protein

Removal of *CNA1* from the zygotic genome resulted in failure of IES excision, failure of DNA replication in the macronuclear anlagen, and arrest of cells in the two-MIC, two-MAC exconjugant phenotype (Cui and Gorovski, 2006).

9. The One-MIC, Two-MAC Exconjugant Phenotype. RAD51, a DNA Repair Gene

A mutation in *RAD51* leading to defective DNA repair produces an early meiotic arrest phenotype and a later, zygotically expressed "one-MIC, two-MAC" exconjugant phenotype. Cells lacking a functional *RAD51* gene in their somatic macronucleus show progressive micronuclear aneuploidy during vegetative development (Marsh *et al.*, 2000). This is no doubt due to the need to employ *RAD51* during, or shortly after MIC DNA replication. When mated, these cells behave "star"-like. This may be due not just to the absence of*RAD51* activity during meiosis, but to the MIC damages accumulated during vegetative growth. In an attempt to circumvent this, rad51-knockouts were mated twice: once to replace their germ-line MIC by outcrossing to a wild-type, diploid partner, and a second time to test the phenotype in a *rad51*-X *rad51*-mating pair. The result was a conjugal arrest phenotype that resembled a prophase I arrest (cells rarely passed beyond diakinesis). This meiotic defect was clearly under the control of the parental MAC gene expression. The authors also examined the phenotype associated with loss of zygotic gene expression (Marsh *et al.*, 2001). These cell lines have wild-type *RAD51* in their somatic MAC, but are *rad51-null* mutants in the germline MICs. When such cells are mated, development proceeds normally (under the control of the parental MAC) up until pair separation. At this point, pairs separate, the parental MAC is resorbed, and one MIC is eliminated. Cells survive, although they are now expressing a *rad51-null* MAC. The resulting *rad51-null* exconjugants replicate their macronuclear anlagen DNA up to the 128C amount typically seen in wild-type exconjugants just prior to their first exconjugant cell division (although at a slower pace). Then, *rad51-null* exconjugants arrest, unable to initiate the first exconjugant cell division and re-enter the vegetative development pathway (although they can survive for weeks).

ASI2, a Potential Cell-Signal Gene with the "One MIC, Two MAC" Mutant Phenotype

In a different approach, genes whose transcription was up-regulated during macronuclear anlagen development were identified. One, in particular *ASI2* (Anlagen-Stage Induced # 2), was cloned, its expression profile determined, and a knockout-phenotype characterized (Li *et al.*, 2006). When *asi2* knockouts were mated, pairs separated, IES excision was normal, but DNA replication was halted in the developing macronuclear anlagen. The result: another one-MIC, two-MAC phenotype. This gene shows a set of predicted coding domains resembling membrane-spanning signal transduction proteins. Its actual role in development, though clearly essential, remains mysterious.

VII. Overview of the Developmental Logic of Conjugation

Mutations affecting conjugal development focus attention on three major transitions: the successful completion of meiosis resulting in "selection" of one meiotic product (Fig. 9), the successful assembly of the pronuclear exchange configuration

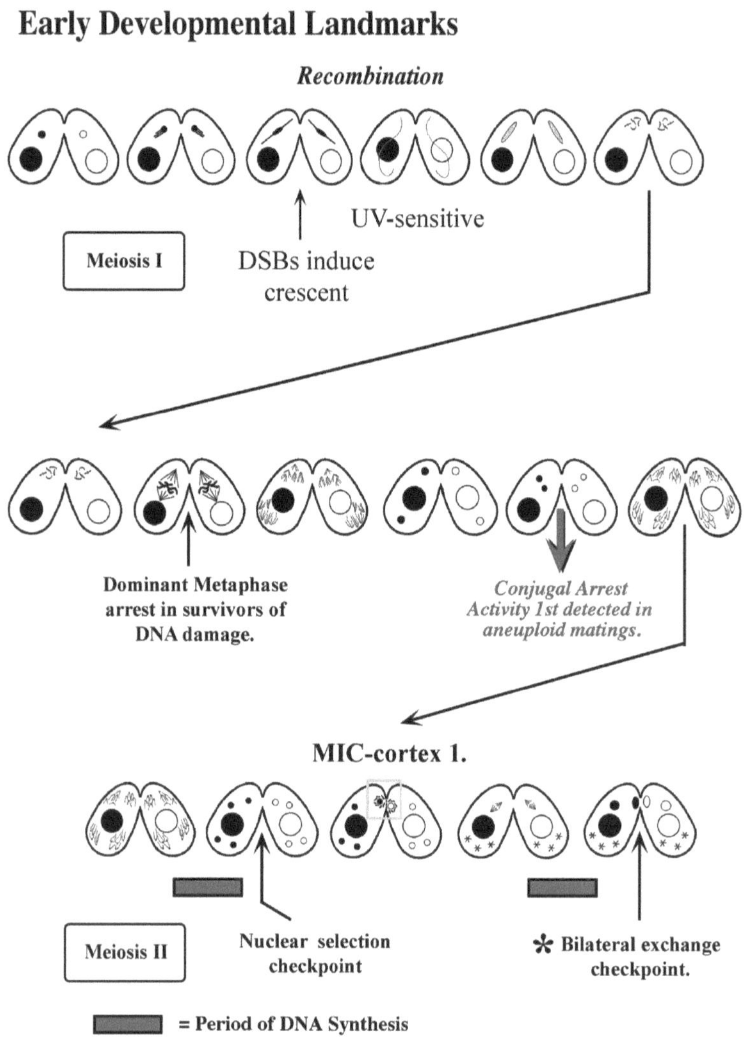

Fig. 9 Early developmental landmarks. Although chromosomes are indicated after MII, these are typically only seen when cells are "spread" for karyotype analysis, not in live observations. Arrows indicate the point at which double-stranded DNA breaks must be present to induce crescent formation; the UV-sensitive period (at which UV induces selection failure); the point at which DNA damage (induced by aphidicolin or etoposide in earlier generations) can induce a dominant-metaphase I arrest; the point at which electrofusion demonstrates a detectable conjugal arrest activity; the point of nuclear selection, and the bilateral exchange junction checkpoint (regulating entry into the post-zygotic developmental program). Gray boxes indicate periods of DNA synthesis (associated with MIC anaphase). Green box indicates the first place in which nuclear association with the cell cortex drives nuclear fate: here cortical tethering (selection) rescues an MIC from programmed nuclear degradation. (See color plate.)

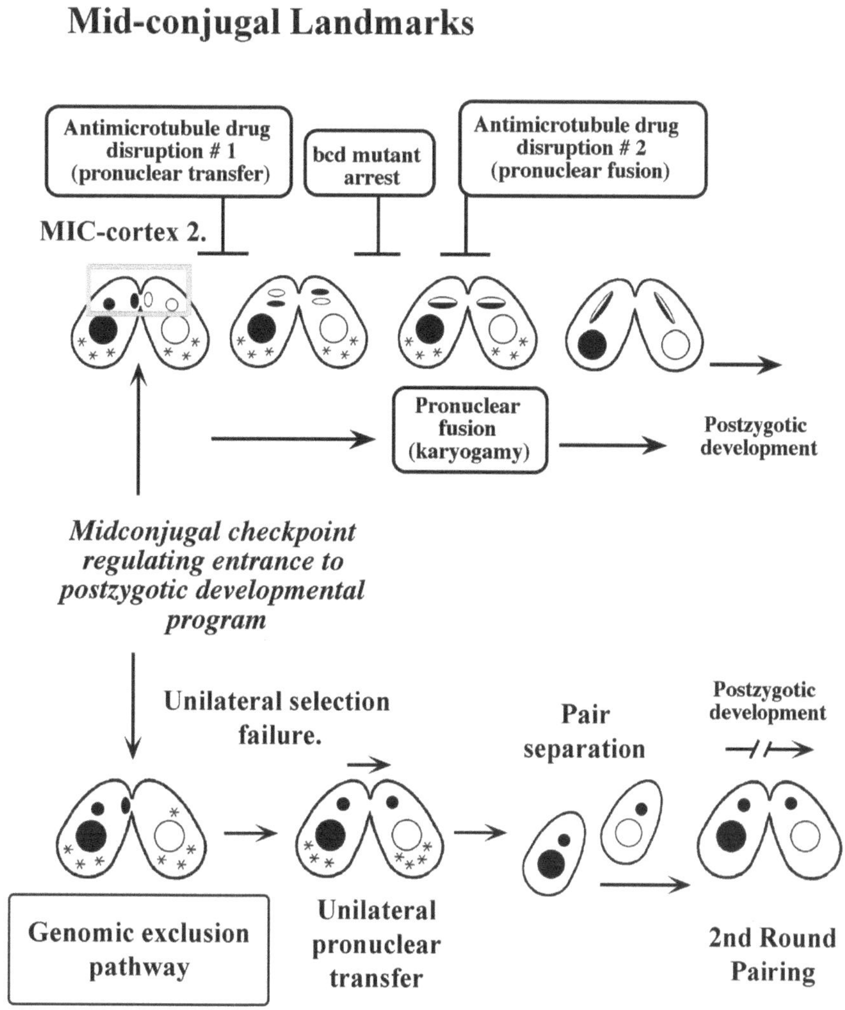

Fig. 10 Mid-conjugal developmental landmarks. Lines indicate where anti-microtubule drugs can disrupt pronuclear transfer and pronuclear fusion (fertilization). Green box indicates the second place in which nuclear association with the cell cortex drives nuclear fate: here a bilateral nuclear/cortical configuration is necessary to trigger post-zygotic development. This figure also shows what happens when one partner of a pair fails to execute nuclear selection (genomic exclusion). Such cells separate and can re-pair leading to a complete developmental program. (See color plate.)

triggering initiation of the post-zygotic developmental program (Fig. 10), and the successful completion of macronuclear anlagen differentiation that serves as the gateway from conjugal to vegetative development (Fig. 11). These studies also reveal a rich dialog occurring between the nucleus and specific regions of the cell cortex,

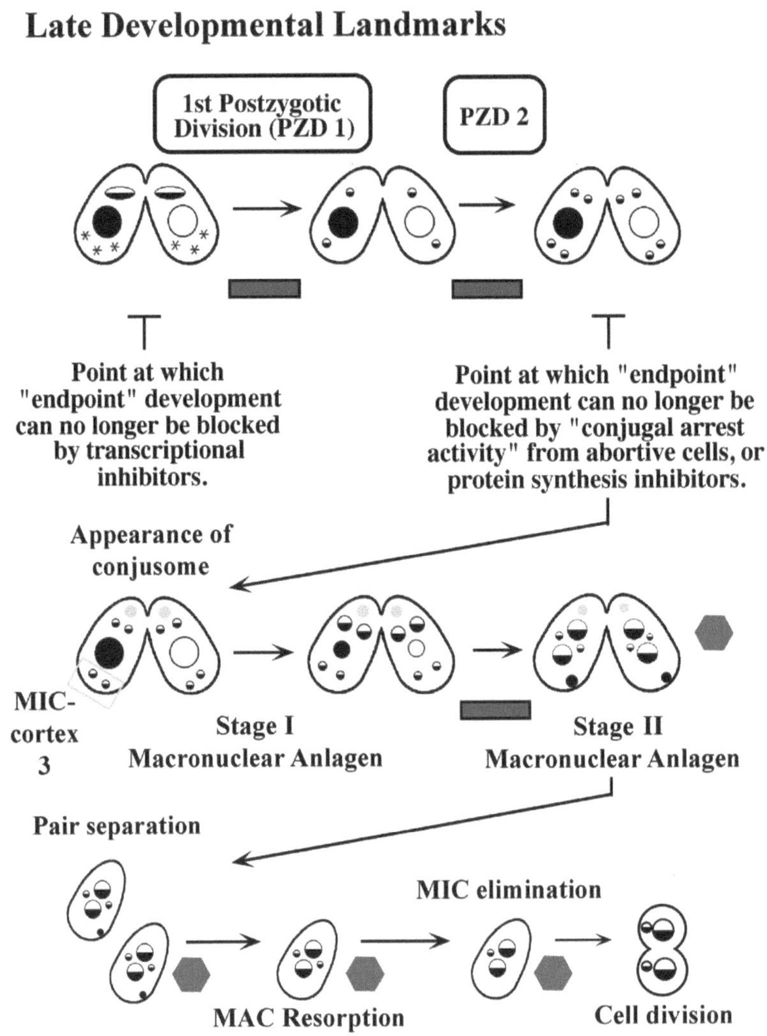

Fig. 11 Late developmental landmarks. Lines indicate the first stages at which transcriptional and translational inhibitors, as well as "conjugal arrest activity" from dividing or genomic exclusion pairs can no longer prevent developmental progression toward "endpoint." Green circles represent the "conjusome" organelle. Green box indicates the third place in which nuclear association with the cell cortex drives nuclear fate: here cortical tethering shields nuclei from signals driving somatic, macronuclear differentiation. It should be noted that tethering actually occurs just prior to the second postzygotic division (the second post-zygotic division spindle is anchored to the posterior cortex, see live-cell observations diagrammed in Fig. 7). Gray boxes indicate periods of DNA synthesis. Red hexagons indicate a variety of arrest phenotypes (see text). (See color plate.)

driving differential nuclear fate. In particular: (1) cytoskeletal tethering of a meiotic product to the nuclear exchange junction appears to shield it from cytoplasmic signals triggering PND; (2) bilateral tethering of gametic pronuclei to the nuclear exchange junction appears to be necessary (though perhaps not sufficient) to trigger entry into the postzygotic developmental program; and (3) cytoskeletal tethering of postzygotic nuclei to the posterior cell cortex appears to shield them from cytoplasmic signals triggering macronuclear anlagen differentiation. Detailed cytological observations of mating cells identify a host of subtle nuclear movements and cortical/cytoskeletal associations that remain to be explored.

Acknowledgments

This chapter was supported by grants from NSF (0817993): "The Gene Stream II" and NIH (1-R15-HD-064147-01): "The Developmental Genetics of *Tetrahymena thermophila*" awarded to E. S. Cole. Authors are grateful to Dr. M. Endo for his help in constructing Fig. 2, and to Dr. Joseph Frankel, Dr. Carolyn Price, and Dr. Douglas Chalker for editorial assistance and valuable conversations and communications.

References

Abraham, R. T. (2001). Cell cycle checkpoint signaling through the ATM and ATR kinases. *Genes Dev.* **15**, 2177–2196.

Allen, S. L. (1967a). Genomic exclusion: a rapid means for inducing homozygous diploid lines in Tetrahymena pyriformis, syngen 1. *Science* **155**, 575–577.

Allen, S. L. (1967b). Cytogenetics of genomic exclusion in Tetrahymena. *Genetics* **55**, 797–822.

Allen, S. L., File, S. K., and Koch, S. L. (1967). Genomic exclusion in *Tetrahymena*. *Genetics* **55**, 823–837.

Aronica, L., Bednenko, J., Noto, T., DeSouza, L. V., Siu, K. W. M., Loidl, J., Pearlman, R. E., Gorovsky, M. A., and Mochizuki, K. (2008). Study of an RNA helicase implicates small RNA-noncoding RNA interactions in programmed DNA elimination in *Tetrahymena*. *Genes Dev.* **22**, 2228–2241.

Aufderheide, K. (1982). An improvement in the protargol procedure of Ng and Nelsen. *Trans. Am. Microsc. Soc.* **101**, 100–104.

Bruns, P. J., and Brussard, T. B. (1974). Pair formation in *Tetrahymena pyriformis,* an inducible developmental system. *J. Exp. Zool.* **188**, 337–344.

Bruns, P. J., and Palestine, R. F. (1975). Costimulation in *Tetrahymena pyriformis*: a developmental interaction between specially prepared cells. *Dev. Biol.* **42**, 75–83.

Bruns, P. J. (1986). The Fenetic Organization of *Tetrahymena.* The Molecular Biology of Ciliated Protozoa. J. Gall, pp. 27-44. Academic Press, New York.

Bruns, P. J., Brussard, T. B., and Merriam, E. V. (1983). Nullisomic *Tetrahymena*: a set of nullisomics define the germinal chromosomes. *Genetics* **104**, 257–270.

Cameron, I. L., and Jeter, J. R. (1970). Synchronization of the cell cycle of *Tetrahymena* by starvation and refeeding. *J. Eukaryot. Microbiol.* **17**, 429–431.

Campbell, C., and Romero, D. P. (1998). Identification and characterization of the RAD51 gene from the ciliate *Tetrahymena thermophila*. *Nucleic Acids Res.* **26**, 3165–3172.

Cervantes, M. D., Xi, X., Vermaak, D., Yao, M. C., and Malik, H. S. (2006). The CNA1 histone of the ciliate *Tetrahymena thermophila* is essential for chromosome segregation in the germline micronucleus. *Mol. Biol. Cell* **17**, 485–497.

Chalker, D. L., Fuller, P., and Yao, M. C. (2005). Communication between parental and developing genomes during *Tetrahymena* nuclear differentiation is likely mediated by homologous RNAs. *Genetics* **169**, 149–160.

Chilcoat, N. D., and Elde, N. C., *et al.* (2001). An antisense approach to phenotype-based gene cloning in Tetrahymena. *Proc. Natl. Acad. Sci. U.S.A.* **98**, 8709–8713.

Cleffmann, G. (1980). Chromatin elimination and the genetic organisation of the macronucleus in Tetrahymena thermophila. *Chromosoma (Berlin)* **78**, 313–325.

Cole, E. S. (1991). Conjugal blocks in *Tetrahymena* pattern mutants and their cytoplasmic rescue. I. Broadened cortical domains (bcd). *Dev. Biol.* **148**, 403–419.

Cole, E. S., and Frankel, J. (1991). Conjugal blocks in *Tetrahymena* pattern mutants and their cytoplasmic rescue. II. *janusA*. *Dev. Biol.* **148**, 420–428.

Cole, E. S., and Bruns, P. J. (1992). Uniparental cytogamy: a novel method for bringing micronuclear mutations of *Tetrahymena* into homozygous macronuclear expression with precocious sexual maturity. *Genetics* **132**, 1017–1031.

Cole, E. S., Cassidy-Hanley, D., Hemish, J., Tuan, J., and Bruns, P. J. (1997). A mutational analysis of conjugation in *Tetrahymena themophila* 1. Phenotypes affecting early development: meiosis to nuclear selection. *Dev. Biol.* **189**, 215–232.

Cole, E. S., and Soelter, T. A. (1997). A mutational analysis of conjugation in *Tetrahymena themophila* 2. Phenotypes affecting middle and late development: third prezygotic division, pronuclear exchange, pronuclear fusion and postzygotic development. *Dev. Biol.* **189**, 233–245.

Cole, E. S., Virtue, M. A., and Stuart, K. R. (2001). Development in electrofused conjugants of *Tetrahymena themophila*. *J. Eukaryot. Microbiol.* **48**, 266–279.

Cole, E. S. (2006). The Tetrahymena conjugation junction. *In* "Cell–cell channels," (F. Baluska, D. Volkmann, P. W. Barlow, eds.), pp. 39–62. Springer, New York.

Coyne, R. S., Nikiforov, M. A., Smothers, J. F., Allis, C. D., and Yao, M. C. (1999). Parental expression of the chromodomain protein Pdd1p is required for completion of programmed DNA elimination and nuclear differentiation. *Mol. Cell* **4**, 865–872.

Cui, B. W., and Gorovsky, M. A. (2006). Centromeric histone H3 is essential for vegetative cell division and for DNA elimination during conjugation in *Tetrahymena thermophila*. *Mol. Cell. Biol.* **26**, 4499–4510.

Davis, M. C., Ward, J. G., Herrick, G., and Allis, C. D. (1992). Programmed nuclear death; apoptotic-like degradation of specific nuclei in conjugating *Tetrahymena*. *Dev. Biol.* **154**, 419–432.

Ding, Y. Z., and Weijie, C. Y. (1987). Fluctuations of cyclic AMP contents of *Tetrahymena* SI during vegetative growth and conjugation. *Acta. Zool. Sinica* 1987-2003.

Doerder, F. P., and DeBault, L. E. (1975). Cytofluorimetric analysis of nuclear DNA during meiosis, fertilization, and macronuclear development in the ciliate *Tetrahymena thermophila*. *J. Cell Sci.* **17**, 471–493.

Doerder, F. P., and Shabatura, S. K. (1980). Genomic exclusion in *Tetrahymena thermophila*: a cytogenetic and cytofluorimetric study. *Dev. Gen.* **1**, 205–218.

Eisen, J. A., and Coyne, R. S., *et al.* (2006). Macronuclear genome sequence of the ciliate *Tetrahymena thermophila,* a model eukaryote. *PLoS Biol.* **4**, e286.

Endo, M., and Sugai, T. (2005). Division-specific globular chromatin in the macronucleus of *Tetrahymena thermophila*. *Zool. Sci.* **22**, 1509.

Finley, M. J., and Bruns, P. J. (1980). Costimulation in *Tetrahymena*. II. A non-specific response to heterotypic cell–cell interactions. *Dev. Biol.* **79**, 81–94.

Francis, J. T., and Hennessey, T. M. (1995). Chemorepellents in *Paramecium* and *Tetrahymena*. *J. Eukaryot. Microbiol.* **42**, 78.

Frankel, J. (1962). The effects of heat, cold, and p-fluorophenylalanine on morphogenesis in synchronized *Tetrahymena pyriformis* GL. *C. R. Trav. Lab. Carlsberg* **33**, 1–52.

Frankel, J. (1964). Cortical morphogenesis and synchronization in *Tetrahymena pyriformis* GL. *Exp. Cell Res.* **35**, 349–360.

Frankel, J. (1967a). Studies on the maintenance of development in *Tetrahymena pyriformis* GL-C. I. An analysis of the mechanism of resorption of developing oral structures. *J. Exp. Zool.* **164**, 435–460.

Frankel, J. (1967b). Critical phases of oral primordium development in *Tetrahymena pyriformis* GL-C: an analysis employing low temperature treatments. *J. Protozool.* **14**, 639–649.

Frankel, J. (1969a). Participation of the undulating membrane in the formation of oral replacement primordia in Tetrahymena pyriformis. *J. Protozool.* **16**, 26–35.

Frankel, J. (1969b). The relationship of protein synthesis to cell division and oral development in synchronized Tetrahymena pyriformis GL-C: an analysis employing cycloheximide. *J. Cell. Physiol.* **74**, 135–148.

Frankel, J. (1970). The synchronization of oral development without cell division in *Tetrahymena pyriformis* GL-C. *J. Exp. Zool.* **173**, 79–99.

Frankel, J. (2008). What do genic mutations tell us about the structural patterning of a complex single-celled organism? *Eukary. Cell* **7**, 1617–1639.

Frankel, J., and Heckmann, K. (1968). A simplified Chatton–Lwoff silver impregnation procedure for use in experimental studies with ciliates. *Trans. Am. Microsc. Soc.* **87**, 317–321.

Frankel, J., Jenkins, L. M., Nelsen, E. M., and Doerder, F. P. (1976a). Mutations affecting cell division in *Tetrahymena pyriformis*. I. Selection and genetic analysis. *Genetics* **83**, 489–506.

Frankel, J., Jenkins, L. M., and DeBault, L. E. (1976b). Causal relations among cell cycle processes in *Tetrahymena pyriformis*: an analysis employing temperature-sensitive mutants. *J. Cell Biol.* **71**, 242–260.

Frankel, J., Nelsen, E. M., and Jenkins, L. M. (1977). Mutations affecting cell division in *Tetrahymena pyriformis,* syngen 1. II. Phenotypes of single and double homozygotes. *Dev. Biol.* **58**, 255–275.

Frankel, J., Mohler, J., and Frankel, A. W. K. (1980a). Temperature-sensitive periods of mutations affecting cell division in *Tetrahymena thermophila*. *J. Cell Sci.* **43**, 59–74.

Frankel, J., Mohler, J. D., and Frankel, A. K. (1980b). The relationship between the excess-delay phenomenon and temperature-sensitive periods in *Tetrahymena thermophila*. *J. Cell Sci.* **43**, 75–91.

Frankel, J., Jenkins, L. M., Bakowska, J., and Nelsen, E. M. (1984). Mutational analysis of patterning of oral structures of *Tetrahymena*. I. Effects of increased size on organization. *J. Embryol. Exp. Morphol.* **82**, 41–66.

Frankel, J., and Nelsen, E. M. (2001). The effects of supraoptimal temperatures on population growth and cortical patterning in *Tetrahymena pyriformis* and *Tetrahymena thermophila*: a comparison. *J. Eukaryot. Microbiol.* **48**, 135–146.

Frisch, Q., Levkowitz, H., and Loyter, A. (1976). Inhibition of conjugation and cell division in *Tetrahymena pyriformis* by tunicamycin: a possible requirement of glycoprotein synthesis for induction of conjugation. *Biochem. Biophys. Res. Commun.* **72**, 138–145.

Frisch, A., and Loyter, A. (1977). Inhibition of conjugation in *Tetrahymena pyriformis* by ConA: Localization of ConA-binding sites. *Exp. Cell Res.* **110**, 337–346.

Furukawa, Y., Yamada, R. H., and Kohno, M. (1979). Presence of two DNA polymerases in *Tetrahymena pyriformis*. *Nucleic Acids Res.* **7**, 2387–2398.

Gaertig, J., Seyfertand, H. -M., and Kaczanowski, A. (1986). Post-meiotic DNA synthesis in nocodazole-blocked nuclei during conjugation of *Tetrahymena thermophila*: induction of polyploidy in the micronucleus. *Exp. Cell Res.* **164**, 562–567.

Gaertig, J., and Kaczanowski, A. (1987). Correlation between the shortened period of cell pairing during genomic exclusion and the block in post–transfer nuclear development in *Tetrahymena thermophila*. *Dev. Growth Differ.* **29**, 553–562.

Gaertig, J., Kiernowska, M., and Iftode, F. (1988). Induction of cybrid strains of *Tetrahymena* by electrofusion. *J. Cell Sci.* **89**, 253–261.

Gaertig, J., and Iftode, F. (1989). Rearrangement of the cytoskeleton and nuclear transfer in *Tetrahymena thermophila* cells fused by electric field. *J. Cell Sci.* **93**, 691–703.

Gaertig, J., and Fleury, A. (1992). Spatiotemporal reorganization of intracytoplasmic microtubules is associated with nuclear selection and differentiation during developmental process in the ciliate *Tetrahymena thermophila*. *Protoplasma* **167**, 74–87.

Gaertig, J., and Cole, E. S. (2000). The role of cortical geometry in the nuclear development of *Tetrahymena thermophila*. *J. Eukaryot. Microbiol.* **47**, 590–596.

Gavin, R. H. (1965). The effects of heat and cold on cellular development in *Tetrahymena pyriformis* WH-6. *J. Protozool.* **12**, 307–318.

Hamilton, E. P., and Suhr-Jessen, P. B. (1980). Autoradiographic evidence for self–fertilization in *Tetrahymena thermophila*. *Exp. Cell Res.* **126**, 391–396.

Hamilton, E. P. (1984). *Dissection of Fertilization and Development in Tetrahymena using Anti-Microtubule Drugs*. Division of Biological Sciences, University of California, Santa Barbara, p. 229.

Hamilton, E. P., Suhr-Jessen, P. B., and Orias, E. (1988). Pronuclear fusion failure – an alternate conjugational pathway in *Tetrahymena thermophila*, induced by vinblastine. *Genetics* **118**, 627–636.

Haremaki, T., Sugai, T., and Takehashi, M. (1995). The vegetative micronucleus has a critical role in maintenance of cortical structure in *Tetrahymena thermophila*. *Cell Struct. Funct.* **20**, 239–244.

Haremaki, T., Sugai, T., and Takahashi, M. (1996). Involvement of active cellular mechanisms on the disorganization of oral apparatus in amicronucleate cells in *Tetrahymena thermophila*. *Cell Struct. Funct.* **21**, 73–80.

Hartwell, L. H., and Weinert, T. A. (1989). Checkpoints: controls that ensure the order of cell cycle events. *Science* **246**, 629–634.

Holz, G. G., Scherbaum, O. H., and Williams, N. (1957). The arrest of mitosis and stomatogenesis during temperature induction of synchronous division in *Tetrahymena pyriformis,* mating type 1, variety 1. *Exp. Cell Res.* **13**, 618–621.

Iwamoto, M., and Nakaoka, Y. (2002). External GTP binding and induction of cell division in starved *Tetrahymena thermophila*. *Eur. J. Cell Biol.* **81**, 517.

Iwamoto, M., Sugai, T., and Nakaoka, Y. (2004). Cell division induced by mechanical stimulation in starved Tetrahymena thermophila: cell cycle without synthesis of macronuclear DNA. *Cell Biol. Int.* **28**, 503–509.

Kaczanowska, J., Buzanska, L., and Ostrowski, M. (1993). Relationship between spatial patterning of basal bodies and membrane skeleton (epiplasm) during the cell cycle of *Tetrahymena: cdaA* mutant and anti-membrane immunostaining. *J. Eukaryot. Microbiol.* **40**, 747–754.

Kaczanowska, J., Joachimiak, E., Buzanska, L., Krawczynska, W., Wheatley, D. N., and Kaczanowski, A. (1999). Molecular subdivision of the cortex of dividing *Tetrahymena* is coupled with the formation of the fission zone. *Dev. Biol.* **212**, 150–164.

Kaczanowska, J., Kaczanowski, K., Kiersnowska, M., Fabczak, H., Tulodziecka, K., and Kaczanowski, A. (2008). Acquisition of cell polarity during cell cycle and oral replacement in *Tetrahymena*. *Int. J. Dev. Biol.* **52**, 249–258.

Kaczanowski, A. (1978). Gradients of proliferation of ciliary basal bodies and the determination of the position of the oral primordium in *Tetrahymena*. *J. Exp. Zool.* **204**, 417–430.

Kaczanowski, A. (1992). Mutation affecting cell separation and macronuclear resorption during conjugation in *Tetrahymena thermophila*: early expression of the zygotic genotype. *Dev. Genet.* **13**, 58–65.

Kaczanowski, A., Gaertig, J., and Kubiak, J. (1985). Effect of the antitubulin drug nocodazole on meiosis and post–meiotic development in *Tetrahymena thermophila*: induction of achiasmatic meiosis. *Exp. Cell Res.* **158**, 244–256.

Kaczanowski, A., Gornicka, I., and Cleffmann, G. (1989). Arrest of micronuclear DNA replication during genomic exclusion in *Tetrahymena* produces haploid strain. *Genetics* **121**, 37–45.

Kaczanowski, A., Ramel, M., Kaczanowska, J., and Wheatley, D. (1991). Macronuclear differentiation in conjugating pairs of *Tetrahymena* treated with the antitubulin drug nocodazole. *Exp. Cell Res.* **195**, 330–337.

Kaczanowski, A., and Kiersnowska, M. (1996a). The differentiation of postzygotic posterior nuclei into macronuclei in Tetrahymena thermophila is not affected by electrofusion with the anterior pole of another conjugant. *Eur. J. Protistol.* **32**, 62–67.

Kaczanowski, A., and Kiersnowska, M. (1996b). Induction of blocks in nuclear divisions and over-condensation of meiotic chromosomes with cycloheximide during conjugation of Tetrahymena thermophila. *J. Eukaryot. Microbiol.* **43**, 380–388.

Kaczanowski, A., Kiersnowska, M., and Kaczanowska, J. (2004). Isolation of a *Tetrahymena thermophila* strain which induced metaphase i meiotic arrest: new pathway of abortive conjugation. *J. Eukaryot. Microbiol.* **51**, 351–363.

Kaczanowski, A., and Kiersnowska, M. (2011 Oct). Inactivation of a macronuclear intraSphase checkpoint in *Tetrahymena thermophila* with caffeine affects the integrity of the micronuclear genome. *Protist* **162**(4), 616–636.

Karrer, K. M. (2000). *Tetrahymena* genetics: two nuclei are better than one. *Meth. Cell. Biol.* **62**, 128–189.

Kaney, A. R., and Speare, V. J. (1983). An amicronucleate mutant of *Tetrahymena thermophila*. *Exp. Cell Res.* **143**, 461–467.

Karrer, K., Stein-Gavens, S., and Allitto, B. A. (1984). Micronucleus-specific DNA sequences in an amicronucleate mutant of *Tetrahymena*. *Dev. Biol.* **105**, 121–129.

Kiersnowska, M., Kaczanowski, A., and Morga, J. (2000). Macronuclear development in conjugants of *Tetrahymena thermophila,* which were artificially separated at meiotic prophase. *J. Eukaryot. Microbiol.* **47**, 139–147.

Kirk, K. E., Christ, C., McGuire, J. M., Paul, A. G., Vahedi, M., Stuart, K. R., and Cole, E. S. (2008). Abnormal micronuclear telomeres lead to an unusual cell cycle checkpoint and defects in *Tetrahymena* oral morphogenesis. *Eukaryot. Cell* **7**, 1712–1723.

Kobayashi, T., and Endoh, H. (1998). Abortive conjugation induced by UV-B irradiation at meiotic prophase in *Tetrahymena thermophila*. *Dev. Genet.* **23**, 151–157.

Lambert, S., and Carr, A. M. (2005). Checkpoint responses to replication fork barriers. *Biochimie* **87**, 591–602.

Lansing, T. J., Frankel, J., and Jenkins, L. M. (1985). Oral ultrastructure and oral development in the misaligned undulating membrane mutant of *Tetrahymena thermophila*. *J. Protozool.* **32**, 126–139.

Li, S., Yin, L., Cole, E. S., Udani, R. A., and Karrer, K. M. (2006). Progeny of germ line knockouts of ASI2, a gene encoding a putative signal transduction receptor in *Tetrahymena thermophila,* fail to make the transition from sexual reproduction to vegetative growth. *Dev. Biol.* **295**, 633–646.

Loidl, J., and Mochizuki, K. (2009). *Tetrahymena* meiotic nuclear reorganization is induced by a checkpoint kinase-dependent response to DNA damage. *Mol. Biol. Cell* **20**, 2428–2437.

Malone, C. D., Anderson, A. M., Motl, J. A., Rexer, C. H., and Chalker, D. L. (2005). Germ line transcripts are processed by a Dicer-like protein that is essential for developmentally programmed genome rearrangements of *Tetrahymena thermophila*. *Mol. Cell. Biol.* **25**, 9151–9164.

Marsh, T. C., Cole, E. S., Stuart, K. R., Campbell, C., and Romero, D. P. (2000). *RAD51* is required for propagation of the germinal nucleus in *Tetrahymena thermophila*. *Genetics* **154**, 1587–1596.

Marsh, T. C., Cole, E. S., and Romero, D. P. (2001). The transition from conjugal development to the first vegetative cell division is dependent on RAD51 expression in the ciliate *Tetrahymena thermophila*. *Genetics* **157**, 1591–1598.

Matsuda, A., Shieh, A. W. Y., Chalker, D. L., and Forney, J. D. (2010). The conjugation-specific Die5 protein is required for development of the somatic nucleus in both *Paramecium* and *Tetrahymena*. *Eukaryot. Cell* **9**, 1087–1099.

McDonald, B. B. (1962). Synthesis of deoxyribonucleic acid by micro- and macronuclei of *Tetrahymena pyriformis*. *J. Cell Biol.* **13**, 193–203.

McDonald, B. B. (1966). The exchange of RNA and protein during conjugation in *Tetrahymena*. *J. Protozool.* **13**, 277–285.

Melia, S. M., Cole, E. S., and Turkewitz, A. P. (1998). Mutational analysis of regulated exocytosis in *Tetrahymena*. *J. Cell Sci.* **111**, 131–140.

Mochizuki, K., Fine, N. A., Fujisawa, T., and Gorovsky, M. A. (2002). Analysis of a piwi-related gene implicates small RNAs in genome rearrangement in *Tetrahymena*. *Cell* **110**, 689–699.

Mochizuki, K., and Gorovsky, M. A. (2004). Small RNAs in genome rearrangement in *Tetrahymena*. *Curr. Opin. Genet. Dev.* **14**, 181–187.

Mochizuki, K., Novatchkova, M., and Loidl, J. (2008). DNA double-strand breaks, but not crossovers, are required for the reorganization of meiotic nuclei in *Tetrahymena*. *J. Cell Sci.* **121**, 2148–2158.

Mohammad, M., York, R. D., Hommel, J., and Kapler, G. M. (2003). Characterization of a novel origin recognition complex-like complex: implications for DNA recognition, cell cycle control, and locus-specific gene amplification. *Mol. Cell. Biol.* **23**, 5005–5017.

Morrison, T. L., Yakisich, J. S., Cassidy-Hanley, D. M., and Kapler, G. M. (2005). *TIF1* represses rDNA replication initiation, but promotes normal S phase progression and chromosome transmission in *Tetrahymena. Mol. Biol. Cell* **16**, 2624–2635.

Naduparambil, K. J., Skopp, R., and Price, C. M. (2001). G-overhang dynamics at *Tetrahymena* telomeres. *EMBO J.* **20**, 4299–4308.

Nachtwey, D. S. (1967a). Cold-shock-induced division delay of synchronized Tetrahymena: temperature–duration relationship. *Res. Dev. Tech. Rep.* **5**, 1–13.

Nachtwey, D. S., and Dickinson, W. J. (1967b). Actinomycin D: blockage of cell division of synchronized Tetrahymena pyriformis. *Exp. Cell Res.* **47**, 581–595.

Nachtwey, D. S., and Giese, A. C. (1968). Effects of ultraviolet light irradiation and heat–shocks on cell division in synchronized *Tetrahymena*. *Exp. Cell Res.* **50**, 167–176.

Nakagawa, T., Fujiu, K., Cole, E. S., and Numata, O. (2008). Involvement of a 25 kDa Tetrahymena Ca2 +–binding protein in pronuclear exchange. *Cell Struct. Func.* **33**, 151–162.

Nanney, D. L. (1953). Nucleocytoplasmic interactions during conjugation in *Tetrahymena*. *Biol. Bull.* **105**, 133–148.

Nanney, D. L. (1975). Patterns of basal body addition in ciliary rows in *Tetrahymena*. *J. Cell Biol.* **65**, 503–512.

Nanney, D. L. (1976). Calcium chloride effects on nuclear development in *Tetrahymena*. *Genet. Res. Camb.* **27**, 297–302.

Nanney, D. L., and McCoy, J. W. (1976). Characterization of the species of the *Tetrahymena pyriformis* complex. *Trans. Am. Microsc. Soc.* **95**, 6640682.

Nelsen, E. M. (1970). Division delays and abnormal oral development produced by colchicine in *Tetrahymena*. *J. Exp. Zool.* **175**, 69–84.

Nelsen, E. M. (1981). The undulating membrane of *Tetrahymena:* Formation and reconstruction. *Trans. Am. Microsc. Soc.* **100**, 285–295.

Nelsen, E. M., and DeBault, L. E. (1978). Transformation in *Tetrahymena pyriformis*—Description of an inducible phenotype. *J. Protozool.* **25**, 113–119.

Ng, S. F. (1986). The somatic function of the micronucleus of ciliated protozoa. *Progr. Protistol.* **1**, 215–286.

Ng, S., and Nelsen, E. M. (1977). The protargol staining technique: an improved version for *Tetrahymena pyriformis*. *Trans. Am. Microsc. Soc.* **96**, 369–376.

Nikiforov, M., Smothers, J., Gorovsky, M., and Allis, C. (1999). Excision of micronuclear-specific DNA requires parental expression of Pdd2p and occurs independently from DNA replication in *Tetrahymena thermophila*. *Genes Dev.* **13**, 2852–2862.

Orias, E., and Flacks, M. (1973). Use of genomic exclusion to isolate heat-sensitive mutants in Tetrahymena. *Genetics* **73**, 543–549.

Orias, E., and Bruns, P. J. (1976). Induction and isolation of mutants in *Tetrahymena*. *Methods Cell Biol.* **13**, 247–283.

Orias, E., and Hamilton, E. P. (1979). Cytogamy: an inducible, alternate pathway of conjugation in *Tetrahymena thermophila*. *Genetics* **91**, 657–671.

Orias, E., Hamilton, E. P., and Flacks, M. (1979). Osmotic shock prevents nuclear exchange and produces whole genome homozygotes in conjugating *Tetrahymena*. *Science* **203**, 660–663.

Orias, J. D., Hamilton, E. P., and Orias, E. (1983). A microtubule meshwork associated with gametic pronucleus transfer across a cell–cell junction. *Science* **222**, 181–184.

Orias, E. (1986). Ciliate conjugation. *In* "The Molecular Biology of Ciliated Protozoa," (L. G. Gall, ed.), pp. 45–94. Academic Press, Orlando.

Pitts, R. A. (1979). Age-associated micronuclear defects of *Tetrahymena thermophila*: genetic and cytogenetic studies. Thesis, University of Pittsburgh, p. 64.

Pitts, R. A., and Doerder, F. P. (1988). Genomic exclusion and other micronuclear anomalies are common in genetically defective clones of *Tetrahymena thermophila*. *Genetics* **120**, 135–143.

Roberts Jr., C. T., and Orias, E. (1973). A cycloheximide-resistant mutant of *Tetrahymena pyriformis*. *Exp. Cell Res.* **81**, 312–316.

Roberts Jr., C. T., and Orias, E. (1974). On the mechanism of adaptation to protein synthesis inhibitors by Tetrahymena. Facilitation, cross adaptation, and resensitization. *J. Cell Biol.* **62**, 707–716.

Sakai, A., and Watanabe, Y. (1982). DNA polymerases of *Tetrahymena pyriformis*. I. Characterization of two N-ethylmaleimide-sensitive DNA polymerases from exponentially growing cells. *J. Biochem.* **91**, 845–853.

Scherbaum, O. H., and Zeuthen, E. (1954). Induction of synchronous cell division in mass cultures *of Tetrahymena pyriformis*. *Exp. Cell Res.* **6**, 221–227.

Scherbaum, O. H., Louderback, A. L., and Jahn, T. L. (1958). The formation of subnuclear aggregates in normal and synchronized protozoan cells. *Biol. Bull.* **115**, 269–275.

Slade, K., Freggiaro, S., Cottrell, K. A., Smith, J. J., and Wiley, E. A. (2011). Sirtuin-mediated nuclear differentiation and programmed degradation in Tetrahymena. *BMC Cell Biology* **12**(1), 40.

Smith, J. J., Cole, E. S., and Romero, D. P. (2004a). Transcriptional control of RAD51 expression in the ciliate Tetrahymena thermophila. *Nucleic Acids Res.* **32**, 4313–4321.

Smith, J. J., Yakisich, J. S., Kapler, G. M., Cole, E. S., and Romero, D. P. (2004). A beta-tubulin mutation selectively uncouples nuclear division and cytokinesis in *Tetrahymena thermophila*. *Eukaryot. Cell* **3**, 1217–1226.

Song, X., Gjoneska, E., Ren, Q. H., Taverna, S. D., Allis, C. D., and Gorovsky, M. A. (2007). Phosphorylation of the SQ H2A.X motif is required for proper meiosis and mitosis in *Tetrahymena thermophila*. *Mol. Cell. Biol.* **27**, 2648–2660.

Stover, N. A., and Rice, J. D. (2011). Distinct cyclin genes define each stage of ciliate conjugation. *Cell Cycle.* **15**, 1699–1701.

Stone, G. E. (1968a). Synchronized Cell Division in *Tetrahymena pyriformis* Following Inhibition with Vinblastine. *J. Cell Biol.* **39**, 556–563.

Stone, G. E. (1968b). Method for reversible inhibition of cell division in *Tetrahymena pyriformis* using vinblastine sulfate. *Methods Cell Physiol.* **3**, 161–170.

Sugai, T., and Hiwatashi, K. (1974). Cytologic and autoradiographic studies of the micronucleus at meiotic prophase in Tetrahymena pyriformis. *J. Protozool.* **21**, 542–548.

Tamura, S., Tsuruhara, T., and Watanabe, Y. (1969). Function of nuclear microtubules in macronuclear division of *Tetrahymena pyriformis*. *Exp. Cell Res.* **55**, 351–358.

Tang, L., Adl, S. M., and Berger, J. D. (1997). A CDC2-related kinase is associated with macronuclear DNA synthesis in *Paramecium tetraurelia*. *J. Eukaryot. Microbiol.* **44**, 269–275.

Thormar, H. (1959). Delayed division in *Tetrahymena pyriformis* induced by temperature changes. *C.R. Trav. Lab. Carlsberg.* **31**, 207–226.

Virtue, M. A., and Cole, E. S. (1999). A cytogenetic study of devolopment in mechanically disrupted pairs of *Tetrahymena thermophila*. *J. Eukaryot. Microbiol.* **46**, 597–605.

Ward, J. G., and Herrick, G. (1994). Effects of the transcription inhibitor actinomycin D on postzygotic development *of Tetrahymena thermophila* conjugants. *Dev. Biol.* **173**, 174–184.

Ward, J. G., Davis, M. C., Allis, C. D., and Herrick, G. (1995). Effects of nullisomic chromosome deficiencies on conjugation events in *Tetrahymena thermophila:* insufficiency of the parental macronucleus to direct postzygotic development. *Genetics* **140**, 989–1005.

Watanabe, S., Toyohara, A., and Suzaki, T. (1988). The relation of concanavalin A receptor distribution to the conjugation process in *Tetrahymena thermophila*. *J. Protozool.* **28**, 171–175.

Wellnitz, W. R., and Bruns, P. J. (1979). The pre-pairing events in *Tetrahymena thermophila*: analysis of blocks imposed by high concentrations of Tris-Cl. *Exp. Cell Res.* **119**, 175–180.

Wellnitz, W. R., and Bruns, P. J. (1982). The pre-pairing events in Tetrahymena. II. Partial loss of developmental information upon refeeding starved cells. *Exp. Cell Res.* **137**, 317–328.

Williams, N. E. (1964). Relations between temperature sensitivity and morphogenesis in *Tetrahymenapyriformis* GL. *J. Protozool.* **11**, 566–572.

Williams, N. E. W., and Williams, R. J. (1976). Macronuclear division with and without microtubules in *Tetrahymena*. *J. Cell Sci.* **20**, 61–77.

Williams, N. E., and Macey, M. (1991). Is cyclin Zeuthen's "division protein"? *Exp. Cell Res.* **197**, 137–139.

Williams, S. A., Hosein, R. E., Garcés, J. A., and Gavin, R. H. (2000). *MYO1*, a novel, unconventional myosin gene affects endocytosis and macronuclear elongation in *Tetrahymena thermophila*. *J. Eukaryot. Microbiol.* **47**, 561–568.

Wolfe, J. (1974). Reciprocal induction of cell division by cells of complementary mating types in *Tetrahymena*. *Exp. Cell Res.* **87**, 39–46.

Wolfe, J. (1976). G1 arrest and the division/conjugation decision in *Tetrahymena*. *Dev. Biol.* **54**, 116.

Wolfe, J., and Grimes, G. W. (1979). Tip transformation in Tetrahymena: a morphogenetic response to interactions between mating types. *J. Protozool.* **26**, 82–89.

Wolfe, J. (1982). The conjugation junction of *Tetrahymena*: its structure and development. *J. Morphol.* **172**, 159–178.

Wolfe, J. (1985). Cytoskeletal reorganization and plasma membrane fusion in conjugating *Tetrahymena*. *J. Cell Sci.* **73**, 69–85.

Wolfe, J., Pagliaro, L., and Fortune, H. (1986). Coordination of concanavalin-A-receptor distribution and surface differentiation in *Tetrahymena*. *Differentiation* **31**, 1–9.

Woodard, J., Kaneshiro, E., and Gorovsky, M. A. (1972). Cytochemical studies on the problem of macronuclear subnuclei in Tetrahymena. *Genetics* **70**, 251–260.

Worthington, D. H., Salamone, M., and Nachtwey, D. S. (1976). Nucleocytoplasmic ratio requirements for initiation of DNA replication and fission in *Tetrahymena*. *Cell Tissue Kinet.* **9**, 119–130.

Yakisich, J. S., Sandoval, P. Y., Morrison, T. L., and Kapler, G. M. (2006). TIF1 activates the intra-S-phase checkpoint response in the diploid micronucleus and amitotic polyploid macronucleus of *Tetrahymena*. *Mol. Biol. Cell* **17**, 5185–5197.

Yanagida, M. (1998). Fission yeast cut mutations revisited: control of anaphase. *Trends Cell Biol.* **8**, 144–149.

Zeuthen, E. (1964). *In* "*Synchrony in Cell Division and Growth*," (E. Zeuthen, ed.), pp. 99–158. Interscience, New York.

Zhang, H., Huang, X., Tang, L., Zhang, Q. -J., Frankel, J., and Berger, J. D. (2002). A cyclin-dependent protein kinase homologue associated with the basal body domains in the ciliate *Tetrahymena thermophila*. *Biochim. Biophys. Acta* **19**, 1–3.

Zweifel, E., and Smith, J., *et al.* (2009). Nested genes CDA12 and CDA13 encode proteins associated with membrane trafficking in the ciliate *Tetrahymena thermophila*. *Euk. Cell* **8**, 899–912.

PART III

Operating Principles

CHAPTER 8

Tetrahymena in the Laboratory: Strain Resources, Methods for Culture, Maintenance, and Storage

Donna M. Cassidy-Hanley

Department of Microbiology and Immunology, College of Veterinary Medicine, Cornell University, Ithaca, New York, USA

METHODS IN CELL BIOLOGY, VOL 109

0091-679X/10 $35.00
DOI 10.1016/B978-0-12-385967-9.00008-6

Abstract

The ciliated protozoan *Tetrahymena thermophila* has been an important model system for biological research for many years. During that time, a variety of useful strains, including highly inbred stocks, a collection of diverse mutant strains, and wild cultivars from a variety of geographical locations have been identified. In addition, thanks to the efforts of many different laboratories, optimal conditions for growth, maintenance, and storage of *Tetrahymena* have been worked out. To facilitate the efficient use of *Tetrahymena*, especially by those new to the system, this chapter presents a brief description of many available *Tetrahymena* strains and lists possible resources for obtaining viable cultures of *T. thermophila* and other *Tetrahymena* species. Descriptions of commonly used media, methods for cell culture and maintenance, and protocols for short- and long-term storage are also presented.

I. Introduction

The increasing use of *Tetrahymena* for both research and educational purposes has been facilitated by the ease with which it can be grown and maintained in a wide range of conditions, from single cells in hanging drops to multiliter cultures grown in large bioreactors. Sexual reproduction is dependably controlled by transfer to non-nutritive media, and simple selection schemes are available for the identification of sexual progeny. Under optimal conditions, *Tetrahymena* has a rapid growth rate, with a doubling time of less than 2 h. However, slowly growing vegetative cultures can be maintained on the bench for several months with very limited loss of function or fertility, and strains can be stored for years in liquid nitrogen. A number of mutant and inbred strains of *Tetrahymena thermophila,* the species most commonly used for physiological, biochemical, and molecular research, are readily available. A number of other *Tetrahymena* species, many of which can be maintained under conditions similar to those used to culture *T. thermophila*, are also easily obtainable. This chapter provides basic information on the most frequently utilized *T. thermophila* strains, current sources for obtaining *T. thermophila* strains and other *Tetrahymena* species, and methods for growing, maintaining, mating, and storing *Tetrahymena* cultures.

II. *Tetrahymena thermophila* Strains

T. thermophila provides a rich resource of useful strains, including highly inbred stocks derived from wild isolates over half a century ago (Allen and Gibson, 1973; and Chapter 2 in this volume), a collection of diverse mutant strains, and wild cultivars from a variety of geographical locations. *T. thermophila* strains are generally named according to location of origin. Wild-type isolates are given a two-letter prefix based on location of origin (e.g., WH for isolates originally collected at

Woods Hole), while strains developed in individual laboratories are given a two-letter prefix representing the location of the lab of origin followed by a strain number (e.g., CU428 is Cornell University strain number 428). Basic standards for describing micronuclear and macronuclear genotypes and phenotypes can be found in Allen (2000) and Allen *et al.* (1998). A more extensive revised version of the current preferences for *Tetrahymena* nomenclature can be found on the *Tetrahymena* Stock Center website (http://tetrahymena.vet.cornell.edu/extras/revised_tetrahymena_nomenclature.doc). A brief description of frequently utilized *T. thermophila* strain types is provided below. More detailed information about individual *T. thermophila* strains can be found on the *Tetrahymena* Stock Center website (http://tetrahymena.vet.cornell.edu).

A. Inbred Wild-Type Strains

The *T. thermophila* strains most commonly used in research labs are highly inbred B lines derived from fertile natural isolates collected in Woods Hole, MA, in the early 1950s by Elliott and co-workers (Nanney and Simon, 2000; and Chapter 2 in this volume). Inbred strains derived from other mating-type families are also available, but are not generally used for physiological, genetic, or molecular research. One exception is inbred strain C3. Naturally occurring genetic polymorphisms inherent in inbred strains B and C3 have proven to be useful genetic tools for genome mapping (Brickner *et al.*, 1996; Lynch *et al.*, 1995; Orias, 1998) and for investigation of amplification, replication, and maintenance of the rDNA (Larson *et al.*, 1986; Lovlie *et al.*, 1988; Luehrsen *et al.*, 1987). The American Type Culture Collection (ATCC; http://www.atcc.org) houses representative strains from inbred mating-type families A and B, derived from cells collected at Woods Hole, Massachusetts, family C, derived from a strain originally collected in Vermont, and family D, derived from a strain originally isolated in Michigan (Nanney and Simon, 2000). ATCC *T. thermophila* strains are identified by mating-type allele family, inbreeding history, and mating type. In many cases, multiple mating types of the same inbreeding cross are represented. ATCC *Tetrahymena* cultures are stored frozen in liquid nitrogen freezers and are shipped frozen on dry ice. A smaller collection of strain B, C, and D inbred lines are also available from the *Tetrahymena* Stock Center at Cornell University. *Tetrahymena* Stock Center cultures are stored in liquid nitrogen freezers, and shipped as viable cultures in proteose peptone media.

B. Star Strains

Star strains are specialized strains of *T. thermophila* that lack a functional germinal nucleus. The vestigial genetic material remaining in the micronucleus cannot contribute to the formation of viable sexual progeny. Matings involving star strains follow an alternative developmental program known as "genomic exclusion" (Allen, 1967a, 1967b). During the mating of a star cell and a cell with a functional micronucleus, mating partners separate prematurely, following unidirectional exchange of a gametic pronucleus from the nonstar cell to the star partner. The results of this initial round of mating (RdI) are two cells with identical 100% homozygous micronuclei derived from one meiotic product from the nonstar parent. Both RdI exconjugant cells retain their original macronucleus, and thus their original phenotypes, including mating type. The RdI exconjugants can immediately enter into a second round of mating. Since both partners now contain identical functional micronuclei, the second round of mating proceeds normally, giving rise to progeny that are whole genome homozygotes. Star strains are a valuable genetic tool, useful for an array of genetic applications, including the construction of homozygous strains (Allen, 1967b), the creation of functional heterokaryons including gene knock-out (KO) strains (Dave *et al.*, 2009; Hai *et al.*, 2000), conjugation rescue (Satir *et al.*, 1986), short circuit genomic exclusion (Bruns *et al.*, 1976), and uniparental cytogamy (Cole and Bruns, 1992).

Star lines derived from inbred strain A, B, and C genetic backgrounds are available in several different mating types (A*III, A*V, C*III, B*VI, and B*VII) from the *Tetrahymena* Stock Center. Any star strain can be used to construct homozygous strains following two rounds of mating. However, star strains show different utility for use in other genetic procedures. C*III is the strain of choice for short circuit genomic exclusion and can be used for uniparental cytogamy. A*III is not ideal for short-circuit genomic exclusion or uniparental cytogamy progeny but is excellent for making homozygous heterokaryons and a good choice for conjugation rescue (Satir *et al.*, 1986). A*V and B*VI are the best star strains for use in experiments involving uniparental cytogamy, and B*VII can also be used effectively in that procedure (Cole and Bruns, 1992). It should be noted that A*III carries resistance to 6-methylpurine in its macronucleus.

C. Mutant Strains

In *Tetrahymena,* the physical and functional separation of germinal and somatic nuclei (nuclear dimorphism; see Prescott, 1994; Chapter 3 in this volume) permits the creation of cells that are genetically different in the germinal micronucleus and the somatic macronucleus. Nuclear dimorphism, in conjunction with allelic assortment in the somatic macronucleus, has fostered the creation of genetically useful strains that carry a nonexpressed mutant or modified allele in the germinal nucleus but express only the nonmutant allele in the somatic nucleus (functional heterokaryons; Bruns and Brussard, 1974b). Such lines can be either homozygous or heterozygous in the micronucleus. Potentially lethal genetic constructs, including homozygous knockouts of essential genes (Dave *et al.*, 2009; Hai *et al.*, 2000), lethal mutations, and chromosome modifications such as deletions and loss of one or more entire micronuclear chromosomes (nullisomics; Bruns and Brussard, 1981; Bruns *et al.*, 1983) can only be maintained as heterokaryons. Functional heterokaryons carrying such potentially lethal constructs in the micronucleus can produce viable progeny when mated with cells carrying the equivalent wild-type sequence (except in the case of dominant lethal genes), although in many cases such matings exhibit a somewhat lower frequency of progeny production.

All work with mutant strains must take into consideration the natural phenotypic assortment that occurs in the macronucleus during vegetative growth. The macronucleus is highly polyploid, containing ~45 copies of each macronuclear chromosome (Doerder, 1979; Doerder *et al.*, 1992; Larson *et al.*, 1991), with the exception of the rDNA palindromic chromosome, which is present at about 9000 copies (Kapler, 1993; Yao and Gorovsky, 1974). Since assortment of macronuclear chromosomes is random, over time cultures initially heterozygous in the macronucleus will produce daughter cells that are homozygous for a given allele. Such assortment can result in the loss of a mutant allele within a clone if the cells are maintained under conditions that provide any selective advantage for cells expressing the wild-type allele. This phenomenon is used to advantage to produce heterokaryons, but can lead to problems if continued expression of the mutant allele is desired. A similar problem can occur if wild-type revertants arise in mutant cells maintained in conditions that favor the growth of wild-type cells. Mutant strains should be maintained under the most restrictive conditions possible, and frozen in liquid nitrogen as soon as possible (Bruns *et al.*, 2000; Cassidy-Hanley *et al.*, 1995; Section VI.B). When working with stock cultures maintained on the bench, mutant phenotypes should be verified before undertaking any experimental protocol that requires the expression of a mutant allele that is heterozygous in the macronucleus to eliminate the possibility of allelic loss due to phenotypic assortment. If loss of the mutant phenotype occurs, new working stocks should be established from frozen cultures.

The *Tetrahymena* Stock Center houses a variety of *T. thermophila* mutant strains, including those carrying naturally occurring and induced mutations, defined chromosomal modifications, and genetically engineered modifications. A brief description of basic types of mutant and modified strains available to researchers is listed below. A complete listing of available strains, updated as new strains become available, can be found at http://tetrahymena.vet.cornell.edu/strains.php.

1. Drug Resistant Functional Heterokaryons

Among the most frequently utilized *T. thermophila* mutants are a series of functional heterokaryons homozygous for mutations conferring resistance to either cycloheximide, 6-methylpurine (Byrne, 1978; Byrne *et al.*, 1978), or paromomycin (Bruns *et al.*, 1985) in the germinal micronucleus but expressing the wild-type (drug sensitive) allele in the somatic macronucleus. Paromomycin and cycloheximide heterokaryons are available in both B and C3 backgrounds, while the 6-methylpurine mutation is currently limited to the B strain. Resistance heterokaryons greatly simplify genetic analyses in *Tetrahymena*, allowing for direct selection of progeny cells in a mass mating (Bruns and Brussard, 1974b).

2. Exocytosis Mutants

T. thermophila strains carrying a variety of mutations that affect regulated mucocyst secretion have been characterized (Gutierrez and Orias, 1992; Haddad and Turkewitz, 1997; Melia *et al.*, 1998; Orias *et al.*, 1983). These strains have been used in the study of secretory granule biogenesis and regulated exocytosis (Turkewitz, 2004). Mutants blocked in exocytosis have also proven useful for the efficient purification of cell organelles and macromolecules (Dentler, 1995; Johnson, 1986; Lombillo *et al.*, 1993; Williams, 2000) since secretion of the sticky mucocyst contents can interfere with the purification of cellular components (Tiedtke, 1985).

3. Temperature Sensitive Mutations

Temperature sensitive (ts) mutations affecting pathways as diverse as phagocytosis (Suhr-Jessen and Orias, 1979), morphological development (Frankel *et al.*, 1993; Williams and Honts, 1987), and cell division (Frankel *et al.*, 1976, 1980) have been described in *T. thermophila*. Many of the original ts mutants developed by Joseph Frankel (University of Iowa) and Eduardo Orias (University of California, Santa Barbara) are available through the *Tetrahymena* Stock Center.

4. Conjugation Mutants

Conjugation, the sexual stage in the *Tetrahymena* life cycle, is a complex developmental program that is highly conserved in a number of ciliate species (Raikov, 1976). In *Tetrahymena*, conjugation involves an array of activities including mating-type recognition, pair attachment, and cell fusion, as well as nuclear events including meiosis, mitosis, nuclear transfer, fertilization, and developmental nuclear modification (Chapter 3 and Chapter 7 in this volume). A series of mutant strains have been developed which affect early, middle-, and late-stage events in conjugation, from chromatin condensation to macronuclear anlagen development (Cole and Soelter, 1997; Cole *et al.*, 1997). Morphological pattern mutants arising as a result of problems in conjugation have also been identified (Cole and Frankel, 1991; Cole, 1991). These strains provide a unique and valuable resource for examining factors influencing prezygotic, postzygotic, and exconjugant developmental.

5. Chromosomal Modifications

Although the *Tetrahymena* germinal micronucleus is not expressed (Gorovsky and Woodard, 1969; Mayo and Orias, 1981), and many *Tetrahymena* species apparently lack a micronucleus (Elliot and Hayes, 1955), the *T. thermophila* strains commonly used in the lab appear to require at least a vestigial micronucleus for cell viability. With one possible exception (Kaney and Speare, 1983; Karrer *et al.*, 1984), complete loss of the micronucleus in *T. thermophila* is lethal. However, cells missing large portions of the micronuclear genome are viable when grown vegetatively. A collection of single and multiple nullisomic strains missing both copies of one or more micronuclear chromosomes have been created (Bruns and Brussard, 1981; Bruns *et al.*, 1983), as well as strains unisomic (containing only a single micronuclear chromosome) for each of the micronuclear chromosomes, in both a B and a C3 genetic background. A series of overlapping deletions have also been created for each of the five micronuclear chromosomes. Since monosomic and hemizygous strains containing a single copy of part or all of any micronuclear chromosome are viable, deletion and nullisomic strains are very useful for genetic mapping (Altschuler and Bruns, 1984).

D. Meiotic Segregation Panels

Panels of B-C3 meiotic segregants and terminal assortants derived from heterozygous progeny of matings of inbred B and C3 strains, developed by the Orias lab for use in mapping genes to micronuclear and macronuclear chromosomes (Brickner *et al.*, 1996; Lynch *et al.*, 1995), are available from the *Tetrahymena* Stock Center. These strains are useful for localizing mutant genes to the micronucleus to help determine relationships among mutants with similar phenotypes and for identifying the macronuclear location associated with a specific phenotype to facilitate identification and cloning of mutant genes (Hamilton and Orias, 2000).

E. Genetically Engineered Lines

With the development of facile techniques for gene manipulation, genetic engineering of new *Tetrahymena* strains has become routine. Gene disruptions, gene replacements, knockouts, and knock-ins can easily be accomplished in the micronucleus by biolistic transformation (Bruns and Cassidy-Hanley, 2000a; Cassidy-Hanley *et al.*, 1997), and in the macronucleus by biolistic transformation, electroporation (Gaertig *et al.*, 1994a, 1994b; Gaertig and Kapler, 2000), and microinjection (Chalker *et al.*, 2000; Tondravi and Yao, 1986). Genetically modified strains are available from the labs of origin, or increasingly, through the *Tetrahymena* Stock Center. Among the *T. thermophila* strains currently available are significant portions of the original strain collections developed in the laboratories of Joseph Frankel, University of Iowa (IA strains), Peter Bruns (CU strains), Eduardo Orias (SB strains), and Martin Gorovsky (various genetically modified transformant strains), as well as a variety of genetically engineered strains developed in other labs (for a current list see http://tetrahymena.vet.cornell.edu/strains.php).

III. Other *Tetrahymena* Species

Although, as discussed above, *T. thermophila* is the primary species of choice for *Tetrahymena* research, significant work has been carried out using other *Tetrahymena* species. *Tetrahymena ssp.* are useful indicators for ecotoxicity tests (Gerhardt *et al.*, 2010), and *Tetrahymena pyriformis* GL, an amicronucleate strain used for much of the early *Tetrahymena* research, is still frequently used for toxicological studies (Artemenko *et al.*, 2011; Sauvant *et al.*, 1995, 1997, 1999). Ribosomal RNA-based phylogenies of various *Tetrahymena* species have been developed (Nanney *et al.*, 1989; Preparata *et al.*, 1989). In addition, phylogenetic relationships of many species have been analyzed by bar coding using the cytochrome c oxidase subunit 1 (COX1) (Kher *et al.*, 2011), and the small subunit ribosomal RNA (SSrRNA) genes (Chantangsi *et al.*, 2007; Chantangsi and Lynn, 2008). Phylogenetic relationships have also been examined using comparisons of telomerase RNA (Ye and Romero, 2002).

The American Type Culture Collection houses a collection of 38 different *Tetrahymena* species, including *T. thermophila,* and 4 unidentified species. The *Tetrahymena* Stock Center houses a smaller collection of known *Tetrahymena* species, and a large collection of unknown *Tetrahymena* species derived from wild isolates collected by Paul Doerder, Cleveland State University. Fourteen different *Tetrahymena* species, as well as six *T. thermophila* strains, are available from the Culture Collection of Algae and Protozoa (CCAP) in the UK (http://www.ccap.ac.uk/index.htm). However, all of the *Tetrahymena* strains in the CCAP collection are maintained solely by serial subculture, and normally micronucleate strains are likely to have become germline-senescent and not useful for any research requiring integrity of the germline. Although fertility is not an issue, nonetheless, amicronucleate strains may also be adversely affected by continued serial transfer over long periods of time and may exhibit decreased viability and eventual die-off of clones.

IV. Cell Culture Media

Originally cultured in bacterized hay or vegetable matter infusions, *Tetrahymena* was the first animal-like eukaryotic cell to be grown axenically (Lwoff, 1923). *Tetrahymena* has two separate nutrient uptake systems; phagocytosis, which in *Tetrahymena* involves intake of particulate matter via a highly specialized oral apparatus and subsequent nutrient digestion in food vacuoles, and a surface uptake system that transports nutrients in solution into the cell (Orias and Rasmussen, 1976; Rasmussen and Orias, 1975; Chapter 6 in this volume). Each of these uptake mechanisms is sufficient to support normal growth in appropriate media. In most standard media, phagocytosis is essential for cell growth (Rasmussen and Kludt, 1970; Rasmussen and Modeweg-Hansen, 1973). However, complete chemically defined media (CDM) (Hagemeister *et al.*, 1999; Szablewski *et al.*, 1991) can support normal growth without phagocytosis. Axenic proteose peptone-based media are currently the most common choice for laboratory culture, but, as discussed below, *Tetrahymena* can be successfully grown in a wide variety of media, including bacterized peptone, bacterized infusions of lettuce or rye leaves (Cerophyll), skim milk-based media, and chemically defined media.

A. Glassware

Tetrahymena is sensitive to even very low levels of some types of impurities in the media. To prevent problems, high purity distilled and/or deionized water should be used in making all media, and dedicated glassware should be set aside solely for use in making and storing media. It is critical to ensure that no soap or acid residue remains on the surface of the glassware following washing. Although some automated dish washing protocols may yield adequate glassware, it is highly advisable to wash and rinse bottles and flasks used for making and storing media by hand, ending with several careful rinses with high-purity distilled/deionized water. To prevent build-up of water residue, glassware should be thoroughly drained in an inverted position before drying.

B. Proteose Peptone-Based Media

Proteose peptone (PP), an enzymatic digest of animal protein high in proteoses, is the traditional basis for most media used for growing *Tetrahymena* in the laboratory. In rich axenic media, PP is often supplemented by varying concentrations of yeast extract, glucose, and some form of iron. Bacto proteose peptone and Bacto yeast extract, originally manufactured by Difco, are now available directly from BD Sciences (BD Diagnostic Systems No.:211684 and BD Diagnostic Systems No.:212750, respectively) and through retailers like Fisher Scientific. It should be noted that not all grades of proteose peptone are suitable for making *Tetrahymena* media, and care must be taken in selecting the appropriate peptone. PP media is occasionally supplemented with liver extract, primarily to maintain *Tetrahymena* species newly isolated from the wild, or other species that are difficult to maintain in standard PP media (Doerder, personal communication). Liver fraction L, which is no longer available, was originally used for *Tetrahymena* culture. A possible substitute is Sigma 03077 Liver Hydrolysate, which has proven useful in the culture of *Trichomonas* and other difficult to culture protozoa. The compositions of several common PP-based media are shown in Table I. Regardless of the PP media chosen, a few basic considerations must be kept in mind.

1. Composition

Growth in all PP media is limited by iron. The need for iron can be met by supplementation with iron salts ($FeCl_3$ is most commonly used) at a final concentration as low as 10 μM, or chelated iron salts like Fe-EDTA (Ethylenediaminetetraacetic acid iron(III) sodium salt hydrate, 12–14% Fe, Sigma # 03650), or sequestrene (Becker Underwood, Dayton, Ohio or Trilon B Fe 13% powder, BASF Corp., Mount Olive, NJ). Yeast extract contains some iron but it is a good policy to supplement media containing yeast extract with additional iron in one of the above forms. PP media containing liver extract generally does not require added iron. The addition of ferric or ferrous chloride can produce an iron precipitate if added prior to autoclaving. Although the precipitate does not affect growth, it can interfere with some downstream operations like electronic cell counting or collection of cells by high-speed centrifugation. Precipitation can be prevented by filter sterilizing a concentrated iron solution separately and adding it to the autoclaved media, immediately after it has cooled or just prior to use. A simpler method is to add the appropriate amount of iron chloride from a 1000 × sterile concentrated stock solution (stored at 4 °C) to about one-fourth of the

Table I
Rich axenic nutrient media

Medium	Recipe
PP[a]	2% Proteose peptone
	10 μM $FeCl_3$ or 90 μM sequestrene (Fe-EDTA)
SP210[b]	2% Proteose peptone
	10 μM $FeCl_3$ or 90 μM sequestrene (Fe-EDTA)
	250 ug/mL streptomycin sulfate
	250 ug/mL penicillin G
Modified Neff[c]	0.25% Proteose peptone
	0.25% Yeast extract
	0.5% Glucose
	33.3 μM $FeCl_3$
	(To avoid precipitate formation, the $FeCl_3$ is first dissolved in one quarter of the final H_2O volume, and the glucose, yeast and PP are added and dissolved next. The remainder of the H_2O is then added, media is bottled and autoclaved.)
SSP[d]	2% Proteose peptone
	0.1% Yeast extract
	0.2% Glucose
	0.003% Sequestrene (Fe-EDTA) (can be replaced with 33 μM $FeCl_3$)
PPY[e]	1% Proteose peptone
	0.15% Yeast extract
	0.01 mM $FeCl_3$
PPYG[f]	0.4% Proteose peptone
	0.2% Yeast extract
	1.0% Glucose
PPYS[e]	1% Proteose peptone
	0.15% Yeast extract
	0.01 mM $FeCl_3$
	0.2 M NaCl
EPP[g]	2% Proteose peptone
	2 mM Na_3 citrate $2H_20$
	1 mM $FeCl_3$
	30 uM $CuSO_4$ $5H_20$
	1.7 uM Folinic acid, Ca salt
Liver peptone[h]	1.5% Proteose peptone
	0.1% Yeast extract
	0.25% Bactotryptone
	0.25% Liver fraction L (liver hydrolysate, Sigma 03077)
	0.5% Glucose
	0.1% KH_2PO_4 (7.35 mM)
	0.1% Na_2HPO_4 (7.04 mM)

(Continued)

Table I *(Continued)*

Medium	Recipe
Skim milk medium[i]	2% Skimmed milk 0.5% Yeast extract 0.1% Ferrous sulfate chelate solution 1% Glucose
MYE skim milk medium[j]	1%(w/v) Skim milk 1%(w/v) Yeast extract

[a] Orias *et al.* (2000).
[b] Diaz *et al.* (2007).
[c] Cassidy-Hanley *et al.* (1997).
[d] Gorovsky *et al.* (1975).
[e] Smith, and Doerder (1992).
[f] Mori *et al.* (2011).
[g] Orias and Rasmussen (1976).
[h] (P. Doerder, personal communication.)
[i] Weide *et al.* (2006)
[j] De Coninck *et al.* (2004).

final volume of water, stir well before dissolving the other ingredients one at a time in the solution, bring up to final volume, and autoclave. Filter sterilization of any PP media is not recommended since some particulate matter appears to be necessary to induce the formation of food vacuoles (Rasmussen and Kludt, 1970; Rasmussen and Modeweg-Hansen, 1973). Alternatively, Fe-EDTA can be used as an iron source. Although somewhat more expensive, Fe-EDTA generally does not form a precipitate when added directly to the media before autoclaving. PP media should be autoclaved at 121 °C and 15 psi for 30 min. Excessive autoclaving will decrease the ability of the media to support optimal growth.

Yeast extract (0.1–0.25%) and glucose (0.1–1%) are frequently included in rich axenic media used for growing *Tetrahymena*, with higher concentrations of each generally combined with lower concentrations of PP (Table I). In 2% PP, both yeast extract and glucose can be omitted with little effect on the growth rate. For routine work, cells may be grown in a simple 2% PP media supplemented with 10 μM $FeCl_3$ or 90 μM Fe-EDTA. In media with lower PP concentrations like Neff (0.25% PP), the addition of yeast extract is recommended to maintain optimal growth rates. For short-term expansion of cultures for nongenetic work, 1% PP supplemented with 10 μM $FeCl_3$ will support normal doubling times, but long-term maintenance in this media results in a greatly increased rate of infertility, and it is not recommended for work involving genetic analysis (Orias *et al.*, 2000).

SPP (Gorovsky *et al.*, 1975) and modified Neff's medium (Cassidy-Hanley *et al.*, 1997) represent the PP concentration extremes among commonly used PP media (Table I). Both efficiently support rapid growth in cell culture, even though there is an eightfold difference in PP concentration, perhaps because the lower PP concentration in Neff is counterbalanced by a two and a half fold increase in the concentration of both yeast extract and glucose. There are, however, slight differences in the growth curve generated by cells growing in the two media. Cells double somewhat faster in SPP than in Neff (~2.5 h vs. ~3 h at 30 °C in shaking cultures), but cells plateau at slightly higher concentrations in Neff than in SPP (2–3 × 10^6 vs. 1–2 × 10^6, respectively). Once stationary phase is reached, Neff cultures maintain longer before crashing than equivalent SPP cultures, holding up to 2 weeks even in microtiter plates kept at room temperature. The longer holding time makes Neff an ideal media for maintaining cultures in stock tubes (Section V.B.1.), while SPP may be preferable for expanding cultures if a more rapid growth rate is required. For settings where cost is a major consideration (e.g., educational use), Neff media is a good all round media that is significantly less expensive than other PP-based media. It should be noted that *T. thermophila* strains with different genetic backgrounds may grow differently in various PP media growth, and that many other *Tetrahymena* species grow more quickly and to higher densities in richer PP media.

2. Storage

PP media can be stored in various ways. Concentrated media can be made, aliquoted, and immediately stored frozen at −20 °C until needed, at which time it is thawed, diluted to the appropriate concentration, and autoclaved (Orias *et al.*, 2000). If larger volumes are needed on a regular basis, sterile 1 × PP media can also be stored in bottles for months at room temperature, although care should be taken to limit long-term exposure to bright light since some required vitamins exhibit light sensitivity over time.

C. Bacterized Media

Bacterized media has historically been used as a method for inducing mating without the need to physically manipulate the *Tetrahymena* cells involved. Currently, growth in axenic media followed by replication into nonnutritive starvation media (Section IV.H) is the preferred protocol for inducing mating, but bacterized media can be useful when the mating of multiple clones without the need for further cell manipulation is desired, for example, matings in 96-well microtiter plates. Bacterized media may also be useful for maintaining fastidious newly isolated wild *Tetrahymena* species that do not thrive in PP-based media. The maximum *Tetrahymena* cell concentration achievable in bacterized media is relatively low as compared to standard PP media, ~ 2 × 10^4 cells/mL (Ducoff *et al.*, 1964).

1. Bacterized PP

Although no longer commonly used, *Tetrahymena* can be maintained in bacterized peptone. Bacterized peptone can be prepared by inoculating PP medium (essentially any 1% to 2% PP media will suffice), prepared without added antibiotics, with *Klebsiella pneumoniae* (formerly *K. aerogenes*), and shaking the culture at ~200 rpm overnight at either 30 °C or 37 °C (Orias *et al.*, 2000). This 100% bacterized PP medium (100% BP medium) can be maintained at 4 °C for up to a week. The 100% BP medium is diluted 1:50 to 1:100 with sterile water just prior to use. If the bacterized medium is to be used for mating (Orias and Flacks, 1973; Simon and Whang, 1967), it is important to keep the final concentration of PP at a level that will allow starvation-induced mating once the bacteria have been consumed. To do this, the *Tetrahymena* culture used for inoculation must also be diluted at least 50-fold in sterile water. Following growth in bacterized media, any remaining bacteria can be eliminated without affecting the *Tetrahymena* cells by the addition of penicillin and streptomycin (250 μg/mL each) to the media.

2. Bacterized Cereal Grass (Cerophyll) or Lettuce Infusions

Bacterized infusions of lettuce (derived from early Paramecium culture media; Nanney, 1953; van Wagtendonk and Hackett, 1949) or rye leaves (Cerophyll; Nanney, 1953; Simon and Nanney, 1979) were commonly used in early work with *Tetrahymena*. Although rarely used for general culture or maintenance at present, cereal grass based media are occasionally useful for specific experimental purposes. The original Cerophyll is no longer available, but Ward's Natural Science Hay Medium Solution, provided as a sterile 2 × solution, is similar. If dry rye or cereal grass products are used, 0.6 g should be added to 500 mL of boiling distilled H_2O, boiled for 2 min, filtered through Whatman filter paper #1 while still warm, and autoclaved. The sterile infusion media is innoculated with *Klebsiella pneumoniae* and incubated overnight at 37 °C.

D. Chemically Defined Media

Consistent growth of *Tetrahymena* cells in synthetic, chemically defined media was first shown by Kidder and Dewey (1951) and provided a controlled means of examining cell nutritional requirements. The modified chemically defined media (CDM, Table II), initially described by Szablewski *et al.* (1991), support rapid growth similar to that observed in PP media (2 h doubling time at 37 °C), and cell concentrations of up to 10^6 cells/mL. Amino acids in group A are required for growth, but minimal chemically defined media (minimal CDM) can be prepared by omitting amino acids contained in solutions B-E in the CDMA recipe shown in Table II. Minimal CDM is useful for working with auxotrophic mutants (Sanford and Orias, 1981), but cell doubling time in minimal CDM is slightly slower than that in complete CDM (~ 2.5 h at 30 °C). CDM will support indefinite propagation by serial transfer provided that the initial inoculum concentration is at least 500 cells/mL. However, lower concentrations of cells transferred to CDM will not survive, perhaps as a result of a need for a critical initial cell density to condition the medium with required autocrine factors (Christensen *et al.*, 1995, 2001; Rasmussen *et al.*, 1996). If an inoculum of less than 2500 cell/mL CDM medium is necessary, the medium should be supplemented with 7.5 μM hemin to ensure cell viability. Hemin can be prepared as a stock solution by dissolving in 0.01 N NaOH and autoclaving

Table II

Chemically defined synthetic media: composition of stock solutions and preparation protocol.

	CDMA[a] (mg/mL)	CDMC[b] (mg/mL)		CDMA[a] (mg/mL)	CDMC[b] (mg/mL)
Amino acid solution A			**Salts and chelator solution**		
L-Arg-HCl	12	2.4	$K_2HPO_4 \bullet 3H_2O$	25	5
L-His-HCl • H_2O	8	1.6	KH_2PO_4	25	5
L-Ile	8	8	$MgSO_4 \bullet 7H_2O$	50	10
L-Leu	8	8	$CaCl_2 \bullet 2H_2O$	1	0.2
L-Lys-HCl	8	16	Tri-potassium citrate	65	13
L-Met	6	6	**Vitamins (Solution A)**		
L-Phe	6	6	Na riboflavin phosphate • $2H_2O$	0.05	0.05
L-Ser	6	12	**Vitamins (Solution B)**		
L-Thr	8	16	DL-6, 8-Thioctic acid	0.01	0.01
L-Trp	6	12	**Vitamins (Solution C)**		
L-Val	4	4	Thiamin-HCl	0.05	0.05
Amino acid solution B			Prydoxal-HCl	0.01	0.01
L-Gln	4	0.8	Nicotinic acid	0.09	0.09
Amino acid solution C			D-Pantothenic acid, hemi Ca-salt	0.08	0.08
L-Asn • H_2O	8	16	**Vitamins (Solution D)**		
L-Pro	8	16	Folinic acid, Ca salt	0.01	0.01
Amino acid solution D			**Trace metals solution**		
L-Ala	6	12	$FeCl_2 \bullet 6H_2O$	1	0.2
L-Asp	8	1.6	$MnSO_4 \bullet 4H_2O$	0.16	0.032
L-Glu	16	0.8	$Co(NO_3)_2 \bullet 6H_2O$	0.05	0.01
Gly	16	32	$ZnSO_4 \bullet 7H_2O$	0.45	0.09
Amino acid solution E			$CuSO_4 \bullet 5H_2O$	0.03	0.006
L-Tyr (Do not prepare ahead)	8	8	$(NH_4)_6 Mo_7O_{24} \bullet 4H_2O$	0.01	0.002
Nucleoside solutions			**Glucose solution**		
Adenosine	0.2	0.2	Glucose	250	250
Cytidine	0.2	0.2			
Guanosine	0.2	0.2			
Uridine	0.2	0.2			

1. **Media preparation**:

Unless otherwise noted, all ingredients are made up as stock solutions in high-purity distilled water, sterilized by filtration, and stored at 4 °C.

- Amino acid solutions A–D are 40-fold concentrated stock solutions. Preparation notes: Solutions A and C: adjust pH to 7 and sterilize by filtration; Solution B: store frozen; Solution D: dissolve aspartic and glutamic acids in water with stirring, keep pH from dropping below 7 with 1N KOH, add alanine and glycine, adjust pH to 7, filter sterilize, and store at 4 °C.
- Nucleoside solutions are 10-fold concentrated stock solutions.
- Salts and Chelator solutions are 100-fold concentrated stock solutions.
- Vitamins are 100-fold concentrated stock solutions and should be stored frozen. To make vitamin solution B, dl-6, 8-thioctic acid is dissolved in 1 mL absolute ethanol, and then diluted in 100 mL H_2O and filter sterilized.
- Trace metals solution are a 100-fold concentrated solution and should be adjusted to approximately pH 2 with 1 N HCl.

Table II *(Continued)*

- Glucose solution is a 50-fold concentrated solution.
  To make the complete media, dissolve Tyr at 60 °C, adjust to give a concentration of 0.2 mg/mL in the final medium, cool, and then add the remaining solutions. The pH may be adjusted as required. The medium may be sterilized by autoclaving or by filtration; if autoclaving is used, glucose should be added aseptically after cooling.

2. **Special-purpose modifications:**

- Minimal defined media. Amino acid solution A contains all required amino acids. A minimal defined medium can be created by omitting amino acid solutions B–E.
- Low cell inoculum in CDMA. When inocula of less than 2500 cells/mL of final medium will be used, CDMA should be supplemented with hemin at a final concentration of 7.5 μM. To prepare a stock solution, dissolve hemin in 0.01 *N* NaOH and autoclave (Christensen and Rasmussen, 1992).
- Phagocytosis deficient cells. For growing phagocytosis-deficient cells, the final concentrations of $FeCl_3$, $CuSO_4$ and folinic acid should be increased to 1 mM, 25 μM, and 1 mg/mL, respectively, (Orias and Rasmussen, 1979).

[a] CDMA is modified chemically defined media that supports rapid growth similar to that observed in PP media (Szablewski et al., 1991).

[b] CDMC is a modification of CDMA that permits growth of low concentration inoculum (including single cells) without additional supplements (Hagemeister *et al.*, 1999).

(Christensen and Rasmussen, 1992). Alternatively, Hagemeister *et al.* (1999) suggested that *Tetrahymena* cell death in CDM inoculated with a low concentration of cells is not apoptotic, but is rather the result of accidental cell lysis at the medium–air interface. They provide an alternate CDM recipe (CDMC, Table II) that permits the growth of low cell inoculum or single cells in the presence of an air–medium interface without the addition of any supplements. However, the doubling time in CDMC is 3.5 h at 36 °C, considerably slower than that in CDM. Initial cell concentration is not an issue in PP and bacterized media, which provide adequate amounts of any growth factors necessary for autocrine regulation of cell survival and support strong cell proliferation growth even in single-cell subcultures.

E. Skimmed Milk Media

PP media, while widely used, are an expensive option for large-scale cultivation of *Tetrahymena*. Increasing interest in the industrial use of *Tetrahymena* (Ethuin *et al.*, 1995; Jayaram *et al.*, 2010; Kiy and Tiedtke, 1992; Weide *et al.*, 2007) spurred the examination of less expensive media for large-scale culture, in particular, the use of skim milk-based media (Table I). In a bioreactor under conditions of high cell density fermentation with cell retention, skim milk-based medium has supported the culture of *Tetrahymena* at cell densities of more than 2.2×10^7 cells/mL, equivalent to 48 g dry weight (Weide *et al.*, 2006). A more dilute skim milk-based medium (MYE medium, 1% (w/v) skim milk, 1% (w/v) yeast extract) has also been successfully used in batch fermenters to support growth up to $\sim 3 \times 10^6$ cells/mL (De Coninck *et al.*, 2004).

F. Media for Phagocytosis-Deficient Cells

Phagocytosis is essential for cell growth in most media (Rasmussen and Kludt, 1970; Rasmussen and Modeweg-Hansen, 1973) and temperature-sensitive mutants defective in phagocytosis cannot survive past two to three doublings in any standard PP medium (Orias and Pollock, 1975). EPP, a modified PP-based axenic medium (Table I), allows indefinite cell growth in the absence of phagocytosis (Orias and Rasmussen, 1976). A specialized chemically defined media (Table II) for phagocytosis deficient cells have also been developed, based on modifications of standard CDM (Orias *et al.*, 2000). Phagocytosis deficient cells cannot grow in bacterized media.

G. Media for Long-Term Stock Culture

In situations where vegetative cells are only used infrequently and/or freezing is not an option, it is sometimes useful to maintain unfrozen *Tetrahymena* cultures long term (e.g., in classroom settings requiring only vegetatively propagated cells). Long-term maintenance of growing cultures also presents an alternative for *Tetrahymena* species that are difficult to maintain in PP media or that are not amenable to freezing in liquid nitrogen. It should be noted that although these methods can successfully maintain viability for long periods, the effects on genetic stability have not been carefully analyzed, and it is likely that genetic deterioration and eventual sterility will occur in micronucleate *Tetrahymena* strains maintained under these conditions (Simon and Nanney, 1979).

1. Bean Medium

The simplest media for prolonged storage in stock tube cultures uses a whole soy or garbanzo bean as a nutrient source (Sweet and Allis, 2010; Williams *et al.*, 1980). To prepare the media, place a single bean in 10 mL distilled water and autoclave or boil ~5 min in a capped culture tube. Any tubes that show signs of significant evaporation or that become cloudy within 24 h should be discarded. Streptomycin and penicillin (250 μg/mL each) and 0.25 μg/mL Amphotericin B (Fungizone; Fisher BioReagents, #BP2645-20) can be added after the media cools, but for general use neither are necessary. After inoculation with sterile cells, add 1–2 mL of sterile paraffin oil to the tubes to prevent evaporation, cap lightly, and store at 15–20 °C. Cells should be transferred every 6–8 months. Direct transfer from bean medium tube to bean medium tube is possible, but it is much preferable to provide an intervening passage in PP media between bean tube transfers.

2. Rat Gut Medium

A second method with much more limited utility uses rat intestine as the primary nutrient source (Williams *et al.*, 1980). While crude, this method is useful for the maintenance of *Tetrahymena* species that are difficult to freeze or maintain in other media. To make the growth media, place a 1 cm section of cleaned rat intestine in 8 mL of distilled water in a culture tube, layer with 8 mm of heavy paraffin oil, and autoclave. When the media is cool, *Tetrahymena* can be added by inoculating through the oil layer with a sterile small-bore pipette. Cultures in rat gut medium can be kept at room temperature for at least a year. Cells should be briefly grown in a rich PP media like SPP between yearly rat gut tube transfers. Three sequential transfers in the PP media over several days will reinvigorate the culture and prepare it for re-inoculation into the rat gut media, where it can be maintained for another year.

H. Starvation Media

A number of protocols, especially those involving mating reactivity, require starvation in nonnutritive media. Starved cells undergo rapid and extensive physiological, biochemical, and molecular changes, including starvation-induced proteolysis (Grinde and Jonassen, 1987), and changes in ribosome biosynthesis (Hallberg and Bruns, 1976) and gene expression (Miao *et al.*, 2009; Song and Gorovsky, 2007; Xiong *et al.*, 2011). Cells can be maintained in starvation media for several days, although viability is decreased if the cells are held at high density or in low surface to volume culture vessels that do not provide adequate aeration (e.g., culture tubes). Low ionic strength salt-based media (see Table III for recipes) like Dryl's medium (Dryl, 1959), 10 mM Tris, pH 7.5 (Bruns and Brussard, 1974a), or NKC media (Sugai and Hiwatashi, 1974) are routinely used to induce sexual reactivity. Starvation without induction of mating reactivity can be accomplished by starving cells in 50–70 mM Tris, pH 7.5. In this medium, cells starve but fail to undergo initiation, the first step in the activation of the sexual cycle, and do not become mating reactive (Bruns and Brussard, 1974a).

Table III
Starvation media

Medium	Recipe
Dryl's [a]	0.59 g of Na citrate-$2H_2O$ (2 mM)
	0.14 g of $NaH_2PO_4 \bullet H_2O$ (1 mM)
	0.14 g of Na_2HPO_4 (1 mM)
	0.13 g of $CaCl_2$ (1.5 mM)
	(To avoid precipitation of the CA phosphate, the $CaCl_2$ solution is autoclaved separately from the mixture of sodium salts, and the two solutions are mixed aseptically after cooling.)
Tris buffer[b]	10 mM Tris HCl, pH 7.5
NKC solution[c]	0.2% NaCl
	0.008% KCl
	0.12% $CaCl_2$

[a] Dryl (1959).
[b] Bruns and Brussard (1974).
[c] Sugai and Hiwatashi (1974).

If large numbers of individual matings must be performed, for example, in microtiter plates, initial growth in Neff medium simplifies the starvation process. Since Neff contains only 0.25% PP, a simple 1:10 dilution decreases nutrient availability to the point where cells become mating reactive. Replicating clones grown in Neff from microtiter plates into microtiter plates containing enough starvation medium to ensure at least a 1:10 dilution of the original Neff will produce mating reactive cells in the replicate plates. Growth in bacterized PP medium (Section IV. C.1) can also be used to carryout matings since once bacteria in the medium have been consumed, *Tetrahymena* will undergo starvation and become mating reactive.

V. Cell Culture

A. Basic Information

Tetrahymena can be easily cultured using a wide variety of media, containers, and conditions, as long as basic requirements for nutrition, aeration, temperature, and cell concentration are met. Culture vessels must be kept meticulously clean, and the use of dedicated flasks for cell growth is strongly recommended. The same stringent criteria described above (Section IV.A) for preparing glassware used in making media should be applied to all culture vessels. High surface-to-volume ratios should be maintained in standing cultures. If volume requirements preclude the use of shallow conditions, culture vessels should be shaken or rotated to provide necessary aeration. When very large volumes are required, as in a fermenter or multiliter bottle, forced aeration and agitation must be supplied to ensure sufficient oxygenation and gas exchange for optimal growth. If aeration and/or agitation are violent enough to cause foaming, an antifoam agent (e.g., 0.001% (v/v) Sigma AntiFoam 204, A6426) should be added to the culture to minimize cell damage. Since aeration stimulates growth and maximizes final cell concentration, but excessive agitation can cause cell damage, overall conditions for each large-scale culture unit must be optimized individually (De Coninck *et al.*, 2004).

It is important to note that, regardless of the culture method employed, care must be taken to maintain cell viability when harvesting or manipulating cells. *Tetrahymena* cells are extremely sensitive to the changes in aeration and cell concentration resulting from centrifugation, much more so than bacteria or yeast. The force necessary to efficiently pellet *Tetrahymena* cells may vary depending on culture volume, tube, and centrifuge type. For routine work, wild-type cells can be pelleted in 50 mL conical tubes at ~ 600–1000 g for 1 min with no adverse effects. However, the supernate must be immediately removed from the pellet since *Tetrahymena* are strong swimmers, and significant numbers of cells may be lost from the pellet if the supernate is not swiftly removed. High centrifugal velocities, especially if accompanied by sudden, dramatic temperature changes, can cause massive cell lysis, as can too vigorous resuspension of cell pellets. Vortexing should never be used to resuspend intact *Tetrahymena*. Rapid, massive cell death, accompanied by the release of high amounts of proteases and nucleases, can also occur if cells are allowed to remain in unfavorable conditions (e.g., in a pellet following centrifugation) for more than a few minutes. Cell pellets must be either immediately resuspended in fresh medium and returned to an appropriate culture vessel (if maintenance of viable cells is desired) or processed without delay.

When *Tetrahymena* cells must be cultured for a large number of generations requiring repeated serial transfer, care should be taken that cells are not transferred in a manner that constantly maintains continuous exponential growth resulting from the repeated transfer of a very small initial inoculum into a highly dilute culture. The creation of clones exhibiting reduced growth rate or the establishment of variant cells incapable of net telomere elongation can occur as a result of coordinate telomere lengthening in vegetatively growing log-phase cells. To avoid this possibility, cultures should periodically be allowed to remain at stationary phase between sustained bursts of log-phase growth (Larson *et al.*, 1987). One simple solution is to refrain from transferring cell cultures on weekends to provide an opportunity for cells to regularly undergo a brief transition to stationary phase. Maintaining cultures at room temperature will prevent the long-telomere phenotype, but also extends the time necessary for long-term growth procedures like maturation of immature clones or terminal phenotypic assortment of mixed macronuclei.

B. Methods of Cell Culture

1. Stock Cultures

For routine daily use, *Tetrahymena* cultures are generally maintained out of direct light, between 18 and 20–24 °C (room temperature) in slow growing stock tube cultures. Stock tubes then provide a constant source of cells from which to establish larger working cultures. Neff medium is especially useful for stock tubes since cell cultures generally maintain longer in Neff than in richer PP media. Serial transfer of a small cell inoculum into fresh stock tubes every 2–4 weeks is an efficient, economical way to maintain cultures for routine short-term use (up to 6 months). However, stock tubes are not suitable for longer-term maintenance of cell lines in which stable genetic characteristics are required. Serial transfer of viable cells in liquid culture should only be used for long-term maintenance (>6 months) if the integrity of the germline is not essential for downstream use. Genetically important strains should be stored frozen in liquid nitrogen (Section VI.B), and new stock tubes established from frozen cultures about every 6 months to ensure genetic stability within working clones. If freezing in liquid nitrogen is not an option, fresh clones of many strains can be obtained from the *Tetrahymena* Stock Center at a relatively inexpensive cost.

Methods of maintaining stock tubes vary among labs (e.g., Orias *et al.*, 2000 provides an alternate approach), but the following method, used by the *Tetrahymena* Stock Center, works well for general maintenance. Ten milliliters of Neff medium, with no antibiotics, are pipetted into 18 mm by 150 mm culture tubes, loosely capped, and autoclaved. Tubes are stored out of direct light and used as needed. A sterile 9-inch

Pasteur pipette (Fisher Scientific #136786B) is used to inoculate fresh stock tubes with a few drops (~100–300 μL) of cells taken from the top region of the original stock tube. It is preferable not to disturb the bottom of a stock tube when removing samples since the upper two-thirds of a well-grown stock tube culture contains healthy, very slowly dividing cells, but over time the bottom third becomes densely littered with dead and dying cells. Long Pasteur pipettes are ideal for transfer between stock tubes since the extra length prevents any inadvertent contact of unsterile surfaces with the inside of the stock tube, which can be a problem if 5-inch Pasteur pipettes or micropipettors and tips are used. It is good practice to keep at least two sequential transfer tubes of every clone to insure against contamination or accidental loss.

When a larger volume of cells (working culture) is needed, a 1:10 dilution of cells taken from the top third of a healthy stock tube culture and inoculated into fresh PP medium (1 mL inoculum to 9 mL PP medium in a culture plate or 100 mL flask) will result in a well-grown mid-log phase culture following overnight (16–20 h) incubation at 30 °C. If larger working cultures are desired, to ensure healthy, rapid growth in the larger volume, it is good practice to first establish a fresh 10 mL culture from a stock tube and use that log-phase culture to seed the larger volume culture.

2. Growth in Liquid Culture

Tetrahymena will thrive under conditions as diverse as standing cultures, shaken cultures, rotated bottles or tubes, industrial fermenters, microtiter plates, or hanging drops. In the lab, working cultures of *Tetrahymena* are generally established from slowly growing cultures maintained in stock tubes at room temperature (Section V. B.1). When grown in stationary containers, maximal growth rates are obtained if the depth of the working culture is limited. As a general rule, standing cultures in flasks should be limited to about 1/10 the flask volume, for example, 100 mL in a one liter flask. Somewhat larger total volumes are possible in Fernbach-style culture flasks designed for culturing organisms requiring a large surface area to volume ratio (e.g., PYREX® 2800 mL Fernbach-Style Culture Flask #4420). In smaller containers like Petri plates, depth should be limited to ~5 mm. If volume requirements make these limitations impractical, then cultures should be shaken or rotated. Cells in PP media can be shaken up to 200 rpm, although generally shaking speeds of ~100–150 rpm provide sufficient aeration to support optimal growth without foaming or bubble formation. Baffled Fernbach flasks (e.g., PYREX® 2800 mL Fernbach-Style Culture Flask with Baffles #4423) can be used to provide maximal oxygen transfer to shaken cultures. The temperature range for sustained growth in liquid media varies considerably among *Tetrahymena* species. *T. thermophila* exhibits sustained growth at 40.7 °C, higher than many other *Tetrahymena* species, and shows very slow but consistent growth at 15 °C (Nyberg, 1981). Optimal doubling time for *T. thermophila* occurs at ~35 °C, with a generation time of about 2 h (Frankel and Nelsen, 2001; Orias *et al.*, 2000), but for routine laboratory use, where factors in addition to doubling time are often important, cultures are most frequently grown between 27 and 32 °C.

Successful large-scale fermentation strategies using low-cost nutrient media have been developed using *T. thermophila* (Hellenbroich *et al.*, 1999; Kiy and Tiedtke, 1992; Noseda *et al.*, 2007). However, conditions for growing *Tetrahymena* in fermenters or bioreactors can vary considerably depending on instrument type and experimental design, and optimal settings need to be worked out specifically for each large-scale setup (De Coninck *et al.*, 2004).

3. Growth in Micro Volumes

A number of procedures frequently used in genetic analysis and the establishment of clonal lines require isolation of single cells into hanging drops and/or the growth of clonal cultures in 96-well microtiter plates. Clonal lines can be initiated by physically separating single cells into individual hanging drops, letting the cells replicate to high concentration within the drop, and replicating into microtiter plates. Individual mating pairs and exconjugants arising from individual mating pairs can also be cloned using the hanging drop system (Bruns and Cassidy-Hanley, 2000b). When isolating mating pairs or ex-conjugants, it is important to note that mating type in *T. thermophila* is not inherited across the sexual stage of the life cycle, and that progeny from a single mating pair frequently give rise to clones expressing different mating types. The segregation of different mating types among the descendants of a given pair appears to be correlated with the segregation of new macronuclei to the four karyonides produced by the first postzygotic division of the two exconjugant cells (Orias, 1981). The mating type of a *T. thermophila* clone is not uniquely expressed until sexual maturity, ~50–80 fissions after conjugation. If the creation of a pure clone expressing a single mating type is needed, individual cells must be isolated after exconjugant clones become sexually mature and tested for expression of a single mating type using a panel of known mating-type testers (available from the *Tetrahymena* Stock Center). Immature cells cannot mate, adolescent cells can mate, but are not exclusive, that is, they fail to form pairs with more than one mating-type tester strain. Sexually mature clones express a unique mating type, forming pairs with all but one mating type tester strain. All of the necessary subcloning, growth, and mating can be carried out in hanging drops and microtiter plates, as long as care is taken to prevent contamination. Mature clones expressing a single mating type are stable, and generally will not give rise to cells of different mating types. Clonal lines established in microtiter plates can easily be expanded into stock tubes that can be used to provide working cultures of any volume, and cells for freezing in liquid nitrogen.

Drop Plate Culture

Drop plates can be created in 100 × 15 mm standard Petri plates in several ways. Regardless of the method used, the drop array should match the wells of half of a 96-well microtiter plate, that is, a 6 × 8 grid array spaced to correspond to microtiter plate wells (Fig. 1). The use of a "drop maker" device is the simplest method, especially if the creation of drop plates is a frequent, routine procedure in the lab but requires the initial construction of drop maker (Bruns and Cassidy-Hanley, 2000b). This device (Fig. 2) consists of 48 aluminum prongs, each 6-mm in diameter, separated by 9-mm center to center (matching the arrangement of wells in a half of a 96-well microtiter plate), and can be easily constructed by a local machine or metal working shop. To make the drops, the drop maker is sterilized by dipping in distilled water to remove any residual material clinging to the prongs, blotting the prongs on several layers of paper towels to completely remove the water, and dipping the prongs into acetone (or 95% alcohol) placed in a large (15 × 150 mm) covered glass petri dish, preferably placed on a separate metal cart outside the hood and away from any possible contact with the Bunsen burner. The acetone should be deeper than the depth of the medium that will be used to form the drops to ensure sterility. The glass lid should be replaced on the Petri plate containing the acetone as soon as the drop maker is removed to protect the contents from accidental contact with flames or hot liquid. The acetone covered prongs are briefly flamed in a Bunsen burner to remove all residual acetone (even a slight residue is potentially lethal to *Tetrahymena* cells) and allowed to cool. When completely cool, the drop maker is dipped into 25–30 mL of sterile PP media containing penicillin and streptomycin (250 μg/mL each) and 0.25 μg/mL Amphotericin B in a 15 × 100 mm Petri plate, rapidly lifted straight up out of the medium and immediately touched down on the inside of a fresh sterile Petri

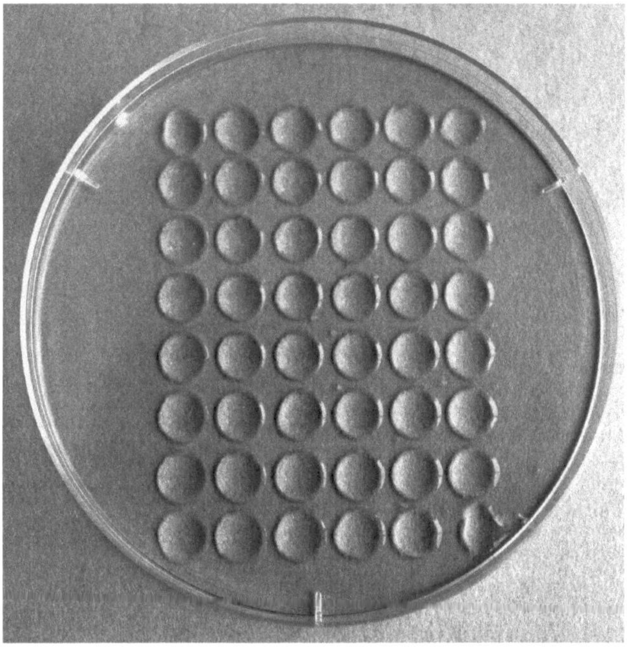

Fig. 1 Drop plate array. The drop array shown is created in a standard 100 mm × 15 mm Petri using a using a 6 × 8 grid drop maker (Fig. 2). The 6 × 8 drop array matches half of a 96 well microtiter plate. Each drop is about 50 μl.

Fig. 2 **Drop maker**. The drop maker shown consists of 48 aluminum prongs, each 6-mm in diameter, separated by 9-mm center to center, matching the arrangement of wells in one half of a 96 well microtiter plate. The device can be easily constructed by a local machine or metal working shop.

plate, where each prong deposits a 40–50 μL drop of culture medium. As long as standard precautions are taken, and work is done in a sterile hood, multiple drop plates can be made from the same medium without resterilizing the drop maker. When finished, the drop maker should be rinsed in distilled water to prevent the build-up of medium on the surface and dried on a paper towel. Drops made using this protocol are quite flat, optically clear, and very stable, adhering more firmly to the plate surface than drops made using other methods. If a drop maker is not available, or if only occasional drop plates are needed, drops can be placed on a Petri plate in the appropriate array by hand using a Pasteur pipette or a single or multichannel pipettor. Drop plates should be stored in plastic boxes (TriState Plastics, #079C or #195C) on a raised platform above a small amount of water to decrease the rate of evaporation. Clean, empty microtiter plates work well for supporting the drop plates above the water level.

A drawn-out glass micropipette is used to transfer single cells into individual drops while viewing the transfer under a dissecting microscope. Micropipettes of appropriate size can be made by hand by drawing the end of a Pasteur pipette or thin glass tubing through the flame of a Bunsen burner or by using a micropipette puller. Suction to control cell deposition can be controlled manually using a pipette bulb or by mouth using tubing to connect a mouthpiece to the pipette (Orias and Bruns, 1976). A much more easily controlled pipette system utilizes braking pipettes (Bruns and Cassidy-Hanley, 2000b), in which thin pieces of capillary tubing (Kimble Chase Capillary tube, # 34500-99, 1.5–1.8 × 100 mm) are drawn out at both ends to an inner diameter of ~0.1–0.15 mm (Fig. 3a). Both ends are clipped with fine forceps under a dissecting microscope to make sure they are open, and one end is carefully inserted about half way into an aspirator tube assembly (Sigma #A5177)(Fig. 3b). The attached mouthpiece is used to control flow, which is fairly easy since pulling out both ends of the capillary tube gives a pipette with neutral action with the back constriction acting as a brake. Once the delivery end is filled with liquid by capillary action, the rest of the tube does not fill, providing much more exact control than that provided by pipettes with a single drawn end. In all cases, the pipette tip is sterilized by repeatedly drawing in and expelling boiling distilled water maintained on a hot

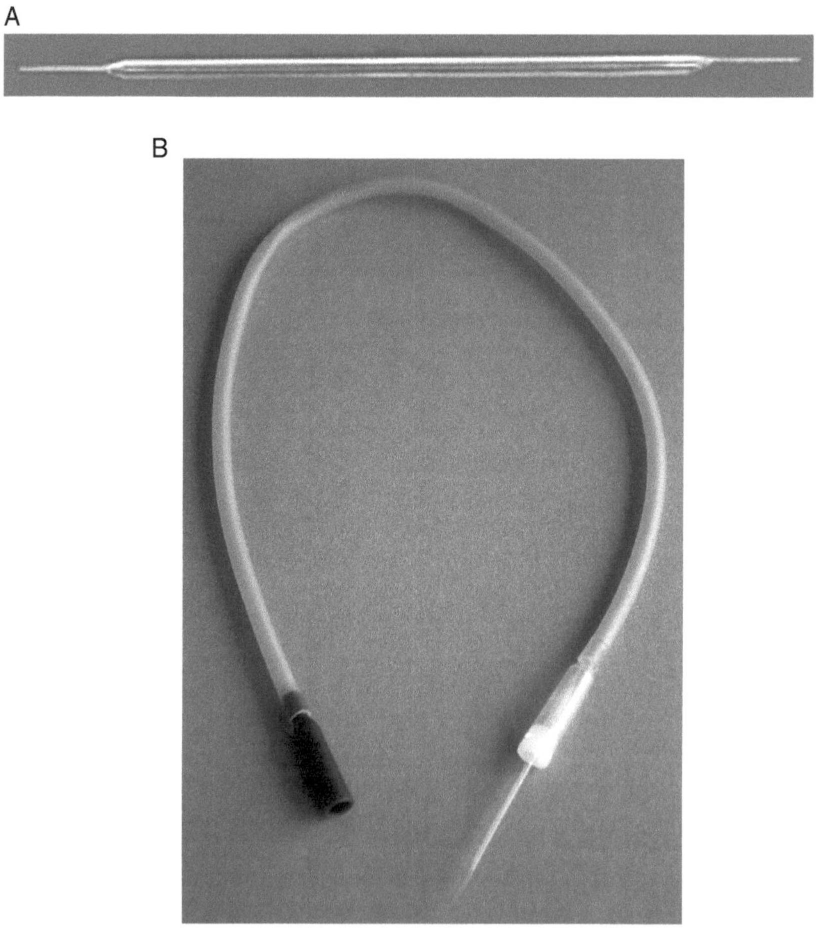

Fig. 3 a. Braking pipette. To make the braking pipette shown, thin pieces of capillary tubing (Kimble Chase Capillary tube, # 34500-99, 1.5–1.8 mm × 100 mm) are drawn out at both ends to an inner diameter of ~0.1–0.15 mm. Both ends are clipped with fine forceps under a dissecting microscope to make sure they are open. **b. Braking pipette in aspirator tube assembly**. To assemble the aspirator, one end of a braking pipette is carefully inserted about half way into the aspirator tube assembly (Sigma #A5177) as shown.

plate adjacent to the microscope, and allowed to cool before use. A few microliters of the cell culture from which cells are to be isolated are placed in a marked corner drop, and under the microscope 50 or so cells are picked up in the pipette tip. Individual cells are deposited into separate drops by gently moving the drop plate under the microscope to position each drop sequentially in the field of view, carefully monitoring deposition to ensure that only one cell is released into each drop. When isolating pairs, the initial cell inoculum placed in the source drop should result in a dilute cell concentration to make it easier to pick out mating pairs and avoid single cells. With a little practice, it is relatively easy to isolate cells into 10 or more drop plates in an hour. Once cells have been transferred, the drop plates should be placed in a humid chamber (covered plastic box with distilled water on the bottom and a raised platform to hold the plates above the water) to prevent drying out, and placed at 30 °C in a stationary incubator for 2–3 days, or until the cells have grown enough to allow replication to microtiter plates.

Microtiter Plate Culture

Tetrahymena is well adapted to rapid growth in very small volumes, for example, 100 μL cultures in 96-well microtiter plates. Clear U-bottom microtiter plates with 100 μL PP medium per well are used for most routine procedures, although for fixation and cytology, flat bottom plates give better optics. Two types of replicators are useful for manipulating *Tetrahymena* grown in microtiter plate cultures: a 48-prong replicator, used to transfer cultures from drop plates to microtiter plates, and a 96-prong replicator for transfer from microtiter plate to microtiter plate. Replicators can be made in a variety of ways, using either a wooden block with straight metal rods ~ 1/16 in diameter arrayed to match either a half (48-prong) or full (96-prong) microtiter well array (Orias *et al.*, 2000) or an aluminum assembly similar to the drop maker described above, except that the diameter of each prong is 4 mm, to easily fit into microtiter plate wells (Bruns and Cassidy-Hanley, 2000b). When using the larger pronged aluminum replicator, care must be taken to ensure that drops do not run together when touched by the prongs. This is best done by first dipping the sterile replicator into fresh, sterile growth medium, so that each prong tip is covered by a drop of medium, and then touching the ends of the prongs to the drops on the drop plate. This ensures that each prong transfers a reasonable cell sample and decreases the risk of drop-to-drop contact during the transfer process. A 96-prong replicator (Fig. 4) is available commercially (Nunc replicator #250520, available from Fisher Scientific). For occasional use with small numbers of samples, a pipettor (multichannel or single channel) can be used for both transfer from drop plates and replication between microtiter plates, but for large numbers of samples the process is labor and cost intensive. Replicators are sterilized using the same technique described for drop makers, paying particular attention to rinsing the replicators in distilled water between uses to prevent the build-up of peptone residue on the prongs. When large numbers of transfers must be done, it is useful to have several replicators of each size that can be used in an alternating fashion. One replicator is rinsed, sterilized, flamed, and then

Fig. 4 96-prong replicator. The 96 prongs of the replicator shown match the well distribution on a 96 well microtiter plate. The replicator can be used to transfer small volumes (~5 μl) of cells to new microtiter plates for screening purpose, to set up small volume plate matings, and for serial plate transfer.

allowed to cool while a second replicator is used, rinsed, sterilized, and flamed. This process ensures that repeated flaming will not cause the replicator to become too hot.

It is possible to reuse microtiter plates, as long as they are not allowed to dry out before washing. Dried-on cells and medium can be difficult to remove, and may complicate subsequent analyses. Plates should be decontaminated by submerging in 25% Clorox (1.3% sodium hypochlorite) for 5 min, followed by washing in a dishwashing machine or by hand. After rinsing to ensure removal of all traces of bleach, plates can also be sonicated for 10 min to ensure removal of all organic material from the wells. In all cases, after washing, the plates must be carefully rinsed by hand in high-purity distilled water. All wells must be completely filled and emptied by vigorous shaking with the plate inverted over a sink at least three times. Covers should be washed and rinsed in a similar manner. Plates are dried inverted standing on edge at an angle to optimize draining. Following air drying, the open plates and lids can be sterilized by exposure to 20 min of UV irradiation at a distance of 15 cm from a bank of six 40-watt sunlamps (Bruns and Cassidy-Hanley, 2000b), or by a 1 h exposure to a germicidal UV lamp 6490 (Orias *et al.*, 2000). In both cases, UV lamps are enclosed in a light tight container to protect eyes and skin form UV exposure. Since UV lamp output can decrease with use, UV lamps should occasionally be monitored using a UV dosimeter. Following irradiation, lids are placed on the plates, and closed plates are stored until needed.

Cells grown in microtiter plates at 30 °C can generally be replicated after 1–3 days. For longer-term maintenance (up to a week), plates should be kept at room temperature out of direct light. If maximum growth rate is desired, for example, when attempting to quickly mature progeny clones, plates can be replicated daily, but long-term continuous daily replication can sometimes cause clonal loss (Section V.A). For optimal clonal health, it is advisable to allow the clones to periodically reach stationary stage, most simply by replicating daily Monday through Friday but letting the clones rest each weekend.

4. Growth on Solid Medium

Although *Tetrahymena* is generally grown in liquid culture, in certain circumstances (e.g., the isolation of mutant clones following mutagenesis or cytotoxicity assays), it is useful to grow cells on solid medium. Several methods have been developed for obtaining clones on solid medium. Gardonio *et al.* (1973) developed the following agar plate method using 1.5% bottom agar and a thin layer of 0.3% top agar, both made up in either 2% PP and 0.1% liver extract, or a defined medium supplemented with 0.04% PP, and containing 250 μg each of penicillin and streptomycin sulfate. Plates are dried for 2 days at 37 °C or a week at room temperature before adding cells. To inoculate the plate, about 0.5 mL of a very dilute *Tetrahymena* culture (~100–150 cells) is placed on the dried agar surface. After the liquid is partially absorbed by the agar, sterile G-25 fine Sephadex is sprinkled onto the plate. Individual clones, established around individual Sephadex beads, can then be isolated into liquid culture. Dobra and Ehret (1980) developed an alternate technique for culturing *Tetrahymena* as continuous monolayers on the surface of nutrient agar. Cells grown at low densities on solid agar are motile and have normal structural characteristics, including production of food vacuoles and an oral apparatus. At higher densities ($2 - 5 \times 10^5/cm^2$), cells are relatively nonmotile and form a tightly packed continuous monolayer. A modification of this method has also been used to cultivate *Tetrahymena* on the surface of sterile cellulose nitrate filters placed on top of PP agar.

C. Culture Contamination: Prevention and Treatment.

Basic sterile techniques should be employed whenever possible when working with all *Tetrahymena* cultures. All glass and plasticware should be sterile, media should be monitored for possible signs of contamination like clouding prior to each use, and whenever feasible cell work should be carried out in a sterile hood. However, some techniques like isolation of cells into hanging drops are more easily done under a microscope on an open lab bench. In these cases, contamination can be minimized by judicious use of antibiotics and fungizone. Additionally, anything coming into direct contact with the cells or media (pipettes, drop makers, etc.) must be carefully sterilized, and exposure time of open plates minimized. With these basic precautions, contamination of drops or microtiter wells is rarely a problem.

If contamination occurs in working cultures or stock tubes, the simplest response is to discard the contaminated cells and re-establish the culture from a back-up stock tube or a frozen stock. Contaminated cultures should immediately be autoclaved and discarded to prevent the spread of potential contaminants within the lab area. Since most contaminants grow more quickly than *Tetrahymena*, contaminated cultures are generally easy to spot fairly early on by the macroscopic appearance of the culture. Bacterial contaminants make flask cultures look cloudier and denser than pure *Tetrahymena* cultures, and generally release an off odor when the flask is opened. Fungal and mold contaminants generally produce obvious mats of growth. Contamination in individual wells of a microtiter plate presents a special problem since contamination can quickly spread within a plate. At the first sign of contamination, the contents of the affected well should be removed by aspiration with a sterile pipette tip and the well cleaned with a cotton swab soaked in 95% ethanol. Any remaining alcohol must be aspirated from the well, and the plate left uncovered in a sterile hood until the alcohol is completely evaporated. If the plate is covered before the alcohol evaporates completely, alcohol vapors will kill the entire contents of the plate. If multiple wells are affected, the plate should be decontaminated and discarded. If a contaminated culture is irreplaceable, possible methods for generating healthy, clean cultures are discussed below.

1. Antibiotics and Fungizones

Tetrahymena cultures grown in the laboratory using basic sterile techniques generally do not require the addition of antibiotics or fungizones to maintain sterility. However, when cells must be knowingly exposed to potentially contaminating conditions, penicillin G and streptomycin sulfate should be routinely added to the PP media at a final concentration of 250 μg/mL each. A 1000 × stock solution containing both antibiotics can be prepared by filter sterilization and stored in 1 mL aliquots in cryovials at −20 °C. One microliter of the drug cocktail is added aseptically per mL culture media immediately prior to use. The fungizone Amphotericin B can also be added to minimize fungal contamination, which is often a major problem in *Tetrahymena* cultures exposed to unsterile conditions. *Tetrahymena* is quite resistant to Amphotericin B and concentrations from 0.025 to 25 μg/mL have been used without problems. For general preventative use, 0.25 μg/mL is usually sufficient.

If preventative measures fail, and cultures become contaminated, if possible they should be autoclaved and discarded, and new cultures established from the freezer or a noncontaminated stock tube (Section V.B.1). Penicillin/streptomycin treatment is generally not effective in cleaning up already contaminated cultures. If a contaminated culture must be rescued, 100 μg/mL neomycin, kanamycin, or tetracycline can be used to try to eliminate bacterial contamination. These drugs may also be useful in cleaning up newly collected isolates since most *Tetrahymena* collected from the wild grow well even at these elevated levels (Clifford Brunk, personal communication; Chapter 9 in this volume). However, the overall most effective, albeit expensive, drug treatment for retrieving contaminated cultures is Normocin (Invivogen, Cat. # ant-nr-1), which can be used at 2 μL/mL culture media directly as shipped. Normocin can also be used in conjunction with penicillin and streptomycin with no adverse effects on *Tetrahymena* and has proved remarkably effective in eliminating both bacterial and fungal contamination. In extreme cases, the Normocin concentration can be increased two to threefold without affecting the *Tetrahymena*. To prevent the development of Normocin resistance, routine use of Normocin in standard cultures is not recommended.

2. Serial Subculture

If all else fails, as may be the case with some fungal or antibiotic resistant bacterial contaminants, serial dilution (Lwoff, 1923) can be used to create contamination free clones. Individual cells are placed in the top eight drops of a 6 × 8 drop array in several drop plates. Every 30 min a single cell is moved with a minimum of media into the drop directly below, for a total of five transfers. Since some bacteria can survive in *Tetrahymena* food vacuoles (Berk *et al.*, 2008; Gourabathini *et al.*, 2008; Meltz Steinberg and Levin, 2007; Rehfuss *et al.*, 2011), the timing between transfers must be long enough to ensure that original food vacuole contents are fully excreted prior to the final transfer so as to avoid contaminating the final drop. Drops are incubated at 30 °C overnight, and drops containing growing *Tetrahymena* and visually free of contamination are transferred to individual stock tubes for further incubation at 30 °C. After 2 days, stock tubes with no sign of contamination can be further tested by spotting a small sample of the tube contents on a PP nutrient agar plate and incubating for 2 days. Since *Tetrahymena* can form small colony-like plaques on the agar surface of agar plates, the agar plate should be examined under a dissecting scope to differentiate bacterial colonies, which are generally rounded, dense, and opaque, from *Tetrahymena* plaques, which are more translucent, flat, and irregular in shape.

VI. Long-Term Storage

A. Serial Transfer

Tetrahymena can be maintained for years by serial transfer, provided that a reasonable cell inoculum (a minimum of ~1000 cells) is used for each transfer. However, prolonged vegetative growth can lead to both micronuclear and macronuclear genetic changes over time. In micronucleate strains like *T. thermophila*, the transcriptionally inactive germinal micronucleus is not subject to direct selection and can accumulate chromosomal changes, including whole chromosome loss, deletions, and lethal point mutations, that have no effect on vegetative viability but eventually result in the inability to produce true sexual progeny (clonal sterility; Allen *et al.*, 1984; Nanney, 1974; Simon and Nanney, 1979). It is important to note that sterile cells generally maintain the ability to form physical pairs that may give rise to viable exconjugants. However exconjugants arising from matings between normal and sterile cells are not true progeny in that both exconjugant cells retain their parental macronuclei and parental phenotypes, and contain micronuclei of unknown genotype. Loss of fertility within a clone occurs gradually, and timing will vary with different strains and maintenance conditions.

Changes during prolonged vegetative growth can also effect macronuclear composition. Strains in which the macronucleus is not homozygous for alleles of interest will undergo random macronuclear phenotypic assortment during vegetative growth, eventually leading to the production of cells that are homozygous for a given allele (Bruns and Brussard, 1974b). In terms of strain utility, the importance of these changes depends on the type of strain and the intended use. Wild-type strains not intended for genetic use can be maintained by serial transfer, as can strains homozygous in the macronucleus for a specific mutation of interest, as long as downstream use does not require fertility (e.g., for expression of a specific gene product in vegetative cells). However, the macronuclear composition of strains with macronuclei that are heterozygous for a particular gene of interest cannot be guaranteed to remain stable following prolonged serial vegetative transfer. To ensure continued fertility and optimize genetic stability, important clones should be stored frozen in a liquid nitrogen freezer as soon as possible after construction. It should be noted that maintenance of a mixed genotype in the macronucleus cannot be absolutely guaranteed even if the clone is stored frozen in a liquid nitrogen freezer. During thawing, only a relatively small population of cells is recovered, and there is a chance that the subpopulation that survives thawing may not be representative of the prefreezing population, especially if assortment gives rise to a subclone better able to withstand the freezing process, as may be the case for some wild-type clones assorting from a deleterious mutation in a heterozygous macronucleus. It is therefore good practice to check the relevant phenotype of clones heterozygous in the macronucleus after thawing if retention of a mixed genotype expressing a specific phenotype or gene product is essential.

B. Storage in Liquid Nitrogen

There are several protocols available for freezing *Tetrahymena* (Flacks, 1979; Orias *et al.*, 2000; Simon, 1982). A number of factors shown to affect successful freezing of eukaryotic cells, including the physiological state of the cells (Rauen *et al.*, 1994), cryoprotectant (Anchordoguy *et al.*, 1987), rate of cooling (Farrant and Morris, 1973), storage temperature, and rate and temperature of thawing (McGann and Farrant, 1976a, 1976b), have been optimized for *Tetrahymena* (Cassidy-Hanley *et al.*, 1995). The method described below is based on these optimized parameters (Bruns *et al.*, 2000) and is used routinely by the *Tetrahymena* Stock Center. This technique has proven successful with multiple *T. thermophila* strains as well as a wide variety of other *Tetrahymena* species. All cell work is carried out in a sterile hood, and all equipment and materials used in the procedure are sterilized.

1. Growth of Cultures for Freezing

Cells designated for freezing should be grown at 30 °C to log phase (100 mL @ $\sim 5 \times 10^5$/mL) on a shaking incubator. It is important to start with a fresh, healthy, uncontaminated culture to ensure consistently high rates of recovery. For optimal aeration and culture health, it is recommended that the flask culture conditions recommended above (Section V.B.2) be used.

2. Starvation

To ensure efficient freezing, the cells must be starved prior to freezing. Unstarved cells give very poor recovery following freezing. Starve 100 mL of cells at ~ 1.5–2×10^5 per mL in sterile 10 mM Tris pH 7.5 for 2–3 days at 30 °C in a flask with a high surface area to volume ratio for adequate aeration. Cell concentration and starvation temperature are important for optimal recovery of live cells.

3. Freezing

Concentrate 100 mL of starved cells by centrifugation (2 min at ~1100 g at room temperature) followed by aspiration of the supernate to 1 mL total volume (cells plus Tris). Immediately add 4 mL of high-quality 10% DMSO (Fisher #D1281) and gently resuspend cells into a total volume of 5 mL (final DMSO concentration 8%). Dispense 0.3 mL of the cell:DMSO mixture to individual cryovials (e.g., Nalgene® #5000-0020 or #5000-0012). The number of tubes frozen is a matter of individual preference, but in all cases, one additional tube should be frozen for a test thaw. Incubate the cryovials at room temperature ~30 min, transfer the cryovials to a Nalgene® Cryo 1 °C controlled rate freezing container (Nalgene #5100-0001) and place the container in a −80 °C freezer overnight. Cryovials should not be left in the low-temperature freezer for extended periods, and care should be taken that the temperature in the freezer does not fluctuate or rise above −70 °C since either occurrence can decrease recovery of viable cells. Cells maintained long term in a −80 °C freezer often exhibit increasingly poor recovery over time, perhaps as a result of temperature changes when the freezer is opened and closed. Following initial freezing in an −80 °C freezer, cells should be transferred to a liquid nitrogen freezer, using a nitrogen dewar or dry ice to prevent the cells from warming up during the transfer process. Do not use wet ice for transfer. For safety reasons, it is recommended that cryovials be frozen in the vapor phase of the liquid nitrogen tank to reduce the risk of cryotube explosion during thawing. If tubes must be frozen in the liquid phase, it is recommended that tubes to be thawed be moved into the vapor phase of the freezer for at least 24 h before thawing. One possibility is to leave an empty box in the top slot of one of the freezer racks, above the liquid level. Move the tubes to be thawed to that box at least 24 h before thawing, being very careful to not let them warm up.

4. Thawing

Thaw tubes individually in a 42 °C water bath, going directly from liquid nitrogen or dry ice into the waterbath. After ~15 s, using a sterile glass transfer (Pasteur) pipette, add ~1 mL of NEFF prewarmed to 42 °C. Gently move the tube in the water bath to speed thawing. When the pellet is dissolved, pipette the contents of the cryovial into a Petri plate containing 10 mL of NEFF plus penicillin and streptomycin (250 μg/mL each) and 0.25 μg/mL Amphotericin B, prewarmed to 30 °C. Keep the pipette tip under the liquid to avoid bubbles, and swirl the plate contents gently. Culture the cells at 30 °C. Live cells can often be observed within 30–60 min and should be visible within 24 h. Once the culture is established, transfer to a stock tube or use as needed.

Acknowledgments

D.C.H. is supported by NIH NCRR SEPA grant 5 R25 RR025126-03 and by NIH Grant No. P40 RR019688-01A1. The *Tetrahymena* Stock Center is funded by NIH Grant No. P40 RR019688-01A1. Past support by NIH, NSF, and USDA is gratefully acknowledged for providing funding for the development of many of these techniques. DCH is a founding member of Tetragenetics, Inc., Ithaca, NY, a company developing and utilizing *Tetrahymena* as a platform for the manufacture of biotechnological products. Neither this connection nor any financial, personal, or professional interest of the author has influenced this paper.

References

Allen, S. L. (1967a). Cytogenetics of genomic exclusion in tetrahymena. *Genetics* **55**(4), 797–822.

Allen, S. L. (1967b). Genomic exclusion: a rapid means for inducing homozygous diploid lines in *Tetrahymena pyriformis*, syngen 1. *Science* **155**(762), 575–657.

Allen, S. L. (2000). Genetic nomenclature rules for *Tetrahymena thermophila*. *Methods Cell Biol.* **62**, 561–563.

Allen, S. L., Altschuler, M. I., Bruns, P. J., Cohen, J., Doerder, F. P., and Gaertig, J., *et al.* (1998). Proposed genetic nomenclature rules for *Tetrahymena thermophila*, *Paramecium primaurelia* and *Paramecium tetraurelia*. The seventh international meeting on ciliate molecular biology genetics nomenclature. *Genetics* **149**(1), 459–462.

Allen, S. L., Ervin, P. R., McLaren, N. C., and Brand, R. E. (1984). The 5S ribosomal RNA gene clusters in *Tetrahymena thermophila*: strain differences, chromosomal localization, and loss during micronuclear ageing. *Mol. Gen. Genet.* **197**(2), 244–253.

Allen, S. L., and Gibson, I. (1973). Genetics of tetrahymena. *In* "Biology of Tetrahymena," (A. M. Elliott, ed.), pp. 307–373. Dowden, Hutchinson & Ross, Stroudsburg, PA.

Altschuler, M. I., and Bruns, P. J. (1984). Chromosome-designated mutation selection in *Tetrahymena thermophila*. *Genetics* **106**(3), 387–401.

Anchordoguy, T. J., Rudolph, A. S., Carpenter, J. F., and Crowe, J. H. (1987). Modes of interaction of cryoprotectants with membrane phospholipids during freezing. *Cryobiology* **24**(4), 324–331.

Artemenko, A. G., Muratov, E. N., Kuz'min, V. E., Muratov, N. N., Varlamova, E. V., and Kuz'mina, A. V., *et al.* (2011). QSAR analysis of the toxicity of nitroaromatics in *Tetrahymena pyriformis*: structural factors and possible modes of action. *SAR QSAR Environ. Res.* **22**(5–6), 575–601.

Berk, S. G., Faulkner, G., Garduno, E., Joy, M. C., Ortiz-Jimenez, M. A., and Garduno, R. A. (2008). Packaging of live legionella pneumophila into pellets expelled by tetrahymena spp. does not require bacterial replication and depends on a Dot/Icm-mediated survival mechanism. *Appl. Environ. Microbiol.* **74**(7), 2187–2199.

Brickner, J. H., Lynch, T. J., Zeilinger, D., and Orias, E. (1996). Identification, mapping and linkage analysis of randomly amplified DNA polymorphisms in *Tetrahymena thermophila*. *Genetics* **143**(2), 811–821.

Bruns, P. J., and Brussard, T. B. (1974a). Pair formation in *Tetrahymena pyriformis*, an inducible developmental system. *J. Exp. Zool.* **188**(3), 337–344.

Bruns, P. J., and Brussard, T. B. (1974b). Positive selection for mating with functional heterokaryons in *Tetrahymena pyriformis*. *Genetics* **78**(3), 831–841.

Bruns, P. J., and Brussard, T. B. (1981). Nullisomic tetrahymena: eliminating germinal chromosomes. *Science* **213**, 549–551.

Bruns, P. J., Brussard, T. B., and Kavka, A. B. (1976). Isolation of homozygous mutants after induced self-fertilization in tetrahymena. *Proc. Natl Acad. Sci. U.S.A.* **73**(9), 3243–3327.

Bruns, P. J., Brussard, T. B., and Merriam, E. V. (1983). Nullisomic tetrahymena: II. *A set of nullisomics define the germinal chromosomes. Genetics* **104**(2), 257–270.

Bruns, P. J., and Cassidy-Hanley, D. (2000a). Biolistic transformation of macro- and micronuclei. *Methods Cell Biol* **62**, 501–512.

Bruns, P. J., and Cassidy-Hanley, D. (2000b). Methods for genetic analysis. *Methods Cell Biol* **62**, 229–240.

Bruns, P. J., Katzen, A. L., Martin, L., and Blackburn, E. H. (1985). A drug-resistant mutation in the ribosomal DNA of tetrahymena. *Proc. Natl Acad. Sci. U.S.A.* **82**(9), 2844–2846.

Bruns, P. J., Smith, H. R., and Cassidy-Hanley, D. (2000). Long-term storage. *Methods Cell Biol.* **62**, 213–218.

Byrne, B. C. (1978). Induced resistance to 6-methylpurine and cycloheximide in tetrahymena: II. Variation within vegetative cultures of micronucleate *T. thermophila* and amicronucleate *T. pyriformis*. *C. R. Seances Soc. Biol. Fil.* **172**(2), 314–319.

Byrne, B. C., Brussard, T. B., and Bruns, P. J. (1978). Induced resistance to 6-methylpurine and cycloheximide in tetrahymena: I. Germ line mutants of *T. thermophila*. *Genetics* **89**(4), 695–702.

Cassidy-Hanley, D., Bowen, J., Lee, J. H., Cole, E., VerPlank, L. A., and Gaertig, J., *et al.* (1997). Germline and somatic transformation of mating *Tetrahymena thermophila* by particle bombardment. *Genetics* **146**(1), 135–147.

Cassidy-Hanley, D., Smith, H. R., and Bruns, P. J. (1995). A simple, efficient technique for freezing *Tetrahymena thermophila*. *J. Eukaryot. Microbiol.* **42**(5), 510–515.

Chalker, D. L., Ward, J. G., Randolph, C., and Yao, M. C. (2000). Microinjection of *Tetrahymena thermophila*. *Methods Cell Biol.* **62**, 469–484.

Chantangsi, C., and Lynn, D. H. (2008). Phylogenetic relationships within the genus tetrahymena inferred from the cytochrome c oxidase subunit 1 and the small subunit ribosomal RNA genes. *Mol. Phylogenet. Evol.* **49**(3), 979–987.

Chantangsi, C., Lynn, D. H., Brandl, M. T., Cole, J. C., Hetrick, N., and Ikonomi, P. (2007). Barcoding ciliates: a comprehensive study of 75 isolates of the genus tetrahymena. *Int. J. Syst. Evol. Microbiol.* **57**, 2412–2425.

Christensen, S. T., and Rasmussen, L. (1992). Evidence for growth factors which control cell multiplication in *Tetrahymena thermophila*. *Acta Protozoologica* **31**(4), 215–219.

Christensen, S. T., Sorensen, H., Beyer, N. H., Kristiansen, K., Rasmussen, L., and Rasmussen, M. I. (2001). Cell death in *Tetrahymena thermophila*: new observations on culture conditions. *Cell Biol. Int.* **25**(6), 509–519.

Christensen, S. T., Wheatley, D. N., Rasmussen, M. I., and Rasmussen, L. (1995). Mechanisms controlling death, survival and proliferation in a model unicellular eukaryote *Tetrahymena thermophila*. *Cell Death Differ.* **2**(4), 301–308.

Cole, E. S. (1991). Conjugal blocks in tetrahymena pattern mutants and their cytoplasmic rescue: I. Broadened cortical domains (bcd). *Dev. Biol.* **148**(2), 403–419.

Cole, E. S., and Bruns, P. J. (1992). Uniparental cytogamy: a novel method for bringing micronuclear mutations of tetrahymena into homozygous macronuclear expression with precocious sexual maturity. *Genetics* **132**(4), 1017–1031.

Cole, E. S., Cassidy-Hanley, D., Hemish, J., Tuan, J., and Bruns, P. J. (1997). A mutational analysis of conjugation in *Tetrahymena thermophila*: 1. Phenotypes affecting early development: meiosis to nuclear selection. *Dev. Biol.* **189**(2), 215–232.

Cole, E. S., and Frankel, J. (1991). Conjugal blocks in tetrahymena pattern mutants and their cytoplasmic rescue: II. Janus A. *Dev. Biol.* **148**(2), 420–428.

Cole, E. S., and Soelter, T. A. (1997). A mutational analysis of conjugation in *Tetrahymena thermophila*: 2. Phenotypes affecting middle and late development: third prezygotic nuclear division, pronuclear exchange, pronuclear fusion, and postzygotic development. *Dev. Biol.* **189**(2), 233–245.

Dave, D., Wloga, D., and Gaertig, J. (2009). Manipulating ciliary protein-encoding genes in *Tetrahymena thermophila*. *Methods Cell Biol.* **93**, 1–20.

De Coninck, J., Leclercq, B., Exbrayat, J. M., and Duyme, F. (2004). Factorial designs: an efficient approach to choosing the main factors influencing growth and hydrolase production by *Tetrahymena thermophila*. *J. Ind. Microbiol. Biotechnol.* **31**(5), 204–208.

Dentler, W. L. (1995). Isolation of cilia from *Tetrahymena thermophila*. *Methods Cell Biol.* **47**, 13–15.

Dobra, K. W., and Ehret, C. F. (1980). Growth kinetics of three species of tetrahymena on solid agar*. *J. Eukaryot. Microbiol.* **27**(2), 226–230.

Doerder, F. P. (1979). Regulation of macronuclear DNA content in *Tetrahymena thermophila*. *J. Protozool.* **26**, 28–35.

Doerder, F. P., Deak, J. C., and Lief, J. H. (1992). Rate of phenotypic assortment in *Tetrahymena thermophila*. *Dev. Genet.* **13**(2), 126–132.

Dryl, S. (1959). Antigenic transformation in paramecium aurelia after homologous antiserum treatment during autogamy and conjugation. *J. Protozool.* **6**(Supplement), 25.

Ducoff, H. S., Williams, S. M., Syn, T. B., and Butler, B. B. (1964). Quantitative studies of metabolism in phagotrophic ciliates: I. Growth of *Tetrahymena pyriformis* W on viable and on killed bacteria. *J. Protozool.* **11**(3), 309–313.

Elliot, A. M., and Hayes, R. E. (1955). Tetrahymena from mexico, panama, and colombia, with special reference to sexuality. *J. Eukaryot. Microbiol.* **2**(2), 75–80.

Ethuin, P., De Coninck, J., Dhulster, P., and Guillochon, D. (1995). Comparison of complex organic media for the cultivation of the temperature-sensitive mutant *Tetrahymena thermophila* SJ180. *Enzyme Microb. Technol.* **17**(11), 998–1002.

Farrant, J., and Morris, G. J. (1973). Thermal shock and dilution shock as the causes of freezing injury. *Cryobiology* **10**(2), 134–140.

Flacks, M. (1979). Axenic storage of small volumes of tetrahymena cultures under liquid nitrogen: a miniaturized procedure. *Cryobiology* **16**(3), 287–291.

Frankel, J., Jenkins, L. M., and DeBault, L. E. (1976). Causal relations among cell cycle processes in *Tetrahymena pyriformis*. An analysis employing temperature-sensitive mutants. *J. Cell Biol.* **71**(1), 242–260.

Frankel, J., Jenkins, L. M., Nelsen, E. M., and Stoltzman, C. A. (1993). Hypoangular: a gene potentially involved in specifying positional information in a ciliate, *Tetrahymena thermophila*. *Dev. Biol.* **160**(2), 333–354.

Frankel, J., Mohler, J., and Frankel, A. K. (1980) Temperature-sensitive periods of mutations affecting cell division in *Tetrahymena thermophila*. *J. Cell Sci.* **43**, 59–74.

Frankel, J., and Nelsen, E. M. (2001). The effects of supraoptimal temperatures on population growth and cortical patterning in *Tetrahymena pyriformis* and *Tetrahymena thermophila*: a comparison. *J. Eukaryot. Microbiol.* **48**(2), 135–146.

Gaertig, J., Gu, L., Hai, B., and Gorovsky, M. A. (1994a). High frequency vector-mediated transformation and gene replacement in tetrahymena. *Nucleic Acids Res.* **22**(24), 5391–5398.

Gaertig, J., and Kapler, G. (2000). Transient and stable DNA transformation of *Tetrahymena thermophila* by electroporation. *Methods Cell Biol.* **62**, 485–500.

Gaertig, J., Thatcher, T. H., Gu, L., and Gorovsky, M. A. (1994b). Electroporation-mediated replacement of a positively and negatively selectable beta-tubulin gene in *Tetrahymena thermophila*. *Proc. Natl Acad. Sci. U.S.A.* **91**(10), 4549–4553.

Gardonio, E., Crerar, M., and Pearlman, R. E. (1973). Growth and cloning of *Tetrahymena pyriformis* on solid meduium. *J. Bacteriol.* **116**(3), 1170–1176.

Gerhardt, A., Ud-Daula, A., and Schramm, K. (2010). Tetrahymena spp. (protista, ciliophora) as test species in rapid multi-level ecotoxicity tests. *Acta Protozool.* **49**(4), 271–280.

Gorovsky, M. A., and Woodard, J. (1969). Studies on nuclear structure and function in *Tetrahymena pyriformis*: I RNA synthesis in macro- and micronuclei. *J. Cell Biol.* **42**(3), 673–682.

Gorovsky, M. A., Yao, M. C., Keevert, J. B., and Pleger, G. L. (1975). Isolation of micro- and macronuclei of *Tetrahymena pyriformis*. *Methods Cell Biol.* **9**(0), 311–327.

Gourabathini, P., Brandl, M. T., Redding, K. S., Gunderson, J. H., and Berk, S. G. (2008). Interactions between food-borne pathogens and protozoa isolated from lettuce and spinach. *Appl. Environ. Microbiol.* **74**(8), 2518–2525.

Grinde, B., and Jonassen, T. O. (1987). Methionine is a regulator of starvation-induced proteolysis in tetrahymena. *Exp. Cell Res.* **173**(2), 496–503.

Gutierrez, J. C., and Orias, E. (1992). Genetic characterization of *Tetrahymena thermophila* mutants unable to secrete capsules. *Dev. Genet.* **13**(2), 160–166.

Haddad, A., and Turkewitz, A. P. (1997). Analysis of exocytosis mutants indicates close coupling between regulated secretion and transcription activation in tetrahymena. *Proc. Natl Acad. Sci. U.S.A.* **94**(20), 10675–10680.

Hagemeister, J. J., Grave, M., Kristiansen, T. B., Assaad, F., and Hellung-Larsen, P. (1999). Interface-mediated death of unconditioned tetrahymena cells: effect of the medium composition. *J. Eukaryot. Microbiol.* **46**(1), 6–11.

Hai, B., Gaertig, J., and Gorovsky, M. A. (2000). Knockout heterokaryons enable facile mutagenic analysis of essential genes in tetrahymena. *Methods Cell Biol.* **62**, 513–531.

Hallberg, R. L., and Bruns, P. J. (1976). Ribosome biosynthesis in *Tetrahymena pyriformis*. Regulation in response to nutritional changes. *J. Cell Biol.* **71**(2), 383–394.

Hamilton, E. P., and Orias, E. (2000). Genetically mapping new mutants and cloned genes. *Methods Cell Biol.* **62**, 265–280.

Hellenbroich, D., Valley, U., Ryll, T., Wagner, R., Tekkanat, N., and Kessler, W., *et al.* (1999). Cultivation of *Tetrahymena thermophila* in a 1.5-m^3 airlift bioreactor. *Appl. Microbiol. Biotechnol.* **51**(4), 447–455.

Jayaram, J., Papoyan, A., Zhang, X., Bisharyan, Y., Colussi, P., and Cassidy-Hanley, D., *et al.* (2010). An alternative platform for rapid production of effective subunit vaccines. *BioPharm. Int.* **23**(10 Suppl.), 6–13.

Johnson, K. A. (1986). Preparation and properties of dynein from tetrahymena cilia. *Methods Enzymol.* **134**, 306–317.

Kaney, A. R., and Speare, V. J. (1983). An amicronucleate mutant of *Tetrahymena thermophila*. *Exp. Cell Res.* **143**(2), 461–547.

Kapler, G. M. (1993). Developmentally regulated processing and replication of the tetrahymena rDNA minichromosome. *Curr. Opin. Genet. Dev.* **3**(5), 730–735.

Karrer, K., Stein-Gavens, S., and Allitto, B. A. (1984). Micronucleus-specific DNA sequences in an amicronucleate mutant of tetrahymena. *Dev. Biol.* **105**(1), 121–129.

Kher, C. P., Doerder, F. P., Cooper, J., Ikonomi, P., Achilles-Day, U., and Kupper, F. C., *et al.* (2011). Barcoding tetrahymena: discriminating species and identifying unknowns using the cytochrome c oxidase subunit I (cox-1) barcode. *Protist* **162**(1), 2–13.

Kidder, G. W., and Dewey, V. C. (1951). The biochemistry of ciliates in pure culture. *In* "Biochemistry and physiology of protozoa," (A. Lwoff, and S. H. Hutner, eds.), pp. 323–400. Academic Press, New York.

Kiy, T., and Tiedtke, A. (1992). Continuous high-cell-density fermentation of the ciliated protozoon tetrahymena in a perfused bioreactor. *Appl. Microbiol. Biotechnol.* **38**(2), 141–146.

Larson, D. D., Blackburn, E. H., Yaeger, P. C., and Orias, E. (1986). Control of rDNA replication in tetrahymena involves a cis-acting upstream repeat of a promoter element. *Cell* **47**(2), 229–240.

Larson, D. D., Spangler, E. A., and Blackburn, E. H. (1987). Dynamics of telomere length variation in *Tetrahymena thermophila*. *Cell* **50**(3), 477–483.

Larson, D. D., Umthun, A. R., and Shaiu, W. L. (1991). Copy number control in the tetrahymena macronuclear genome. *J. Protozool.* **38**(3), 258–263.

Lombillo, V. A., Coue, M., and McIntosh, J. R. (1993). *In vitro* motility assays using microtubules tethered to *Tetrahymena pellicles*. *Methods Cell Biol.* **39**, 149–165.

Lovlie, A., Haller, B. L., and Orias, E. (1988). Molecular evidence for somatic recombination in the ribosomal DNA of *Tetrahymena thermophila*. *Proc. Natl Acad. Sci. U.S.A.* **85**(14), 5156–5160.

Luehrsen, K. R., Baum, M. P., and Orias, E. (1987). A restriction fragment length polymorphism in the 5' non-transcribed spacer of the rDNA of *Tetrahymena thermophila* inbred strains B and C3. *Gene* **55** (2–3), 169–178.

Lwoff, A. (1923). Sur la nutrition des infusoires. *C.R. Acad. Sci.* **176**, 928–930.

Lynch, T. J., Brickner, J., Nakano, K. J., and Orias, E. (1995). Genetic map of randomly amplified DNA polymorphisms closely linked to the mating type locus of *Tetrahymena thermophila*. *Genetics* **141**(4), 1315–1325.

Mayo, K. A., and Orias, E. (1981). Further evidence for lack of gene expression in the tetrahymena micronucleus. *Genetics* **98**(4), 747–762.

McGann, L. E., and Farrant, J. (1976a). Survival of tissue culture cells frozen by a two-step procedure to −196 degrees C: I. Holding temperature and time. *Cryobiology* **13**(3), 261–268.

McGann, L. E., and Farrant, J. (1976b). Survival of tissue culture cells frozen by a two-step procedure to −196 degrees C: II. Warming rate and concentration of dimethyl sulphoxide. *Cryobiology* **13**(3), 269–273.

Melia, S. M., Cole, E. S., and Turkewitz, A. P. (1998). Mutational analysis of regulated exocytosis in tetrahymena. *J. Cell Sci.* **111**(Pt 1), 131–140.

Meltz Steinberg, K., and Levin, B. R. (2007). Grazing protozoa and the evolution of the escherichia coli O157: H7 shiga toxin-encoding prophage. *Proc. R. Soc. Biol. Sci. Series B* **274**(1621), 1921–1929.

Miao, W., Xiong, J., Bowen, J., Wang, W., Liu, Y., and Braguinets, O., *et al.* (2009). Microarray analyses of gene expression during the *Tetrahymena thermophila* life cycle. *PloS one* **4**(2), e4429.

Nanney, D. L. (1953). Nucleo-cytoplasmic interaction during conjugation in tetrahymena. *Biol. Bull.* **105** (1), 133–148.

Nanney, D. L. (1974). Aging and long-term temporal regulation in ciliated protozoa. A critical review. *Mech. Ageing Dev.* **3**(2), 81–105.

Nanney, D. L., Meyer, E. B., Simon, E. M., and Preparata, R. M. (1989). Comparison of ribosomal and isozymic phylogenies of tetrahymenine ciliates. *J. Protozool.* **36**(1), 1–8.

Nanney, D. L., and Simon, E. M. (2000). Laboratory and evolutionary history of *Tetrahymena thermophila*. *Methods Cell Biol.* **62**, 3–25.

Noseda, D. G., Gentili, H. G., Nani, M. L., Nusblat, A., Tiedtke, A., and Florin Christensen, J., *et al.* (2007). A bioreactor model system specifically designed for tetrahymena growth and cholesterol removal from milk. *Appl. Microbiol. Biotechnol.* **75**(3), 515–520.

NYBERG, D. (1981). Three new "biological" species of tetrahymena (*T. hegewischi* n. sp., *T. sonneborni* n. sp., *T. nipissingi* n. sp.) and temperature tolerance of members of the "pyriformis" Complex1. *J. Eukaryot. Microbiol.* **28**(1), 65–69.

Orias, E. (1981). Probable somatic DNA rearrangements in mating type determination in *Tetrahymena thermophila*: a review and a model. *Dev. Genet.* **2**(2), 185–202.

Orias, E. (1998). Mapping the germ-line and somatic genomes of a ciliated protozoan, *Tetrahymena thermophila. Genome Res.* **8**(2), 91–99.

Orias, E., and Bruns, P. J. (1976). Induction and isolation of mutants in tetrahymena. *Methods Cell Biol.* **13**, 247–282.

Orias, E., and Flacks, M. (1973). Use of genomic exclusion to isolate heat-sensitive mutants in tetrahymena. *Genetics* **73**(4), 543–549.

Orias, E., Flacks, M., and Satir, B. H. (1983). Isolation and ultrastructural characterization of secretory mutants of *Tetrahymena thermophila. J. Cell Sci.* **64**, 49–67.

Orias, E., Hamilton, E. P., and Orias, J. D. (2000). Tetrahymena as a laboratory organism: useful strains, cell culture, and cell line maintenance. *Methods Cell Biol.* **62**, 189–211.

Orias, E., and Pollock, N. A. (1975). Heat-sensitive development of the phagocytotic organelle in a tetrahymena mutant. *Exp. Cell Res.* **90**(2), 345–356.

Orias, E., and Rasmussen, L. (1976). Dual capacity for nutrient uptake in tetrahymena: IV. Growth without food vacuoles and its implications. *Exp. Cell Res.* **102**(1), 127–137.

Orias, E., and Rasmussen, L. (1979). Dual capacity for nutrient uptake in tetrahymena. V. utilization of amino acids and proteins. *J Cell Sci* **36**, 343–353.

Preparata, R. M., Meyer, E. B., Preparata, F. P., Simon, E. M., Vossbrinck, C. R., and Nanney, D. L. (1989). Ciliate evolution: the ribosomal phylogenies of the tetrahymenine ciliates. *J. Mol. Evol.* **28**(5), 427–441.

Prescott, D. M. (1994). The DNA of ciliated protozoa. *Microbiol. Rev.* **58**(2), 233–267.

Raikov, I. B. (1976). Evolution of macronuclear organization. *Annu. Rev. Genet.* **10**, 413–440.

Rasmussen, L., Christensen, S. T., Schousboe, P., and Wheatley, D. N. (1996). Cell survival and multiplication. The overriding need for signals: from unicellular to multicellular systems. *FEMS Microbiol. Lett.* **137**(2–3), 123–128.

Rasmussen, L., and Kludt, T. A. (1970). Particulate material as a prerequisite for rapid cell multiplication in tetrahymena cultures. *Exp. Cell Res.* **59**(3), 457–463.

Rasmussen, L., and Modeweg-Hansen, L. (1973). Cell multiplication in tetrahymena cultures after addition of particulate material. *J. Cell Sci.* **12**(1), 275–286.

Rasmussen, L., and Orias, E. (1975). Tetrahymena: growth without phagocytosis. *Science* **190**(4213), 464–465.

Rauen, U., Noll, T., Piper, H. M., Lauchart, W., Becker, H. D., and De Groot, H. (1994). Endothelial cell toxicity of preservation solutions: comparison of endothelial cells of different origin and dependence on growth state. *Cryobiology* **31**(2), 144–153.

Rehfuss, M. Y., Parker, C. T., and Brandl, M. T. (2011). Salmonella transcriptional signature in tetrahymena phagosomes and role of acid tolerance in passage through the protist. *ISME J.* **5**(2), 262–273.

Sanford, Y. M., and Orias, E. (1981). Phenylketonuric tetrahymena: phenylalanine hydroxylase mutants and other tyrosine auxotrophs. *Proc. Natl Acad. Sci. U.S.A.* **78**(12), 7614–7618.

Satir, B. H., Reichman, M., and Orias, E. (1986). Conjugation rescue of an exocytosis-competent membrane microdomain in *Tetrahymena thermophila* mutants. *Proc. Natl Acad. Sci. U.S.A.* **83**(21), 8221–8225.

Sauvant, M. P., Pepin, D., Bohatier, J., Groliere, C. A., and Guillot, J. (1997). Toxicity assessment of 16 inorganic environmental pollutants by six bioassays. *Ecotoxicol. Environ. Saf.* **37**(2), 131–140.

Sauvant, M. P., Pepin, D., Groliere, C. A., and Bohatier, J. (1995). Effects of organic and inorganic substances on the cell proliferation of L-929 fibroblasts and *Tetrahymena pyriformis* GL protozoa used for toxicological bioassays. *Bull. Environ. Contam. Toxicol.* **55**(2), 171–178.

Sauvant, M. P., Pepin, D., and Piccinni, E. (1999). *Tetrahymena pyriformis*: a tool for toxicological studies. A review. *Chemosphere* **38**(7), 1631–1669.

Simon, E. M. (1982). Breeding performance of *Tetrahymena thermophila* following storage for 5 to 6 years in liquid nitrogen. *Cryobiology* **19**(6), 607–612.

Simon, E. M., and Nanney, D. L. (1979). Germinal aging in *Tetrahymena thermophila. Mech. Ageing Dev.* **11**(4), 253–268.

Simon, E. M., and Whang, S. W. (1967). Tetrahymena: effect of freezing and subsequent thawing on breeding performance. *Sci. (New York, N.Y.)* **155**(763), 694–696.

Song, X., and Gorovsky, M. A. (2007). Unphosphorylated H1 is enriched in a specific region of the promoter when CDC2 is down-regulated during starvation. *Mol. Cell. Biol.* **27**(5), 1925–1933.

Sugai, T., and Hiwatashi, K. (1974). Cytologic and autoradiographic studies of the micronucleus at meiotic prophase in tetrahymena pyriformis. *J. Protozool.* **21**(4), 542–548.

Suhr-Jessen, P. B., and Orias, E. (1979). Mutants of *Tetrahymena thermophila* with temperature sensitive food vacuole formation: II. Physiological and morphological studies. *Exp. Cell Res.* **124**(2), 317–327.

Sweet, M. T., and Allis, C. D. (2010). Long-term storage of unfrozen tetrahymena cultures in soybean medium. *Cold Spring Harbor Protoc.* **2006**(3) pdb.prot4472.

Szablewski, L., Andreasen, P. H., Tiedtke, A., Florin-Christensen, J., Florin-Christensen, M., and Rasmussen, L. (1991). *Tetrahymena thermophila*: growth in synthetic nutrient medium in the presence and absence of glucose. *J. Eukaryot. Microbiol.* **38**(1), 62–65.

Tiedtke, A. (1985). Isolation of pure pellicles containing intact basal bodies of *Tetrahymena pyriformis*. *J. Cell Sci.* **77**, 155–165.

Tondravi, M. M., and Yao, M. C. (1986). Transformation of *Tetrahymena thermophila* by microinjection of ribosomal RNA genes. *Proc. Natl Acad. Sci. U.S.A.* **83**(12), 4369–4373.

Turkewitz, A. P. (2004). Out with a bang! tetrahymena as a model system to study secretory granule biogenesis. *Traffic* **5**(2), 63–68.

van Wagtendonk, W. J., and Hackett, P. L. (1949). The culture of paramecium aurelia in the absence of other living organisms. *Proc. Natl Acad. Sci. U.S.A.* **35**(3), 155–159.

Weide, T., Bockau, U., Rave, A., Herrmann, L., and Hartmann, M. W. W. (2007). A recombinase system facilitates cloning of expression cassettes in the ciliate *Tetrahymena thermophila*. *BMC Microbiol.* **7**, 12.

Weide, T., Herrmann, L., Bockau, U., Niebur, N., Aldag, I., and Laroy, W., *et al.* (2006). Secretion of functional human enzymes by *Tetrahymena thermophila*. *BMC Biotechnol.* **6**, 19.

Williams, N. E. (2000). Preparation of cytoskeletal fractions from *Tetrahymena thermophila*. *Methods Cell Biol.* **62**, 441–447.

Williams, N. E., and Honts, J. E. (1987). The assembly and positioning of cytoskeletal elements in tetrahymena. *Development* **100**(1), 23–30.

Williams, N. E., Wolfe, J., and Bleyman, L. K. (1980). Long-term maintenance of tetrahymena spp. *J. Protozool.* **27**(3), 327.

Xiong, J., Lu, X., Lu, Y., Zeng, H., Yuan, D., and Feng, L., *et al.* (2011). Tetrahymena gene expression database (TGED): a resource of microarray data and co-expression analyses for tetrahymena. *Sci. China Life Sci.* **54**(1), 65–67.

Yao, M. C., and Gorovsky, M. A. (1974). Comparison of the sequences of macro- and micronuclear DNA of *Tetrahymena pyriformis*. *Chromosoma* **48**(1), 1–18.

Ye, A. J., and Romero, D. P. (2002). Phylogenetic relationships amongst tetrahymenine ciliates inferred by a comparison of telomerase RNAs. *Int. J. Syst. Evol. Microbiol.* **52**(Pt 6), 2297–2302.

CHAPTER 9

Natural Populations and Inbred Strains of *Tetrahymena*

F. Paul Doerder[*] and Clifford Brunk[†]

[*]Department of Biological, Geological, and Environmental Sciences, Cleveland State University, 2121 Euclid Avenue, Cleveland, OH, USA

[†]Department of Ecology and Evolutionary Biology, University of California, Los Angeles, 621 Charles E. Young Drive So., Los Angeles, CA, USA

Abstract

Tetrahymena typically is found in freshwater lakes, ponds, and streams in association with submerged or emergent vegetation. The genus consists of numerous breeding species with micronuclei and many asexual species without micronuclei. In summer months when most populations are at their peak, 30–50% of water samples may yield one or more species of *Tetrahymena*. This chapter describes both bulk and trapping procedures for collecting *Tetrahymena* and also evaluates barcode methods for species identification. The history and inbreeding of the laboratory model

0091-679X/10 $35.00
DOI 10.1016/B978-0-12-385967-9.00009-8

Tetrahymena thermophila is also discussed. There are numerous unresolved questions about *Tetrahymena* evolution and biogeography that may be solved by additional collecting.

I. Introduction/Background

Tetrahymena is such a well-known laboratory organism that it is sometimes easy to forget that it exists in nature, where it is relatively common and easily isolated from ponds and streams. The best-studied member of the genus, *Tetrahymena thermophila*, is but one of about 40 named species and numerous unnamed species identified by DNA sequence analysis. In this respect, *Tetrahymena* is like many ciliates in which a morphospecies consists of many reproductively isolated populations that are morphologically indistinguishable. In this chapter, we examine the reasons for studying natural populations and provide details as to methods.

Tetrahymena became a laboratory favorite in 1923 when Andre Lwoff grew a natural isolate called *Glaucoma pyriformis* in axenic media (Lwoff, 1923). Capable of growing to high densities, these *G. pyriformis*, later *Tetrahymena pyriformis* (Furgason, 1940) quickly became an important "animal" cell model, particularly for biochemistry and physiology (reviewed by Hill, 1972). In the next decades, as investigators isolated more tetrahymenas from nature, it became evident that there were important strain differences. Most significantly some had a germinal micronucleus typical of ciliates, whereas others, such as Lwoff's isolate, did not. There were also biochemical differences and, in a few instances, there were sufficient morphological differences (e.g., cell shape, cyst formation, and macrostome transformation) to warrant designations as separate species. However, most tetrahymenas were so similar that they were simply called *T. pyriformis* and given a stain designation such as "GL" or "W" or "S."

A turning point in the history of *Tetrahymena* was the discovery in 1952 of complementary mating types among micronucleate isolates from Eel Pond at Woods Hole MA (Elliott and Gruchy, 1952). Micronucleate strains are capable of conjugation, whereas amicronucleates, such as Lwoff's *T. pyriformis*, are not. Alfred Elliott and his students had begun a project to collect wild tetrahymenas, and matings among these Woods Hole (WH) strains yielded, for the first time, viable progeny. As domesticated by Nanney, the WH descendants initiated a new era of genetic analysis similar to what happened earlier following Sonneborn's (1937) discovery of mating types in *Paramecium*. Now there were two genetically tractable ciliates, each making important contributions.

Elliott's group discovered 11 other clusters of complementary mating types within the *T. pyriformis* complex. Consistent with the practice in other ciliates, the clusters were designated as "syngens," the equivalent of species. The WH strains were assigned to *T. pyriformis* syngen 1. By 1973, there were 14 named *Tetrahymena* species, including the morphospecies *T. pyriformis*, a complex of micronucleate syngens and numerous amicronucleate strains. It was widely acknowledged that some species were dubious, reflecting the difficulty of distinguishing *Tetrahymena* species. With the advent of isozyme technology, species could be identified without maintaining stocks of living reference strains, and Latin binomials were given to the syngens (Nanney and McCoy, 1976). *T. pyriformis,* syngen 1 became *T. thermophila* because it was among the most temperature tolerant of the *Tetrahymena* species. Nanney and McCoy named eight additional micronucleate (from Elliott's collection) and four amicronucleate species. Over the years, additional species have been named, for a total of approximately 40 named species (see Chapter by Lynn and Doerder in this volume). For further historical details, the interested reader is referred to chapters in the *Biology of Tetrahymena* (Elliott, 1973).

II. The Species Problem

Among protists, including ciliates, the biological species concept has been problematic (Nanney, 1999; Sonneborn, 1957; Schlegel and Meisterfeld, 2003). While there are indeed many reproductively isolated populations readily identifiable as biological species, there are exceptions that complicate species assignment. Many ciliates are obligate inbreeders (e.g., selfers), some, particularly in the genus *Tetrahymena,* are amicronucleate and hence asexual, and some simply appear not to mate, perhaps because the complementary mating type has not been found. Another complicating factor is that many species are morphologically indistinguishable, existing as complexes of cryptic species. A given morphospecies, such as *T. pyriformis* or *Paramecium aurelia*, might consist of dozens of biological species. This causes considerable difficulty in identification of unknowns. First, it requires testing unknowns with living reference strains, a rather formidable housekeeping task for large numbers of species. Second, the conditions, which maximize mating, are unknown for most species, and since new species are based on negative reactions with other species, isolates may be erroneously assigned to new species. For instance, as recounted by Nanney and McCoy (1976), the species *T. pyriformis* syngens 6 and 8 were established on the basis of mating tests done in distilled water, a procedure that works well with *T. thermophila*. However, upon mating cells in bacterized cerophyll, it was found that strains of syngens 6 and 8 yielded fertile F1 and F2 generations, indicating they were the same species, now known as *T. pigmentosa*. Third, many ciliates, including tetrahymenas, have an immaturity period following conjugation during which they cannot mate. About 20% of *T. thermophila* isolated from nature are sexually immature. Lack of mating therefore does not necessarily indicate separate species. An added complication is that the lengths immaturity periods are unknown for most species. For *T. thermophila*, the immaturity period of inbred strains is probably shorter than that of wild isolates, perhaps due to selection during inbreeding.

It was problems such as these that led to the search for molecular markers to identify species without reference to living strains. The first of these were isozyme mobilities (reviewed by Meyer and Nanney, 1987), which quickly led to the assignment of Latin binomials to the syngens (Nanney and McCoy, 1976). Nearly 40 years later, the most successful of the molecular markers for *Tetrahymena* appears to be mitochondrial cytochrome oxidase subunit 1 barcodes (Chantangsi *et al.*, 2007; Chantangsi and Lynn, 2008; Kher *et al.*, 2011), which resolve species that have identical small ribosomal subunit (SSU) rDNA (also see discussion below). The average *cox*1 sequence difference among *bone fide* species is 10%, and for nonproblematic species intraspecific difference is ~0–2%. Thus, most unknowns are easily identified. However, there are several problems with *cox*1 barcodes that remain to be resolved, primarily because the barcodes suggest that some reference strains are either mislabeled or misidentified, with most requiring further breeding tests to resolve the ambiguities (Chantangsi and Lynn, 2008; Kher *et al.*, 2011). DNA barcodes for *Tetrahymena* therefore should be considered as work in progress.

Amicronucleate tetrahymenas present additional problems as they do not mate and hence cannot be assigned to a species based on sexuality; moreover, they lack the germinal micronucleus, thus precluding assessment of F1 and F2 fertility. It should be noted that aged micronucleate strains, though readily conjugating, cannot form gametic nuclei and hence cannot be used to assess fertility. Based on isozyme differences that rivaled those of micronucleate species, several amicronucleates were assigned Latin names (Nanney and McCoy, 1976).

The emerging consensus, as based on the work described above and supported by our own unpublished work (see below), is that DNA barcodes are an effective way to identify most species of *Tetrahymena*. The rate of discovery of new species suggests that there are many more. The *Tetrahymena* species may be quite informative regarding species evolution in ciliates.

III. *Tetrahymena* Collections

Though numerous investigators have collected *Tetrahymena* (or taxa now thought to be *Tetrahymena*), it was Elliott and his students who made the first systematic survey of *Tetrahymena* (Elliott, 1970). In all they collected from around the world nearly 1300 water samples and examined 7115 clones. Those that mated fell into 12 reproductively isolated groups or syngens of *T. pyriformis*, later named as species as mentioned above. This work also provided the first insights into the biogeography of this speciose genus. According to Elliott, 39% of "habitats" examined contained *Tetrahymena*, and many species appeared to have limited distribution. For instance, *T. thermophila* was found only in North America, *T. australis* only in Australia, and *T. hyperangularis* only in Europe. Other species had broader, perhaps global distributions, such as *T. pigmentosa*, which was found in North America, Europe, and Africa.

The next major collection spanning roughly 1960–1990 was made by David Nanney and Ellen Simon. Simon also assembled information on many smaller collections made by others (e.g., Nyberg, 1981). The reference strains used by Chantangsi *et al.* (2007) and Kher *et al.* (2011) came primarily from their collection as donated by Ellen Simon to the American Type Culture Collection. A summary of the collection data is available at http://www.life.illinois.edu/nanney/tetrahymena/biogeography.html#Anchor-Species-A-49575. This database contains results from approximately 600 sites, of which ~500 are in North America and ~100 from the rest of the world. The tetrahymenas were variously identified by mating reactions, isozyme mobilities, and sequences of the D2 region of the large subunit ribosomal RNA (LSU rRNA). This database contains ~30 species, many unknowns, and some suspected subspecies. It also contains 17 micronucleate (breeding) species, of which *T. americanis* is the most abundant, followed by *T. borealis* and *T. elliotti*. This database extends the range of many species, for example *T. australis* and *T. hyperangularis* to North America (Simon *et al.* 2008).

In recent years, one of us (FPD) has extensively sampled for *T. thermophila*, primarily in northeastern United States by challenging wild isolates with all seven of its mating types. This collection has examined 832 sites, ~11,000 samples, and ~21,500 isolates. Collaterally, other species of *Tetrahymena* were identified (see below). 46.6% of sites and 55% of samples yielded tetrahymenas, though the latter is biased by repeated sampling of sites with resident populations of *T. thermophila* in Pennsylvania and New Hampshire. A map of sites yielding *Tetrahymena* in the Nanney/Simon and Doerder databases is shown in Fig. 1. Further details on *T. thermophila*, which remains confined to North America, primarily in the Northeast USA (Fig. 2), are described below.

The distribution of *Tetrahymena* species as indicated by various collections is relevant to fundamental questions of ecology and biogeography. It has been argued that microorganisms are "everywhere" with most species globally distributed (Finlay and Fenchel, 1999; but see also Foissner, 2006; Nanney, 2004). According

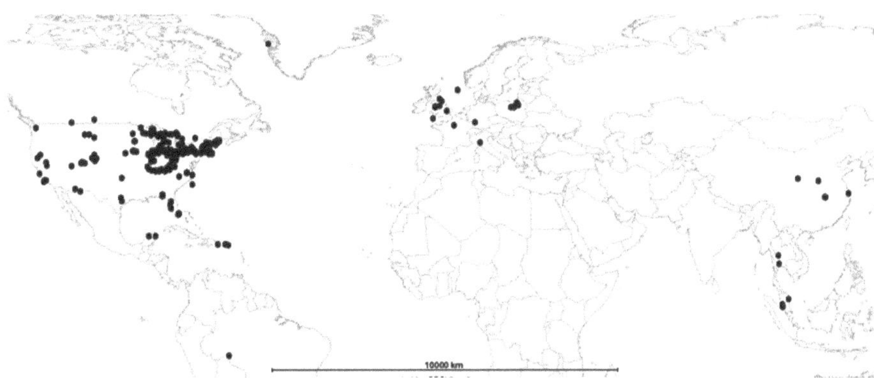

Fig. 1 Map of locations in Nanney/Simon and Doerder collections yielding *Tetrahymena*. Each dot, many overlapping, represents a single collecting site. Some sites yielded multiple *Tetrahymena* species. Much of the world remains to be sampled.

Fig. 2 Map of locations yielding *T. thermophila*. Each dot indicates a site for which *T. thermophila* has been reliably reported.

to this, hypothesis ciliate species, including tetrahymenas, do not have biogeographies. The global distribution of some ciliates and tetrahymenas supports this view, whereas the limited distribution of other species, e.g., *T. thermophila* with strong evidence of population structure, does not (Foissner *et al.*, 2008; Katz *et al.*, 2006). Further, the collections are showing that many tetrahymenas are sympatric, sometimes with many species found in the same pond or stream, suggesting that the taxa occupy distinct ecological niches. Food sources and predation are the two major selective pressures resulting in speciation. For most ciliates, and particularly *Tetrahymena*, neither the preferred bacterial food sources nor their common predators are currently known. As a suggestion for future research, it might be possible to determine food preference by single-cell PCR on specimens caught in the field, on the assumption that food vacuoles contain undigested DNA.

IV. Species of *Tetrahymena*

A. *Tetrahymena thermophila* and Its Inbred Stains

Because mating among the WH strains isolated by Elliott was the first such mating that yielded viable progeny, *T. thermophila* became the *Tetrahymena* genetic model of choice. Nanney's group demonstrated that it has seven mating types controlled by complex alleles specifying frequency distributions of mating types rather than fixed mating types (Nanney *et al.*, 1955). They also showed that mating-type determination is karyonidal, where each new macronucleus is independently determined for mating type (Nanney and Caughey, 1953), and, using rare selfers (intraclonal conjugation), they discovered the phenomenon of macronuclear assortment (Allen and Nanney, 1958).

Inbred strains were established from the WH strains and two other isolates through a series of sib matings to establish homozygosity. These strains were given letter designations (A–F) as based on the mating-type allele. Strains A and B were both derived from WH6XWH14, whereas strains C–E were derived from wild isolates UM226 (Vermont, strain C), ALP-4 (Michigan, strains D and F), and IL-12 (Michigan; strain E) and one or more crosses to Strains A or B. All inbred strains have identical *cox1* sequence (Chantangsi *et al.*, 2007), suggesting that their mitochondria are derived from the common WH ancestor. Numerical designations such as B2 or C3 indicate that the strain has the same mating-type gene as other strains with the same letter but differ at other loci due to crosses to other strains early in the inbreeding process. The article by Allen and Gibson (1973) contains information on the derivation of the strains. Strain B has become the standard strain for genetic analysis and is deposited at ATCC (http://www.atcc.org/) and at the National *Tetrahymena* Stock Center (http://tetrahymena.vet.cornell.edu/) along with other tetrahymenas.

While *T. thermophila* strain B is the most commonly used laboratory strain, most investigators have prefaced their strains with letters designating their university (e.g., CU = Cornell University; SB = Santa Barbara; UR = University of Rochester) followed by numbers designating an internal code specifying a specific genotype. All of these strains had their origin in David Nanney's laboratory and nearly all represent over 28 generations of inbreeding. Complete strain descriptions, including the laboratory of their origin, in publications help trace their history. Despite its being highly inbred, it should not be assumed that all Strain B cells are identical. For instance, the *mat* locus mapped in two labs to different chromosome arms, the apparent result of a translocation (Lynch *et al.*, 1995). The complete macronuclear sequence of *T. thermophila* SB210 (Eisen *et al.*, 2006) makes this particular derivative of Strain B the standard reference.

As the micronuclear and macronuclear genomes can be independently manipulated (a useful tool) and modified (e.g., nullisomy; transformation), it cannot be assumed that the micronuclear and macronuclear genotypes are identical nor can it be assumed that strains with the same name contain the same set of genes or that the genes are in the same linkage arrangement. In other words, the history and provenance of an individual strain is critical. Occasionally strains other than B (e.g., C and D) are employed for specific reasons. Strain C3, for example, provided an especially useful set of polymorphisms that allowed the mapping of Lynch *et al.* (1995). Because strains C–E and those with numbers are derived in part from wild isolates other than the WH strains, it should not be assumed that these strains have the properties associated with strain B, even though they might have the Strain B mating-type locus (e.g, strain B2). It should be kept in mind that inbred strains are but snapshots of the *T. thermophila* gene pool.

Studies of the inbred strains (A–F) revealed genetic polymorphisms for surface (immobilization) antigens, enzyme mobilities, membranelle number, and numerous RFLPs. Studies of wild strains of *T. thermophila* found additional mating-type alleles as well as additional surface antigen loci and alleles. With the exception of mating-type alleles, which can only be revealed by crosses, only the macronuclear genomes of wild isolates are characterized. Almost certainly, their micronuclear genomes will reveal additional polymorphisms.

T. thermophila appears to have a relatively limited distribution (Fig. 2). It is found in many ponds in New England (MA, ME, NH, and VT) and appears to be resident in ponds of both western and eastern PA. Although it was found in MI by Nanney and Allen, extensive resampling (FPD) of both the upper and lower peninsulas has failed to find additional isolates. In addition, *T. thermophila* was found in two ponds in FL, but has not been found in OH, KY, or NY (of the states that have been more extensively sampled). Since the greatest variation of *cox1*is among New England samples, it is likely that *T. thermophila* originated there. The mechanism of dispersal is unknown, though it may be postglacial and influenced by waterfowl as well as humans, as *T. thermophila* is found in numerous man-made ponds. It is an interesting problem in biogeography, likely to be remedied by additional sampling, that *T. malaccensis*, the closest relative to *T. thermophila*, so far has been found only in Malaysia.

T. thermophila has been found primarily in ponds, typically small ones, including roadside ditches. It is sometimes found in streams, downstream from ponds with resident populations. The preferred habitat appears to be near shore among either floating or emergent vegetation. Samples from open water are rarely positive for this species. Multiple mating types, including all seven mating types, have been found in single samples. It is probable that *T. thermophila* is attracted to bacterial food sources where it reaches densities critical for mating, which occurs upon exhaustion of the food. To date, *T. thermophila* has not been collected in ponds where it is resident if the water is <13 °C. The mechanism of overwintering, where the temperature is well below 13 °C for extended periods, is unknown.

The number of mating types appears to be fixed at seven as was found among progeny of the very first crosses of the WH strains (Nanney and Caughey, 1953). No new mating types have appeared in any subsequent collection or their descendants, including ~7600 isolates tested for mating type in the Doerder lab. There is, however, evidence for numerous mating-type (*mat*) alleles that determine the frequencies with which specific mating types are chosen during macronuclear development of each karyonide (Arslanyolu and Doerder, 2000; Doerder *et al.*, 1995). Among the first inbred strains, Nanney's group identified two qualitatively different mating-type alleles (referred to as A- and B-type alleles). The A-type allele specifies mating-types I, II, III, V, and VI; the B-type allele specifies mating-types II–VII. In addition to the critical mating-types I, IV, and VII that distinguish between the two alleles, these alleles differ quantitatively in the frequencies with which the common mating types appear. It was subsequently determined that environmental cues, especially temperature, influence the frequency of mating types. In no case has an allele been described which specifies all seven mating types, nor has an allele been found which specifies both I and IV or I and VII. It will be interesting to characterize the molecular nature of the A- and B-type alleles to determine the cause of apparent suppression of recombination.

In all ponds that have been sufficiently sampled, all seven mating types are found, indicating that the founding population contained both A- and B-type alleles. Theoretically, all seven mating types should be equally frequent in a pond, maximizing compatibility of potential mates. However, though the overall distribution of mating types in an area may be close to equal, observed frequencies in individual ponds are frequently skewed (Doerder *et al.*, 1995; Doerder, unpublished). To determine which *mat* alleles are present in individual ponds, genomic exclusion (Allen, 1967) was used to create instant, whole genome homozygotes, which permit assessment of environmental influences in the absence of genetic variation (Arslanyolu and Doerder, 2000). Surprisingly, nearly every new genotype appeared to contain a new *mat* allele, which differed both qualitatively and quantitatively from each other and the alleles in the inbred strains. All alleles were either A- or B-type (missing IV and VII or I), but some were also missing mating-types II and/or III, and all differed markedly in the frequencies of expressed mating types. A theoretical study (Paixão *et al.*, 2011), concludes that such multiple alleles explain mating-type frequencies observed in natural populations if the effective population size is small. A more complete understanding of mating type in natural populations requires molecular characterization of the *mat* locus as well as reliable estimates of population parameters.

B. Other Species of *Tetrahymena*

The "gold standard" for identification of biological species is sexual compatibility. By mating wild isolates in all possible combinations, Elliott described 12 species (syngens). Using representatives of Elliott's collection as mating-type testers, subsequent investigators identified additional species. However, as the number of species grew, so did the burden of maintaining stocks and performing the necessary matings, not all of which are as straightforward as *T. thermophila* matings. Thus, molecular methods have been applied, including isozyme mobility, histone H4 gene, the intergenic region between histone H3 and H4 genes, 5S and 5.8S rRNA, telomerase RNA, and the D2 region of the LSU rDNA and the *cox*1 gene. Though useful, none proved completely satisfactory, both because of limited databases and their inability to distinguish among all taxa. The nuclear SSU rDNA, often of great utility over large evolutionary distances, lacks resolving power, particularly among *Tetrahymena* species, some of which have identical SSU rDNA sequences (Chantangsi *et al.*, 2007; Sogin et al., 1986).

Among the most useful barcodes (see also below) is the mitochondrial *cox*1 sequence. It has the largest number of sequences (Kher *et al.*, 2011) and is useful in identifying unknowns isolated from nature and errors in labeling strains (which isozyme analysis also revealed). Based on *cox1* barcode differences of >5%, Doerder (unpublished) has identified >30 new species of *Tetrahymena*. Some of these are represented many times in the collection, and others occur only once. It would appear that there are numerous species of *Tetrahymena*, and, given the relatively limited geographic area sampled, many more species are awaiting discovery.

C. Amicronucleates

Among ciliates, *Tetrahymena* is especially rich in amicronucleates, which account for some 10–70% of isolates. Among 2093 wild isolates that were not *T. thermophila* in Doerder's collection, 26.9% were amicronucleate. In contrast to the prevalence of amicronucleates in nature and their vigorous growth when brought into the laboratory, construction of amicronucleate strains in the laboratory by a variety of manipulations all fail to produce viable clones (Ng, 1986). Manipulated and aged strains may become hypodiploid, as in the classical C* and A*III strains used in genomic exclusion, but these are not true amicronucleate strains. The only exception is an amicronucleate strain of *T. thermophila* that appeared after mutagenesis; though this strain grows well, it dies upon conjugation (Kaney and Speare, 1983). This strain is the only amicronucleate *Tetrahymena* known to mate.

With the widespread occurrence of asexual amicronucleate tetrahymenas in natural populations, it was suggested that they may represent a "senile" phase of the life cycle (Sonneborn, 1957). This has been doubted, however, primarily because amicronucleates arising in the laboratory inevitably die, while those from nature reproduce at rates indistinguishable from micronucleate strains. Though requiring additional investigation, a more likely scenario is that amicronucleates arise through errors in macronuclear development and micronuclear misdivision. Early isozyme studies showed that amicronucleates likely were of multiple origin but failed to associate any amicronucleate with a micronucleate species (Borden *et al.*, 1977). Subsequently, the amicronucleate species *T. elliotti*, an amicronucleate named by Nanney and McCoy (1976), was found to have micronucleate counterparts (Simon *et al.*,1985). *T. elliotti* is the second most common amicronucleate in North America, and, interestingly, most selfers are micronucleate *T. elliotti* (Doerder, unpublished). The most abundant amicronucleate is *T. borealis*, whose micronucleate forms do not self.

*Cox*1 barcodes have shown that the status of some named amicronucleates should be revisited (Chantangsi *et al.*, 2007; Kher *et al.*, 2011). *T. pyriformis* and *T. setosa* have the same *cox*1 sequence and are likely synononomous, as earlier suggested by Sadler and Brunk (1992). Similarly, *T. lwoffi* and *T. furgasoni* are likely synononomous. None of these currently has a micronucleate counterpart. Analysis of *cox*1 sequences of >200 amicronucleates (Doerder, unpublished) indicate that they are distributed widely across the *Tetrahymena* phylogenetic tree, with the exception of the "*australis*" clade. Only about half have micronucleate counterparts, and these appear to be of recent origin as their *cox*1 sequences are either identical or differ by only 1–3 nucleotides from micronucleate species. Amicronucleate counterparts of *T. thermophila* have been isolated from multiple sources (Doerder, unpublished). Given the relatively small sampling region worldwide, it is likely that there are dozens more amicronucleate *Tetrahymena* species.

V. Methods for Collecting and Identification of *Tetrahymena*

A. Choice of Habitat Locations

Ciliates, particularly *Tetrahymena*-like taxa, can be found in most fresh-water habitats. We concentrate here on free-living species, but note that some species are parasitic in snails, insect larvae, etc. (see chapter by Lynn and Doerder). Ponds with reeds and vegetation along the margins are preferred sites (Hersha *et al.,* 2009), though streams and rivers, particularly eddies have also yielded *Tetrahymena*. Cells can be recovered from most locations within a water source; however, there are dramatic differences in the probability of obtaining cells. The bottom of a pond at the base of reeds, a likely site of decaying vegetation and bacteria, is the best site for collecting cells. Collection of cells from surface water, even by vegetation, is less probable, and collection of cells from open water has the lowest probability. In one pond, we found ~ 1 cell per 75 mL in benthic samples next to reeds, while there was ~ 1 cell per 200 mL in surface water adjacent to the reeds, but surface samples several meters from the reeds had only ~1 cell/L. Ponds or lakes suitable for collecting are found in a myriad of sizes and venues, and even quite small ponds, including vernal ponds, are good collection sites. Metropolitan parks with ponds are excellent collection sites easily comparable to rural ponds (but see below). Figure 3 shows typical *Tetrahymena* habitats.

The exact ecology of most *Tetrahymena* species is unknown, and thus it is difficult to offer guidance for finding specific species. The results of any collection, regardless of "preferred" locations in the water column, will provide important information. This applies to parasitic *Tetrahymena* species as well, since a species discovered as a guppy parasite in Israel (Leibowitz and Zilberg, 2009) has been found in an OH stream (Doerder, unpublished).

The best time to collect that maximizes recovery of tetrahymenas is during the summer months from mid-May to mid-October. This is particularly true for *T. thermophila*, which has not been recovered in water <13 °C. There is some

Fig. 3 Typical collection sites. Upper left: reeds in the fishpond in Frazier Park, CA. Upper right: a city park in Chiana, Crete, Greece. Lower left: reeds in Fern Lake, Pine Mountain Club, CA. Lower left: the bank of the Danube at Regensberg, Germany. (See color plate.)

evidence that other species may be preferentially found at lower temperatures, but this needs more systematic investigation. The time of day at which samples are taken does not seem to affect collection efficiency.

Some cautions about collecting. It is important to be aware that collecting water samples in national parks and on private land requires special permission. In some states (check natural resource, hunting, fishing, and boating websites), a fishing license and/or a permit (possibly free) may be required. It is also important to be aware that wet gear, wading shoes, and clothing not only could transport tetrahymenas but also transport invasive species such as the diatom *Didymosphenia geminata* or Eurasian milfoil *Myriophyllum spicatum*. Natural resource experts recommend disinfectants and removal of plant fragments. Lastly, it should be emphasized that safety is of utmost importance. Shores may be slippery or treacherously rocky, and rivers and streams may have high banks or dangerous currents. All of these hazards have caused us to abandon collecting at a specific site. We have also abandoned potential collecting sites because of lack of parking, heavy traffic, gaggles of geese, and poison ivy.

B. Bulk Collecting

Collecting apparatus need not be elaborate nor should there be undue concern about sample volume. Nearly any type of bottle or bag can be used, including recycled soda or water bottles, and volumes as small as 25 mL have yielded tetrahymenas. The important thing is to get a sample when the opportunity presents itself. For more systematic studies, both the number and volume of samples must be considered, as well as the location of the samples (e.g., benthic or surface). For the standard Doerder collection procedure, an inflated labeled one-quart Ziploc™ bag is inserted into the end of a golf-ball retriever, which is then used to collect a 200–500 mL sample (Fig. 4). Because of the retriever telescopes, suitable water usually can be reached without wading. Typically, samples are collected in 15–300 cm of water with emergent, floating, or benthic vegetation as such sites have a greater likelihood of containing tetrahymenas. To maximize recovery of ciliates likely feeding on the bottom, the water and substrate should be mixed by dragging the bag. The presence of "muck" and decaying vegetation in the bag is probably a plus. After the sample is obtained, the bulk of air is removed from the bag as it is closed (less bulky for transport) and placed a large plastic box or bucket for transport to the laboratory. In the field, the date, highway location, latitude, longitude, and water temperature are recorded. It is important neither to refrigerate the samples nor to allow them to become too hot. Though tetrahymenas survive the winter without forming cysts, they do not tolerate rapid change in temperature due to refrigeration. Although *T. thermophila* tolerates temperatures up to about 41 °C, many other tetrahymenas have much lower temperature tolerances.

Fig. 4 Collecting water samples in 1-quart bags using a telescoping golf-ball retriever. The retriever allows the collection of a bulk sample while also stirring the substrate. (For color version of this figure, the reader is referred to the web version of this book.)

Either in the field or in the laboratory, samples are processed to enhance the recovery of *Tetrahymena*. The bags are opened and proteose peptone and antibiotics are added. Typically, 50–150 mg of powdered proteose peptone is sprinkled from a tube (much like feeding aquarium fish), and ~0.3 mL antibiotics (from concentrated stock of 0.1 g/mL each of penicillin G sodium and streptomycin) are added with a plastic pipette. Precision is not critical. The proteose peptone encourages *Tetrahymena* growth, while the antibiotics tend to reduce the presence of flagellates and other ciliates, perhaps by interfering with their food sources. After 3–5 days at room temperature, a few drops from each bag are examined with a dissecting microscope (25×) for the presence of *Tetrahymena*-like organisms. Assuming that the sample initially contained a single *Tetrahymena* that divides every 6 h (most divide faster), this waiting period allows for sufficient doublings to produce a high density of *Tetrahymena*; if multiple mating types are present conjugating pairs may also be seen. Though many species of *Tetrahymena* have been isolated in this way because they grow to high density, it is suspected that some tetrahymenas do not grow under these conditions. In rare instances, *Tetrahymena*-like cells are seen at low density, even after 2–3 more days, and these typically fail to grow when transferred to bacterized laboratory medium. Environmental PCR procedures might provide useful information to determine whether there are tetrahymenas that are difficult or impossible to grow in the lab. In addition to tetrahymenas, it is not unusual also to find *Halteria*, *Vorticella*, *Paramecium*, and various hypotrichs when samples are examined.

For samples that contain *Tetrahymena*-like organisms, it is essential to clone cells by single-cell isolation prior to any DNA purification or cytology as samples may contain multiple mating types (e.g., all seven for *T. thermophila*) and/or multiple species. Though samples can be diluted, the surest way to isolate single cells is with a micropipette pulled from a heated Pasteur pipette. Though this procedure takes some practice, an experienced investigator can isolate numerous single cells in a short amount of time. Though cells can be isolated directly into antibiotic containing PPY (1% proteose peptone, 0.15% yeast extract; 10 mg/mL each of penicillin G sodium and streptomycin), they can also be isolated into Cerophyll (or rye grass) inoculated with *Klebsiella pneumoniae* as the bacterial food source. For bacterized medium, 24-well cell culture plates typically are used. Cerophyll has the advantage that many species will conjugate in this medium, whereas no tetrahymena will conjugate in axenic proteose peptone. If cells are isolated into antibiotic containing PPY, it is essential to dilute the bacteria (which can be harbored in food vacuoles) by serially transferring cells through four to seven changes of medium. Typically, this is done by placing drops of PPY with antibiotic in a petri plate and transferring cells with a micropipette at intervals of 10–30 min through four to five successive drops and placing a single cell in the final drop. Isolates that die in this process are often species of the related genus *Glaucoma*, but they may also be *T. borealis*, *T. candadensis,* or several of the new species of *Tetrahymena*. Often, but not always, these can be grown in the much richer liver-peptone medium (Phillips, 1967). Cells failing to grow in axenic media usually grow in bacterized Cerophyll so that sufficient quantities can be obtained for breeding tests, cytology, and DNA purification.

C. Specialized Collecting and Attractant Traps

Surface samples can be collected by simply scoping water using a jar or bag. Taking benthic samples is more challenging. An easy method of benthic collection is to use a syringe or "turkey baster" to extract a sample from a specific location. A long tube guided by a pole with a pump to draw in samples is also an effective collection mode.

A "fountain" collector developed in the Brunk laboratory can be used in combination with a rod or long pole to collect benthic samples from the bank (Fig. 5). A "fountain" collector has a tube leading from the bottom of the collection vessel to just below the top of the chamber. The collection vessel has a small vent hole in the top, which allows the air in the chamber to slowly escape while water enters from the bottom. Filling the chamber takes over a minute, which allows the vessel to sink to the bottom as the chamber begins to fill; thus, the vast majority of the sample is taken at depth. This allows collection of benthic samples, which is often difficult with jars or bags.

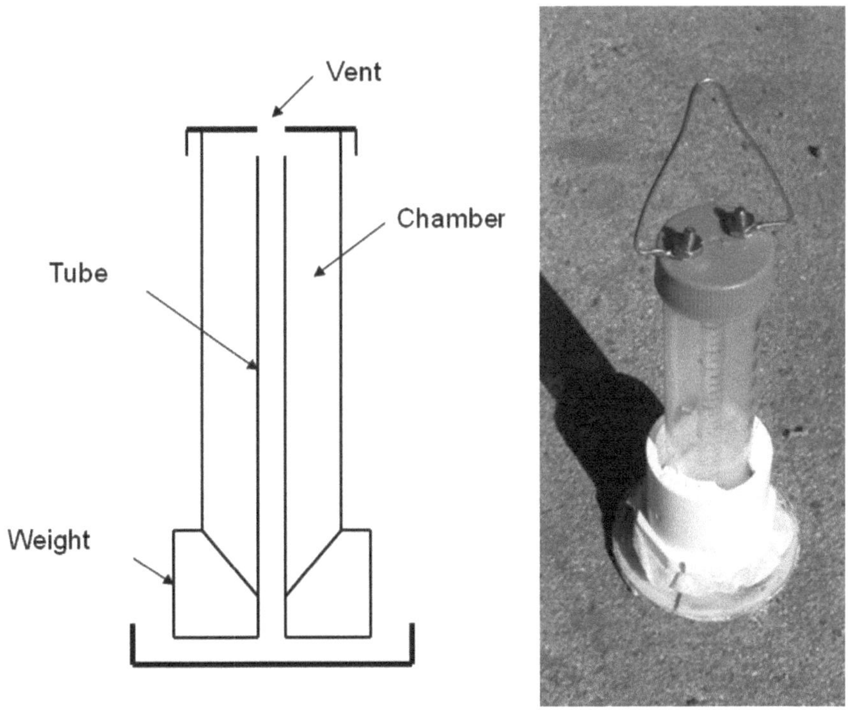

Fig. 5 The drawing at left shows the elements of the "fountain" collector. The sample enters at the bottom and flows up the tube filling the chamber. The vent in the top allows air to escape slowly from the chamber extending the collection time. When retrieved, the sample remains in the chamber. A "fountain" collector constructed from a 50 mL disposable centrifuge tube and a 10 mL disposable pipette is shown at right. (For color version of this figure, the reader is referred to the web version of this book.)

Bulk samples can be added directly to PYG media (2% proteose peptone, 0.1% yeast extract, 0.5% dextrose) or concentrated by centrifugation (1000 g for 15 min) prior to culturing. The addition of antibiotic is essential; 100 mg/L each of penicillin, streptomycin, neomycin, kanamycin, and tetracycline works well. All pond water we have encountered has a significant number of bacteria many of which are resistant to one or more of these antibiotics.

The use of "traps" dramatically enhances the collection of *Tetrahymena*-like cells (Leick and Lindemose, 2007). Small vial traps (2 mL freezer vials) filled with an attractant and closed by a filter works well. Cells easily pass through a common coffee filter; however, the filter keeps dirt and larger organisms out of the vial. A 25-mm disk of coffee filter is secured by a ring of silicon (thin section cut from a silicon tube) in the upper portion of the vial (Fig. 6A). The vials are filled with 5% proteose peptone and antibiotics using a syringe with a fine needle. The screw caps are placed on the vials for storage and transport. For collection, the vials are uncapped and placed in a pond for 10–20 h. A number of vials can be place at a single site. The vials must be weighted (they are buoyant) and some means of retrieval attached to the vials. The retrieval mechanism can be as simple as a string leading to the bank; however, leaving the vials overnight subjects them to vandalism, thus attaching a bail to the vials makes them less conspicuous and allows them to be easily fish out of water. The filter also prevents spills; actually the vials can be inverted without loss of the sample. After retrieval the water above the filter is removed (blotted with a tissue) and the caps returned. The cells in the vials remain viable for several days in capped vials.

A very effective collection trap is a 96-well microtiter plate (450 μL wells) overlaid by a coffee filter and a cover with holes that match the microtiter wells (Fig. 6B). A seal lid is placed over the cover with holes for storage and transport. The plates are deployed like the vials (they must also be weighted). Once the plate is retrieved, the water above the filter is removed (blotted with a tissue) and then the plate can be sealed with a solid lid placed over the cover with holes. Plates are usually deployed for 10–20 h. When using the plates, it is common to have 40–90 wells containing cells. Usually there is a single cell entering a well, but they reproduce, in the proteose peptone, during the collection period. Traps dramatically enhance cell collection efficiency, but require significant deployment time. We have collected cells using plates in as short as 2 h, but fewer wells contain cells (usually 10–20).

D. Growth Conditions

The proteose peptone and PYG are strong selective agents favoring *Tetrahymena*-like cells. Cells in samples concentrated by centrifugation when mixed with PYG and observed microscopically show that the media is actually toxic to many non-*Tetrahymena*-like cells as evidenced by the fact that they stop moving. This selective pressure of proteose peptone is a large part of the screening for *Tetrahymena*-like cells among the vast number protists present in ponds. In addition to PYG media,

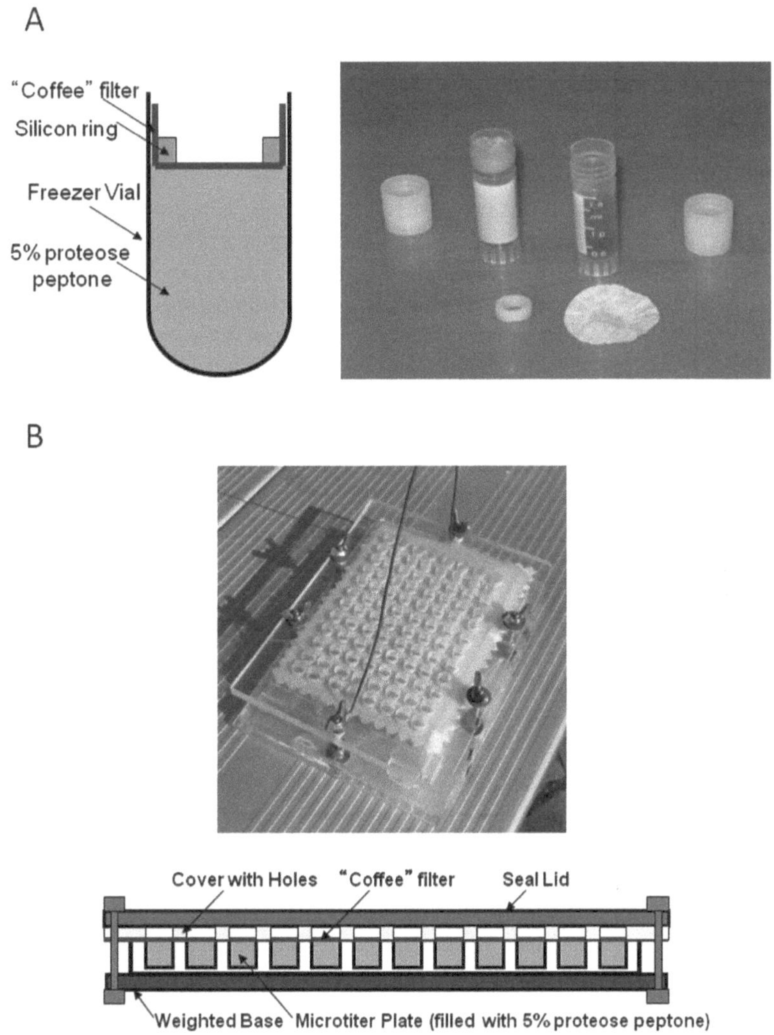

Fig. 6 (A) The drawing at left shows the elements of the vial trap. A 2 mL freezer vial has a 25 mm diameter "coffee" filter disk secured by a silicon ring. These vials with filters in place are autoclaved and then filled with 5% proteose peptone and antibiotic using a syringe. At right an uncapped vial is shown and to its right is a filled vial with the silicon ring and "coffee" filter disk removed. (B) The drawing below shows the elements of the microtiter plate collection device. The upper photo shows a microtiter plate collection device, ready for deployment, with the seal lid removed. (For color version of this figure, the reader is referred to the web version of this book.)

Tetrahymena-like cells grow well in garbanzo bean medium (one garbanzo bean in 10 mL distilled water and autoclaved). As with Cerophyll mentioned earlier, some taxa grow better in bean medium than in PYG. Cell growth in PYG or bean media allows densities of 10^5 to 10^6 cells/mL. Long-term cell maintenance is supported by bean media capped with 3–4 mm of mineral oil, which significantly reduces the oxygen concentration. Under these conditions, cells can persist for a number of years.

E. Taxa Identification

Identifying species by morphology is highly problematic with *Tetrahymena*-like cells as they are virtually indistinguishable. Silver staining and detailed analysis of patterns of cilia requires expertise and is labor intensive for nominal identification (Corliss, 1973). Sequence comparison of specific DNA regions provides a reliable and broad-ranging identification tool with specificity.

PCR amplification followed by conventional Sanger DNA sequencing allows virtually any region of the genome of cultured cells to be determined. DNA suitable for PCR amplification can be easily prepared from 10 to 25 mL of cultured cells. Even cells collected from "traps" often represent more than one taxon. Cells from environmental samples should be cultured from a single-cell isolate to avoid mixed cultures. This can be done by suitable dilution into micro-titer wells or single-cell isolation. Many commercial DNA preparation protocols are available; however, a simple Sarkosyl lysis and phenol-chloroform extraction works well. The cells from a small culture are harvested by brief centrifugation (1000 g, 3 min) and resuspended in 500 μL NET* (0.5 M NaCl, 50 mM Tris, 50 mM EDTA, pH 8.5) in a 1.5 mL microfuge tube. Sarkosyl is added to 1% and the lysate is extracted with phenol-chloroform (1:1 phenol chloroform plus 100 mg/L 8-hydroxyquinoline). The extract is centrifuged (15 kg, 10 min) to separate the phases. The upper phase is collected, and the nucleic acids are precipitated by the addition of an equal volume of iso-propanol, followed by centrifugation (15 kg, 10 min). The nucleic acid pellet is resuspended in 50 μL NET-RNase (100 mM NaCl 10 mM Tris, 10 mM EDTA, 10 units/mL RNase, pH 8.5) and digested at 37 °C for 30 min to remove RNA, which interferes with PCR amplification. The DNA can be precipitated with isopropanol prior to PCR amplification or PCR amplification can be performed directly on the RNase digest.

The small subunit ribosomal RNA (SSU rDNA) sequence has an extensive data-base available. SSU rDNA sequences will identify any organism and place it in a general group. However, this sequence has very limited resolution capacity for closely related *Tetrahymena* species, as mentioned earlier. Regions of the LSU rDNA and the 5S ribosomal RNA sequences have been used for phylogenetic reconstruction, but they also have limited taxa resolution (Nanney *et al.*, 1998, Van Bell, 1985).

Three additional regions hold promise for taxa identification and phylogenetic reconstruction, the mitochondrial gene *cox1*, the 5.8S ITS (intergenic region between the SSU rDNA and the LSU rDNA), and the intergenic region between the histone H3 and histone H4 genes. The factors contributing to the value of each of these regions as an identification sequence include the sequence variability exhib-ited by the region, the existence of an extensive library of sequences from related taxa, and easy with which the region can be PCR amplified.

There is a growing library of *cox*1 sequences from ciliates related to the use of this sequence as a barcode identification region for a broad range of organisms (http://www.boldsystems.org). An ~980-bp region of the *cox1* gene is commonly PCR amplified and sequenced for its 689-bp barcode (Kher *et al.*, 2011). Many standard PCR primer sequences for this region are available. The *cox1* gene is of mitochondrial origin; thus, there are several thousand copies of this region per cell.

The 5.8S ITS region has a substantial database for comparison including many *Tetrahymena*-like species. A number of conserved PCR primer sites located in the SSU rDNA sequence and LSU rDNA sequence are available to amplify the 5.8 S ITS region which is ~ 460 bp long. The highly conserved 5.8 S (~150 bp) is situated roughly in the center of this region. In *Tetrahymena*-like species, the rDNA genes are amplified to ~9000 copies; thus, this region is readily PCR amplified even from small DNA preparations.

The H3H4 intergenic regions is ~350 bp between the histone H3 and H4 genes. PCR primers can be easily located in these highly conserved flanking genes. In *Tetrahymena*-like species the genomic orientation of the histone H3 and H4 genes permits the PCR amplification of this region; however, this orientation does not extend to more distantly related ciliates such as *Paramecium.* To date all of the *Tetrahymena*-like species capable of growing in PYG media have the appropriate orientation of histone H3H4 to allow PCR amplification. The available database for histone H3H4 sequences is limited. *Tetrahymena*-like species have ~45 copies of this region per cell due to macronuclear amplification.

The nucleotide substitution rate in these regions varies substantially (Fig. 7). The region with the greatest variability is the histone H3H4 intergenic region. Comparing the number of nucleotide substitutions at each base position in the most parsimonious phylogenetic tree relating 31 *Tetrahymena*-like species, there are an average of 2.90 nucleotide substitutions per nucleotide position in the histone H3H4 region. A similar analysis of the ITS region (a concatenation of ITS1 and ITS2 with the 5.8S rRNA removed) has an average of 1.40 nucleotide substitutions per base position, while the *cox1* regions has an average of 1.39 nucleotide substitutions per base position.

The histone H3H4 region is by far the most efficient region in terms of identification potential for length of region sequenced. The region is relatively small, which facilitates PCR amplification and sequencing; however, a larger region may be desired for detailed identification. In this case, a concatenation of the histone H3H4 and 5.8S ITS region may be preferable. The *cox1* sequence is the coding regions for an essential protein; thus, it has a relatively low nucleotide substitution rate per length. The *cox1* gene is a mitochondrial gene and yields a phylogeny of the mitochondrial lineage. If mitochondrial exchange occurs between taxa, the species (nuclear) lineage and mitochondrial lineage may not be congruent.

Sequence determination of specific regions is the "gold standard" for taxa identification; however, with collection and culturing of numerous taxa this becomes expensive. We have had success in preliminary identification of taxa by restriction

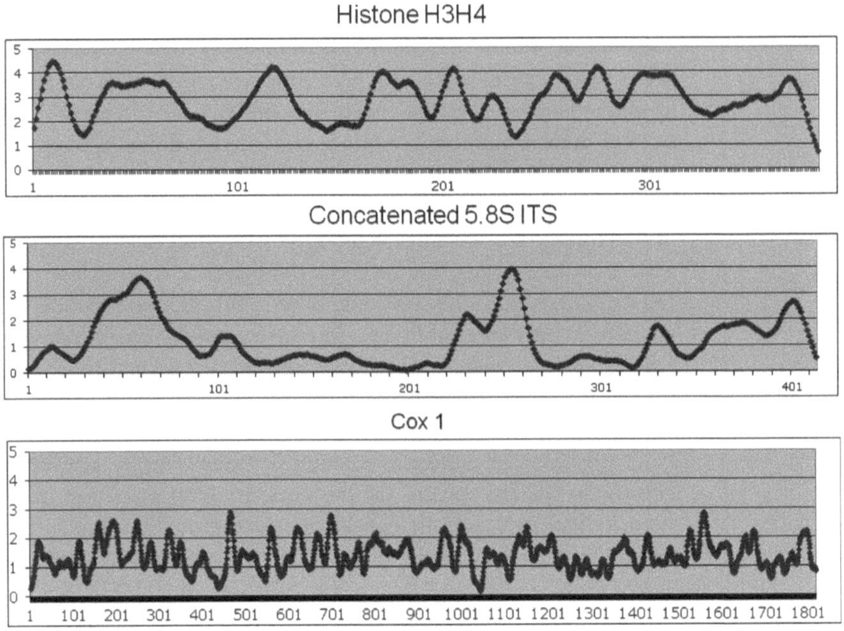

Fig. 7 A series of plots showing the nucleotide substitution frequency per nucleotide position (smoothed) for 31 *Tetrahymena*-like taxa. The 5.8S ITS1 and ITS2 regions are concatenated (removing the highly conserved 5.8S gene). The histone H3H4 region has an average of 2.90 nucleotide substitutions per nucleotide position, while the 5.8 S ITS regions have 1.40 and the *cox-I* gene has 1.39 nucleotide substitutions per nucleotide position. (For color version of this figure, the reader is referred to the web version of this book.)

fragment length polymorphism (RFLP) of PCR products from the histone H3H4 region and the 5.8 S rRNA gene. PCR product are digested with restriction enzymes such as Sau3A I (GATC) and Tsp509 I (AATT) and analyzed on a 3% agarose gel. The RFLP patterns allow identification of similar and different taxa. Well-described taxa can be identified by their pattern and then only taxa with novel patterns need be sequenced, which allows a much higher through put for collected samples.

F. Phylogenetic Analysis

A maximum parsimony phylogenetic tree for 60 recently isolated as well as published *Tetrahymena*-like taxa based on a concatenation of the histone H3H4 and 5.8S ITS sequences is shown in Fig. 8. *Ichthyophthirius multifiliis* makes an ideal out-group for this collection of sequences. A number of the sequences are virtually identical, and they are represented by a single entry in bold followed by the number of identical sequences. These sequences group into six clusters based on a comparison of the average pairwise percentage distance within the group versus the

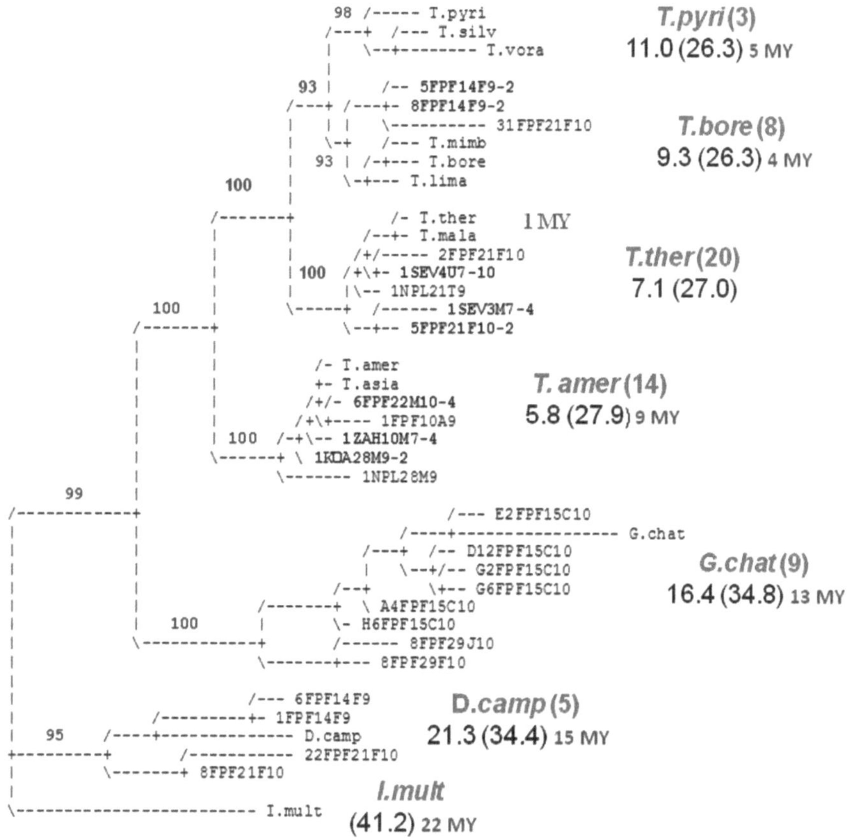

Fig. 8 A phylogenetic tree for 60 *Tetrahymena*-like 5.8S–H3H4 concatenated sequences generated using PAUP (Swofford, 1991). The tree uses 38 recently obtained sequences and 22 published sequences. Where several taxa have identical sequences, the taxon (in bold) is followed by the number of strains. Six clusters are identified (the number of taxa in each cluster) with the average intracluster pairwise percentage distance shown below and the average pairwise percentage distance to all other taxa shown in parentheses. An estimate of the divergence time from *T. thermophila* for the clusters as well as *I. mult* and *T. mala* are shown in MY (green). The bootstrap values for various nodes are shown in blue.. Species abbreviations: *T. pyri.*, *T. pyriformis*; *T. silv*, *T. silvani*; *T. vora*, *T. vorax*; *T. mimb*, *T. mimbres*; *T. bore*, *T. borealis*; *T. lima*, *T. limacis*; *T. ther*, *T. thermophila*; *T. mala*, *T. malaccensis*; *T. amer*, *T. americanis*; *T. asia*; *T. asiatica*; *G. chat*, *Glaucoma chattoni*; *D. camp*, *Dexiostoma campylum*; *I. mult*, *Ichthyophtherius multifiliis*. (For interpretation of the references to color in this figure legend, the reader is referred to the web version of this book.)

average pairwise percentage distance between group members and all other sequences. These clusters are generally similar to previous phylogenetic groupings (Kher *et al.*, 2011). The taxa represented in Fig. 8 include over 20 different sequences of taxa collected from the city fishpond in Frazier Park, CA. This is an indication of the number of *Tetrahymena*-like taxa to be found in a single pond.

An estimate of the divergence time of various sequences from the sequence for *T. thermophila* was made based on the sequence divergence of the 5.8 S ITS sequences of mouse and rat, taken as 30 million years (MY) (Nei *et al.*, 2001; O'hUigin and Li, 1992). The number of indels (insertion or deletions), transitions, and transversions per kilobase (kb) were calculated independently for a comparison of the mouse/rat 5.8 S ITS sequences. A similar comparison of the *T. thermophila* 5.8 S ITS sequence with the 5.8 S ITS sequence from a representative of each group, as well as *Ichthyophthirius multifiliis* and *T. malaccennsis* was also computed. Assuming a 30 MY divergence of mouse and rat, the divergence time of each group from *T. thermophila* was calculated independently for indels, transitions, and transversions and these estimated times were averaged (the similarity in these three estimates was relatively close). The *Tetrahymena*-like taxa are known to accumulate nucleotide substitutions at an accelerated rate; thus, these estimates are maximum divergence times (Sadler & Brunk, 1992; Katz *et al.*, 2004). These estimated divergence times are shown in Fig. 8.

The techniques presented here will allow a substantial expansion of the environmental isolations of *Tetrahymena* and *Tetrahymena*-like taxa. As additional environmental isolations of *Tetrahymena*-like taxa are characterized, the phylogenetic tree for these organisms is expected to become more bush-like. It will be of great interest to see if the clustering of these sequences reflects deep evolutionary divergences or is an artifact of limited collections. The current geographic distribution of *Tetrahymena*-like taxa is certain to be modified by more extensive collection.

References

Allen, S. L. (1967). Genomic exclusion: a rapid means for inducing homozygous diploid lines in *Tetrahymena pyriformis*, syngen 1. *Science* **155**, 575–577.

Allen, S. L., and Gibson, I. (1973). Genetics of Tetrahymena. *In* "Biology of Tetrahymena," (A. M. Elliott, ed.), pp. 307–373. Dowden, Hutchinson & Ross, Stroudsburg, PA.

Allen, S. L., and Nanney, D. L. (1958). An analysis of nuclear differentiation in the selfers of *Tetrahymena*. *Am. Nat.* **92**, 139–160.

Arslanyolu, M., and Doerder, F. P. (2000). Genetic and environmental factors affecting mating type frequency in natural isolates of *Tetrahymena thermophila*. *J. Euk. Microbiol.* **47**, 412–418.

Borden, D., Miller, E. T., Whitt, G. S., and Nanney, D. L. (1977). Electrophoretic analysis of evolutionary relationships in *Tetrahymena*. *Evolution* **31**, 91–102.

Chantangsi, C., Lynn, D. H., Brandl, M. T., Cole, J. C., Hetrick, N., and Ikonomi, P. (2007). Barcoding cilitaes: a comprehensive study of 75 isolates of genus Tetrahymena. *Int. J. Syst. Evol. Microbiol.* **57**, 2412–2425.

Chantangsi, C., and Lynn, D. H. (2008). Phylogenetic relationships within the genus *Tetrahymena* inferred from the cytochrome c oxidase subunit 1 and the small subunit ribosomal RNA genes. *Mol. Phyl. Evol.* **49**, 979–987.

Corliss, J. O. (1973). History, taxonomy, ecology and evolution of species of Tetrahymena. *In* "Biology of Tetrahymena," (A. M. Elliott, ed.), pp. 1–55. Dowden, Hutchinson & Ross, Stroudsberg, PA.

Doerder, F. P., Gates, M. A., Eberhardt, F. P., and Arslanyolu, M. (1995). High frequency of sex and equal frequencies of mating types in natural populations of *Tetrahymena thermophila*. *Proc. Natl Acad. Sci. U.S.A.* **92**, 8715–8718.

Eisen, J. A., Coyne, R. S., Wu, M., Wu, D. Y., Thiagarajan, M., Wortman, J. R., Badger, J. H., Ren, Q. H., Amedeo, P., Jones, K. M., Tallon, L. J., Delcher, A. L., Salzberg, S. L., Silva, J. C., Haas, B. J., Majoros, W. H., Farzad, M., Carlton, J. M., Smith, R. K., Garg, J., Pearlman, R. E., Karrer, K. M., Sun, L., Manning, G., Elde, N. C., Turkewitz, A. P., Asai, D. J., Wilkes, D. E., Wang, Y. F., Cai, H., Collins, K., Stewart, A., Lee, S. R., Wilamowska, K., Weinberg, Z., Ruzzo, W. L., Wloga, D., Gaertig, J., Frankel, J., Tsao, C. C., Gorovsky, M. A., Keeling, P. J., Waller, R. F., Patron, N. J., Cherry, J. M., Stover, N. A., Krieger, C. J., del Toro, C., Ryder, H. F., Williamson, S. C., Barbeau, R. A., Hamilton, E. P., and Orias, E. (2006). Macronuclear genome sequence of the ciliate *Tetrahymena thermophila*, a model eukaryote. *PLOS Biol.* **4**, 1620–1642.

Elliott, A. M. (1970). The distribution of *Tetrahymena pyriformis*. *J. Protozool.* **17**, 162–168.

Elliott, A. M. (1973). *Biology of Tetrahymena.* Dowden, Hutchinson & Ross, Inc, Stroudsburg, PA.

Elliott, A. M., and Gruchy, D. F. (1952). The occurrence of mating types in *Tetrahymena*. *Biol. Bull.* **103**, 301.

Finlay, B. J., and Fenchel, T. (1999). Divergent perspectives on protist species richness. *Protist* **150**, 229–233.

Foissner, W. (2006). Biogeography and dispersal of micro-organisms: a review emphasizing protists. *Acta Protozool.* **45**, 111–136.

Foissner, W., Chao, A., and Katz, L. A. (2008). Diversity and geographic distribution of ciliates (Protista: Ciliophora). *Biodiv. Conserv.* **17**, 345–363.

Furgason, W. H. (1940). The significant cytostomal pattern of the Glaucoma-Colpidium group, and a proposed new genus and species, *Tetrahymena geleii*. *Arch. fur Protistenkunde* **94**, 224–266.

Hersha, D. K., Willams, L. R., and Bouchard, V. (2009). Collection strategies for quantifying protist assemblages in temperate headwater streams. *Hydrobiologia* **630**, 115–126.

Hill, D. L. (1972). *The Biochemistry and Physiology of Tetrahymena.* Academic Press, New York.

Kaney, A. R., and Speare, V. J. (1983). An amicronucleate mutant of *Tetrahymena thermophila*. *Exp. Cell. Res.* **143**, 161–167.

Katz, L. A., Bornstein, J. G., Lasek-Nesselquist, E., and Muse, S. V. (2004). Dramatic diversity of ciliate histone H4 genes revealed by comparisons of patterns of substitutions and paralog divergences among eukaryotes. *Mol. Biol. Evol.* **21**, 555–562.

Katz, L. A., Snoeyenbos-West, O. L. O., and Doerder, F. P. (2006). Unusual patterns of molecular evolution at the *SerH* surface antigen locus in *Tetrahymena thermophila*: implication for estimates of effective population. *Mol. Biol. Evol.* **23**, 608–614.

Kher, C. P., Doerder, F. P., Cooper, J., Ikonomi, P., Achilles-Day, U., Kupper, F. C., and Lynn, D. H. (2011). Barcoding *Tetrahymena*: discriminating species and identifying unknowns using the cytochrome c oxidase subunit 1 (*cox-1*) barcode. *Protist* **162**, 2–13.

Leibowitz, M. P., and Zilberg, D. (2009). *Tetrahymena* sp. infection in guppies, *Poecilia reticulata* Peters: parasite characterization and pathology of infected fish. *J. Fish Dis.* **32**, 845–855.

Leick, V., and Lindemose, S. (2007). Chemokinesis by *Tetrahymena* in response to bacterial oligopeptides. *J. Eukaryot. Microbiol.* **54**, 271–274.

Lwoff, A. (1923). Sur la nutrition des infusoires. *C. R. Hebd. Seances Acad. Sci.* **176**, 928–930.

Lynch, T. J., Brickner, J., Nakano, K. J., and Orias, E. (1995). Genetic map of randomly amplified DNA polymorphisms closely linked to the mating type locus of *Tetrahymena thermophila*. *Genetics* **141**, 1315–1325.

Meyer, E. B., and Nanney, D. L. (1987). Isozymes in the ciliated protozoan Tetrahymena. *In* "Isozymes in the Biological and Medical Sciences," (M. C. Rattazzi, J. G. Scandalios, G. S. Whit, eds.), pp. 61–101. A. R. Liss, New York.

Nanney, D. L. (1999). When is a rose? The kinds of *Tetrahymena*. *In* "Species: New Interdisciplinary Essays," (R. A. Wilson, ed.), pp. 93–118. MIT Press, Cambridge, MA.

Nanney, D. L. (2004). No trival pursuit. *BioScience* **54**, 720–721.

Nanney, D. L., and Caughey, P. A. (1953). Mating type determination in *Tetrahymena pyriformis*. *Proc. Natl Acad. Sci. U.S.A.* **39**, 1057–1063.

Nanney, D. L., Caughey, P. A., and Tefankjian, A. (1955). The genetic control of mating type potentialities in *Tetrahymena pyriformis*. *Genetics* **40**, 668–680.

Nanney, D. L., and McCoy, P. A. (1976). Characteristics of the species of the *Tetrahymena pyriformis* complex. *Trans. Am. Microscop. Soc.* **95**, 664–682.

Nanney, D. L., Park, C., Preparata, R., and Simon, E. M. (1998). Comparison of sequence differences in a variable 23S rRNA domain among sets of cryptic species of ciliated protozoa. *J. Eukaryot. Microbiol.* **45**, 91–100.

Nei, M., Xu, P., and Glazko, G. (2001). Estimation of divergence times from multiprotein sequences for a few mammalian species and several distantly related organisms. *Proc. Natl Acad. Sci.* **98**, 2497–2502.

Ng, S. F. (1986). The somatic function of the micronucleus of ciliated protozoa. *In* "Progress in Protislolology," (J. O. Corliss, and D. J. Patterson, eds.), pp. 215–286. Biopress Ltd, Bristol.

Nyberg, D. (1981). Three new "biological" species of *Tetrahymena* (*T. hegewischi* n. sp., *T. sonneborni* n. sp., *T. nipissingi* n. sp.) and temperature tolerance of members of the "pyriformis" complex. *J. Protozool.* **28**, 65–69.

O'hUigin, C., and Li, W. -H. (1992). The molecular clock ticks regularly in muroid rodents and hamsters. *J. Mol. Evol.* **35**, 377–384.

Paixão, T., Phadke, S. S., Azevedo, R. B. R., and Zufall, R. A. (2011). Sex ratio evolution under probabilistic sex determination. *Evolution* **65**, 2050–2060.

Phillips, R. B. (1967). Inheritance of T serotype in *Tetrahymena*. *Genetics* **56**, 667–681.

Sadler, L. A., and Brunk, C. F. (1992). Phylogenetic relationships and unusual diversity in histone H4 proteins within the *Tetrahymena pyriformis* complex. *Mol. Biol. Evolution* **9**, 70–84.

Schlegel, M., and Meisterfeld, R. (2003). The species problem in protozoa revisited. *E. J. Protistol.* **39**, 349–355.

Simon, E. M., Meyer, E. B., and Preparata, R. M. (1985). New wild *Tetrahymena* from southeast Asia, China and North America, including *T. malaccensis*, *T. asiatica*, *T. nanneyi*, *T. caudata*, and *T. silvana* n. spp. *J. Protozool.* **32**, 183–189.

Simon, E. M., Nanney, D. L., and Doerder, F. P. (2008). The "Tetrahymena pyriformis" complex of cryptic species. *Biodiversity Conserv.* **17**, 365–380.

Sogin, M. L., Ingold, A., Karlok, M., Nielsen, H., and Engberg, J. (1986). Phylogenetic evidence for the acquisition of ribosomal RNA introns subsequent to the divergence of some of the major *Tetrahymena* groups. *EMBO J.* **5**, 3625–3630.

Sonneborn, T. M. (1937). Sex, sex inheritance and sex determination in *Paramecium aurelia*. *Proc. Natl Acad. Sci. U.S.A.* **23**, 378–385.

Sonneborn, T. M. (1957). Breeding systems, reproductive methods and species problems in protozoa. *In* "The Species Problem," (E. Mayr, ed.), pp. 155–324. AAAS, Washington, DC.

Swofford, D. L. (1991). *PAUP: Phylogenetic Analysis Using Parsimony Version 3.1.* Illinois Natural History Survey, Champaign, IL.

Van Bell, C. T. (1985). 5S and 5.8S Ribosomal RNA Evolution in the Suborder Tetrahymenina (Ciliophora: Hymenostomatida). *J. Mol. Evol.* **22**, 231–236.

CHAPTER 10

Tetrahymena thermophila Genetics: Concepts and Applications

Eduardo Orias

Department of Molecular, Cellular and Developmental Biology, University of California, Santa Barbara, Santa Barbara, CA 93106, USA

0091-679X/10 $35.00
DOI 10.1016/B978-0-12-385967-9.00010-4

Abstract

The differentiation of germline and somatic genomes in *Tetrahymena thermophila* results in two independent systems of genetic transmission. One is the conserved, sexual Mendelian genetics system of the germline genome. The other is a random genetic assortment mechanism, which operates in the somatic genome during asexual propagation. This chapter describes both systems, their interplay, and how they are exploited to construct useful biological reagents and powerful tools, which can be used to answer a variety of experimental questions.

I. Introduction

As a typical ciliate, *Tetrahymena thermophila* separates germline and soma by way of a diploid germline nucleus, the micronucleus (MIC), and a somatic nucleus, the macronucleus (MAC). It also exhibits the haploid/diploid alternation of germline ploidy typical of eukaryotes, implemented by conserved meiosis and fertilization during sexual reorganization (conjugation). Thus this organism normally displays Mendelian transmission genetics patterns identical to those of multicellular eukaryotes, such as animals and plants. But *Tetrahymena* genetic capabilities are combinatorially enriched by another phenomenon, akin to the genetics of bacterial plasmids: the somatic genome, derived from a mitotic copy of the germline genome, is highly polyploid and divides by random distribution of somatic chromosome copies. This introduces an independent, asexual dimension of transmission genetics in *Tetrahymena*.

A series of discoveries and applications, combined with the latest molecular approaches, have generated a set of remarkably versatile genetic tools for experimental analysis. This chapter covers diverse and useful purposes for which conventional (phenotype-based) genetics has been recruited in the lab to generate these tools. The fundamental concepts of *Tetrahymena* genetics will be covered first – not as a comprehensive review, but rather to provide the foundations for understanding the power and limitations of genetic experimental tools described in the rest of the article. These are grouped under basic genetic operations, genetic mapping and isolating mutants. Box 1 contains a glossary of special terms used in *Tetrahymena* genetics.

II. Fundamental Concepts of *Tetrahymena* Genetics

A. The Two Nuclear Genomes of *Tetrahymena*

The *Tetrahymena* life cycle and the separation of germline (MIC) and soma (MAC) have been reviewed (Orias *et al.*, 2011; Prescott, 1994). They are summarized below and in Fig. 1 because of their fundamental importance for understanding the basic genetic concepts. The MIC is the germline, that is, the store of genetic information for the sexual progeny. It is diploid, contains five pairs of chromosomes

Box 1 Glossary of special terms used in *Tetrahymena* genetics

Anlage (pl. anlagen): diploid products of the second postzygotic mitosis of the fertilization nucleus. Depending on their location in the cell, they differentiate into MACs (anterior) or MICs (posterior).

Assortment: the process by which a heterozygous MAC becomes genetically pure, for one allele or the other, as a consequence of the random distribution of MAC chromosome copies during asexual multiplication.

Coassortment: statistically significant tendency of parental alleles of two loci on the same MAC chromosome to assort together during vegetative multiplication.

Cytogamy: variant of conjugation in which exchange of gamete pronuclei is blocked, leading to self-fertilization. End result: MIC and MACs of each exconjugant are homozygous for the same allele at every locus.

Exconjugant: a cell that results from the separation of the two conjugants of a pair.

Genomic exclusion: variant of conjugation in which one conjugant is a "star" strain. Each exconjugant contains a MIC, which is a diploidized copy of the surviving meiotic product from the normal mate, retains its parental MAC, and remains sexually mature. End result: each MIC is homozygous for the entire genome.

Hemizygote: cell line haploid for a chromosome segment; the MIC contains only one copy of the genes on that segment. A MAC would be hemizygous when generated from a hemizygous MAC anlage.

Heterokaryon: cell line with different alleles at one or more loci in the MAC compared to the MIC.

Homokaryon: cell line with identical MIC and MAC genotypes (allowing for ploidy differences).

Karyonide: one of the two first cell division products of an exconjugant.

Nullisomic: cell line missing both copies of one or more chromosomes. In Tetrahymena, also used to refer to losses of both copies of a putative chromosome arm.

Pronuclear fusion failure: variant of conjugation in which migratory gamete pronuclei are exchanged but their subsequent fusion to stationary pronuclei is blocked. Each pronucleus gives rise to a MIC and a MAC. Because of the programmed destruction of one new MIC in each exconjugant, homokaryon and heterokaryon sister karyonides, with whole-genome homozygous nuclei, are produced.

Star strain: cell line with a defective MIC, incapable of generating meiotic products when it conjugates.

Uniparental cytogamy: variant of cytogamy in which one of the conjugants is a "star" strain. One exconjugant undergoes self-fertilization (as in cytogamy) while the other becomes anucleate and dies. End result: whole-genome homozygous MIC and MACs in the surviving exconjugant.

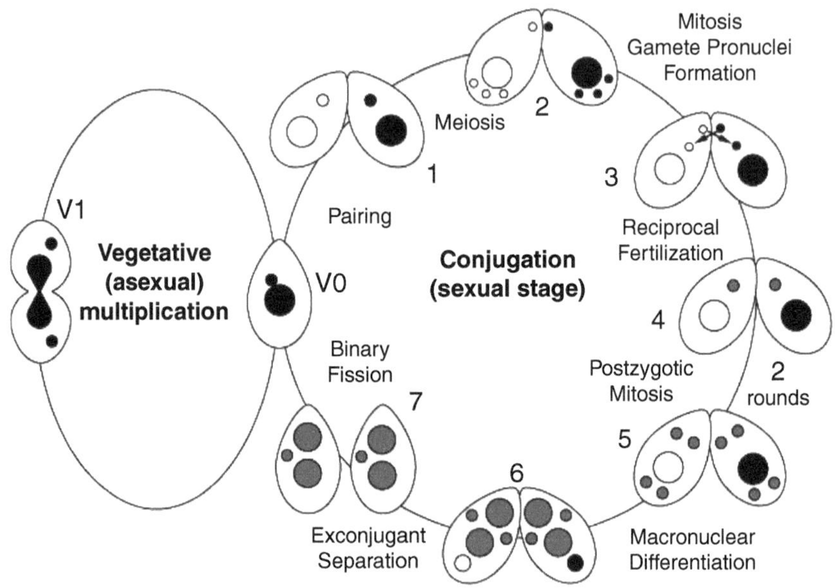

Fig. 1 The *Tetrahymena* life cycle. V0 and V1: vegetative cells. 1–7: conjugating cells. Stage 5: anterior and posterior pairs of nuclei are, respectively, the MAC and MIC anlagen. In each conjugant, the three posterior meiotic products (Stages 2–3), the parental MAC (Stages 5–7), and one of the new MICs (Stages 6–7) selectively undergo programmed nuclear death. Stage 7 is the terminal stage of conjugation in the absence of food. When food is available, each exconjugant generates two daughter cells called "karyonides." Each karyonide receives one of the new MACs and a mitotic daughter of the surviving new MIC. Soluble cytosolic macromolecules are exchanged between the conjugants of a pair. (Figure reproduced from Orias *et al.*, 2011).

and divides mitotically. No genes are known to be expressed in the MIC during vegetative multiplication. The MAC is the somatic nucleus, that is, the nucleus actively expressed during vegetative multiplication, and thus determines the cell's phenotype. The MAC is highly polyploid and divides amitotically, that is, by random distribution of chromosome copies. The MAC genome contains approximately 180 chromosomes (Hamilton, Dear and Orias, mss. in preparation) which, with the sole exception of the rDNA, are maintained at an average G1 copy number of ~45. No MAC DNA is known to be transmitted to the sexual progeny.

B. Normal Conjugation events

Tetrahymena cells conjugate in pairs. Major nuclear events of conjugation are meiosis, haploid gamete nucleus formation, fertilization, two postzygotic mitotic divisions to generate MIC and MAC anlagen, and nuclear differentiation (Fig. 1). The first division products of an exconjugant are the two "karyonides," each inheriting an independently differentiated new MAC. The nuclear events of conjugation have important genetic consequences: (1) as already indicated, the events generate a conventional Mendelian system of transmission of genetic information from parent to sexual offspring, as the diploid MIC undergoes meiosis and the diploid zygote genome (from which progeny MIC and MAC genomes are derived) is formed by the merger of two haploid genomes, one from each parent. Thus classical Mendelian ratios are obtained in crosses. (2) Genetic recombination occurs during meiosis, using conserved eukaryotic mechanisms, which allows the assignment of mutations and natural DNA polymorphisms to MIC linkage groups (see the *Tetrahymena* Genome Database webpage at http://www.ciliate.org; Brickner *et al.*, 1996; Wickert and Orias, 2000). (3) The MIC and MAC anlagen of a conjugant are mitotic descendants of the fertilization nucleus and thus start out genetically identical to one another; therefore, the karyonides are "homokaryons." (4) In a given pair, the two fertilization nuclei are genetically identical; therefore, the eight MIC and MAC anlagen are also identical, no matter how genetically different the parental cells were or what history of crossover events generated the surviving meiotic product in each conjugant. Individual pairs are the units counted for determining genetic ratios among the progeny of a cross. (5) During MAC differentiation, several types of site-specific developmentally programmed DNA rearrangements occur (Coyne *et al.*, 1996). These include chromosome fragmentation and the deletion of internally eliminated sequences (IES), introduced in Chapter 3.

Mitochondria are normally not exchanged between conjugants (Roberts and Orias, 1973). Thus traits determined by mitochondrial DNA differences normally exhibit cytoplasmic inheritance. In contrast, macromolecules and presumably small molecules are rather freely exchanged (MacDonald, 1966).

C. Variations of Normal Conjugation

Useful and well-characterized variants of normal conjugation occur spontaneously or can be induced at will: genomic exclusion, cytogamy (bi- or uniparental self-fertilization), and pronuclear fusion failure (described below). Some of these variant conjugation pathways generate whole-genome homozygotes, that is, cells that are *homozygous for their entire genome*, in either just the MIC (genomic exclusion) or in both MICs and MACs (cytogamy and fusion failure). Cytogamy generates homokaryons, while fusion failure generates both hetero- and homokaryons. Cytogamy allows the efficient isolation of laboratory-induced recessive mutants (see Section V). Pronuclear fusion failure is used for the generation of heterokaryon strains for one or more loci, whose MICs and MACs are both whole genome homozygotes (see Section III).

Genomic exclusion (Allen, 1967) is induced by conjugation with a "star" strain, characterized by having a grossly defective MIC, unable to produce functional meiotic products. Genomic exclusion refers to the inability of the defective MIC of the star strain to contribute any germline genomic DNA to sexual progeny. Instead both conjugants get a new diploid MIC, which is in effect derived from the diploidization of the single surviving haploid meiotic product of the normal mate and is thus homozygous for the entire genome. Subsequent conjugation events are blocked; these nuclei fail to undergo postzygotic mitotic divisions and to differentiate new MACs; instead, the old MAC is retained. The cell line derived from star-strain exconjugant (but not the one from its mate) is unable to maintain a normal MIC during asexual multiplication and soon reverts to the star state (Weindruch and Doerder, 1975). The generation of heterokaryons with a homozygous MIC is the most common application of genomic exclusion, often done after phenotypic assortment (described later) has generated a genetically pure *MAC*.

Because round I exconjugants retain their old MACs, they are sexually mature and keep their original (different) mating types. If the conjugating culture is not refed, the exconjugants immediately undergo a second round of conjugation (round II of genomic exclusion). Since all recently generated round I exconjugants now have a fully diploid MIC, round II is a normal round of conjugation. In the special case that the two exconjugants from *the same round I pair* undergo round II, homokaryons with both MIC and MAC homozygous for the entire genome result.

Additional ways to obtain whole-genome homozygotes involve experimentally induced blocks of gamete pronuclear exchange or fusion, which can occur spontaneously but at low frequency. *Cytogamy* (Orias and Hamilton, 1979) is a self-fertilization that occurs when the exchange of gamete pronuclei is blocked (Fig. 1, Stage 3), for example, by hyperosmotic shock or a pulse of microtubule inhibitors. The two sister gamete pronuclei of each conjugant fuse to one another and generate a diploid fertilization nucleus, which is homozygous for the entire haploid genome of the functional meiotic product. Postzygotic nuclear divisions and differentiation occurs normally, and whole-genome homozygous homokaryons are produced. The two exconjugants will generally be genetically different from one another. *Uniparental cytogamy* combines a cross to a "star" strain with the induction of cytogamy during the first round of genomic exclusion. In contrast to genomic exclusion, the nonstar conjugant does differentiate new macronuclei in round I. The exconjugant from the normal cell is a whole-genome homozygote, while the exconjugant from the star cell dies for lack of nuclei.

Pronuclear fusion failure occurs when gamete pronuclei are exchanged but their fusion is blocked (Fig. 1, Stages 3–4), for example, by a carefully timed pulse of microtubule inhibitors (Hamilton *et al.*, 1988). In each conjugant, the two unfused gamete pronuclei diploidize. Each pronucleus divides once, giving rise to one MIC and one MAC anlagen. In the end, half of the resulting karyonides are heterokaryons and half are homokaryons.

D. Genetics of the Macronucleus

1. Phenotypic Assortment

When cells with a MAC initially heterozygous at a given locus undergo asexual multiplication, subclones that irreversibly express phenotypes associated with either homozygote are generated. This phenomenon was discovered (Allen and Nanney,

1958) in the context of mating-type selfers (see below). Because recessive alleles can come to expression in this way, the phenomenon has been called phenotypic assortment. Starting with a heterozygous cell with a mixed MAC, the steady-state rate at which subclones that are pure for either allele arise is ~0.011/fission (Allen and Nanney, 1958; Doerder *et al.*, 1992). Assortment is attributed to the random distribution of allelic copies in a compound MAC (Allen and Nanney, 1958; Orias and Flacks, 1975). Mathematically, the steady-state rate of assortment of pure MACs from mixed MACs approaches $1/(2N-1)$ per fission for a large N, where N is the number of copies in the G1 MAC (Schensted, 1958). This measure of assortment rate led to the first determination of ~45 as the average G1 ploidy of the MAC. The ploidy of bulk MAC chromosomes was subsequently confirmed by molecular measurements of MIC and MAC DNA amounts and sequence complexity (see discussion in Orias and Flacks, 1975). Phenotypic assortment allows a recessive allele in a heterozygote to come to full expression in some of its vegetative descendants. It also generates descendants in which a single mutant allele, introduced by mutation or by DNA-mediated transformation, can completely replace all ~45 wild-type alleles in the MAC.

2. Coassortment

When two loci lie in the same MAC chromosome, parental alleles in a double heterozygote tend to assort together in the MAC during asexual multiplication; this phenomenon is referred to as coassortment (Longcor *et al.*, 1996). To illustrate, if a MAC start starts with a mixture of AB and *ab* MAC chromosomes and assortment is allowed, most of the vegetative descendants will become pure for either the AB or the ab combinations (i.e., the two parental combinations). Less than 10% of the descendants become pure for either of the recombinant types (Ab or *aB*). The percentage of recombinant assortants does not reliably measure *distance* along the MAC chromosome. Coassortment groups consist of loci whose members coassort with one another. MAC chromosomes are the physical basis of coassortment groups (Wong *et al.*, 2000). Thus, a coassortment group is the MAC genetic analog of a MIC meiotic linkage group. If two loci are on *different* MAC chromosomes, then they assort independently: "terminal assortants" become pure for all four possible combinations of the two alleles at the two loci in comparable frequencies.

Rarely, two loci on the same MAC chromosome assort independently or nearly so (Deak and Doerder, 1998; Hamilton and Orias unpubl. obs.); a hot spot of MAC recombination is suspected to be the cause. Therefore, finding independent assortment means most of the time, but not always, that two loci are on different MAC chromosomes. On the other hand, no exceptions have been found to the rule that two loci that coassort are on the same MAC chromosome.

E. The Mating-Type System and Mating-Type Determination

The basic *Tetrahymena* mating-type phenomena were discovered by David L. Nanney, later joined by Sally L. Allen and other collaborators in the 1950 s and 1960 s and were reviewed in Orias (1981). In order to become sexually reactive and mate, cells must be starved, have different mating types and be "old enough" to mate. Adolescents can mate with fully mature cells but not with other adolescents. Adolescence and full sexual maturity are, respectively, reached approximately after 50–80 and 95–110 fissions after karyonides are generated (Rogers and Karrer, 1985).

Seven mating types were originally described in *T. thermophila* and no additional ones have since been found among several thousand independent isolates from natural populations (Doerder *et al.*, 1995). The MIC encodes the potential for five to seven mating types, depending on the genotype at the mating type (*mat*) locus complex. Nevertheless, a cell generally expresses only one mating type. For example, cells of inbred strain B are homozygous for the *mat-2* allele. Their mating type can be one of the following: II, III, <u>IV</u>, V, VI, or <u>VII</u>. In contrast, inbred strain C3 cells are homozygous for the mat-1 allele; their mating type can be <u>I</u>, II, III, V, or VI. (The mating types diagnostic of each allele have been underlined in each list.). *mat-1/mat-2* heterozygotes (e.g., B/C3 inbred strain hybrids) can express any one of the seven mating types.

The choice of which one of the mating types allowed by MIC genotype is actually expressed is the result of a somatically inherited, *irreversible*, stochastic event that occurs during the differentiation of the new macronucleus. The mating type of a sexual progeny is not correlated with the mating type of its parents or that of the other three karyonides from the same pair. The frequencies with which various mating types arise are not necessarily equal, and they are significantly affected by environmental conditions prevailing at the time of MAC differentiation, for example, temperature and nutritional state (Orias and Baum, 1984). Very recent observations (Cervantes *et al.* unpubl. obs.) show that mating-type determination is accompanied by DNA rearrangement that accounts for its irreversibility. This has recently become an active area under investigation.

Nearly 50% of newly differentiated MACs contain a mixture of two or more mating type determinants but, as a result of phenotypic assortment, most MACs have become pure for a single mating type by the time sexual maturity is reached. Sexually mature cells with mixed MACs give rise to clones whose members can mate with one another when starved; they are called "selfers." Upon continued asexual multiplication, selfers assort subclones with MACs pure for each one of two mating types.

III. Basic Genetic Operations

A fundamental element of genetic analysis is making a cross and deriving conclusions from the fate of parental genetic differences among the progeny. A cross is also important to generate useful strains for experimental biology. As indicated earlier, *T. thermophila* cells lines can be directly crossed if they differ in mating type. If the mating types are unknown, they should be tested as described further below. Once putative progeny of the cross are obtained, it is necessary to ascertain that they have indeed undergone sexual reorganization and to test their relevant phenotypes and genotypes. Sometimes, it is important to derive progeny cell lines that have assorted to a useful phenotype. This section describes important considerations relevant to various genetic procedures, including making a cross, phenotypically characterizing the progeny, making useful genetic constructs and genetically mapping useful DNA features, such as mutations, MIC-limited sequences and DNA polymorphisms.

A. Making a Cross

Synchronous mating is generally useful and is especially important for certain experiments, for example, timing events that occur during conjugation. An acceptable degree of synchrony is readily achieved by first inducing sexual reactivity and then mixing the sexually reactive cells. We set up crosses as follows in our lab. We grow 15 mL overnight cultures of two strains that are sexually mature and of different mating types. Cells are grown in growth medium in plastic Petri plates to a density between 2E5 (2×10^5) and 5E5 cells/mL. After cell density is measured, the growth medium is replaced with starvation medium (Dryl's medium or 10 mM Tris buffer) by two cycles of centrifugation ($600 \times g$) and pellet resuspension in starvation medium. It is important to do this step as quickly as possible, as cells deteriorate while sitting in the pellet (high density and low surface to volume ratio for gas exchange). They also deteriorate if centrifuged too fast. The volume of the final resuspension is adjusted to give a final cell density of ~2.5E5 cells/mL. The cultures, in standard Petri plates, are then returned to 30 °C and incubated overnight; this allows them to become sexually reactive ("initiated"; Bruns and Brussard, 1974a). (Three hours of starvation are sufficient for initiation; overnight incubation is just for scheduling convenience.)

To start the cross, the density of each starved culture is measured; if necessary it is diluted with starvation medium to achieve a final concentration of 2E5 cells per mL. The cultures are then mixed. (If every cell were to pair, the resulting density would be 1E5 pairs/mL.) After about 1 h of "costimulation" (Bruns and Palestine, 1975), pairs begin forming and conjugants remain paired for ~12 h at 30 °C. (Conjugants that abort nuclear differentiation and retain parental MACs often separate much earlier.) Exconjugants must be refed in order to undergo the first cell division, which generates the karyonides.

B. Establishing the Progeny of a Cross

This section describes how to initiate progeny cell lines from conjugating pairs. The next section describes how to test the progeny and detect "false" progeny that failed to undergo MAC differentiation.

Isolating the progeny of a cross can proceed in different ways, depending on the purposes of the experiment:

1. To keep track of progeny from individual conjugating pairs, pairs are isolated into drops of nutrient medium in a Petri plate arrayed in 96-well half-plate format; three days later they are replica-plated to nutrient medium in 96-well plates for further testing.

2. Conjugating pairs can be mixed with nutrient medium and then distributed into 96-well plates at a desired number of pairs per well. They are then allowed to grow for three days. This is done when one need not keep track of every pair and can subsequently eliminate parental cells by selecting for the progeny by drug resistance phenotype. Distribution to many wells maximizes the diversity of progeny obtained.
3. The mating culture can be kept in the Petri plate, refed by the addition of nutrient medium and allowed to grow for one more day. This is done when one simply wants to select progeny of a particular phenotype and does not care about characterizing the rest of the progeny.
   Please note that in Cases 2 and 3, cells that paired are still mixed with parental cells that failed to mate; the latter can start growing much sooner after nutrient medium is added and, absent selection against them, can easily become the majority of the cells in the refed culture.
4. Occasionally it is important to isolate and compare the genotype and phenotype of individual exconjugants or karyonides of a pair. Usually this is done when exconjugants or karyonides may be genetically different from one another because of an induced variation of conjugation or because alternative genome rearrangements arise during the independent differentiation of the new MACs in a pair. In this case pairs are isolated as in case (1) above, and it becomes critical to isolate the exconjugants before they undergo their first division, which seldom occurs much before 24 h after mixing at 30 °C. When the separated exconjugants are observed under a dissecting microscope, they are individually transferred to new, separate drops of nutrient medium. When it is necessary to isolate karyonides, the exconjugants are observed every 2 h until they divide. At this point, the two cells are individually transferred to new, separate drops of nutrient medium. Because frequent observation under the microscope is needed, it is important to keep the plates in moisture chambers as much as possible so the drops do not dry up. Three days later, when the drop cultures have grown up, they are treated as in (1) above. It is important at this time to also examine the drops transiently inhabited by the pair – and the exconjugants when isolating karyonides. If they are not empty, that indicates a mistake and the descendants of that pair are discarded. (Generally those cases turn out to be unwanted anyway, as they represent mistakes or pairs where the exconjugants retained their parental MAC and started dividing quickly.)
5. Highly parallel crosses can be set up when one wishes to cross many cell lines to the same strain. The crosses are set up in bacterized media exactly as described for the highly parallel mating-type tests (see below), where the strain shared in every cross is treated as a mating-type tester. The addition of penicillin and streptomycin sulfate to the refeeding medium kills any bacteria remaining in the starving conjugating cultures.
6. Situations occasionally arise where one needs to cross two particular cell lines to one another but both have the same mating type. This problem can be circumvented by setting up a three-way cross; the two strains to be crossed are mixed

with a star strain, which has a different mating type, each at ~2E5 cells per mL, in a 1:1:2 ratio, respectively. They are then allowed to undergo two rounds of genomic exclusion (see Section II) by not refeeding them until 2 days later. The genetic results will be identical in every respect as if the two parental cell lines had expressed different mating types and had been crossed to one another.

C. Sorting and Testing the Progeny of a Cross

Not every conjugating pair generates viable exconjugants; those deaths may go undetected unless single pairs are tracked. Furthermore, even when apparent progeny are obtained, it is critical to determine that they indeed completed sexual reorganization and differentiated a new macronucleus. The reason is that conjugating pairs have an "escape" response, such that they can abort postzygotic mitotic divisions and nuclear differentiation; these exconjugants retain the old macronuclei. The trigger is not well understood but may include sensing something abnormal at some step between meiosis and gamete pronucleus exchange (Scholnick and Bruns, 1982). Therefore, some of the resulting exconjugants may have retained their original phenotype and remain sexually mature, as in exconjugants of round I of genomic exclusion. These were originally called "nonconjugants" and later "MAC retainers," which is a less ambiguous name. The original tests to distinguish true progeny with newly differentiated MACs from MAC-retainers relied on screening either for a nonparental phenotype or for sexual immaturity. Progeny that have differentiated new macronuclei are immature if they have undergone less than ~40 fissions and are unable to mate. The immaturity test is conducted by mixing, as early as feasible, starved progeny cell lines with a starved tester culture expressing a nonparental mating type and then looking for pairing. In the near future, it may become possible to use PCR amplification-based tests to immediately screen progeny, even if sexually immature, for possession of nonparental mating-type determinants.

With the advent of heterokaryons (Bruns and Brussard, 1974b), it became common practice to cross cell lines that are homozygous heterokaryons for different dominant drug resistance markers. Unlike parental cells and "MAC retainers," which are sensitive, true progeny are resistant to both drugs and can easily be selected. (The technical difference between screening and selecting, as used in genetic work, is that the former involves individually testing every progeny for the wanted trait; the latter involves killing every unwanted progeny, so that only the wanted ones survive, which is much more efficient.)

D. Passaging Progeny to Sexual Maturity

Once true progeny have been identified, they are ready to be phenotypically characterized. If they will be used for subsequent crosses, it is necessary to serially passage them until they reach sexual maturity – usually for at least 80–100 fissions. Serial passaging is done by replica plating cells to 96-well plates with axenic nutrient medium. A convenient week-day passaging schedule is Monday, Tuesday, Thursday, and Friday. If each replica-plating dilutes the culture ~100-fold, four passages represents a total of ~25 fissions per week. An extra passage can be added during the weekend. (Cultures could be passaged faster, but keeping cells under constant exponential growth leads to telomere lengthening and eventually to lowered growth rate (Larson *et al.*, 1987) and selection for short-telomere somatic mutants (Ahmed *et al.*, 1998). Because sexually mature cell lines, even if derived from a single karyonide, can include cells with different mating types, it is important to make a final subcloning prior to testing for mating type.

While *Tetrahymena* cell lines are somatically immortal, spontaneous mutations, and cytogenetic defects tend to accumulate in the MIC, which lead to a progressive (sometimes catastrophic) loss of *sexual* fertility (Nanney, 1974). The absence of gene expression in the MIC precludes removal of affected cells by natural selection. To protect the genetic integrity of the MIC, it is therefore essential, as soon as practicable, to freeze (Cassidy-Hanley *et al.*, 1995) and maintain cell lines under liquid nitrogen.

E. Mating-Type Testing

It is important to know that two cell lines to be crossed have different mating types. In general, the mating type of useful cell lines that have reached maturity and have been subcloned is determined as follows. Samples from a starved culture of the unknown are separately mixed with samples of starved cells of all the mating types (the "mating-type testers"). Pairing usually becomes clearly visible by about 2 h at 30 °C, and pairs will remain abundant for at least the next 6 h. The unknown should mate with every tester strain but one – the strain with its mating type. Having approximately equal numbers of cells in each mixture makes the results easiest to score, but is not essential. In scoring the results, seeing at least one good pair is sufficient to score the test as positive, but it is necessary to make sure that the two cells in the pair move and swim as a mechanical unit (true pair) rather than as independent single cells engaging in casual contact.

An essential control is to mix samples of all the testers in every possible pair-wise combination, including with self. Pairs should be seen every combination, except the self-mixtures. If this is not observed, something is wrong with the tester strains or with this mating-type test. With cultures of seven mating types to dispense, utmost care is required to make sure that mating-type cultures are not inadvertently exchanged while setting up the test.

A few percent of the cultures of recently matured progeny show pairs with every mating-type tester. In this case, pairs are also observed if an unmixed control of the culture was set up at the same time. These are selfers. The most common basis for selfing is that the recently matured cell line has not yet assorted for mating-type purity (see concepts section). Selfer cell lines are not normally used in a controlled cross; if necessary they are further assorted until pure for a mating type.

Highly parallel (96-plex) mating-type tests can be set up when having to type many cultures, for example, the sexually mature progeny of a cross. In this case, the cultures to be mating type tested are grown in nutrient medium in 96-well plates, and then replica plated to wells containing 50–100 μL of 2% BP medium. Seven replica plates, one for each mating type to be tested, are made for each plate of unknowns. At the same time, cultures of the seven mating type-testers are used to inoculate seven flasks containing 2% bacterized peptone (BP) medium (Phillips, 1967), with a 100-fold dilution of stock cultures of the mating-type testers. All the cultures are incubated 2 days at 30 °C. During this period, the *Tetrahymena* cells will multiply at the expense of the bacteria; when the latter are exhausted, the former automatically starve and become sexually reactive. In the morning of the second day, an equal volume of the corresponding starved testers is added to the wells of the seven replica plates, and incubation at 30 °C is continued. The same controls are included as described above. Cell pairing is comfortably scored in the afternoon.

F. Isolating and Testing Assortants

Assortants are cells whose MAC was initially heterozygous but have become pure for one allele or the other by assortment during successive cell divisions (see concepts section). Several situations occur when it is important to isolate assortants, for example, to get heterokaryon strains whose MAC is pure for a drug-sensitive allele while the MIC is still heterozygous or to obtain MAC-transformed cells whose MAC has become pure for a transgene, such as a gene KO.

While assortment intrinsically is selectively neutral, one can bias the generation of fully assorted cell lines by selection: cells that assort toward the allele selected against are killed. Thus the culture becomes progressively enriched for cells predominantly containing the allele under selection. In some cases it is essential to be sure that assortment is complete, because having a few copies of the undesired allele may not show phenotypically but may distort experimental results. The more dominant the desired allele, the easier it is for copies of the recessive allele to go phenotypically undetected. The only way to *genetically* detect for impurity of assortment is to look for back-assortment to the alternative phenotype. When making drug sensitive assortants, once a fully assorted candidate is obtained, it is customary to make 48 single-cell isolations, allow them to grow (and back assort to resistance, if any resistant alleles remain) and then test their phenotype again. If only fully sensitive cultures are seen, the cell line is declared to have fully assorted. In cases where the alleles have a molecular phenotype, it is possible to test for complete assortment by Southern blot hybridization. In this case, molecular signal from both alleles in the still heterozygous MIC may confuse the answer as to whether the MAC is fully assorted. Tests based on gene expression are immune to this confusion because there is no expression from the MIC in vegetatively growing cells.

G. Test-Crossing Progeny Cell Lines

Because of dominance, phenotypic tests of young cultures are generally insufficient to directly determine their genotype, and testcrosses are required after the cells have become sexually mature. A general, simple, and convenient way to testcross a cell line is to cross it either to a nullisomic strain (see below) missing the DNA segment that includes the locus (Lynch *et al.*, 1995) or to a star strain for two rounds of genomic exclusion. Progeny expressing either allele are produced, which are directly detected phenotypically. Unlike other diploid organisms, a testcross involving multiple loci does not require multiple homozygous recessive strains because the star cross generates both homozygotes at every locus.

H. Making a Whole-Genome Homozygote

Whole genome homozygotes are cell lines where both the diploid MIC and the polyploid MAC are homozygous and pure, respectively, for the same allele at every locus in the genome. Whole-genome homozygotes (inbred cell lines) were originally obtained in *Tetrahymena* by the slow procedure of doing the equivalent of about 20 serial brother–sister crosses (Allen and Gibson, 1973). Since then, ways were discovered to reliably make whole-genome homozygotes in a single step: genomic exclusion, cytogamy, and uniparental cytogamy (see Section II). One-step generation of whole-genome homozygotes has proven extremely valuable for generating recessive mutants, after mutagenesis, nearly as efficiently as if the germline were haploid. It also has myriad uses in generating genetically marked, experimentally useful cell lines.

I. Making Homozygous Heterokaryons

Heterokaryons are cell lines in which the MIC and MAC have different genotypes. Assortment generates heterokaryons in initially heterozygous cell lines. But it is also possible to make homozygous heterokaryons, in which the MIC is homozygous for one allele, while the MAC is homozygous for the alternative allele at the same locus. In these strains, homozygous lethal mutations can be maintained in the transcriptionally silent germline, because they are "covered" by MACs pure for the wild-type allele. Types of lethal mutations that are so maintained include point mutations, knock-outs of essential genes, and deletions of various lengths, including one or more whole MIC chromosomes. Strains that have lost both copies of a MIC chromosome are called nullisomic. (In *Tetrahymena*, cells that have lost both MIC copies of a putative chromosome *arm* are also called nullisomics.) Nullisomic and deletion homozygous strains are invaluable for mapping recessive mutations to MIC chromosomes, to chromosome arms, or to deletion intervals defined by a set of smaller deletions.

Homozygous heterokaryons for a single locus are usually made in three steps. (1) A heterozygous cell line is generated. (2) The heterozygous cell line is passaged until descendants fully assorted for the desired MAC allele are cloned. (3) The assortant is crossed to a "star" strain (one round) and the exconjugants are test-crossed; those having the desired MIC genotype and MAC phenotype and the mating type of the original assortant are saved. (It is important to avoid saving an exconjugant clone derived from the star strain because its MIC will quickly deteriorate and make the strain useless for further crosses; see concepts section.) Adaptation of this method to make a hereditary gene KO using DNA-mediated transformation is explained in Chapter 11.

Heterokaryons having a MIC and MAC that are whole-genome homozygous for one or more genetic differences can be made in a single step by pronuclear fusion failure (PFF). PFF is induced by a pulse treatment of the microtubule inhibitor vinblastine around the time of pronuclear exchange (Mayo and Orias, 1981). Karyonides have to be isolated and test-crossed to screen for those that have the desired genotype. The time window for pronuclear fusion is narrow: earlier treatment blocks pronuclear exchange and leads to self-fertilization while later treatment inhibits postzygotic mitoses. Because of the narrow window and because only a fraction of the PFF progeny will have the desired genotype, the frequency of whole genome heterokaryons is low. But one attempt is usually sufficient to find at least one cell line with the desired genotype.

Whole-genome homozygous B-C3 heterokaryons (Hamilton, unpublished observation; available at the *Tetrahymena* Stock Center) are a remarkable pair of genetic constructs. One cell line has a MIC which is whole-genome homozygous inbred strain C3 and a MAC, which is whole genome homozygous for inbred strain B. The other has the reciprocal genotype. MACs and MICs of each of these two strains have alternative alleles at hundreds of thousands of distinct genomic sites. These strains are useful for timing certain rearrangements in the new MAC during conjugation, by assaying for the appearance of a polymorphic allele present in the parental MICs but absent in the parental MACs.

IV. Genetic Mapping

Tetrahymena is the only ciliate for which germline linkage maps, based on the nearly century old approach of exploiting germline meiotic recombination, have been constructed. This approach provides quantitative genetic distance information, which is used to construct genetic linkage maps. In addition, in *Tetrahymena*, genome-wide deletion mapping can be used to order DNA polymorphisms with respect to one another, by a single set of parallel crosses. This provides a very efficient approach to mapping a DNA polymorphism to a unique segment of the germline genome, defined by the two nearest mapped deletion ends. Finally, as described earlier, the random distribution of somatic chromosome copies during asexual multiplication results in genetic assortment. This phenomenon has been exploited to answer the question of whether two mutations are, or most likely are not, located in the same MAC chromosome. An orientation to the various types of genetic mapping available for *Tetrahymena* is given below. Relevant genetic and genomic resources are listed in Orias *et al*. (2011, Supplementary Materials). In addition, a first draft of the MIC genome sequence assembly has recently been released by the Broad Institute (http://www.broadinstitute.org/annotation/genome/Tetrahymena.1/MultiHome.html).

A. Germline Deletion Mapping

Deletion mapping is a process by which a DNA feature (mutation, DNA polymorphism, or MIC-limited DNA segment) can be genetically mapped to a segment of a MIC chromosome, defined by deletion ends, using a panel of deletion homozygotes. These are strains in which both copies of a DNA segment have been deleted from the MIC. Available deletions range in size up to entire MIC chromosomes (nullisomics), which were the first deletion type to be used for mapping in *Tetrahymena* (Bruns *et al*., 1983). Once obtained, MIC deletion homozygotes are maintained as heterokaryons without any special effort in *Tetrahymena* because their expressed MAC contains the complete genome.

Since both copies of a particular DNA segment are missing from the MIC, deletion mapping has a very simple conceptual basis. When the MIC can be queried without interfering signal from the MAC DNA (e.g., a MIC-limited sequence), the presence or absence of the DNA feature can be detected directly in the deletion homozygotes. The presence or absence of any polymorphism allele can be detected after transmission to an F1 MAC, which becomes hemizygous for the corresponding DNA segment. The answer to a deletion mapping test is an unambiguous "yes" or "no": a DNA feature is or is not within the segment deleted in the deletion homozygote. On the other hand, unlike linkage mapping, deletion mapping cannot measure distance between loci. Thus, resolution is limited by the density of the available deletion ends represented in the panel. A *cre-lox*-based method for inducing relatively small MAC deletions in *Tetrahymena* has been developed (Busch *et al*., 2010). Attempts to use the *cre-lox* method to custom-induce deletions in the MIC are underway (Cassidy-Hanley, personal communication).

Deletion mapping of a recessive mutation is carried out by crossing the homozygous mutant in parallel to every strain of the homozygous deletion panel. For any given deletion cross, F1 progeny show the mutant phenotype only when they are hemizygous for the mutation because it lies within the deleted MIC chromosome segment; otherwise heterozygous progeny with wild-type phenotype are obtained. The locus can then be uniquely assigned to a chromosome arm or smaller deletion interval.

To facilitate mapping, most deletion strains are obtained as heterokaryons for drug resistance, generally cycloheximide: they are homozygous in the MIC for the resistance allele, while the MAC is pure for the sensitive allele. True (cross-fertilized) progeny can be easily selected for: the drug resistance derived from the nullisomic strain kills any nonmating parental cells, any cells that mated but retained their parental MAC, and cytogamous progeny derived from the mutant strain, while deletion homozygosity kills cytogamous progeny of the nullisomic strain. It is important that the mutant strain not be expressing the drug resistance encoded in the MIC of the nullisomic strains; otherwise, true progeny cannot be selected. One additional advantage of deletion mapping is that only hemizygous progeny are obtained when the mutation is in the deleted region, so that in mass crosses the informative, mutant progeny cannot be outgrown by wild-type cells before being phenotypically tested, even if homozygosity (or hemizygosity) for the mutation is deleterious and lowers the growth rate.

Deletion mapping of a dominant mutation is less straightforward, because the progeny of all deletion strains, whether heterozygous or hemizygous, show the mutant phenotype. But the two types can be distinguished based on assortment to the wild-type phenotype: heterozygous progeny can assort while hemizygotes cannot, thus showing that the mutation lies in the segment missing in the deletion strain. Because the steady-state rate of assortment to pure MACs is about 1% per fission, assortants can usually be readily detected well before heterozygotes have undergone 100 fissions. If the dominant mutation confers growth disadvantage, wild-type assortants will show up sooner.

Other DNA features can be deletion-mapped. MIC limited segments, even if not polymorphic, can be mapped to deletion intervals. In this case, whole cell DNAs from the homozygous deletion panels are PCR-amplified using primers designed to amplify only from MIC DNA (Cassidy-Hanley *et al.*, 1994). Molecularly identified DNA B-C3 polymorphisms (see below) can also be mapped to deletion intervals (Brickner *et al.*, 1996).

B. Germline Linkage Mapping by Meiotic Recombination Frequency

As in other eukaryotes, mutations are genetically mapped to MIC linkage groups in individual chromosomes by exploiting meiotic crossing-over. This conserved eukaryotic mechanism generates recombinant genotypes with a frequency which, in the linear portion of the range, is more or less proportional to physical distance. Two inbred strains (B and C3) have been used as a source of natural DNA polymorphisms. Major progress has been made in constructing a solid framework for a genetic map of the *Tetrahymena* MIC genome. The MAC genome of inbred strain C3 has recently been sequenced at low coverage, creating a rich source of potential polymorphisms. A variety of genetically mapped, genome-sequence-related DNA polymorphisms (RAPDs, tandem repeat polymorphisms, and "snip-SNPs," which are single nucleotide polymorphisms that create a restriction site polymorphism) are now available for most MAC chromosomes and thus for most of the germline genome (Hamilton and Orias, manuscript in preparation). Many gene mutations and DNA polymorphisms are listed in http://www.lifesci.ucsb.edu/~genome/*Tetrahymena*/GenomMaps.htm.

C. Somatic Genetic Mapping by Coassortment

Loci that coassort reside on the same MAC chromosome (see concepts section). Thus, a coassortment test allows one to experimentally answer the question of whether two loci (of any type) end up in the same MAC chromosome after MAC differentiation. To test for coassortment, one starts with around 36 progeny that are double- or multiple heterozygotes. These cultures are propagated in parallel for many fissions so that assortment to pure genotypes has occurred at most loci. (300–500 fissions, which takes 12–20 weeks, have been used in the past; it is likely that fewer transfers are needed if one can test pools of wild type and mutant phenotypes (analogous to "bulked segregant analysis"; Michelmore *et al.*, 1991). The cultures are subcloned at the end of the assortment period, and the individual subclones (panel of "terminal assortants") are tested for their phenotype or DNA polymorphism genotype (Longcor *et al.*, 1996). Coassortment makes it possible to map a mutation to a MAC chromosome purely by genetic means, without knowledge of the molecular basis of the mutation or the phenotype. This is done indirectly by detecting its genetic coassortment with a physically mapped, sequence-related DNA polymorphism.

V. Isolating and Using Mutants: Identifying the Mutant Gene

Mutants are useful for answering experimental questions and contributing mechanistic information. This process starts with the time-honored approach of isolating a mutant with a phenotypic defect in some biological mechanism of one's choosing. As gene sequencing became possible, this approach was extended to identifying the mutant gene in order to better understand the biological and molecular basis of the phenotype; this is mechanistically the most useful way to exploit the value of mutants. The term "forward genetics" was then coined to contrast the process to "reverse genetics". The latter is a counterintuitive name for the process whereby an already sequenced gene is purposefully mutated in order to carry out a structure/function investigation of the mechanism by which it determines the wild-type phenotype. Unlike most uses of reverse genetics, forward genetics is a blind, totally preconception-free way to disrupt a biological mechanism and thus in principle it is capable of contributing totally unexpected novel information. The fields of molecular, cell, and developmental biology contain many examples of fundamental discoveries made by the application of forward genetics.

Description of the basic operations of forward genetics in *Tetrahymena* is divided into several sections. The first ones deal with isolating mutants with a phenotype of interest and sorting a mutant collection. The last section deals with the identification of a mutant gene of interest.

A. Mutant Induction and Isolation

Because spontaneous mutants are rare, it is usually necessary to use mutagenesis. The easiest mutants to isolate are those whose phenotype confers selective advantage relative to wild-type cells. Otherwise, it is necessary to screen progeny one-by-one for the mutant phenotype. A new mutation must be present in the MAC in order to be expressed and be detected by the different phenotype it confers. There are three interesting cases of mutant induction to consider: somatic only mutants, dominant germline mutants, and recessive germline mutants. The latter are the most common type of mutation sought or found and require the most stringent approach to find them, because a germline mutation must be induced in the MIC and immediate sexual reorganization is required to express the mutation in the MAC of the progeny. They will be described first, as the other types merely require simplifications of this approach.

1. Germline Recessive Mutations

The initial detection of a *recessive* mutant requires not only that the mutation be in the MAC, but that, without need for lengthy assortment, most if not all of the ~45 copies of the locus carry the mutant allele. This is best accomplished by inducing self-fertilization immediately after mutagenesis to generate a MAC homozygous for the recessive mutation. In this way, two identical copies of the *same* mutation are brought together in the fertilization nucleus and transmitted to the new MICs and new expressed MACs. Self-fertilization makes the mutant frequency proportional to the mutation frequency, as if the germline were haploid. (Under normal conditions of cross-fertilization, the probability that two mutations of independent origin affecting the same locus would come together in an F1 would be proportional to the square of the mutation frequency. This is an impractically small frequency because recessive mutants rarely confer selective advantage, so the mutagenized progeny have to be screened one-by-one.) The efficient induction of cytogamy (self-fertilization) was originally used to isolate recessive mutants (e.g., Sanford and Orias, 1981). Since then uniparental cytogamy (UPC) has become the method of choice for two reasons: unlike cytogamy, the loss of the MAC kills the unwanted (nonmutagenized) exconjugant and the frequency of self-fertilization among recovered progeny is 100% (Cole and Bruns, 1992). An additional, serendipitous advantage is that UPC progeny reach sexual maturity earlier than progeny from normal crosses. Consequently UPC is particularly well adapted for the isolation of mutants with recessive conjugation defects (Cole *et al.*, 1997).

Isolation of recessive mutants uses the following steps, described in the above-cited references:

1. *Mutagenize a cell line*: This cell line would normally be a heterokaryon, with a homozygous drug-resistant MIC and a drug sensitive MAC, to allow selection for the self-fertilized progeny. The mutant frequency is directly related to the strength of the mutagenic treatment. The optimal treatment is a trade-off: strong enough to make it feasible to find rare mutants but not so strong that it induces too many spurious mutations, which can cause infertility and/or compound the difficulty of identifying the mutation responsible for the mutant phenotype.
2. Immediately transfer the mutagenized cells to starvation medium.
3. Mix with starved cells of the star strain.

4. At the appropriate time induce self-fertilization by osmotic shock.
5. Distribute the conjugating cells to nutrient medium in 96-well plates before the exconjugants undergo their first division. Distribution ensures the independent origin of mutant isolates.
6. Select the progeny for drug resistance
7. Screen the progeny for the desired mutant phenotype.

In addition to the germline mutations, the mutagenic treatment will probably generate somatic (macronuclear) mutations. However, these will not be recovered because the mutagenized MACs are destroyed during the sexual reorganization that brings the germline mutations to expression among the drug-selected progeny.

2. Dominant Germline Mutations

These are simpler to get. Although sexual reorganization is required to bring the germline mutation to expression, a single mutant copy in the fertilization nucleus is sufficient to generate heterozygous progeny expressing the dominant phenotype (Roberts and Morse, 1980). Thus, mutant frequency will be proportional to the mutation frequency. While the stringent protocol above will allow the detection of dominant germline mutants as well, wild-type cells can be used instead of the star strain at step 3, making it is unnecessary to induce self-fertilization (step 4) – and even counterproductive if the dominant mutation is homozygous-lethal.

Recessive mutations *with selective advantage* (e.g., recessive drug resistance) can also be isolated by this protocol, provided that a period of assortment is allowed before selection – to increase in some descendants the fraction of mutant allele copies above the threshold required for expression of the mutant phenotype.

3. Somatic Mutations

A mutation in the MAC can directly be expressed. Indeed a sexual reorganization step after mutagenesis should be *avoided* because the parental MAC having the induced mutation will be destroyed during conjugation. A special feature of somatic mutant isolation is that, in all likelihood, only a single copy of the wild-type locus (out of the ~45) will have been mutated. Therefore, it is necessary to allow a period of cell multiplication so that assortment generates descendants possessing enough copies of the mutant allele for the cell to express the mutant phenotype. The more recessive the mutation, the more fissions should be allowed before testing or attempting to select for the mutant phenotype. Please note that somatic mutations cannot be bred into other cell lines, as the parental MAC is destroyed during conjugation and contributes no DNA to sexual progeny.

To isolate somatic mutations, the protocol above can be greatly simplified, as follows (e.g., Orias and Newby, 1975). (1) Mutagenize any cell line. (2) Wash away the mutagen and immediately dilute the mutagenized cells with nutrient medium and distribute to 96-well plates, to ensure the isolation of mutants of independent origin.

(3) Allow a period of asexual multiplication to allow assortment to the mutant phenotype. (4) Screen or select for the mutants being sought.

B. Purifying a Mutation

Mutagenesis is dirty. A mutant with a desired mutation will also carry additional mutations – and become homozygous for them when the recessive germline mutation protocol is used. Some of these spurious mutations can have deleterious effects on the growth rate and/or cause infertility in subsequent crosses. Occasionally the observed mutant phenotype will actually be dependent on synergy between two independent mutations. Therefore, it becomes important to "purify" a new mutation, that is, to segregate spurious mutations away from it and to ascertain that it is solely responsible for the mutant phenotype.

This process is accomplished by repeated cycles of the following pair of crosses:

1. a backcross to a wild-type cell line to make heterozygous F1 and (2) crosses of sexually mature F1 s to one another to regenerate homozygotes for the mutation (1/4 of the progeny). It becomes a matter of judgment and experimental need how many cycles of backcrossing to carry out. The fertility of the mutant homozygotes should increase with just a few backcrossing cycles and provides one objective end-point to what otherwise becomes an open-ended process. The backcrosses can be used to advantage by introducing drug resistance alleles that will facilitate the recognition of true progeny in downstream crosses and genetic analysis.

C. Sorting out a Mutant Collection

Any wild-type phenotypic trait requires the synergistic function of protein products of many different genes. Therefore, in a large collection of mutants with changes in that phenotype, many mutant genes are generally represented – some by multiple independent mutants. To sort the different mutants into functional groups, with each group likely representing a different gene, complementation tests are usually done. Pair-wise crosses of the mutants are made in order to determine whether the F1 progeny have the mutant or wild phenotype. The results are tentatively interpreted according to the following rationale. If two *recessive* mutations are in different genes, the simplest expectation is that the F1 will have wild-type phenotype; the wild-type genes complement one another, as the double heterozygote has one wild-type allele at each locus. If the double heterozygote has the mutant phenotype, it is concluded that the two mutations are in the same gene. (For a *Tetrahymena* example, see Frankel *et al.*, 1976.)

It is useful to sort out the mutant collection as quickly as possible, so that subsequent experimental work can be focused on one representative of each complementation group. However, the primary mutant isolates may show low fertility when crossed, which manifests itself as a high fraction of dead progeny or MAC retainers. If the double heterozygote has the wild-type phenotype, clearly that is a nonparental phenotype – a positive result indicative of complementation. If only mutant progeny are seen in the cross, they may be MAC retainers rather than true progeny, and thus be a "no test." Therefore, it is important to make sure that one is looking at the phenotype of cross-fertilized progeny before accepting a negative result as real. If necessary, the tests can be repeated after backcrossing to wild type has increased the fertility of the mutants and has introduced drug resistance alleles that can be used to select cross-fertilized progeny.

Two approaches can, in principle, speed up the grouping of mutants. (1) The collection of mutants may be presorted by deletion mapping (Section IV). This can frequently be done directly with primary mutant isolates because, even if the cross has low fertility, the rare progeny can be selected. Clearly mutants that map to different deletion intervals must be in different genes. If the mutant collection is large, this presorting can significantly scale down, as a square function, the number of required pair-wise complementation tests. (2) If the phenotype can be scored in conjugating cells, the cytoplasmic exchange that occurs during conjugation (see concepts section) can lead to "instant complementation" if two mutants with different mutated loci conjugate; this happens because the mixed cytosol will transiently contain wild-type gene products from both loci. A good example is mutations that affect exocytosis (Satir *et al.*, 1986).

D. Identifying a Mutant Gene

Genetic mapping procedures alone can greatly narrow down the list of candidate genes, but very seldom lead to the identification of the target mutation – the one responsible for the mutant phenotype. One possible general solution considered for *Tetrahymena* was to construct a complementation library, in which the whole genome is covered by plasmid inserts, each of which is large enough to include a gene and its cis-acting control elements (R. S. Coyne, personal communication). While the basic technology is available, the main difficulty so far has been the high proportion of A+T in the *Tetrahymena* genome (Eisen *et al.*, 2006; Fass *et al.*, 2011), which precludes stable maintenance of large cloned inserts in *E. coli*, such as would be needed to ensure the inclusion of promoter regions. As the cost of DNA sequencing continues to decrease, the most promising alternative for directly identifying the target mutation is by deep whole-genome DNA sequencing. This method has been successful in other eukaryotic organisms which, like *Tetrahymena*, have medium size genomes (Sarin *et al.*, 2008).

A major problem inherent in identifying the target mutation by DNA sequencing are the many additional spurious mutations that mutagenesis generates. The great majority of those will not map near the target mutation and can thus be eliminated from consideration by deletion mapping (see Section IV). If needed, an additional way to filter out spurious mutations exploits meiotic recombination to separate them from the target mutation. The mutant is backcrossed to a wild-type strain; mutant and wild-type meiotic segregants of the F1s, obtained as described (Lynch *et al.*, 1995), are pooled and sequenced. This method ("bulked segregant analysis"; Michelmore *et al.*, 1991) eliminates all but the most tightly linked spurious mutations. When successfully implemented, these methods for mutant gene identification will add the awesome power of unbiased forward genetics to the other powerful experimental tools (Turkewitz *et al.*, 2002) employed for studies of molecular, cell, and developmental biology utilizing this model organism.

Acknowledgments

I gratefully acknowledge Drs. Eileen Hamilton's and Kathleen Collins suggestions, which improved the manuscript, and the U.S. National Science Foundation for my current research support (grant MCB-1025069).

References

Ahmed, S., Sheng, H., Niu, L., and Henderson, E. (1998). Tetrahymena mutants with short telomeres. *Genetics* **150**, 643–650.

Allen, S. L. (1967). Genomic exclusion: a rapid means for inducing homozygous diploid lines in *Tetrahymena pyriformis*, syngen 1. *Science* **155**, 575–577.

Allen, S. L., and Gibson, I. (1973). Genetics of Tetrahymena. *In* "Biology of Tetrahymena," (A. M. Elliott, ed.), pp. 307–373. Dowden, Hutchinson and Ross, Stroudsburg, PA.

Allen, S. L., and Nanney, D. L. (1958). An analysis of nuclear differentiation in the selfers of Tetrahymena. *Am. Nat.* **92**, 139–160.

Brickner, J. H., Lynch, T. J., Zeilinger, D., and Orias, E. (1996). Identification, mapping and linkage analysis of randomly amplified DNA polymorphisms in *Tetrahymena thermophila*. *Genetics* **143**, 811–821.

Bruns, P. J., and Brussard, T. B. (1974a). Pair formation in Tetrahymena pyriformis, an inducible developmental system. *J. Expt. Zool.* **188**, 337–344.

Bruns, P. J., and Brussard, T. B. (1974b). Positive selection for mating with functional heterokaryons in Tetrahymena pyriformis. *Genetics* **78**, 831–841.

Bruns, P. J., Brussard, T. B., and Merriam, E. V. (1983). Nullisomic Tetrahymena: II. a set of nullisomics define the germinal chromosomes. *Genetics* **104**, 257–270.

Bruns, P. J., and Palestine, R. F. (1975). Costimulation in *Tetrahymena pyriformis*: a developmental interaction between specially prepared cells. *Dev. Biol.* **42**, 75–83.

Busch, C. J., Vogt, A., and Mochizuki, K. (2010). Establishment of a Cre/loxP recombination system for N-terminal epitope tagging of genes in Tetrahymena. *BMC Microbiol.* **10**, 191–202.

Cassidy-Hanley, D., Smith, H. R., and Bruns, P. J. (1995). A simple, efficient technique for freezing *Tetrahymena thermophila*. *J. Eukaryot. Microbiol.* **42**, 510–515.

Cassidy-Hanley, D., Yao, M. C., and Bruns, P. J. (1994). A method for mapping germ line sequences in *Tetrahymena thermophila* using the polymerase chain reaction. *Genetics* **137**, 95–106.

Cole, E. S., and Bruns, P. J. (1992). Uniparental cytogamy: a novel method for bringing micronuclear mutations of Tetrahymena into homozygous macronuclear expression with precocious sexual maturity. *Genetics* **132**, 1017–1031.

Cole, E. S., Cassidy-Hanley, D., Hemish, J., Tuan, J., and Bruns, P. J. (1997). A mutational analysis of conjugation in *Tetrahymena thermophila*: 1. Phenotypes affecting early development: meiosis to nuclear selection. *Dev. Biol.* **189**, 215–232.

Coyne, R. S., Chalker, D. L., and Yao, M. C. (1996). Genome downsizing during ciliate development: nuclear division of labor through chromosome restructuring. *Annu. Rev. Genet.* **30**, 557–578.

Deak, J. C., and Doerder, F. P. (1998). High frequency intragenic recombination during macronuclear development in *Tetrahymena thermophila* restores the wild-type SerH1 gene. *Genetics* **148**, 1109–1115.

Doerder, F. P., Deak, J. C., and Lief, J. H. (1992). Rate of phenotypic assortment in *Tetrahymena thermophila*. *Dev. Genet.* **13**, 126–132.

Doerder, F. P., Gates, M. A., Eberhardt, F. P., and Arslanyolu, M. (1995). High frequency of sex and equal frequencies of mating types in natural populations of the ciliate *Tetrahymena thermophila*. *Proc. Natl Acad. Sci. U.S.A.* **92**, 8715–8718.

Eisen, J. A., Coyne, R. S., Wu, M., Wu, D., Thiagarajan, M., Wortman, J. R., Badger, J. H., Ren, Q., Amedeo, P., Jones, K. M., Tallon, L. J., Delcher, A. L., Salzberg, S. L., Silva, J. C., Haas, B. J., Majoros, W. H., Farzad, M., Carlton, J. M., Smith, R. K., Garg, J., Pearlman, R. E., Karrer, K. M., Sun, L., Manning, G., Elde, N. C., Turkewitz, A. P., Asai, D. J., Wilkes, D. E., Wang, Y., Cai, H., Collins, K., Stewart, B. A., Lee, S. R., Wilamowska, K., Weinberg, Z., Ruzzo, W. L., Wloga, D., Gaertig, J., Frankel, J., Tsao, C. C., Gorovsky, M. A., Keeling, P. J., Waller, R. F., Patron, N. J., Cherry, J. M., Stover, N. A., Krieger, C. J., Del Toro, C., Ryder, H. F., Williamson, S. C., Barbeau, R. A., Hamilton, E. P., and Orias, E. (2006). Macronuclear genome sequence of the ciliate *Tetrahymena thermophila*, a model eukaryote. *PLoS Biol.* **4**, e286.

Fass, J. N., Joshi, N. A., Couvillion, M. T., Bowen, J., Gorovsky, M. A., Hamilton, E. P., Orias, E., Hong, K., Coyne, R. S., Eisen, J. A., Chalker, D. L., Lin, D., and Collins, K. (2011). Genome-scale analysis of programmed DNA elimination sites in *Tetrahymena thermophila*. *Genes, Genomes, Genetics* **1**, 515–522.

Frankel, J., Jenkins, L. M., Doerder, F. P., and Nelsen, E. M. (1976). Mutations affecting cell division in *Tetrahymena pyriformis*: I. Selection and genetic analysis. *Genetics* **83**, 489–506.

Hamilton, E. P., Suhr-Jessen, P. B., and Orias, E. (1988). Pronuclear fusion failure: an alternate conjugational pathway in *Tetrahymena thermophila*, induced by vinblastine. *Genetics* **118**, 627–636.

Larson, D. D., Spangler, E. A., and Blackburn, E. H. (1987). Dynamics of telomere length variation in *Tetrahymena thermophila*. *Cell* **50**, 477–483.

Longcor, M. A., Wickert, S. A., Chau, M. F., and Orias, E. (1996). Coassortment of genetic loci during macronuclear division in *Tetrahymena thermophila*. *Eur. J. Protistol.* **32**(Suppl. 1), 85–89.

Lynch, T. J., Brickner, J., Nakano, K. J., and Orias, E. (1995). Genetic map of randomly amplified DNA polymorphisms closely linked to the mating type locus of *Tetrahymena thermophila*. *Genetics* **141**, 1315–1325.

MacDonald, B. (1966). The exchange of RNA and protein during conjugation in Tetrahymena. *J. Protozool.* **13**, 277–285.

Mayo, K. A., and Orias, E. (1981). Further evidence for lack of gene expression in the Tetrahymena micronucleus. *Genetics* **98**, 747–762.

Michelmore, R. W., Paran, I., and Kesseli, R. V. (1991). Identification of markers linked to disease-resistance genes by bulked segregant analysis: a rapid method to detect markers in specific genomic regions by using segregating populations. *Proc. Natl Acad. Sci. U.S.A.* **88**, 9828–9832.

Nanney, D. L. (1974). Aging and long-term temporal regulation in ciliated protozoa. A critical review. *Mech. Ageing Dev.* **3**, 81–105.

Orias, E. (1981). Probable somatic DNA rearrangements in mating type determination in *Tetrahymena thermophila*: a review and a model. *Dev. Genet.* **2**, 185–202.

Orias, E., and Baum, M. P. (1984). Mating type differentiation in *Tetrahymena thermophila*: strong influence of delayed refeeding of conjugating pairs. *Dev. Genet.* **4**, 145–158.

Orias, E., Cervantes, M. D., and Hamilton, E. P. (2011). *Tetrahymena thermophila*: a unicellular eukaryote with separate germline and somatic genomes. *Res. Microbiol.* **162**, 578–586.

Orias, E., and Flacks, M. (1975). Macronuclear genetics of Tetrahymena: I. Random distribution of macronuclear gene copies. *Genetics* **79**, 187–206.

Orias, E., and Hamilton, E. P. (1979). Cytogamy: an inducible alternate pathway of conjugation in *Tetrahymena thermophila*. *Genetics* **91**, 657–671.

Orias, E., and Newby, C. J. (1975). Macronuclear genetic of Tetrahymena: II. Macronuclear location of somatic mutations in Tetrahymena pyriformis syngen 1. *Genetics* **80**, 251–262.

Phillips, R. B. (1967). Inheritance of T serotypes in Tetrahymena. *Genetics* **56**, 667–681.

Prescott, D. M. (1994). The DNA of ciliated protozoa. *Microbiol. Rev.* **58**, 233–267.

Roberts Jr., C. T., and Morse, D. E. (1980). Galactokinase-deficient mutants of Tetrahymena: selection and characterization. *Mol. Gen. Genet.* **180**, 129–134.

Roberts Jr., C. T., and Orias, E. (1973). Cytoplasmic inheritance of chloramphenicol resistance in Tetrahymena. *Genetics* **73**, 259–272.

Rogers, M. B., and Karrer, K. M. (1985). Adolescence in *Tetrahymena thermophila*. *Proc. Natl Acad. Sci. U.S.A.* **82**, 436–439.

Sanford, Y. M., and Orias, E. (1981). Phenylketonuric Tetrahymena: phenylalanine hydroxylase mutants and other tyrosine auxotrophs. *Proc. Natl Acad. Sci. U.S.A.* **78**, 7614–7618.

Sarin, S., Prabhu, S., O'Meara, M. M., Pe'er, I., and Hobert, O. (2008). Caenorhabditis elegans mutant allele identification by whole-genome sequencing. *Nat. Methods* **5**, 865–867.

Satir, B. H., Reichman, M., and Orias, E. (1986). Conjugation-rescue of an exocytosis-competent membrane microdomain in *Tetrahymena thermophila* mutants. *Proc. Natl. Acad. Sci. U.S.A.* **83**, 8221–8225.

Schensted, I. V. (1958). Appendix: model of subnuclear segregation in the macronucleus of ciliates. *Am. Nat.* **92**, 161–170.

Scholnick, S. B., and Bruns, P. J. (1982). A genetic analysis of tetrahymena that have aborted normal development. *Genetics* **102**, 29–38.

Turkewitz, A. P., Orias, E., and Kapler, G. (2002). Functional genomics: the coming of age for *Tetrahymena thermophila*. *Trends Genetics* **18**, 35–40.

Weindruch, R. H., and Doerder, F. P. (1975). Age-dependent micronuclear deterioration in *Tetrahymena pyriformis*, syngen 1. *Mech. Ageing Dev.* **4**, 263–279.

Wickert, S., and Orias, E. (2000). Tetrahymena micronuclear genome mapping: a high resolution map of chromosome 1L. *Genetics* **154**, 1141–1153.

Wong, L., Klionsky, L., Wickert, S., Merriam, V., Orias, E., and Hamilton, E. P. (2000). Autonomously replicating macronuclear DNA pieces are the physical basis of coassortment groups in *Tetrahymena thermophila*. *Genetics* **155**, 1119–1125.

CHAPTER 11

Transformation and Strain Engineering of *Tetrahymena*

Douglas L. Chalker

Department of Biology, Washington University in St. Louis, St. Louis, MO, USA

Abstract

Transformation of *Tetrahymena* by microinjection of DNA was established 25 years ago. This rather labor-intensive technique has since been shelved, replaced by less time consuming and more efficient methods, electroporation and biolistics. Conjugative electroporation is the method of choice for introducing autonomously replicating, rDNA-based vectors into *Tetrahymena*. These are maintained as high-copy linear mini-chromosomes. Versatile expression cassettes in these vectors facilitate expression of most genes. Transformation efficiencies are sufficiently high to permit screens using expression libraries. Biolistic transformation is primarily used to introduce DNA for integration into the genome by homologous recombination. This technique has greatly enhanced strain engineering of *Tetrahymena* through facilitating the disruption of genes

0091-679X/10 $35.00
DOI 10.1016/B978-0-12-385967-9.00011-6

(creating targeted knockout cell lines) or epitope-tagging coding regions, allowing researchers to take full advantage of the sequenced genome. The presence of both germline and somatic nuclei in these cells requires different strategies to target DNA to the desired compartment. This presents challenges, including the need to engineer the polygenic macronuclear genome, which has nearly 50 copies of each gene. However, separate manipulation of functionally distinct genomes provides experimental opportunities, especially for the analysis of essential genes, by modifying the silent micronucleus then subsequently examining phenotypes in the next sexual generation. The flexibility to engineer strains as needed makes *Tetrahymena* a facile system with which to answer many biological questions.

I. Introduction

Tetrahymena emerged as an experimental organism largely due to its interesting biology. However, its utility for exploring many diverse scientific questions is greatly facilitated by the availability of many tools for molecular genetic analysis and strain engineering. Creation of gene knockouts or expression of tagged proteins is relatively simple due to the ability to transform *Tetrahymena*. Expression vectors have been developed to facilitate these studies. Homologous recombination allows any region of the genome to be targeted for manipulation. The relative ease of creating useful engineered strains for phenotypic analyses has been critical to promoting *Tetrahymena* as a major model system for investigation of cellular processes.

Transformation in *Tetrahymena* was first achieved by microinjection of purified rDNA (Tondravi and Yao, 1986), and this technique was later used to introduce autonomously replicating vectors (Godiska and Yao, 1990; Yao and Yao, 1989; Yu and Blackburn, 1989) and to direct homologous recombination (Yao and Yao, 1991; Yu *et al.*, 1988). Microinjection is rarely used today as it requires expensive micromanipulators and is a very time-consuming technique, both to learn and execute; thus, it will not be discussed further. For those wanting to employ this technique, a very detailed description is available (Chalker *et al.*, 1999). The development of transformation by electroporation (Gaertig and Gorovsky, 1992; Gaertig *et al.*, 1994a) and particle bombardment (or biolistics) (Cassidy-Hanley *et al.*, 1997) have replaced microinjection, both due to their ease and increased efficiency relative to microinjection. Electroporation is primarily used for introduction of rDNA-based replicating vectors while biolistics is the method of choice for molecular genetics approaches, such as creating gene knockouts, requiring homologous recombination. At the end of this chapter, we describe the execution of these methods in some detail. Before that, we will discuss various strategies for which transformation of a DNA construct is utilized to produce a desired experimental result.

II. Strategies for Manipulating the Genome of *Tetrahymena* Strains

Of course, when working with *Tetrahymena*, one cannot escape considering its nuclear dimorphism. The macronucleus is the site of all gene expression, whereas the micronucleus contains the only genome that will be passed on to the next generation. Consequently, the genome that needs to contain an expression construct or be lacking a gene of interest for knockout studies is the one housed in the somatic macronucleus. However, if a strain has a knockout allele or expression construct (integrated or on a replicating vector) present only in the macronucleus, it will be lost upon conjugation. Furthermore, if one is examining processes that occur during postzygotic differentiation, all expression will come from the micronucleus-derived developing macronucleus. Thus, for some experimental questions, one must engineer the micronucleus.

Macronuclear transformation is easier to achieve. For one, available autonomously replicating vectors are maintained solely in the macronucleus. When engineering strains by recombining a DNA construct into the genome, the macronucleus is the easy target. It may be that the macronucleus more readily uptakes introduced DNA as micro- and macronuclei have nuclear pores with different selectivity (Iwamoto *et al.*, 2009; Malone *et al.*, 2008). The nuclear envelope does not break down during division (Chapter 7), so DNA must be transported into the nucleus. Even if DNA enters the micronucleus, the frequency of homologous recombination outside of meiosis has not been investigated. Furthermore, when DNA is introduced into cells, selection for transformation requires that the DNA be in the macronucleus for expression of the marker gene. Thus, for both technical and likely biological reasons, getting engineered DNA into the macronucleus is a much simpler task.

Germline (micronuclear) transformation requires that the cells be transformed during prophase of meiosis (Cassidy-Hanley *et al.*, 1997). However, transformation of *Tetrahymena* during this developmental window does not ensure incorporation of the introduced DNA into the micronucleus. The macronucleus is still a much bigger target, and most transformants recovered only have this somatic nucleus modified. Again, the possible biological reasons for this are numerous. To increase the success rate of micronuclear transformation, one must start with highly fertile strains and be efficient with the biolistic procedure as mating cells readily abort conjugation without making new macronuclei if they are roughly handled during early conjugation (more on this below). The use of heterokaryon parental cell lines (Chapter 10) for the transformation, which have a drug resistant allele (usually 6-methylpurine resistance) in the micronucleus, but a sensitive allele in the macronucleus, allows for selection of progeny among the recovered transformants. Cells that completed the conjugation program (i.e., sexual progeny) are the best candidates to have had their germline modified.

Germline transformants that emerge from an initial transformation are heterozygous at the locus targeted. This is because recombination into the genome occurs prior to formation of the zygotic genome and typically into only one of the two mating partners. Thus, the zygotic genome will get one engineered genome from one mating partner and a wild-type allele from the other. To make homozygous mutants, one can simply mate two heterozygous lines (if germline transformants of different mating types are obtained). Alternatively, one can cross the heterozygous transformants to "star" strains that have defective micronuclei (Allen, 1967) (Chapter 10). These genomic exclusion crosses produce strains with homozygous micronuclei (it is equally likely to get strains homozygous for the transformed allele or the wild-type allele). The exconjugants emerging from such crosses that are homozygous for engineered allele can be crossed, and their progeny will have exclusively that allele in both their micro- and macronuclei.

When manipulating the macronuclear genome, one major challenge is the fact that it is polyploid, containing roughly 50 copies of each chromosome (Chapter 3). This increases the time from the initial transformation to that when strains are ready for phenotypic analysis. Upon initial introduction and selection of an integrating construct into the macronucleus, it is typical to obtain strains that have about half of the wild-type copies replaced with the introduced DNA. This can be sufficient for expression of tagged constructs if one is not concerned about competition with the endogenous untagged protein. If one is aiming to knockout a gene, all copies must be eliminated to obtain the loss-of-function phenotype. This can be achieved, taking advantage of the random segregation of macronuclear chromosomes (Chapter 3), by growing transformants in increasingly higher concentrations of the selection drug together with periodic subcloning of individual cells (Chapters 8 and 10). It is widely accepted for evidence that a gene is essential for rapid growth if, after several rounds of subcloning and growth with selection, one cannot obtain cells lacking all copies of the wild-type allele. This can be shown more convincingly by disrupting the gene from the micronucleus, followed by mating these germline transformants, which results in replacement of the existing macronucleus with one derived from the micronucleus. If homozygous mutant cells cannot be obtained, then the gene disrupted must be essential. This type of heterokaryon analysis will be discussed more below.

III. Generating Gene Knockouts

The ability for *Tetrahymena* cells to efficiently incorporate introduced DNA into the genome by homologous recombination enables targeted disruption of nearly any gene (or genes) of interest. This is commonly performed by replacing the coding sequence with a paromomycin resistance *neo* cassette consisting of a *Tetrahymena* promoter driving expression of the *neo* gene (described more below) (Gaertig *et al.*, 1994a; Mochizuki, 2008; Shang *et al.*, 2002). Typically, about 1 kbp of sequence from both upstream and downstream of the coding region to be targeted is cloned flanking the *neo* gene to make a knockout construct that is ready to introduce into cells (Fig. 1). Shorter regions of homology can be used, but can decrease

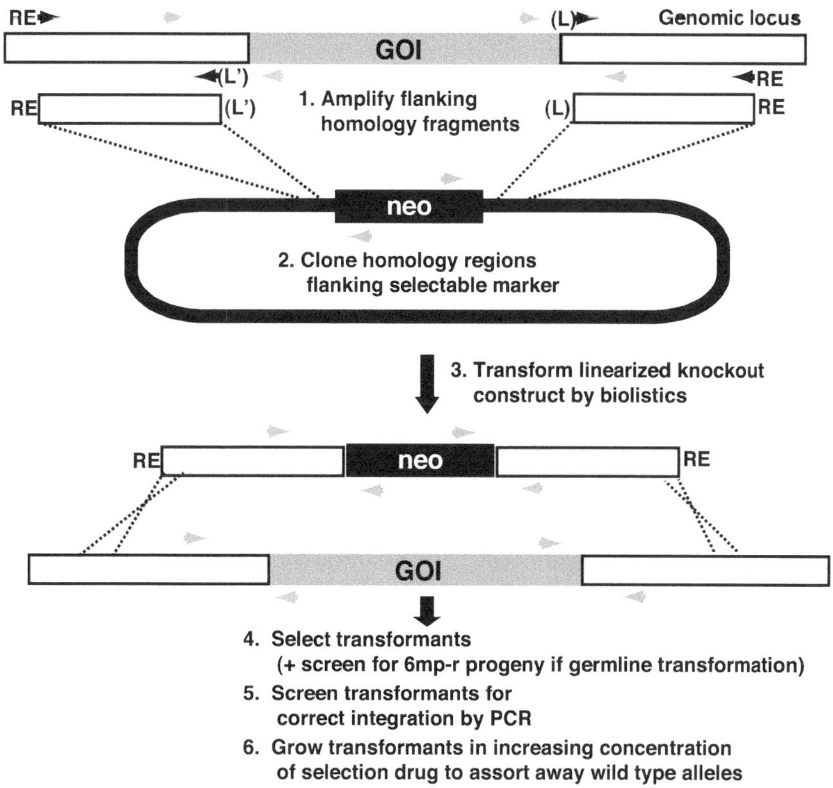

Fig. 1 Gene knockout strategy. The gray box depicts the coding region of a gene of interest (GOI) to be knocked out of the genome. Genomic sequences flanking the gene (open boxes) are amplified using oligonucleotide primers (black arrowheads); RE, unique restriction endonuclease recognition sites added to the gene distal primers; L and L′, optional complementary linkers sequences added to assist in cloning. These fragments are cloned on each side of a *neo* selectable marker (black box) in a suitable vector. The resulting gene knockout construct is linearized at the added RE sites, coated on gold particles, and introduced into cells. Subsequent steps to identify desired knockout lines are listed. Gray arrowheads, primers used to verify correct insertion of construct into the genome and to track assortment of wild-type and mutant alleles (see text).

transformation efficiency. We have achieved high-efficiency transformation when one homology arm is short (~400 bp) and the other is long (1.5–2 kbp). This may be an advantageous configuration for some genes given that intergenic sequences in *Tetrahymena* are routinely short (about 1 kb on average), and it is important to be sure that your targeting strategy does not affect expression of a neighboring gene. It is permissible to place the end of the targeting homology within an upstream or downstream gene, but in that case, one may want to clone and sequence the region of the nearby gene into which recombination initiates to ensure that no mutations were created.

One potential complication, if one plans to introduce the construct into micronuclear genome, may be the presence of an internal eliminated sequence (IESs) in the region of homology used to target the marker into the genome. Five to six thousand IESs are found in the germline genome, all of which are eliminated during macronuclear differentiation. Currently the locations of these are not noted on the TGD genome browser, but many have been mapped on a supplementary browser (http://bioshare.bioinformatics.ucdavis.edu/Data/k6pd8efvnv/all_windows.html) (Fass *et al.*, 2011). If an IES is present in the flanking region of a gene of interest, it will interrupt the genome at the site being targeted and could lower the efficiency of targeting the locus in the micronucleus. In such a case, it is probably best to avoid spanning that region in the design of the knockout construct, as the effect of an IES on targeting efficiency is unexplored.

Two basic cloning strategies are routinely used to make targeting constructs. One is to independently clone the upstream and downstream homology regions separately on each side of the *neo* selectable marker (Fig. 1). These genomic sequences are typically amplified by PCR with unique restriction sites added at the ends distal from the coding region of interest, which will be used to linearize the knockout construct prior to introducing it into cells to direct homologous recombination. While this is a routine molecular biology approach, the A+T rich genome of *Tetrahymena*, especially in the intergenic regions can be challenging to PCR amplify and clone. Careful design of the PCR primers can help to ensure successful generation of one's knockout construct. Another successful cloning strategy involves "stitching" together the upstream and downstream homology regions by sequential PCR. In this approach, a common linker sequence (L and L′ in Fig. 1) is added to the coding region proximal ends of the PCR primers such that the two products can anneal with one another. After the upstream and downstream regions are amplified in separate PCR reactions, the two products are purified and mixed together, and a combined product is amplified using the distal primers of the upstream and downstream regions. The stitched together PCR products are then cloned into a desired vector. The *neo* gene or other selectable marker can then be inserted into the linker sequence used to assist in the stitching reaction, which should be designed to contain restriction enzyme recognition sites that are present on each side of the desired selectable marker.

After creating a knockout construct and using it to transform cells, one should verify that the recombination occurred into the desired locus. A good first indication that homologous recombination occurred as planned is whether both the upstream and downstream borders between *Tetrahymena* DNA and the selectable marker can be detected in the engineered genome. Using genomic DNA from transformants as a template, simply perform PCR with one primer in the upstream flanking region paired with one in the selectable marker (gray arrowheads in Fig. 1). Use the same strategy to detect the downstream border. As the construct does contain other homology to the genome (e.g., the *Metallothionein 1* (*MTT1*) promoter driving *neo* expression) (Shang *et al.*, 2002), it is possible to get aberrant targeting (even occurring by homologous recombination. These PCR primers can also be used to follow assortment of a knockout allele to complete replacement of the wild-type allele by adding in a third primer that is in the region deleted in the knockout allele that will amplify the wild-type allele together with flanking primer used to amplify the border. In this three primer PCR, the products to amplify the knockout allele and wild type are in competition for the one common primer, allowing for semiquantitative assessment of the ratio of these two alleles in the population. Southern blot analysis is useful to finally confirm that a strain lacks all copies of the wild-type allele.

If a gene is essential and all wild-type copies cannot be assorted away by increasing the concentration of the selection drug, it is possible to make a conditional knockout allele. The most straightforward method to achieve this is to create a knock in construct in which the endogenous promoter of the gene is replaced with a cadmium-inducible promoter from one of the *MTT* genes. Once this allele is present in the genome, all copies of the wild-type allele can be assorted away in the presence of cadmium to induce expression of the gene of interest, then cadmium can be removed and the loss of function phenotype examined. This has been successful to study the *POT1* gene (Linger *et al.*, 2011). Alternatively, one can introduce an inducible, epitope-tagged allele at an ectopic locus into a previously generated gene knockout strain, and then assort away the remaining copies of the wild-type allele in the presence of the inducer (e.g., cadmium). While either strategy has great potential for the study of essential genes, we have found that the *MTT1* promoter is too leaky to create conditional alleles for some essential genes (S. McDaniel and D.L Chalker, unpublished data). The *MTT3* promoter may have the lowest basal expression and may be better choice (K. Collins, personal communication).

IV. Engineering Strains for Expression of Tagged Alleles

The ability to express epitope-tagged proteins in cells greatly facilitates biochemical or cytological studies (see Chapters 12 and 13). As *Tetrahymena* can be transformed with autonomously replicating vectors or by homologous recombination targeted to the genome, many options exist to create a desired strain. For many reasons, it is usually best practice to integrate a tagged allele into the endogenous locus. For one, it is the most reliable way to ensure expression of the gene at physiologically normal levels. Also if the tagged allele can completely replace the wild-type allele and no deleterious phenotypes can be detected, it indicates that the tagged protein is functional.

The basic strategy to construct a strain expressing an endogenously tagged allele is similar to making a knockout strain. Placing the tag on the C terminus turns out to a bit more user friendly. The targeting construct to be created should contain at least 1 kbp of the coding region fused to the desired tag followed by a 3′UTR (either the gene's own or one borrowed from another), this is followed by a selectable marker and about 1 kbp of downstream gene sequence (the marker and downstream sequence can be the same as used in the knockout construct). A versatile collection of C-terminal, epitope tagging modules has been generated by K. Mochizuki and colleagues to facilitate such strain engineering (Kataoka *et al.*, 2010).

N-terminal tags can be more challenging to create. If the selectable marker is placed upstream of the gene, it may interrupt the promoter. If the marker is placed downstream of the targeted gene, the marker and the tagged portion of the gene are often sufficiently distant from one another that the recombination can occur between the tag and the construct, leading to selection of a nontagged allele. Busch *et al.* have generated an N-terminus tagging strategy that consists of the *neo4* selectable marker flanked by loxP sites interrupting the introduced tag. This ensures that the tag is incorporated in all transformants. Once transformants are selected, Cre recombinase is expressed in cells to excise the marker, allowing the tagged allele is expressed (Busch *et al.*, 2010).

In addition to tagging genes at their own loci, one can express tagged constructs integrated at ectopic loci. This has several practical merits. Universal expression cassettes have been created that target a tagged allele to specific loci. These expression cassettes typically incorporate the inducible *MTT1* promoter, allowing for controlled expression. All one needs to do is clone the coding region of a gene of interest into the cassette and then introduce the expression vector into cells. Genomic loci that have been used routinely for ectopic expression included regions encoding *BTUI*, *rpL29*, and *MTT1*. The first two loci have selectable alleles available (see Section V), which has made each particularly useful. Since these loci will be targeted by homologous recombination, these expression vectors must be linearized at the upstream and downstream boundaries of the cloned genomic homology to efficiently recombine with their locus upon transformation, so the introduced coding region must be free of the restriction sites used for linearizing the vector. Most designed expression constructs possess G+C-rich recognition sites at the ends of the targeting homology, which are rarely found in the A+T-rich *Tetrahymena* genome, maximizing the compatibility of these vectors with *Tetrahymena* genes.

Eptitope tagged genes can also be expressed when carried on high-copy vectors. The most commonly used autonomously replicating vectors for *Tetrahymena* are based on the micronuclear rDNA. The rRNA gene of *Tetrahymena* is present as a single copy in the micronuclear genome, but is excised from its locus in the developing macronucleus and maintained as a 20-kbp minichromosome (Yao *et al.*, 1979; Yao and Gall, 1977). After excision from the genome, it is converted into a head-to-head palindromic molecule (Yao *et al.*, 1985), with two copies of the rRNA gene with telomeres added at each free end and amplified to ~9000 copies per macronucleus. This allows for high-level expression of transgenes carried on the replicating minichromosome.

Two features of these rDNA-based vectors are important for their overall effectiveness. First, several distinct point mutations have been identified in the rRNA-coding region that confer various antibiotic resistance to cells carrying these vectors (e.g., Paromomycin, Anisomycin, and Hygromycin resistance) (Spangler and Blackburn, 1985). Second, the C3-type replication origin present upstream of the rDNA-coding region exhibits a replication maintenance advantage over the B-type origin (found in the inbred B strains used for most genetic studies performed with *Tetrahymena*) (Larson *et al.*, 1986; Orias *et al.*, 1988). Thus, when the C3-type rDNA vectors are transformed into B-type laboratory strains, the vector-based rDNA, and any expression cassette it carries, out-competes the endogenous rDNA chromosomes and typically becomes the major or only rDNA minichromosome in the transformed cells. The combination of antibiotic selection and the replication advantage ensures that these vectors are maintained stably at high copy number. The pD5H8 plasmid is a widely used vector containing a paromomycin-resistant allele of the 17S rRNA gene and the C3-type replication origin described above (Godiska and Yao, 1990; Spangler and Blackburn, 1985). The pD5H8 plasmid also possesses a polylinker sequence ~400-bp downstream of the rRNA coding region into which expression modules are inserted.

These high-copy vectors have both advantages and limitations as expression vehicles. As for limitations, they are large vectors (~18 kbp) that can make recombinant DNA manipulations challenging. Also, these vectors must be introduced into cells during conjugation to allow the circular *Escherichia coli* vector to be processed into a palindromic minichromosome. This is a problem when using heterokaryon strains that contain an existing mutant allele, such as gene knockout, carried only within the somatic macronucleus, as this nucleus is lost and replaced with a copy of the genome from the germline micronucleus. A further limitation is that only a single version of an rDNA vector can be stably maintained in a given transformant; thus, approaches requiring dual expression of differently tagged constructs (e.g., colocalization studies) require a second vector platform.

As for advantages, the rDNA vectors provide for high-level expression of transgenes. By combining these vectors with the inducible *MTT1* promoter, expression levels can be modulated by varying cadmium concentration in the growth medium. Strong induction can drive overexpression, if desired. Overexpression cannot be achieved for all genes, as *Tetrahymena* appears to be quite adept at post-transcriptional regulation of gene expression (based on our observation of many genes expressed from the same vectors). Transformation of these vectors into *Tetrahymena* cells requires little special training. The only necessary equipment is an electroporator, which is commonly used for other cell types in many laboratories engaged in cellular and molecular biology research (Gaertig and Gorovsky, 1995a, 1995b; Gaertig *et al.*, 1994a; Gaertig and Kapler, 2000). These vectors can support highly efficient transformation (10^4 transformants/μg of DNA), which has allowed them to be used for reverse genetic screening strategies (Chilcoat *et al.*, 2001; Yao *et al.*, 2007).

One version of epitope tagging modules available in both integrating and replicating vectors employ Gateway® recombination-based cloning, which greatly facilitates DNA manipulation and enables high-throughput approaches (Hartley *et al.*, 2000; Walhout *et al.*, 2000). Expression vectors using this recombination-based cloning have facilitated the analysis of large gene families in *Tetrahymena* (Bright *et al.*, 2010; Malone *et al.*, 2008). By using a uniform recombination cloning platform, one's genes of interest, once cloned in a Gateway®-compatible "entry" vector, can be rapidly assembled into any existing "destination" vector with minimal effort. Future development of this resource will greatly aid functional genomic approaches.

V. Selectable Markers and Expression Vectors

Three generations of *neo* selectable cassette have been created to facilitate strain engineering. The original, *neo2*, contained the *neo* gene flanked by the Histone H4 (*HHF1*) promoter and the 3′UTR from the *BTU2* gene (Gaertig *et al.*, 1994a). This cassette was improved by swapping the cadmium-inducible *MTT1* promoter for the *HHFI* promoter to create *neo3* (Shang *et al.*, 2002). The neo-coding sequence was later codon optimized for *Tetrahymena* leading to the generation of *neo4* (Mochizuki, 2008). Each of these improvements has significantly increased the transformation efficiency. Both *neo3* and *neo4* provide sufficient numbers of transformants to perform micronuclear transformation, which is much more demanding than somatic (macronuclear) transformation as described above.

Other selectable markers have been used for strain engineering. The blasticidin S resistance cassette, *bsr1* is a useful alternative to *neo* for macronuclear transformation (Gaertig and Kapler, 2000). As indicated in the next chapter, it has been improved by replacing the *HHF1* promoter with the *MTT1* promoter. A cycloheximide resistance allele of the *rpL29* gene (Yao and Yao, 1991) has also been adapted for selection by replacing its endogeneous promoter with the *MTT1* promoter (J. Bowen and M. Gorovsky and R. Jain and D. Chalker, unpublished data). Its efficiency for transformation has been relatively low, so it has not been widely used, but is available if an additional marker is required.

In addition to these selection cassettes that are useful for targeting a construct to any locus, drug resistant alleles of endogenous genes have served in powerful selection strategies to introduce expression cassettes into the genome.

A. The *BTU1* Locus

This has been a commonly used integration site for ectopic expression of gene products (Gaertig *et al.*, 1994b; Witkin and Collins, 2004). *Tetrahymena* strains containing a paclitaxel sensitive allele of this beta-tubulin gene at the primarily expressed locus (*BTU1*) become resistant to paclitaxel when the *BTU1* locus is disrupted by introduction of the transgene. This results in the upregulation of the paclitaxel resistant beta-tubulin allele expressed from the minor *BTU2* locus. This is a powerful selection but has limitations for strain engineering. All expression constructs targeting the *BTU1* locus must be introduced into paclitaxel sensitive strains such as CU522 or CU727 (both are available from the stock center), which limits the utility of this marker.

B. The *rpL29* Locus

This locus has been developed for efficient expression of introduced transgenes. An allele of the *rpL29* gene that confers cycloheximide resistance (*rpl29-cy-r*) was engineered based on known mutations in yeast (Yao and Yao, 1991). A site just upstream of *rpL29-cy-r* promoter is used for insertion of transgenes back into the genome (Kowalczyk *et al.*, 2006). Both of these loci are excellent sites to insert tagged genes for biochemical and/or protein localization studies (Chapters 12 and 13).

VI. The Use of Heterokaryon Strains for the Study of Essential Genes

Essential genes can be difficult to study. If an essential gene is disrupted in the macronucleus, it is unlikely that complete replacement of the wild-type copies with the selectable marker will ever be achieved. In some cases, one can detect phenotypes in the cells that have the fewest copies of the essential gene remaining, but this takes very careful observation. Generation of germline knockout heterokaryons offers a potentially powerful means to characterize such difficult to study genes. These strains are homozygous for the knockout allele in their silent micronuclei, but are wild type in their macronucleus. When cells homozygous for the gene knockout in their micronuclei are mated, all progeny cells will be complete knockouts. If the gene is essential, then no viable progeny will be recovered. Nevertheless, it is possible to examine the phenotype of the mutant cells as they deplete their maternal load of protein in the cells as they undergo the first few rounds of postconjugative cell division (e.g., Cervantes *et al.*, 2006; Malone *et al.*, 2008).

Germline heterokaryon strains can also be used for mutational analysis. If the germline is homozygous for the knockout, copies of the gene can be introduced during postzygotic development, either on replicating vectors or integrating constructs to assess whether an introduced copy can rescue the knockout phenotype. Epitope-tagged versions of the gene can be introduced to determine whether or not the tag disrupts function. Truncated versions of the gene can be tested to map essential domains of a protein of interest. The major advantage over simply trying to replace the macronuclear gene by assortment is that one can be confident that the only version of the protein expressed is the one introduced during conjugation.

A rapid means of creating two homozygous germline knockout heterokaryons is to mate an existing germline transformant with "star" strains, which have defective micronuclei (e.g., B*VI and B*VII). This genomic exclusion cross-transfers the micronucleus from the knockout line to the star strain. The exconjugant derived from the star parent will have a homozygous micronucleus of the knockout, but retain its wild-type macronucleus. As this was an abortive mating, the strains are mature and ready to mate; therefore, as soon as these cell lines are expanded by growth, they can be used in phenotypic studies.

VII. Executing Biolistics Transformation and Electroporation

In the following section, we describe the methodologies to transform *Tetrahymena* by bombarding cells with DNA-coated particles (biolistics) (Cassidy-Hanley *et al.*, 1997) and electroporation (Gaertig and Gorovsky, 1992; Gaertig *et al.*, 1994a).

Biolistics is the method of choice for carrying out homologous recombination, while conjugative electroporation is preferred for transformation of cells with high-copy rDNA vectors. Electroporation can be used for homologous recombination as well, but the efficiency is much reduced.

One must first select the appropriate cells to use for the transformation experiment. Biolistic transformation works well for most strains. Some strains exhibit higher or lower tolerance to the drugs used for selection, which appears to have an impact on transformation efficiency. It may be worthwhile to titrate a lab stock of drug against the cell lines to be used in the transformation to determine the effective concentration at which the drug will kill. Drug effectiveness can also be cell density dependent. Dense cultures are much more resistant to some drugs (e.g., blasticidin S). If using an expression construct that targets the *BTU1* locus, a paclitaxel sensitive strain such as CU522 must be used. For germline transformation, strains CU428 [*mpr1-1/mpr1-1*; (VII mp-s)] and B2086 (II) are routinely used. These strains produce progeny at high frequency in crosses, which is essential for recovering germline transformants. In addition, the dominant 6-methylpurine resistance allele in the micronucleus allows one to identify progeny cells among the transformants.

A. Biolistics

(*Note*. These instructions provided are for use with the PDS-1000 particle bombardment apparatus from Bio-Rad (Hercules, CA). The standard apparatus uses a single macrocarrier holder. Newer versions have a Hepta adaptor that uses seven separated macrocarrier holders, which disperse the particles more evenly to give higher efficiencies, but will not be discussed)

1. Grow and Starve Cells (see Chapter 8 for media recipes and descriptions)

Cells to be transformed by biolistics need to be starved prior to bombarding with DNA-coated gold particles. Starvation reduces vesicle content inside cells to allow the particles better access to the cytoplasm or nucleus. For each shot, grow 50 mL of cells in 1× SPP to a density of 2×10^5 cells/mL. Collect cells by centrifugation at $1100 \times g$ for 3 min in a conical bottom tube, wash cells in 10 mM Tris–HCl (pH 7.5), one or two times, pellet cells after the final wash, and resuspend in 10 mM Tris–HCl at 2×10^5 cells/mL. Starve cells overnight at 30 °C (at least 6 h). Starvation should be carried out in dedicated flasks that never see detergent or in 150 mm disposable petri dishes (50 mL/dish).

If performing biolistics for germline transformation, starved strains (CU428 and B2086) should be mixed in equal numbers at a cell density of 2×10^5 cells/mL (Bruns and Brussard, 1974). As with starved cell transformation, 50 mL of cells are needed per shot (fewer cells can be used, but the chance of recovering transformants will be diminished). To ensure synchrony of mating, it is useful to allow cells to co-stimulate upon mixing by shaking cells at 200 rpm for 15 min at 30 °C to prevent pairing, allow cells to rest without shaking for 15 min at 30 °C, and then shake again. Once cells are co-stimulated, stop shaking (cells can be either left in flasks or poured into 150-mm petri dishes), and allow cells to mate undisturbed for ~2 h and 10 min, at which point the first 50 mL of cells should be harvested for transformation. Four shots in succession ~10 min apart should be performed to guarantee that the transforming DNA is introduced into the population at the stage of meiosis just prior to when recombination occurs.

2. Linearize DNA Construct

Free DNA ends catalyze homologous recombination. Sufficient DNA should be digested with appropriate restriction enzymes to yield 1–3 μg per shot. The enzyme should be removed by phenol/chloroform extraction and ethanol precipitation or other means of purification and resuspended in ddH_2O at a concentration of 1–3 μg/μL.

3. Coating the Gold Particles with DNA

All steps of the coating should be performed at 4 °C. To prepare particles, they should be washed in sterile 1 mL ddH_2O, followed by washing in 1 mL each of 70% ethanol, 100% ethanol, ddH_2O, and finally resuspended in 50% sterile glycerol at 60 mg/mL and stored at −20 °C in 40 μL aliquots (enough for four shots). Between each step harvested particles by brief centrifugation, the particles are dense and pellet easily. For the coating, to one aliquot of gold add in rapid succession:

- 4 μL (4–12 μg) of linearized DNA
- Add 40 μL 2.5 M $CaCl_2$
- Add 16 μL spermidine.

Vortex for 3 s between each addition. After all components are added, shake at 4 °C for 10 min (a vortex mixer equipped with a tube holder works well for this step).

To prepare DNA/gold for shooting, harvest the particles in a 4 °C microcentrifuge for 6 s (~10,000 g). Remove the supernatant with a pipette, being careful not to lose any gold particles. Wash gold particles with 100 μL of 70% ethanol. Spin again, discard ethanol and wash gold particles in 100 μL of 100% ethanol. Harvest gold, discard wash, and resuspend in 40 μL 100% ethanol. Pipette up and down to resuspend and sonicate gold particles for 2–3 s in a bath sonicator (e.g., Branson model 1510) to disrupt clumping. Put 10 μL of the gold particle solution in the center of each macrocarrier, assembled in a stainless steel holder, being sure to pipette up and down between removing each aliquot, as gold tends to settle quickly. Dry particles on macrocarriers positioned in macrocarrier holders by vacuum dessiccation. Transport dried gold on macrocarrier from dessicator to biolistics apparatus in a petri dish containing drierite or other dessicant. (*Note.* Spermidine and 100% ethanol are sensitive reagents and should be replaced if transformation efficiency is poor.)

4. Preparing the Cells

Place a circular Whatmann 50 filter in each of four 100-mm Petri dishes and moisten with 10 mM Tris–HCl (pH 7.5). Aspirate off excess liquid, filters should be moist, but not wet. Harvest 50 mL of starved (or mating) cells by centrifugation at 1100 × g for 3 min in a conical bottom tube, decant supernatent, and resuspend in 0.5 mL of 10 mM Tris–HCl. Distribute onto filter paper, starting in center and spreading cells over the central three-fourth of dish. Harvest cells just before each shot. During centrifugation, ready the particle bombardment apparatus for biolistics.

5. Shooting

Wet a 900 psi rupture disk in isopropanol, place in its threaded holder, and assemble onto the treaded helium port inside the particle bombardment apparatus, tighten with the Torque wrench provided. Place a stopping screen and a macrocarrier holder with DNA-coated particles into adapter tray. Secure with lock ring and insert tray into the apparatus at highest position. (The gap distance between the rupture disk holder and the lock rings should be ~1 cm – this should be optimized using a construct of known transformation efficiency.) Place a 100-mm Petri dish holding cells, with its lid removed, onto the specimen dish tray placed at the lowest position (we have found both the lowest and next to lowest positions work well). Open the valve on the helium tank and turn on the vacuum compressor. Push the "Vacuum" switch and pull a vacuum to 26.5–27 in. of Hg, quickly switch "Vacuum" to "Hold." Push and hold the "Fire" button, record the pressure at which the rupture disk bursts. Release the "Fire" button and switch vacuum from "Hold" to "Vent." Once the vacuum is release, recover the petri dish and add 1 mL of 30 °C 10-mM Tris–HCl to the cells to prevent cells from drying out. Place shot cells in the 30 °C incubator and complete the remaining shots.

6. Recovery and Selection.

If transforming starved cells for macronuclear transformation, the cells can immediately be returned to 1× SPP growth medium. If the selectable marker used is expressed from a cadmium inducible promoter, 0.5 μg/mL $CdCl_2$ should be added to the medium during recovery to induce marker expression. It is important to test the sensitivity of nontransformed lab strains to selection drugs as different lots of drugs and different medium preparations can alter the cell's response to these antibiotics.

If drugging cells with paromomycin or blasticidin S, cells should be allowed to recover for 5–7 h before increasing $CdCl_2$ to 1 μg/mL and adding either drug to a concentration of 80 μg/mL. Cells can be selected in mass in flasks, but distributing at least part of the culture into 96-well plates allows one to recover individual transformants and to assess transformation efficiency.

If selecting cells with cycloheximide, it is best to let cells recover >10 h (up to 20 h) to allow for accumulation of ribosomes with the resistant version of the rpL29 protein. Cycloheximide should then be added to 12.5 μg/mL to select transformants.

For selection of germline transformants, the drug regimen is the same as with starved cells; however, the cells are not immediately returned to growth medium. They are resuspended in 25 mL of 10 mM Tris–HCl for 12–24 h to allow cells to complete conjugation at 30 °C. This prevents the unmated cells in the population from getting a head start on the mating cells and taking over the population. After this period, 25 mL of 2× SPP is added to the culture (to make the culture 1× SPP), and the transformants are selected as described for the macronuclear transformants, that is, cells must be allowed to outgrow in growth medium prior to adding drug.

B. Conjugative Electroporation

Because rDNA-based vectors must be processed by the chromosomal breakage machinery into a linear minichromosome, they are introduced into *Tetrahymena* during conjugation, just prior to the time that DNA rearrangement in the developing macronucleus occurs (Yao and Yao, 1989). The circular plasmid is then cleaved at breakage sequences flanking the *E. coli* vector sequences (which will fail to replicate), telomeres are added downstream, and inverted repeats present near the upstream breakage site direct palindrome formation of the entire linear chromosome (Yasuda and Yao, 1991). Transformants are selected based on base changes in the 17S rRNA gene (Spangler and Blackburn, 1985) in the vector that confers resistance to paromomycin on the cells. One drawback is that this confers the same resistance as the *neo* gene, limiting dual use of the commonly used rDNA vectors and the standard selectable marker used for disrupting genes.

1. Growth and Preparation of Cells (see Chapter 8 for media recipes and descriptions).

Cells to be transformed by electroporation should be grown and starved as described above for germline biolistics transformation. Cells should be starved overnight 30 °C in 10 mM Tris–HCl at 2 × 10^5 cells/mL. The day that the electroporation is to be performed, equal numbers of cells of two strains of different mating types [eg. CU427 (VI), CU428 (VII), B2086(II)] are mixed and allowed to conjugate for 9–10 h (depending upon the strains). About 15 mL of mating cells is sufficient for each transformation. The co-stimulation steps described above for germline biolistics transformation is usually unnecessary as the synchrony is not as critical for electroporation, and the timing can be adjusted based on tracking the progression of mating. A good rule of thumb is to plan to harvest cells for the electroporation ~1.5 h after 50% of the mating population has reach the Mac II developmental stage and formed developing macronuclei (anlagen) (Martindale *et al.*, 1982) (Chapter 7). This can be determined by DAPI staining DNA and visualizing nuclei. A simpler method to visualize anlagen is to treat mating cells in Schaudin's fixative (one part ethanol and two parts saturated mercuric chloride) (Wenkert and Allis, 1984). This makes the anlagen easy to see by phase contrast microscopy. Mating cells can even be fixed directly on a microscope slide by mixing 10 μL of cells with 1–2 μL of fixative and immediately mounting under a coverslip. Three nuclei (two anlagen and the parental macronucleus) in each mating partner of a pair will be quite obvious when cells have reached the correct (Mac II) developmental stage; the two micronuclei are hard to see. One should start monitoring cells for the percentage of cells with anlagen about 7 h postmixing.

2. Preparation of DNA

Any standard plasmid DNA preparation can be used. The main challenge is that rDNA vectors are large so many commercial kits provide poor yields. Standard alkaline lysis preparations, followed by a final 8–10% PEG 8000/0.6 M NaCl DNA precipitation to remove RNA oligomers and other impurities can yield DNA of high quality for electroporation, but in some hands, this preparation yields inhibitory concentrations of endotoxin that lowers transformation efficiency. We have observed large differences in the efficiency of electroporation based on the method used to isolate the plasmids. Between 10 and 30 μg of plasmid DNA is optimal for each electroporation, so medium-scale plasmid isolation is needed to obtain sufficient quantities of DNA for electroporation. Just prior to electroporation, 10–30 μg of DNA should be mixed with enough 10 mM HEPES (pH 7.4) to achieve a total of 50 μL of DNA solution.

3. Executing the Electroporation

Mating cell populations, 9–10 h postmixing, should harvested in 50 mL conical centrifuge tubes by centrifugation for 3 min at 1100 × g. Decant supernatant quickly and gently resuspend in a volume of 10 mM HEPES (pH 7.4) equal to the starting culture volume. Allow cells to equilibrate in HEPES at room temperature for 5 min. Harvest cells again by centrifugation, decant supernatant, and resuspend cells in 200 μL of 10 mM HEPES buffer for each 15 mL of mating cells collected. Adjust the electroporator (e.g., BTX model ECM 630) to the following settings:

Capacitance = 275 μF
Resistance = 25 Ω
Voltage = 250 V

Mix 200 μL of the mating cells to the 50 μL of DNA, transfer to an electroporator cuvette (0.2 cm gap), and pulse in the electroporation chamber. Pulse lengths should be between 6 and 6.5 ms.

4. Recovery and Selection of Transformants

Wait 1 min after the electroporation and then recover cells from the cuvette by adding 1 mL of 1× SPP growth medium with a pasteur pipette. Transfer cells from the cuvette into 30 mL of 1× SPP and distribute into 96-well plates (100 μL/well). Allow cells to complete mating for 12–20 h at 30 °C. Select transformants by adding 100 μL of 1× SPP containing 200 μg/mL paromomycin to each well in the 96-well plate with a multichannel pipettor (final concentration 100 μg/mL paromomycin).

If the cells to be transformed have reduced pairing efficiency, electroporation can still be successful, but it may be helpful to resuspend cells after the electroporation in 15 mL of 10 mM Tris–HCl (pH 7.5) and allow conjugation to complete for 8–12 h prior to adding growth medium. This will prevent the unpaired cells from overpopulating the culture. To select transformants, add 15 mL of 2× SPP, plate into 96-well plates and allow cells to grow vegetatively for 8–12 h before adding the selection drug as described in the preceding paragraph.

Transformants should be apparent as healthy, vigorously growing cells after 3 days of selection at 30 °C. By this time, they will have fully amplified the rDNA vector and will carry the expression cassette at high copy. They can be rapidly expanded in culture and used for the experiments planned. However, the transformants obtained by electroporation are new progeny and therefore will not be mature to mate until after 1.5–2 weeks of growth (80–100 fissions). This lengthens the time required until one can analyze the new transformants during conjugation.

Anecdotally, it appears that expression constructs carried on rDNA-based vectors are prone to either silencing or loss. The mechanism underlying loss of expression has not been carefully examined. It is prudent to freeze transformants for long-term storage shortly after the culture is expanded if one wishes to preserve lines for future analysis.

VIII. Summary

The multiple ways to transform *Tetrahymena* makes this organism particularly amenable to molecular genetic manipulation. Biolistics is used primarily in experiments that require the introduced DNA to be targeted to the genome by homologous recombination. Genes can be knocked out, epitope-tagged alleles can be created, and even foreign DNA can be incorporated into the genome at any selected locus. One cannot understate how this ability to engineer the genome significantly contributes to the current and future successes of *Tetrahymena* as a major model system. Even a large-scale gene knockout project is underway (R.S. Coyne, personal communication), taking advantage of the annotated genome. Such efforts should greatly enhance access to new experimental questions in this ciliate. Electroporation is used to introduce high-copy vectors. These are expression vehicles of choice when high transformation efficiency is needed (e.g., enabling screening of libraries) and when very high-level (over)expression is desired. With only a small number of the nearly 25,000 genes studied, *Tetrahymena* and its genome are ripe for exploitation to uncover new and interesting biology.

Acknowledgements

All of us in the Tetrahymena research community owe much gratitude to Meng-Chao Yao, Jacek Gaertig and Martin Gorovsky, and Donna Cassidy-Hanley and Peter Burns who devoted extensive effort to develop transformation vectors and strategies.

References

Allen, S. L. (1967). Cytogenetics of genomic exclusion in Tetrahymena. *Genetics* **55**, 797–822.

Bright, L. J., Kambesis, N., Nelson, S. B., Jeong, B., and Turkewitz, A. P. (2010). Comprehensive analysis reveals dynamic and evolutionary plasticity of Rab GTPases and membrane traffic in *Tetrahymena thermophila*. *PLoS Genet* **6**(10), e1001155.

Bruns, P. J., and Brussard, T. B. (1974). Pair formation in *Tetrahymena pyriformis*, an inducible developmental system. *J. Exp. Zool.* **188**, 337–344.

Busch, C. J., Vogt, A., and Mochizuki, K. (2010). Establishment of a Cre/loxP recombination system for N-terminal epitope tagging of genes in Tetrahymena. *BMC Microbiol.* **10**, 191.

Cassidy-Hanley, D., Bowen, J., Lee, J. H., Cole, E., VerPlank, L. A., Gaertig, J., Gorovsky, M. A., and Bruns, P. J. (1997). Germline and somatic transformation of mating *Tetrahymena thermophila* by particle bombardment. *Genetics* **146**, 135–147.

Cervantes, M. D., Coyne, R. S., Xi, X., and Yao, M. C. (2006). The condensin complex is essential for amitotic segregation of bulk chromosomes, but not nucleoli, in the ciliate *Tetrahymena thermophila*. *Mol. Cell Biol.* **26**, 4690–4700.

Chalker, D. L., Ward, J. G., Randolph, C., and Yao, M. -C. (1999). Microinjection of *Tetrahymena thermophila*. *In* "Tetrahymena thermophila," (D. J. Asai, and J. D. Forney, eds.), pp. 469–484. Academic Press, New York.

Chilcoat, N. D., Elde, N. C., and Turkewitz, A. P. (2001). An antisense approach to phenotype-based gene cloning in Tetrahymena. *Proc. Natl. Acad. Sci. U.S.A.* **98**, 8709–8713.

Fass, J., Joshi, N., Couvillion, M., Bowen, J., Gorovsky, M., Hamilton, E., Orias, E., Hong, K., Coyne, R., Eisen, J., Chalker, D., Lin, D., and Collins, K. (2011). Genome-scale analysis of programmed DNA elimination sites in *Tetrahymena thermophila*. *G3: Genes, Genomes, and Genetics* **1**, 515–522.

Gaertig, J., and Gorovsky, M. A. (1992). Efficient mass transformation of *Tetrahymena thermophila* by electroporation of conjugants. *Proc. Natl. Acad. Sci. U.S.A.* **89**, 9196–9200.

Gaertig, J., and Gorovsky, M. A. (1995 a). DNA-mediated transformation in Tetrahymena. *Methods Cell Biol.* **47**, 559–569.

Gaertig, J., and Gorovsky, M. A. (1995 b). Gene transfer by electroporation in Tetrahymena. *Methods Mol. Biol.* **47**, 331–348.

Gaertig, J., Gu, L., Hai, B., and Gorovsky, M. A. (1994 a). High frequency vector-mediated transformation and gene replacement in Tetrahymena. *Nucleic Acids Res.* **22**, 5391–5398.

Gaertig, J., and Kapler, G. (2000). Transient and stable DNA transformation of Tetrahymena thermophila by electroporation. *Methods Cell Biol.* **62**, 485–500.

Gaertig, J., Thatcher, T. H., Gu, L., and Gorovsky, M. A. (1994 b). Electroporation-mediated replacement of a positively and negatively selectable beta-tubulin gene in Tetrahymena thermophila. *Proc. Natl. Acad. Sci. U.S.A.* **91**, 4549–4553.

Godiska, R., and Yao, M. C. (1990). A programmed site-specific DNA rearrangement in *Tetrahymena thermophila* requires flanking polypurine tracts. *Cell* **61**, 1237–1246.

Hartley, J. L., Temple, G. F., and Brasch, M. A. (2000). DNA cloning using in vitro site-specific recombination. *Genome Res.* **10**, 1788–1795.

Iwamoto, M., Mori, C., Kojidani, T., Bunai, F., Hori, T., Fukagawa, T., Hiraoka, Y., and Haraguchi, T. (2009). Two distinct repeat sequences of Nup98 nucleoporins characterize dual nuclei in the binucleated ciliate tetrahymena. *Curr. Biol.* **19**, 843–847.

Kataoka, K., Schoeberl, U. E., and Mochizuki, K. (2010). Modules for C-terminal epitope tagging of Tetrahymena genes. *J. Microbiol. Methods* **82**, 342–346.

Kowalczyk, C. A., Anderson, A. M., Arce-Larreta, M., and Chalker, D. L. (2006). The germ line limited M element of Tetrahymena is targeted for elimination from the somatic genome by a homology-dependent mechanism. *Nucleic Acids Res.* **34**(20), 5778–5789.

Larson, D. D., Blackburn, E. H., Yeager, P. C., and Orias, E. (1986). Control of rDNA replication in Tetrahymena involves a cis-acting upstream repeat of a promoter element. *Cell* **47**, 229–240.

Linger, B. R., Morin, G. B., and Price, C. M. (2011). The Pot1A-associated proteins TPT1 and PAT1 coordinate telomere protection and length regulation in Tetrahymena. *Mol. Biol. Cell.* **22**(21), 4161–4170.

Malone, C. D., Falkowska, K. A., Li, A. Y., Galanti, S. E., Kanuru, R. C., LaMont, E. G., Mazzarella, K. C., Micev, A. J., Osman, M. M., Piotrowski, N. K., Suszko, J. W., Timm, A. C., Xu, M. M., Liu, L., and Chalker, D. L. (2008). Nucleus-specific importin alpha proteins and nucleoporins regulate protein import and nuclear division in the binucleate *Tetrahymena thermophila*. *Eukaryot. Cell* **7**, 1487–1499.

Martindale, D. W., Allis, C. D., and Bruns, P. (1982). Conjugation in *Tetrahymena thermophila*: a temporal analysis of cytological stages. *Exp. Cell Res.* **140**, 227–236.

Mochizuki, K. (2008). High efficiency transformation of Tetrahymena using a codon-optimized neomycin resistance gene. *Gene* **425**(1-2), 79–83.

Orias, E., Larson, D., Hu, Y. F., Yu, G. L., Karttunen, J., Lovlie, A., Haller, B., and Blackburn, E. H. (1988). Replacement of the macronuclear ribosomal RNA genes of a mutant Tetrahymena using electroporation. *Gene* **70**, 295–301.

Shang, Y., Song, X., Bowen, J., Corstanje, R., Gao, Y., Gaertig, J., and Gorovsky, M. A. (2002). A robust inducible-repressible promoter greatly facilitates gene knockouts, conditional expression, and overexpression of homologous and heterologous genes in *Tetrahymena thermophila*. *Proc. Natl. Acad. Sci. U.S.A.* **99**, 3734–3739.

Spangler, E. A., and Blackburn, E. H. (1985). The nucleotide sequence of the 17S ribosomal RNA gene of *Tetrahymena thermophila* and the identification of point mutations resulting in resistance to the antibiotics paromomycin and hygromycin. *J. Biol. Chem.* **260**, 6334–6340.

Tondravi, M. M., and Yao, M. C. (1986). Transformation of Tetrahymena thermophila by microinjection of ribosomal RNA genes. *Proc. Natl. Acad. Sci. U.S.A.* **83**, 4369–4373.

Walhout, A. J., Temple, G. F., Brasch, M. A., Hartley, J. L., Lorson, M. A., van den Heuvel, S., and Vidal, M. (2000). GATEWAY recombinational cloning: application to the cloning of large numbers of open reading frames or ORFeomes. *Methods Enzymol* **328**, 575–592.

Wenkert, D., and Allis, C. D. (1984). Timing of the appearance of macronuclear-specific histone variant hv1 and gene expression in developing new macronuclei of *Tetrahymena thermophila*. *J. Cell Biol.* **98**, 2107–2117.

Witkin, K. L., and Collins, K. (2004). Holoenzyme proteins required for the physiological assembly and activity of telomerase. *Genes Dev.* **18**, 1107–1118.

Yao, M. C., Blackburn, E., and Gall, J. G. (1979). Amplification of the rRNA genes in Tetrahymena. *Cold Spring Harb. Symp. Quant. Biol.* **2**, 1293–1296.

Yao, M. C., and Gall, J. G. (1977). A single integrated gene for ribosomal RNA in a eucaryote. Tetrahymena pyriformis. *Cell* **12**, 121–132.

Yao, M. C., and Yao, C. H. (1989). Accurate processing and amplification of cloned germ line copies of ribosomal DNA injected into developing nuclei of *Tetrahymena thermophila*. *Mol. Cell Biol.* **9**, 1092–1099.

Yao, M. C., and Yao, C. H. (1991). Transformation of *Tetrahymena* to cycloheximide resistance with a ribosomal protein gene through sequence replacement. *Proc. Natl. Acad. Sci. U.S.A.* **88**, 9493–9497.

Yao, M. C., Yao, C. H., Halasz, L. M., Fuller, P., Rexer, C. H., Wang, S. H., Jain, R., Coyne, R. S., and Chalker, D. L. (2007). Identification of novel chromatin-associated proteins involved in programmed genome rearrangements in Tetrahymena. *J. Cell Sci.* **120**, 1978–1989.

Yao, M. C., Zhu, S. G., and Yao, C. H. (1985). Gene amplification in *Tetrahymena thermophila*: formation of extrachromosomal palindromic genes coding for rRNA. *Mol. Cell Biol.* **5**, 1260–1267.

Yasuda, L. F., and Yao, M. C. (1991). Short inverted repeats at a free end signal large palindromic DNA formation in *Tetrahymena*. *Cell* **67**, 505–516.

Yu, G. L., and Blackburn, E. H. (1989). Transformation of Tetrahymena thermophila with a mutated circular ribosomal DNA plasmid vector. *Proc. Natl. Acad. Sci. U.S.A.* **86**, 8487–8491.

Yu, G. L., Hasson, M., and Blackburn, E. H. (1988). Circular ribosomal DNA plasmids transform *Tetrahymena thermophila* by homologous recombination with endogenous macronuclear ribosomal DNA. *Proc. Natl. Acad. Sci. U.S.A.* **85**, 5151–5155.

CHAPTER 12

Biochemical Approaches Including the Design and Use of Strains Expressing Epitope-Tagged Proteins

Mary T. Couvillion and Kathleen Collins

Department of Molecular and Cell Biology, University of California at Berkeley, Berkeley, CA 94720-3200, USA

Abstract

Epitope tagging is a powerful approach used to enable investigations of a cellular component by elucidating its localization, interaction partners, and/or activity targets. Successful tag-based affinity purification yields a mixture of the molecule of interest, associated proteins and nucleic acids, and nonspecific background proteins and nucleic acids, many of which can depend on details of the protocol for enrichment. This chapter provides guidelines and considerations for designing an affinity purification experiment, beginning with construction of a strain expressing a tagged subunit. Common biochemical methods for detecting protein, RNA, and DNA in *Tetrahymena thermophila* are also discussed.

0091-679X/10 $35.00
DOI 10.1016/B978-0-12-385967-9.00012-8

I. Introduction: General Considerations for Strain Construction

Although there can be advantages to using a custom-produced antibody for protein immunoprecipitation, it is generally faster, easier, more reliable, and better experimentally controlled to make a strain expressing a tagged fusion protein. Epitope tagging allows the opportunity for protocol consistency between affinity purifications, which is especially useful when comparing across a family of related proteins (Couvillion *et al.*, 2009).

We most commonly use a tandem epitope tag consisting of two Protein A domains (ZZ), a Tobacco Etch Virus (TEV) protease cleavage site, and triple FLAG peptide (N-terminal ZZtev3XF or C-terminal 3XFtevZZ) (Lee *et al.*, 2009; Min and Collins, 2009), although other tags have also been used successfully (Lee and Collins, 2007; Mochizuki *et al.*, 2002; Yu and Gorovsky, 2000). Since *Tetrahymena thermophila* has a distinct codon usage compared to other model organisms, the construction of *T. thermophila* codon-optimized tag open reading frames (ORFs) should be considered. We have used tag ORFs designed based on *T. thermophila* codon usage as well as ORFs designed for expression in yeast or human cells, but we have not compared different tag versions with an otherwise identical mRNA sequence. Another important consideration relevant when designing N-terminally tagged fusion proteins is the preferred translation start codon context (Salim *et al.*, 2008). The intended start codon can become highly disfavored if placed immediately after a GC-rich restriction site introduced for cloning.

After choosing a tag, the next consideration is how to engineer tagged protein expression from a genomic locus. We have targeted DNA integration in the somatic macronucleus, although similar principles would apply to micronuclear genome targeting as well. The endogenous locus encoding the protein to be tagged is a natural choice for targeting. In this case, in the targeting vector a drug resistance cassette is placed upstream of the gene for an N-terminal tag or downstream for a C-terminal tag. The most commonly used resistance cassettes are those from the *neo* series, which confer resistance to paromomycin (Gaertig *et al.*, 1994a; Mochizuki, 2008). We have recently modified the *bsr1* cassette by replacement of the histone H4-I (*HHF1*) promoter with the *MTT1* promoter to create *bsr2,* which is analogous to *neo3* (*MTT1* promoter/resistance gene/*BTU2* 3′ untranslated region (UTR) and polyadenylation signal) but confers resistance to blasticidin. Expression from the *HHF1* promoter can be too high to efficiently select for complete assortment of the locus (see Chapters 10 and 11), whereas basal expression from the *MTT1* promoter (without induction with cadmium) is low enough to do so in a reliable manner (unpublished data).

Enough of the targeted genomic locus must be cloned between the resistance cassette and the tag to include the promoter and 5′ UTR (if the tag is N-terminal) or to include the 3′ UTR and polyadenylation signal (if the tag is C-terminal). In

T. thermophila, average UTR lengths are on the order of 150 nucleotides, judging from expressed sequence tag mapping of polyadenylation sites (Coyne *et al*., 2008). For a gene without a definitively annotated mRNA, reverse transcription (RT) and PCR with primer walking or RACE should be performed to map the UTR(s). Complete promoter regions are harder to define by rapid experimental analysis, so there is always a risk that insertion of the drug resistance cassette will perturb promoter activity. Another risk is the possibility that the tag will be separated from the resistance cassette and lost due to homologous recombination in the intervening region. Nonetheless, we have been successful using this strategy when tag fusion is not functionally deleterious for a growth-essential protein. Assortment of the targeted chromosome can be easily monitored during selection in polyclonal cell populations using whole-cell PCR (when the PCR works): boil fewer than 15 cells in 20 μL sterile filtered water for 5 min and then add PCR mix and perform PCR as usual. Another method has recently been established for N-terminal tagging using the Cre/loxP system (Busch *et al*., 2010).

Other options include using an ectopic promoter and 5′ UTR or an ectopic 3′ UTR in the tagging cassette, which precludes loss of linkage between the tag and marker by recombination, or integrating a complete transgene at an ectopic locus with or without subsequent disruption of the endogenous gene locus (Lee *et al*., 2009; Min and Collins, 2009). For protein overexpression from an ectopically integrated transgene, we commonly design an ORF to be under expression control of the cadmium-inducible *MTT1* promoter (Shang *et al*., 2002) at the *BTU1* locus of the strain CU522, which allows for integration selection by taxol resistance without requiring a drug resistance cassette in the transgene vector (Gaertig *et al*., 1994b). For induction, cadmium is added to a final concentration in the range of 0.1–1.0 μg/mL. Exact cadmium concentration will depend on the desired level of overexpression and the choice of medium (see Chapter 8); richer media require more cadmium. For example, starving cells (cultured in 10 mM Tris) reach maximum expression from an *MTT1* promoter at a lower cadmium concentration than growing cells (Shang *et al*., 2002). Another variable is the iron source in the media (see Chapter 8). If Sequestrene® is used for the iron source, it will chelate some of the cadmium. *T. thermophila* has four additional metallothionein genes, two others most highly induced by cadmium and two induced by copper (Boldrin *et al*., 2006; Diaz *et al*., 2007). We have successfully used the promoters from *MTT1*, *MTT2*, *MTT3*, and *MTT5* to drive inducible transgene expression.

Overexpression can be a useful strategy to make readily detectable amounts of protein for immunofluorescence or small-scale affinity purification. However, we have recently come to suspect that even relatively minor tagged protein overexpression, corresponding to basal transcription from the *MTT1* promoter integrated at the *BTU1* locus, can alter overall protein distribution compared that in cells with protein expression from the endogenous locus (unpublished data). An alternative approach is to use the *MTT1* promoter at its endogenous locus where basal transcription appears much lower (Shang *et al.*, 2002).

II. Whole-Cell Protein and Nucleic Acid Isolation, Detection, and Quantification

A. Genomic DNA Isolation

Once a transgene construct is transformed, a variety of methods can be used to test for incorporation into the genome, assortment, and expression. Whole-cell PCR, as discussed above, is the fastest way to check for integration of DNA at the targeted locus. However, it does not work reliably in our hands. Alternately, genomic DNA can first be isolated (scale as desired and mix gently but *thoroughly* throughout the following steps): collect 2.5×10^5 cells and concentrate to 50 μL in 10 mM Tris, pH 7.5. Resuspend in residual Tris and add 200 μL 60 °C lysis buffer (10 mM Tris, pH 7.5, 0.5 M EDTA, 1% SDS, pH adjusted to 9.5 at 60 °C). Add 2 volumes water and incubate at 60 °C at least 1 h. Cool to room temperature, add Proteinase K to 50 μg/mL, and incubate at 37 °C overnight. Extract with one volume phenol/chloroform/isoamyl alcohol (PCI) and precipitate with one-tenth volume sodium acetate, pH 5.2, and one volume isopropanol. Wash pellet in 70% ethanol and resuspend in 75 μL 1× TE. Add RNase A to 0.8 μg/μL and incubate at 37 °C for 30 min. PCI extract, precipitate, and wash as before. This method yields ~15 μg total (macronuclear and micronuclear) genomic DNA per 2.5×10^5 cells. Of note, the AT-rich composition of the *T. thermophila* genome generally necessitates using long primers for reliable PCR, where $T_m \sim 55$ °C using the formula $T_m = 64.9\ °C + [41\ °C\ (\text{number of G} + \text{C}-16.4)/N]$, where N is the length of the primer.

B. Southern Blotting

For a quantitative assessment of assortment, Southern blotting is used (Malone *et al.*, 2005). If the method for DNA isolation does not specifically purify macronuclei, a wild-type locus-sized restriction fragment may be detectable even in a strain in which the macronuclear gene copies are fully replaced. This derives from the diploid micronucleus, which is mostly at 4N since it replicates so early relative to the macronucleus and thus can be only ~10 times less abundant in gene content than the polyploid macronucleus (see Chapter 7). If it is ambiguous whether the endogenous locus has been fully replaced, RT-PCR should be used to check for any remaining endogenous transcript expression. We have noticed that some loci do not seem to assort (or back-assort) as quickly as others (unpublished observation). This means that even after allowing clonal populations to back-assort for the standard 14 days of rapid doubling, the remaining wild-type locus genomic locus restriction fragment may still be hard to discern even if assortment is incomplete.

Therefore, we have made it a common practice to test strains that appear fully assorted by RT-PCR for the mRNA, which is only produced from the macronuclear gene locus.

C. Immunoblotting

Transgene product expression levels can be monitored using whole-cell immunoblotting. Collect 2–4 × 10^5 cells, depending on the expected protein expression level. Rinse cells in 10 mM Tris, pH 7.5, and concentrate to 30 μL. Add protease inhibitors (see the following section for details) and then SDS-PAGE loading buffer to a final concentration of 1.5× (5× stock is 0.3 M Tris, pH 6.8, 10% SDS, 20% β-mercaptoethanol, 50% glycerol). Boil immediately for 5 min and freeze solid (not all freezers will prevent sample degradation) or preferably resolve immediately by SDS-PAGE.

D. Northern Blotting

Transcript mRNA levels can be monitored using northern blotting. Typically 20–30 μg total RNA (2–5 × 10^5 growing cell equivalents when isolated using TRIzol® reagent) in a 2 mm × 7 mm × 10 mm well volume is sufficient to detect low to moderately abundant transcripts. For very low abundance transcripts, a poly (A)+ enrichment step can be included to enrich for mRNAs. For a detailed description of Northern blotting and all other RNA methods, see "RNA: A laboratory manual" (Rio *et al.*, 2011).

A major focus in our lab is the study of ribonucleoprotein (RNP) complexes. Therefore, sensitive detection of small, noncoding RNAs is vital. Small RNAs (sRNAs) are detected using a modified northern blotting protocol, with or without a prior filtration-based enrichment step that removes the large rRNAs and mRNAs (Lee and Collins, 2006). For abundant RNAs of less than 200 nucleotides in length, such as tRNAs, starting with less than 1 μg total RNA is sufficient for visualization of size-enriched RNA by SYBR Gold and northern blotting. For low abundance sRNAs like some Piwi-bound sRNAs, up to 5 μg enriched sRNA may be needed to detect a particular sRNA sequence of interest by northern blotting. This is typically obtained from 200–500 μg of total RNA. For very low abundance species, modifications to the northern blotting procedure can be used to improve the hybridization sensitivity (Pall *et al.*, 2007).

E. Subcellular Fractionation

It may be of interest to differentiate nuclear from cytoplasmic complexes using subcellular fractionation before applying the detection techniques discussed above and/or prior to the purification methods discussed below. Robust methods for isolation of macronuclei, micronuclei, developing macronuclei (anlagen), and nucleoli have been developed (Allis and Dennison, 1982; Gocke *et al.*, 1978; Gorovsky *et al.*, 1975), which we have used prior to affinity purification or immunofluorescence.

III. Cell Lysate Preparation and Affinity Purification

Numerous protocol considerations should factor into an experimental optimization of both the starting extract and the subsequent affinity purification. The process below provides working guidelines and typical procedures in point-by-point detail.

1. Grow cells. Grow 10 mL to 1 L cultures of the tagged protein strain and a control wild-type untagged protein strain (mock) to log phase or the desired life cycle stage. All efforts should be made to treat cultures similarly between experiments to be compared, including choice of growth medium, cell density, amount and duration of cadmium, or other treatment, etc.
2. Collect cells. Spin 1500 × g for 3 min (in a braked rotor) or longer for faster swimming starving cells and wash in 10 mM Tris, pH 7.5 (if necessary to remove rich media contaminants that can increase proteolysis). Alternatively wash in PBS or Dryl's (Dryl, 1959) for downstream applications not compatible with Tris. *Note*. Cells may undergo osmotic stress in 1× PBS.
3. Optional *in vivo* crosslinking, for example, with formaldehyde or UV (Dedon *et al.*, 1991).
4. *Lyse cells*: Lysis conditions will vary depending on desired purification stringency, protein complex characteristics, and whether a crosslinking step was included. Always lyse at 4 °C to reduce protease and nuclease activities, usually at 3–5 × 10^6 cell equivalents/mL for 10–15 min. Conditions will have to be empirically determined for each new experiment. Guidelines follow.
   Native (No crosslink): 20 mM Tris pH 7.5, 0.05–1.0 M NaCl, 10% glycerol, 0.1–0.2% Igepal®, 0.1–0.2% Triton X-100 (optional to more efficiently lyse nuclei). Add protease inhibitors fresh just before use. We use a mammalian protease inhibitor cocktail (Sigma) as well as a final concentration of 0.1 mM phenylmethylsulfonyl fluoride (PMSF) made as a 0.1 M stock in isopropanol. Note that PMSF is inactivated in aqueous solutions, so add it fresh. Keep the protease inhibitor addition to less than 1/500th total volume to minimize the concentration of organic solvents, unless the purification is done under denaturing conditions.
   Optional: (a) 1 mM $MgCl_2$ stabilizes some protein–RNA interactions but will also allow Mg^{2+}-dependent enzymes be catalytically active and release their RNA substrates; (b) 1 mM EDTA as a protease and RNase inhibitor, which can also stall enzymes on RNA substrates by Mg^{2+} chelation; (c) reducing agent such as 0.5–1 mM DTT or 10–20 mM β-mercaptoethanol.
   Denaturing (crosslink step included): Radioimmunoprecipitation assay (RIPA) buffer (Harlow and Lane, 1988) is commonly used, with Triton X-100 in place of Igepal® for chromatin applications. Igepal® may inhibit DNA shearing (unpublished observation).

Resuspend to 5×10^6 cells/mL.
Disrupt cells by sonication.

5. *Clear lysate*: Spin at 16,000 × g for 15 min for a crude lysate or 100,000 × g for 1 h, which may remove polyribosomes and other large macromolecular complexes from the lysate.
6. *Optional*: flash freeze lysate and store −80 °C. This may reduce yield in some cases.
7. *Binding*: Add prewashed antibody-conjugated beads to cleared lysate. A good starting point is 4 μL 50% bead slurry/mL lysate. Larger bead volumes can increase recovery, but can also disproportionately increase background. The amount used depends on binding capacity of beads and concentration of lysate. Bind at room temperature for 1–2 h or at 4 °C for 1.5 h to overnight.
8. *Wash beads*: Wash four to six times in at least 20 bead-slurry volumes 5 min at room temperature or 4 °C. Wash buffer should be similar to binding buffer but is often higher stringency: for example, higher detergent concentration or additional detergents, higher salt concentration, and the addition of urea to 2 M (or higher in the case of prior crosslinking).
9. *Elute protein/RNP complex*: Tag-specific elution results in the lowest background and recovery of native (possibly functional) complexes that can be used for activity assays or a second step of immunoprecipitation. Examples are TEV protease and 3XFLAG peptide. Other options include denaturation by SDS, urea, or low pH (0.1 M glycine pH 2.7).

IV. Detection of Recovered Complexes

When beginning an affinity purification experiment, recovery of the protein and/or RNA of interest should be tracked throughout the procedure using Coomassie staining and/or immunoblotting for protein and using SYBR Gold staining and/or northern blotting for RNA. To direct the reiterative process of purification optimization, complexes should be tracked in whole cells, soluble cell lysate (the input extract), postbinding supernatant (the unbound fraction or flow-through), eluate, and purification resin (beads) postelution (and/or an aliquot of pre-elution beads). Pellets can also be tracked but are often difficult to resuspend uniformly.

After optimization, generally only a fraction of the eluate is sacrificed for SDS-PAGE and silver staining to detect the presence of the desired tagged protein and specific co-purifying proteins, as judged by comparison to background proteins recovered in a parallel mock purification. The remainder of the eluate is then used for activity assays, identification of proteins by immunoblot or mass spectrometry, or analysis of co-purified nucleic acids. To extract DNA, add equal volume of phenol/chloroform/isoamyl alcohol (PCI). To extract RNA, add equal volume of PCI or 10× volume of TRIzol® reagent. Caution: remember to reverse crosslink (if relevant and possible) or treat with proteinase K before extracting DNA or RNA. DNA and RNA can be differentiated by treatment with DNase or RNase.

Co-purified nucleic acids can be detected by PCR, RT-PCR, or northern blotting if the sequence is known. A valuable application of affinity purification is in the identification of unknown nucleic acid species associated with a protein. For example, after native affinity purification, stably bound sRNAs can be purified, ligated to adaptors, and reverse transcribed to make a cDNA library for high-throughput sequencing (sRNA-seq) (Couvillion *et al.*, 2009). Similarly, longer RNAs can be immunoprecipitated and prepared for sequencing (RIP-seq) (Zhao *et al.*, 2010). After crosslinking and denaturing purification, DNA from chromatin immunoprecipitation can be prepared for sequencing (ChIP-seq) (Park, 2009). A variety of methods including high-throughput sequencing of RNA isolated by crosslinking immunoprecipitation (HITS-CLIP) (Licatalosi *et al.*, 2008) and photoactivatable-ribonucleoside-enhanced crosslinking and immunoprecipitation (PAR-CLIP) (Hafner *et al.*, 2010) have been developed to improve recovery of RNA that only transiently interacts with a protein.

Acknowledgments

Funding for *T. thermophila* research in the Collins lab is supported by NIH RO1 GM54198.

References

Allis, C. D., and Dennison, D. K. (1982). Identification and purification of young macronuclear anlagen from conjugating cells of *Tetrahymena thermophila*. *Dev. Biol.* **93**, 519–533.

Boldrin, F., Santovito, G., Gaertig, J., Wloga, D., Cassidy-Hanley, D., Clark, T. G., and Piccinni, E. (2006). Metallothionein gene from *Tetrahymena thermophila* with a copper-inducible-repressible promoter. *Eukaryot. Cell* **5**, 422–425.

Busch, C. J., Vogt, A., and Mochizuki, K. (2010). Establishment of a Cre/loxP recombination system for N-terminal epitope tagging of genes in *Tetrahymena*. *BMC Microbiol.* **10**, 191.

Couvillion, M. T., Lee, S. R., Hogstad, B., Malone, C. D., Tonkin, L. A., Sachidanandam, R., Hannon, G. J., and Collins, K. (2009). Sequence, biogenesis, and function of diverse small RNA classes bound to the Piwi family proteins of *Tetrahymena thermophila*. *Genes Dev.* **23**, 2016–2032.

Coyne, R. S., Thiagarajan, M., Jones, K. M., Wortman, J. R., Tallon, L. J., Haas, B. J., Cassidy-Hanley, D. M., Wiley, E. A., Smith, J. J., Collins, K., Lee, S. R., Couvillion, M. T., Liu, Y., Garg, J., Pearlman, R. E., Hamilton, E. P., Orias, E., Eisen, J. A., and Methe, B. A. (2008). Refined annotation and assembly of the *Tetrahymena thermophila* genome sequence through EST analysis, comparative genomic hybridization, and targeted gap closure. *BMC Genomics* **9**, 562.

Dedon, P. C., Soults, J. A., Allis, C. D., and Gorovsky, M. A. (1991). A simplified formaldehyde fixation and immunoprecipitation technique for studying protein–DNA interactions. *Anal. Biochem.* **197**, 83–90.

Diaz, S., Amaro, F., Rico, D., Campos, V., Benitez, L., Martin-Gonzalez, A., Hamilton, E. P., Orias, E., and Gutierrez, J. C. (2007). *Tetrahymena* metallothioneins fall into two discrete subfamilies. *PLoS One* **2**, e291.

Dryl, S. (1959). Antigenic transformation in *Paramecium aurelia* after homologous antiserum treatment during autogamy and conjugation. *J Protozool.* **6**(Suppl.), 25.

Gaertig, J., Gu, L., Hai, B., and Gorovsky, M. A. (1994a). High frequency vector-mediated transformation and gene replacement in Tetrahymena. *Nucleic Acids Res* **22**, 5391–5398.

Gaertig, J., Thatcher, T. H., Gu, L., and Gorovsky, M. A. (1994b). Electroporation-mediated replacement of a positively and negatively selectable beta-tubulin gene in Tetrahymena thermophila. *Proc. Natl Acad. Sci. U.S.A.* **91**, 4549–4553.

Gocke, E., Leer, J. C., Nielsen, O. F., and Westergaard, O. (1978). Transcriptional properties of nucleoli isolated from *Tetrahymena*. *Nucleic Acids Res.* **5**, 3993–4006.

Gorovsky, M. A., Yao, M. C., Keevert, J. B., and Pleger, G. L. (1975). Isolation of micro- and macronuclei of *Tetrahymena pyriformis*. *Methods Cell Biol.* **9**, 311–327.

Hafner, M., Landthaler, M., Burger, L., Khorshid, M., Hausser, J., Berninger, P., Rothballer, A., Ascano Jr, M., Jungkamp, A. C., Munschauer, M., Ulrich, A., Wardle, G. S., Dewell, S., Zavolan, M., and Tuschl, T. (2010). Transcriptome-wide identification of RNA-binding protein and microRNA target sites by PAR-CLIP. *Cell* **141**, 129–141.

Harlow, E., and Lane, D. A. (1988). *Antibodies: A Laboratory Manual.* Cold Spring Harbor Press, New York.

Lee, S. R., and Collins, K. (2006). Two classes of endogenous small RNAs in *Tetrahymena thermophila*. *Genes Dev.* **20**, 28–33.

Lee, S. R., and Collins, K. (2007). Physical and functional coupling of RNA-dependent RNA polymerase and Dicer in the biogenesis of endogenous siRNAs. *Nat. Struct. Mol. Biol.* **14**, 604–610.

Lee, S. R., Talsky, K. B., and Collins, K. (2009). A single RNA-dependent RNA polymerase assembles with mutually exclusive nucleotidyl transferase subunits to direct different pathways of small RNA biogenesis. *RNA* **15**, 1363–1374.

Licatalosi, D. D., Mele, A., Fak, J. J., Ule, J., Kayikci, M., Chi, S. W., Clark, T. A., Schweitzer, A. C., Blume, J. E., Wang, X., Darnell, J. C., and Darnell, R. B. (2008). HITS-CLIP yields genome-wide insights into brain alternative RNA processing. *Nature* **456**, 464–469.

Malone, C. D., Anderson, A. M., Motl, J. A., Rexer, C. H., and Chalker, D. L. (2005). Germ line transcripts are processed by a Dicer-like protein that is essential for developmentally programmed genome rearrangements of *Tetrahymena thermophila*. *Mol. Cell Biol.* **25**, 9151–9164.

Min, B., and Collins, K. (2009). An RPA-related sequence-specific DNA-binding subunit of telomerase holoenzyme is required for elongation processivity and telomere maintenance. *Mol. Cell* **36**, 609–619.

Mochizuki, K. (2008). High efficiency transformation of *Tetrahymena* using a codon-optimized neomycin resistance gene. *Gene* **425**, 79–83.

Mochizuki, K., Fine, N. A., Fujisawa, T., and Gorovsky, M. A. (2002). Analysis of a piwi-related gene implicates small RNAs in genome rearrangement in *Tetrahymena*. *Cell* **110**, 689–699.

Pall, G. S., Codony-Servat, C., Byrne, J., Ritchie, L., and Hamilton, A. (2007). Carbodiimide-mediated cross-linking of RNA to nylon membranes improves the detection of siRNA, miRNA and piRNA by northern blot. *Nucleic Acids Res.* **35**, e60.

Park, P. J. (2009). ChIP-seq: advantages and challenges of a maturing technology. *Nat. Rev. Genet.* **10**, 669–680.

Rio, D. C., Hannon, G. J., Ares Jr, M., and Nilsen, T. W. (2011). *RNA: A Laboratory Manual.* Cold Spring Harbor Press, New York.

Salim, H. M., Ring, K. L., and Cavalcanti, A. R. (2008). Patterns of codon usage in two ciliates that reassign the genetic code: *Tetrahymena thermophila* and *Paramecium tetraurelia*. *Protist* **159**, 283–298.

Shang, Y., Song, X., Bowen, J., Corstanje, R., Gao, Y., Gaertig, J., and Gorovsky, M. A. (2002). A robust inducible-repressible promoter greatly facilitates gene knockouts, conditional expression, and over-expression of homologous and heterologous genes in *Tetrahymena thermophila*. *Proc. Natl. Acad. Sci. U.S.A.* **99**, 3734–3739.

Yu, L., and Gorovsky, M. A. (2000). Protein tagging in *Tetrahymena*. *Methods Cell Biol.* **62**, 549–559.

Zhao, J., Ohsumi, T. K., Kung, J. T., Ogawa, Y., Grau, D. J., Sarma, K., Song, J. J., Kingston, R. E., Borowsky, M., and Lee, J. T. (2010). Genome-wide identification of polycomb-associated RNAs by RIP-seq. *Mol. Cell* **40**, 939–953.

CHAPTER 13

Cytological Analysis of *Tetrahymena thermophila*

Mark Winey*, Alexander J. Stemm-Wolf*, Thomas H. Giddings Jr.* and Chad G. Pearson†

*MCD Biology, University of Colorado at Boulder, 347 UCB, Boulder, Colorado 80309-0347, USA

†Department of Cell and Developmental Biology, University of Colorado – Denver, RC1 South, MS 8108, 12801 E. 17th Ave. Room 12104, Aurora, Colorado 80045, USA

Abstract

Since their first detection in pond water, large ciliates such as *Tetrahymena thermophila,* have captivated school children and scientists alike with the elegance of their swimming and the beauty of their cortical organization. Indeed, cytology – simply looking at cells – is an important component of most areas of study in cell biology and is particularly intriguing in the large, complex *Tetrahymena* cell. Cytological analysis of *Tetrahymena* is critical for the study of the microtubule cytoskeleton, membrane trafficking, complex nuclear movements and interactions,

0091-679X/10 $35.00
DOI 10.1016/B978-0-12-385967-9.00013-X

and the cellular remodeling during conjugation, to name a few topics. We briefly review previously reported cytological techniques for both light and electron microscopy, and point the reader to resources to learn about those protocols. We go on to present new and emerging technologies for the study of these marvelous cells. These include the use of fluorescent-protein tagging to localize cellular components in live cells, as well as for tracking the dynamic behavior of proteins using pulse labeling and fluorescence recovery after photobleaching. For electron microscopy, cellular and antigenic preservation has been improved with the use of cryofixation and freeze-substitution. The technologies described here advance *Tetrahymena* cell biology to the cutting-edge of cytological analysis.

I. Introduction

Tetrahymena cells and their behavior during their life cycle offer a wealth of interesting cytology (Wloga and Frankel, *this volume*). Simply examining the cells in the light or electron microscope can be informative about a variety of cellular processes. In fact, we routinely cite classic *Tetrahymena* cytological studies, such as Dick Allen's description of basal body assembly based on electron microscopy (Allen, 1969); see also http://www5.pbrc.hawaii.edu/allen/ch18/ and also the ASCB cell image library http://www.cellimagelibrary.org/). As detailed below, the previous 2000 edition of this volume contains still relevant chapters describing protocols for light or electron microscopy. However, since that time gene discovery and analysis in *Tetrahymena* has been greatly enhanced by the availability of the genome sequence and facilitated by improved fluorescent protein tagging constructs coupled with efficient means to alter gene function. Given such advances, robust methods for protein localization and structural analyses are necessary to accomplish the outstanding cell biology that can be done in these cells. It has been demonstrated that large-scale proteomic or genomic screens followed by localization of proteins can effectively reveal genes of interest to the investigator (e.g., Bright *et al.*, 2010; Cole *et al.*, 2008; Jacobs *et al.*, 2006; Kilburn *et al.*, 2007). Furthermore, sensitive live-cell and high-resolution electron microscopy techniques are critical for phenotypic analysis of mutant strains. We present methods and reagents for important microscopy techniques that have been implemented over the last 10 years, and we suggest emerging cytological techniques that should be valuable when applied to *Tetrahymena thermophila*.

II. Rationale

As noted above, cytology is an important component of cell biological investigations in *Tetrahymena*, as in other organisms. The major applications for cytological protocols are protein localization and phenotypic analysis. Both of these applications can be pursued in the light microscope or the electron microscope, and some of the protocols for these applications are presented below. Furthermore, the combined use of both types of instruments to determine protein localization or to reveal a mutant phenotype can provide the most robust and revealing description. *Tetrahymena* is a wonderful subject for microscopy, but it does offer some challenges – its large size can make it difficult to effectively fix for EM, and its motility can be an issue for live-cell imaging. Nonetheless, many of the liabilities of doing cytology in *Tetrahymena* can be mitigated with newer reagents and techniques described here.

III. Light Microscopy

Specific model systems in cell biology are used because of the advantages that they possess. *Tetrahymena* is a good cytological model system because cells are large, highly organized, and several cellular structures are amplified, allowing the researcher to more easily identify structures and events of interest. Protein tagging strategies for fixed cell protein localization using light microscopy has generated a wealth of novel information that contributes to our understanding of the cortical cytoskeleton, nuclear architecture, and cilia function. While still in its infancy, the utility of *Tetrahymena* and specific live-cell imaging strategies are now being developed for a real time view of these biological events. Such strategies allow us to view cellular events as they occur, the dynamics of proteins and DNA, and the short-lived, transient structural events that lead to a final product. The ability to visualize transient events that may not be captured by fixed time-point studies, because they are short-lived, will help reveal key functional events that happen during the cell cycle.

A. Previous Methods and Resources

In the previous edition, Stuart and Cole (2000) presented protocols for the preparation and imaging of immuno-fluorescently labeled fixed cells. In a separate chapter, these authors (Cole and Stuart, 2000) described classical staining techniques for bright-field microscopy. In a later publication, Cole *et al.* (2002) and others described additional techniques for *in situ* hybridization and for live-cell imaging using devices, such as the rotocompressor, that immobilize cells (e.g., Aufderheide, 2008; Cole *et al.*, 2002; Loidl and Scherthan, 2004; Wolfe and Colby, 1981). All of these techniques are still relevant and applicable to *Tetrahymena*. We present further "low-tech" solutions for live-cell imaging, along with new applications based on the ability to tag genes with various fluorescent proteins.

B. New Methods

1. Live-Cell Imaging

A key technological advance in recent years with *Tetrahymena thermophila* studies is the application of green fluorescent protein (GFP) and its spectral variants,

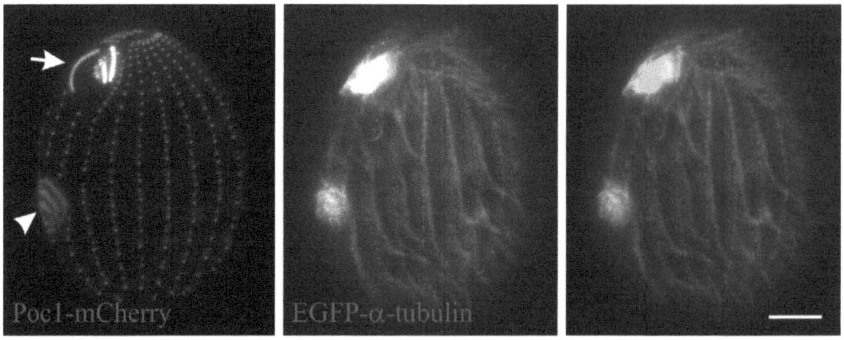

Fig. 1 Live cell, two color colocalization using EGFP and mCherry fusion proteins. Co-localization of Poc1-mCherry (left panel, red) and EGFP-α-tubulin (center panel, green) in a live *Tetrahymena thermophila* cell that is in mitosis. Immobilized cells were imaged for red and green fluorescence using a through volume Z-series. A maximum fluorescence intensity projection was generated using Metamorph Imaging Software (Molecular Devices). Arrow denotes the old oral apparatus and the arrowhead denotes the nascent oral apparatus or oral primordium. For visualization of individual cilia, the image brightness was increased leading to saturation of the oral apparatus GFP-α-tubulin fluorescence. Scale bar, 10 μm. (See color plate.)

which have been used to create fusion proteins labeled with different colors in the same cell (Fig. 1). With this advance comes the ability to visualize the assembly and structural localization of proteins and chromosomes.

Cell Immobilization

Because *Tetrahymena* are vigorous swimmers, live cells must be adhered to a substrate before cellular structures can be followed for imaging. A number of historical studies have identified methods for limiting cell motility during live-cell imaging (Cole *et al.*, 2002). This is still a significant challenge given the hundreds of undulating cilia that propel cells to almost 0.5 mm/s. A range of tactics has been used to slow or stop cell motility (Aufderheide, 2008; Bright *et al.*, 2010). We find that a combination of compression of cells between the glass coverslip and slide in the presence of a high viscosity medium works well for short imaging times of up to 30 min.

While the below methods have worked well for many of our experiments, we expect to develop robust strategies for long time period imaging in the future. We have, so far, been unsuccessful in using optical lithography with microfabricated chambers, and this technology may also provide a means to immobilize cells in the future. Furthermore, an exciting new possibility is the use of temperature-regulated polymers that have already shown promise in the Turkewitz lab (Bright *et al.*, 2010). The techniques described below provide a short-term imaging tactic to follow localization in live cells (Pearson *et al.*, 2009a, 2009b; Fig. 1).

Methods

1. Grow cells to mid-log phase.
2. Spin down ~1 mL of cells at 3000 × g for 30 s.
3. Aspirate supernatant.
4. Gently resuspend cells in ~0.2 mL of *Imaging Media*. Allow the cells to equilibrate for at least 20 min in the viscous *Imaging Media*.
5. Prepare a microscope slide by placing four small (~1 mm^2) spots of silicone vacuum grease on each corner of an 18 mm^2 coverslip.
6. Apply 50 μL of fresh *Imaging Media* on the center of a microscope slide.
7. Apply 20 μL of the resuspended cells on top of the *Imaging Media* pad.
8. Cover with the above prepared coverslip so that media is dispersed and cells become trapped in the viscous solution between the coverslip and slide. The silicone creates a buffer distance between the slide and coverslip so that the cells are not overly compressed or lysed. Use a pipette tip to apply pressure to all four corners of the coverslip. To minimize tilting of the coverslip, it is best to apply pressure to all four corners at the same time.
9. Aspirate or wick away any residual media that is dispersed from the coverslip.
10. Monitor cell motility using a transmitted light microscope at low magnification.
11. If required, apply additional pressure to the corners to inhibit cell motility.
12. Once cell motility is abated, evaporation of the media from the chamber can be minimized by sealing the coverslip to the microscope slide using a thin film of melted *VALAP* around the edges. This allows for air exchange but reduces evaporation.
13. Transfer prepared sample to an appropriate light microscope to visualize cells.

Materials

Imaging Media (modified 1% SPP Cole and Stuart, 2008)

Proteose peptone	1%
Yeast extract	0.1%
Glucose	0.2%
Poly(ethylene oxide) MW 900,000 (PEO, Sigma)	3%

VALAP

Vasolin	30%
Lanolin	30%
Paraffin	30%

Notes

It is important to minimize background autofluorescence to obtain a high signal-to-noise ratio when imaging fluorescent proteins. Media is an unfortunate source of background autofluorescence. We minimize this source of background signal by keeping the proteose peptone at 1% as opposed to the 2% that we use to grow cells in culture. Also, we do not autoclave our media with glucose but rather add sterile glucose at the appropriate concentration (0.2% final) after autoclaving to minimize caramelization of the sugar. Reducing the total media concentration also improves the efficiency with which cells are immobilized on the glass surface.

2. Fluorescent Protein Pulse Experiments

Meselson and Stahl (1958) famously used pulse labeling of DNA to show that DNA is semiconservatively replicated. Additionally, George Palade used strategies to pulse label proteins and define the mechanisms of the cellular secretory pathway (Caro and Palade, 1964; Jamieson and Palade, 1967a, 1967b). These pioneering techniques elucidated the mechanisms of fundamental cellular processes with radioactive markers.

Pulse-chase experiments can now be used in live cells to follow the fate of newly expressed proteins fused to EGFP to assess the assembly and turnover dynamics of both nucleic acids and proteins. Reagents exist for the expression and repression of genes fused to EGFP in *Tetrahymena*. Transient regulation of gene expression can be controlled by promoters that are sensitive to metals in the media (e.g., cadmium and copper, Boldrin *et al.*, 2008; Shang *et al.*, 2002). GFP-tagged proteins are then pulsed on with the addition of metals. Within minutes, EGFP is visualized and the incorporation and dynamics of proteins at their site of activity can be monitored with high temporal resolution. Examples of these experiments are provided in Fig. 2. Furthermore, the ultrastructural localization of the dynamics can be visualized using immuno-EM that is described in the EM section of this chapter and as shown in Pearson *et al.* (2009a).

Methods

1. Grow cells containing MTT-EGFP-Your Favorite Gene (YFG) to mid-log phase or to a cell cycle arrest in SPP media. The constructs for generating these strains are described below.
2. Image live cells to ensure that the EGFP signal is not detectable (this indicates that the promoter is not leaky; see *Notes*).
3. Induce the expression of EGFP-YFG by adding 0.1–1.0 μg/mL $CdCl_2$ to the culture. Concentration will depend both on YFG and the media that you are growing your cells in (less $CdCl_2$ is required in Starvation Media (10 mM Tris pH 7.4)). We use low concentrations of cadmium to minimize deleterious effects from protein overexpression. In addition, high concentrations of $CdCl_2$ (>1.0 μg/mL) affect cell growth rates (Larsen, 1989).
4. After EGFP-YFG is synthesized, cellular expression can be stopped by washing the cells three times with fresh SPP media. Alternatively, EGFP-YFG expression can be constitutively expressed by maintaining the $CdCl_2$ in the media.

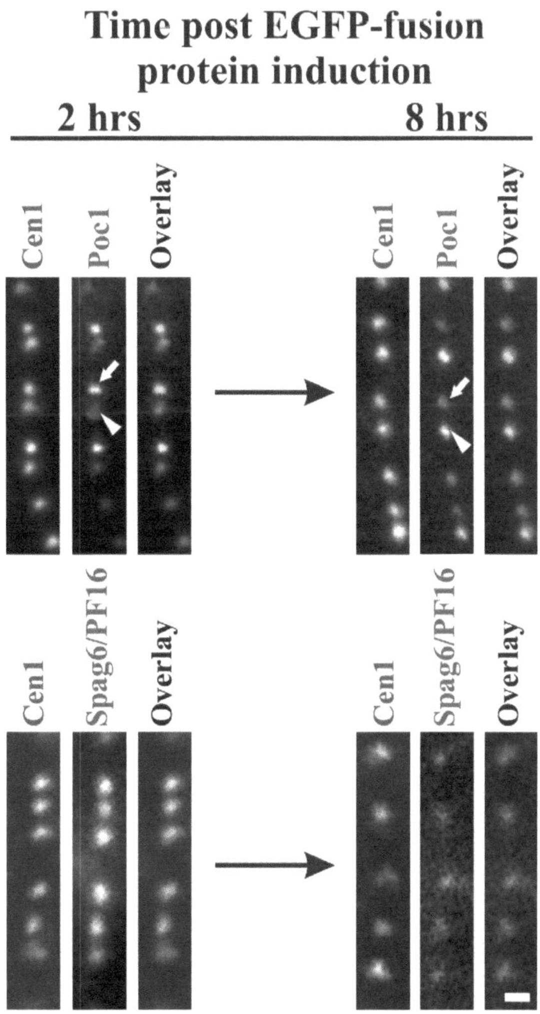

Fig. 2 Pulse-labeling using EGFP fusions. Pulsed expression of EGFP fusion proteins can be used to visualize protein assembly at their binding sites. Expression of basal body proteins Poc1 (top panels) and Spag6/PF16 (bottom panels) controlled by the metallothionein (MTT1) promoter was induced by addition of $CdCl_2$ to the media. All basal bodies were uniformly labeled using α-centrin staining. Fluorescence signal of the EGFP is not observed prior to induction (data not shown). By two hours post-induction basal bodies are labeled in a manner that represents the turnover dynamics of each component. Poc1 labels existing basal bodies with a low level of fluorescence (arrowhead) that represents dynamic protein turnover. Basal bodies that assemble in the presence of EGFP-Poc1 are brightly labeled (arrow) indicating that new basal body assembly is required for complete incorporation of signal. These assembly dynamics correspond to the turnover dynamics measured by FRAP (Fig. 3). In contrast, EGFP-Spag6/PF16 exhibits rapid assembly at all basal bodies shortly after EGFP-Spag6/PF16 induction indicating that it exhibits rapid exchange with its basal body binding sites, independent of new basal body assembly. By eight hours, most basal bodies have assembled in the presence of EGFP tagged proteins and these levels represent steady state incorporation dynamics. (Figure adapted from Pearson *et al.* (2009a, 2009b) Scale bar, 1 μm. (See color plate.)

5. At defined time points after expression, EGFP-YFG localization to the site of activity can be followed by EGFP fluorescence.

Materials

$CdCl_2$Stock	
$CdCl_2$	1 mg/mL
ddH_2O	
Filter sterilize	
Starvation Media	
Tris base	10 mM
dd$H_2$0	
pH to 7.4 and filter sterilize.	

Notes

It is important to monitor the level of EGFP fluorescence signal that is expressed in the absence of $CdCl_2$. The metallothionine (MTT) promoter can exhibit a low level of expression without induction. This is likely due to small amounts of metals in the media that activate the promoter. This can be limited by growing cells in limited media defined by the Gaertig lab (J. Gaertig, personal communication). Finally, new promoter systems are available for use with copper (MTT2; Boldrin *et al.*, 2008). This advancement limits the use of toxic heavy metals.

3. Fluorescence Recovery After Photobleaching

Fluorescence recovery after photobleaching (FRAP) is a powerful strategy to measure protein dynamics and complement pulse-chase studies. The quantitative analysis of protein interactions and dynamics has historically been studied *in vitro*. However, GFP tagged proteins allow for the study of interactions and dynamics in live cells. This technique can be used to measure diffusion rates, dynamics of protein binding, and to measure interactions with other components. Several complementary studies and reviews describe FRAP technology and analysis in more detail (Salmon *et al.*, 1984a, 1984b; Sprague and McNally, 2005; Walczak *et al.*, 2010). We provide a brief introduction to the methods required for photobleaching and live-cell imaging of fluorescence recovery in *Tetrahymena* (Fig. 3).

Methods

1. Grow cells expressing EGFP-YFG to mid-log phase.
2. Prepare immobilized cells as described above.
3. Several methods exist for photobleaching. The most common are to either use a laser scanning confocal to scan and bleach a region of interest (ROI) or to use a

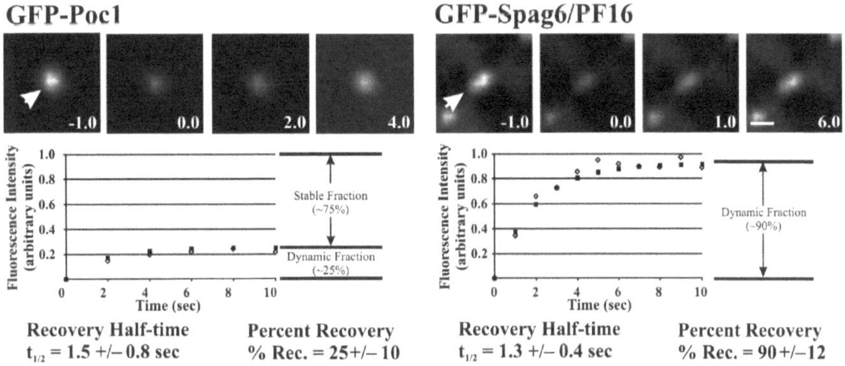

Fig. 3 Fluorescence recovery after photobleaching (FRAP) to visualize protein dynamics. Protein exchange at basal bodies is visualized by photobleaching bound EGFP-tagged protein at basal bodies and visualizing fluorescence recovery over time. Fluorescence recovery (FRAP) represents unbleached molecules in the cytoplasmic pool that replace the bleached GFP molecules. Basal body components exhibit divergent protein exchange. Poc1's binding to basal bodies is bimodal. ~25% of the basal body localized Poc1 protein is dynamic and ~75% is stable and no fluorescence recovery is observed. In contrast, Spag6/PF16 is almost completely dynamic with ~90% of the protein exchanging with rapid kinetics. (Figure adapted from Pearson et al. (2009a, 2009b)) Scale bar, 1 μm.

focused laser beam to bleach a specific region of interest. We will describe the later technique, as we believe this is an effective technique for FRAP studies. We use a Nikon TiE stand with a motorized X–Y stage. Laser light (either 488 nm (for GFP) or 564 nm (for mCherry)) is fiber-optically coupled to the back aperture of the objective, and the collimated laser beam is focused on to the specimen plane as a point. The beam spot size can be expanded by defocusing the laser beam. The site of the beam is then identified as an ROI.

4. Once the cellular target is chosen, the specimen is centered at the laser site (ROI) using the motorized stage driven by Nikon – NIS-Elements Software.
5. A prebleach image is acquired to determine the sample fluorescence intensity prior to photobleaching.
6. The samples are exposed to a short laser pulse (~50 ms) using a shuttered laser light. The laser exposure time, spot size, and intensity is altered depending on the experiment.
7. Immediately following photobleaching, a post-bleach image is then acquired ($t = 0$ s).
8. A time course is then acquired to follow the fluorescence recovery. The time interval between acquisition time points is critical for obtaining appropriate recovery curves to determine the protein turnover kinetics. Often this is achieved by trial and error.
9. Following FRAP image acquisition, the data are analyzed (Salmon *et al.*, 1984a, 1984b; Sprague and McNally, 2005; Walczak *et al.*, 2010).

Notes

The quantification of FRAP is an important aspect to define the turnover dynamics. For simplicity, we refer the reader to prior publications (Salmon *et al.*, 1984a, 1984b; Sprague and McNally, 2005; Walczak *et al.*, 2010). The key parameters to correct for are background fluorescence and the photobleaching that occurs from excitation light exposure during image acquisition. It is also important to ensure that the photobleaching laser irradiation is not causing cellular damage and affecting the biological event of interest.

C. Fluorescent Protein Tagging Strategies

Several technical advances have made creating fluorescent fusion proteins in *Tetrahymena* increasingly efficient. These include the availability of the *Tetrahymena* macronuclear genome sequence, use of metal-sensitive promoters, an increasing number of fluorescent proteins, a choice of drug resistance markers, and development of new vector systems designed to target gene fusions either to exogenous or endogenous loci. These strategies are equally applicable to other kinds of tags, but the focus here is on fluorescent tags for live-cell imaging.

Initially *Tetrahymena* GFP constructs were based on rDNA processing vectors, which, when introduced into the *Tetrahymena* macronucleus following mating, provide resistance to paromomycin. As rDNA sequences are processed and amplified in the maturing macronucleus, the transforming sequence is greatly amplified (Tondravi and Yao, 1986). Two vectors designed to create N-terminal GFP fusions were built into this system, pVGF-1 and pIGF-1 (Table I, Malone *et al.*, 2005; Wiley *et al.*, 2000; Yao *et al.*, 2007), the primary difference being in the promoters used to drive the constructs. pVGF-1 utilizes the rpL29 promoter, which is constitutive during vegetative growth, whereas pIGF employs the MTT1 promoter, which is responsive to the addition of $CdCl_2$ to the media. These constructs accept the gene in sites engineered just 3′ of the GFP-coding region. rDNA processing vectors have been modified further to employ YFP and CFP (Yellow and Cyan Fluorescent Protein, respectively), and the Gateway recombinase cloning system (Invitrogen), as well as to create C-terminal fusions (Cole *et al.*, 2008; Malone *et al.*, 2005; Yao *et al.*, 2007). Because these constructs are highly amplified, one must be cautious in interpreting data due to overexpression of the fusion protein (Stemm-Wolf *et al.*, 2005).

GFP-fusions have also been targeted to the rpL29 locus by using an rpL29 allele that confers resistance to cycloheximide (Matsuda *et al.*, 2010; Yao and Yao, 1991). These constructs take advantage of the Gateway cloning system and are controlled by the MTT1 promoter. Additionally, variants have been made replacing GFP with monomeric Cherry (C. G. Pearson, unpublished). These constructs have several advantageous properties: because they are not reliant on rDNA processing, the gene copy number is considerably lower, and vegetative cells can be transformed by biolistic bombardment. Furthermore, the use of cycloheximide as a selectable marker leaves paromomycin resistance, encoded by any

Table I
Vectors for fluorescent tagging of proteins in *Tetrahymena*.

Vector type	Vector name	*Tetrahymena* selection	Description	Reference
rDNA processing vectors	pVGF-1	Paromomycin	N-terminal EGFP, rpL29 promoter	Wiley *et al.*, 2000 Yao *et al.*, 2007
	pIGF-1	Paromomycin	N-terminal EGFP, MTT1 promoter	Malone *et al.*, 2005
	pIGF-gtw	Paromomycin	N-terminal EGFP, MTT1 promoter	Yao *et al.*, 2007
	pICC-gtw	Paromomycin	C-terminal CyanFP, MTT1 promoter	
	pICY-gtw	Paromomycin	C-terminal YellowFP, MTT1 promoter	Cole *et al.*, 2008 Malone *et al.*, 2005
rpL29 exogenous	pBS-MTT-GFP-gtw	Cycloheximide	N-terminal EGFP, MTT1 promoter	Matsuda *et al.*, 2010
	pBS-MTT-mCherry-gtw	Cycloheximide	N-terminal mCherry, MTT1 promoter	Pearson unpublished
Endogenous	pEGFP-NEO4	Paromomycin	C-terminal EGFP, codon optimized	Kataoka *et al.*, 2010
	pmCherry-NEO4	Paromomycin	C-terminal mCherry, codon optimized	
	ploxP-NEO4-loxP	Paromomycin	N-terminal EGFP, codon optimized. Requires abortive mating to CRE556 or some CRE expressing strain.	Busch *et al.*, 2010
	pmCherryLAP-NEO2	Paromomycin	C-terminal S-peptide-PreScission protease site-mCherry, codon optimized	Stemm-Wolf unpublished
	pNEO2-MTT1pr-mCherryLAP	Paromomycin	N-terminal mCherry-PreScission protease site-RGS6HIS, codon optimized, MTT1 promoter	
	pNEO2-MTT1pr-GFP	Paromomycin	N-terminal EGFP, MTT1 promoter	

number of NEO cassettes engineered for *Tetrahymena*, available for additional vegetative transformations.

Because *Tetrahymena* executes high-fidelity homologous recombination (Dave *et al.*, 2009; Yao and Yao, 1991), fluorescent tags can be targeted directly to the endogenous locus, and systems have been devised to allow expression from the endogenous promoter for both C- and N-terminal fusions. Furthermore, the entire gene does not have to be cloned into these vectors, as sequence is required only to promote homologous recombination at the target locus. This can be a tremendous advantage when studying proteins encoded by large genes. PCR strategies have been employed that entirely bypass the need for cloning new gene specific vectors for transformation (Kataoka *et al.*, 2010). Furthermore, cassettes have been developed that optimize codon usage for expression in *Tetrahymena* (Kataoka *et al.*, 2010). Endogenous C-terminal tagging is straightforward as a drug resistance marker can be inserted downstream of the fluorescent tag, but tagging the N-terminus requires either the addition of an exogenous promoter (such as the MTT1 promoter) or the subsequent removal of the selectable marker after transformation in order for the fusion protein to be expressed. This has been accomplished by introducing the Cre recombinase into a transformed strain that has the selectable marker flanked by loxP sites (Busch *et al.*, 2010).

Once a strain has been constructed, it is sometimes necessary to observe the fluorescent tag following cell fixation for antibody staining of a different protein. In such cases, it is important to minimize the extent of the fixation in order to preserve the fluorescent protein signal. We have effectively used the "Double Fix" (Cole and Stuart, 2000), which employs a short formaldehyde fix followed by an ethanol fix, and a 30 min 2% formaldehyde fix has been reported to be effective as well (Matsuda *et al.*, 2010).

Now that a variety of vector systems are well established, new developments in fluorescence microscopy can be easily incorporated into *Tetrahymena* research. Amongst these are Localization and Purification (LAP) tags which pair a fluorescent protein with a second tag well suited for protein purification and function similarly to TAP tags (Cheeseman and Desai, 2005; Puig *et al.*, 2001). Newer fluorescent proteins, such as Dendra, which can be converted from green fluorescence to red fluorescence by blue or UV light, have the potential to distinguish between unactivated (green) and activated (red) populations of the same protein within the cell (Gurskaya *et al.*, 2006). SNAP and CLIP-tags are flexible tags that can bind a number of fluorescent substrates whose use in *Tetrahymena* is just now being explored (New England Biolabs).

IV. Electron Microscopy

Electron microscopy (EM) is an important tool that allows the cell biologist to peer into the cell and directly image the structures of interest at a resolution of 2 nm or better. Indeed, EM investigation of ciliates has been invaluable in the understanding of basal body assembly, the identification of structures associated with cortical patterning, and the description of cellular membrane systems. Despite outstanding advances in live-cell imaging using light microscopy electron microscopy remains the only means of discerning the ultrastructure of the cell at the macromolecular level. The combination of modern electron and light microscopy techniques provides a powerful approach to the study of cellular processes, protein localization, and correlation of structure and function through the investigation of mutant phenotypes.

A. Previous Methods and Resources

Traditional chemical fixation of *Tetrahymena* cells with aqueous glutaraldehyde and osmium tetroxide generated a wealth of structural information about these cells. In the previous edition, Dentler (2000) and Gavin *et al.* (2000) presented techniques for the chemical fixation of *Tetrahymena* for morphology and for protein localization by antibody staining, respectively. In addition, we have found that simultaneous fixation with a mixture of glutaraldehyde and osmium tetroxide (modified from the method of Orias *et al.*, 1983) yielded particularly well-preserved and stained cellular ultrastructure (described in Giddings *et al.*, 2010).

We describe here methods for cryofixation of *Tetrahymena* by high-pressure freezing and freeze-substitution (HPF/FS) for morphological analysis in thin sections and for the immuno-EM (IEM) localization of proteins. These techniques have been introduced into the analysis of *Tetrahymena* since the previous edition of this manual and have been reviewed elsewhere (Giddings *et al.*, 2010; Meehl *et al.*, 2009). High-pressure freezing, as the name implies, involves rapidly freezing the cells under conditions of high pressure such that the formation of damaging ice crystals is greatly reduced or prevented (Glkey and Staehelin, 1986). Freeze-substitution is the process during which the water in the samples is replaced with an organic solvent, stains, and fixatives at low temperatures in preparation for embedding the cells in plastics. The use of HPF/FS on *Tetrahymena* specimens has resulted in excellent preservation of overall cell structure with very little extraction of material (Meehl *et al.*, 2009; Fig. 4). Immuno-labeling of thin sections (50–70 nm) from similarly prepared HPF/FS samples has been used to localize proteins in cells and to specific domains of cellular structures (Kilburn *et al.*, 2007; Fig. 4). Finally, cells prepared by HPF/FS can be used for electron tomography (ET). ET produces three-dimensional models based on a tilt-series of electron micrographs of semithick sections (~300 nm) to reveal intricate details of cellular structures previously unobserved in traditional thin-section EM. ET is an advanced EM application that generally requires higher voltage TEMs, specialized software and significant expertise. The number of laboratories equipped to perform ET is increasing steadily. We have discussed the application of ET to the study of *Tetrahymena* ultrastructure elsewhere and presented models of basal bodies and other structures (Giddings *et al.*, 2010).

B. High-Pressure Freezing and Freeze-Substitution of *Tetrahymena* Cells

1. High-Pressure Freezing

Small cell pellets are prepared from 8–10 mL of *Tetrahymena* culture by centrifugation in a 15 mL conical centrifuge tube at 500 × g for 2 min. Quick removal of the supernatant from the pellets prevents cells from swimming out of the pellet. Each pellet is gently resuspended in a cryoprotectant solution (500 μL of SPP media supplemented with 15% dextran (Sigma) and 5% Bovine Serum Albumin (BSA) (Sigma)). This cell slurry is centrifuged at 800 – 1000 × g for 4 min, and the

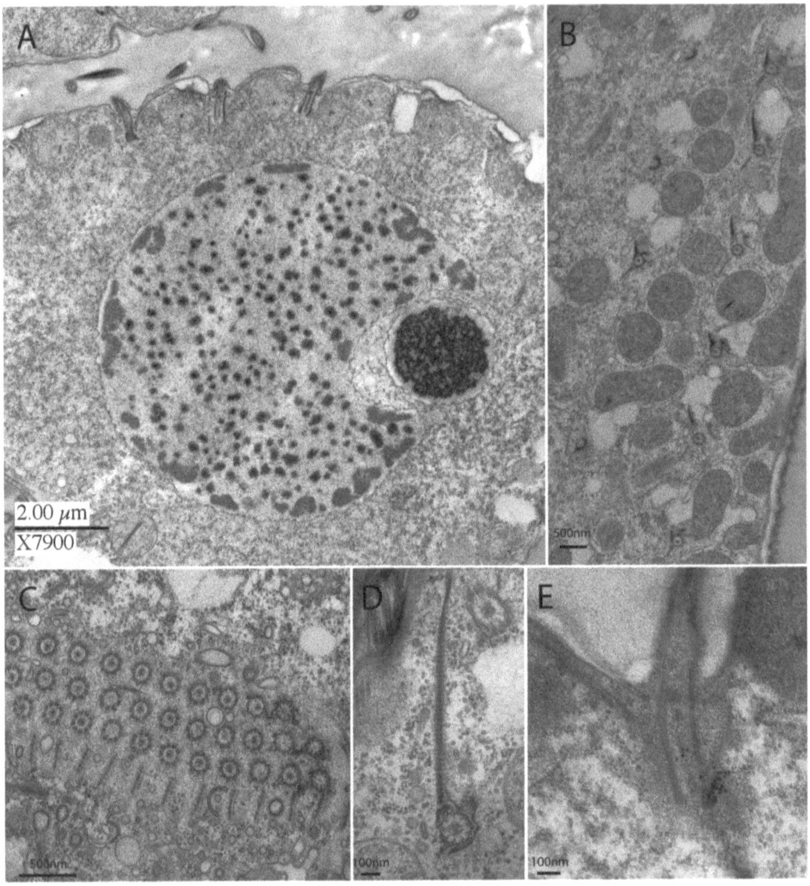

Fig. 4 Thin section transmission electron microscopy and immuno-EM of *Tetrahymena thermophila* prepared by high pressure freezing and freeze substitution. (A) A lower magnification image showing preservation of the two nuclei as well as organelles in the cytoplasm and basal bodies anchored at the cell surface. (B) A lower magnification image of a section near the cortical surface in which two cortical rows of basal bodies are visible. Basal bodies and some of their associated microtubules and structures, as well as mitochondria are visible. (C) Basal bodies and associated microtubule structures in one of the oral apparatus membranelles. Connectors between basal bodies on the right side of the image are visible. (D) A basal body in cross-section along with its associated kinetodesmal fiber. E. A longitudinal section of a basal body that has been stained with anti-Cen1 antibodies followed by a secondary antibody conjugated with 15 nm gold particles. Cen1 is asymetrically localized at the proximal end of the basal body, and is visible at the basal body midzone. The cells in panels A–D were freeze-substituted with osmium/uranyl acetate and embedded in Epon. The cell in panel E were freeze-substituted with glutaraldehyde/uranyl acetate and embedded in Lowicryl HM20.

supernatant is removed leaving a minimal residue of cryoprotectant media with the pellet. The small residue of the cryoprotectant allows for the cells to be somewhat resuspended and separated. Loosely packed *Tetrahymena* cells retain their normal shape, freeze better, and retain more of their cortical cilia. Two to three microliters of the cell preparation are pipetted into the 100 μm deep well (shallow side) of an aluminum Type B specimen carrier (Technotrade International). The samples are then capped with the flat side of a Type A specimen carrier coated with hexadecene (Sigma). The tip of the specimen holder is clamped around the specimen carriers and tightened gently before insertion into the HPF instrument. With the Bal-Tec HPM 010, the freezing process is initiated by simply pressing a button. Immediately after the freezing event and cessation of the audible venting of the freezing chamber, the sample holder is rapidly moved to a tray of liquid nitrogen for unloading the sample. Under liquid nitrogen, the aluminum carrier hats containing the sample are pried apart and transferred to cryovials that contain 1 mL of *FS medium*. The samples will lie on top of the frozen *FS medium*, but sink into it once the vial is warmed sufficiently to initiate freeze-substitution.

2. Freeze-Substitution

We use two different freeze-substitution protocols for fixation and embedding of HPF-prepared *Tetrahymena* cells depending on the experiment (Meehl *et al.*, 2009). To achieve a thorough fixation with strong staining of both membranous and cytoskeletal organelles, we freeze-substitute in 2% osmium tetroxide (OsO_4; Ted Pella) and 0.1% uranyl acetate (UA; Electron Microscopy Sciences, Hatfield, PA) in acetone followed by embedding in Epon-Araldite (Electron Microscopy Sciences, Hatfield, PA). To retain antigenicity for immuno-labeling of plastic-embedded sections, we use a milder fixation with 0.25% glutaraldehyde (Electron Microscopy Sciences, Hatfield, PA) and 0.1% UA in acetone followed by embedding in Lowicryl HM20 (Electron Microscopy Sciences, Hatfield, PA). The Lowicryl low-temperature embedding method has also proven to yield excellent preservation of cellular ultrastructure for high-resolution EM analysis including tomography. Our general practice is to high-pressure freeze enough samples to carry out both FS and embedding protocols on the same batch of cells.

Freeze-Substitution with Osmium Tetroxide for Embedding in Epon-Araldite Epoxy Resin

A metal block cooled to –80 °C is used to hold the cryovials containing the samples in *FS media*. This block with the samples is nestled in a chest of dry ice and placed in a standard –20 °C freezer for 3–4 days. Gradual overnight warming of the samples to –20 °C is achieved by removing the lid from the chest allowing a small residue of the dry ice to evaporate. After remaining at −20 °C for several hours, the metal block containing the samples is moved to 4 °C for 4–6 h and finally to room temperature for 1 h. An alternative method is to use an automated freeze-substitution device. Once the samples have reached room temperature, the *FS media* is removed and the samples are rinsed twice with acetone. The samples are now removed from the aluminum carrier hats prior to embedding. The freeze-substituted cells and cryoprotectant solution typically form a cohesive disk that either falls off or can be removed gently from the aluminum carrier hats by means of dissecting needles or similar tools. It can be worthwhile to work under a dissecting microscope to retrieve any small fragments. Samples are rinsed again in fresh acetone, and then infiltrated with increasing concentrations of Epon-Araldite resin (without DMP30 accelerator, Electron Microscopy Sciences, Hatfield, PA) diluted in acetone. The embedding regimen is 25% Epon in acetone overnight; 50% Epon 8–10 h; 75% Epon overnight; and two changes of 100% Epon during the next day. The samples are then left in Epon with accelerator overnight, transferred to BEEM capsules with fresh embedding resin the next day and placed in a 60 °C oven to polymerize for at least 48 h.

Freeze-Substitution in Glutaraldehyde for Embedding in Lowicryl HM20

As described above, samples of frozen cells were freeze-substituted at −80 °C for 3–4 days followed by gradual warming to –20 °C overnight. The samples are then held at −20 °C for acetone rinses and infiltration with increasing concentrations of Lowicryl HM20 in acetone. After rinsing with acetone chilled to –20 °C, the FS samples were separated from the specimen carriers. The procedure for separating the specimens from the specimen carriers is the same as above except for the use of chilled acetone. Working quickly is best to minimize sample warming that can cause extraction and cause morphological changes. As soon as samples are returned to the cryovial they are immediately rinsed in fresh –20 °C acetone and then infiltrated with increasing concentrations of Lowicryl HM20 diluted in acetone. The infiltration schedule is 25% HM20 in acetone overnight; 50% HM20 for 6–8 h; 75% HM20 overnight. The final incubation of the samples is in 100% HM20 for approximately 1.5 days. During this incubation, four changes with fresh resin are made to ensure the removal of any residual acetone. The samples are transferred to embedding capsules that are half filled with fresh HM20, and then the capsule is filled to the top and capped. Polymerization under UV illumination is carried out at –45 °C in a homemade device (see below).

We have also freeze-substituted high-pressure frozen *Tetrahymena* cells using only 0.1% UA in acetone. The rest of the procedure is identical to the above-described glutaraldehyde/UA FS and HM20-embedding protocol and result in nearly indistinguishable morphology. In other cell types, this has allowed us to obtain significant labeling of aldehyde-sensitive antigens (e.g., Pearson *et al.*, 2009b).

3. Ultramicrotomy and Staining of Sections

Epon or Lowicryl HM20 plastic resin block faces are trimmed to short, wide trapezoids to optimize the number of cells per section while allowing a large number of serial sections per grid. Cells can then be easily tracked from one section to the next in a ribbon. Copper slot grids are used to pick up serial thin sections (50–70 nm).

The sections are stained in 2% UA in 70% methanol for 6 min; rinsed in the same solvent and dried; and then stained in Reynolds lead citrate for 4 min and thoroughly rinsed with water. For immuno-electron microscopy (IEM), staining times for both UA and lead can be reduced to better visualize colloidal gold particles over electron dense structures.

C. Immuno-Labeling Thin Sections

Sectioned *Tetrahymena* cells prepared by HPF/FS and embedded in Lowicryl HM20 as described above are used for immuno-electron microscopy (IEM). We have been successful with either primary antibodies to selected proteins or with antibodies to tags such as GFP fused to your favorite gene (YFP) (Kilburn *et al.*, 2007; Stemm-Wolf *et al.*, 2005). Sections of cells fixed by light chemical fixation and embedded in LR White (Electron Microscopy Sciences, Hatfield, PA) have also been used for immuno-localization studies in *Tetrahymena* (Ueno *et al.*, 2003).

We have previously published IEM methods for the localization of *Tetrahymena* proteins (Meehl *et al.*, 2009). Serial thin sections (50–70 nm) of Lowicryl-embedded cells are collected on Formvar-coated nickel slot grids. The grids are placed, sections side down, onto 15 μL drops of blocking solution for 30 min, followed by 2 h on primary antibody diluted in blocking solution. The grids are then rinsed with a steady stream of Phosphate-Buffered Saline with Tween (PBST) for 20 s before labeling with an appropriate secondary antibody (conjugated to 10 or 15 nm gold) for 1 h. Grids are then rinsed with PBST followed by distilled water, which is removed by careful blotting and air-drying. The visibility of colloidal gold secondary antibody can be improved by using thinner sections and reduced staining times. Expression of GFP-fusion proteins is a valuable technique for the study of *Tetrahymena,* as described earlier. GFP has proven to be robust tag for IEM. To date, we have used two GFP rabbit polyclonal antibodies on *Tetrahymena* cells harboring GFP-tagged proteins (see *Materials* section). Both of these antibodies yield strong signal with low background following this IEM protocol. Of course, the abundance of the given GFP-tagged protein and its concentration at a cellular location contribute to the success of localizing the protein by IEM.

D. Instrumentation and Materials

1. High-Pressure Freezing and Freeze-Substitution Instrumetation

The technology underlying high-pressure freezing and general techniques for its use have been described (e.g., Glkey and Staehelin, 1986; McDonald, 1999). Our instrument is a Bal-Tec HPM 010 (currently available from RMC, Tucson, AZ). Other available models include the Wohlwend HPM 01 (available in the United States through Technotrade International, Manchester, NH) and two models from Leica, the Leica EM PACT2 (McDonald *et al.*, 2007) and the Leica EM HPM100.

We commonly grow, HPF/FS, and embed *Tetrahymena* cells for investigators whose local EM facility lacks a freezer. Once embedded in plastic, the bullets can be easily shipped to the investigator for sectioning, staining, and imaging at their home EM facility. High-pressure freezers and associated technological expertise can be found in numerous electron microscopy core facilities at universities, medical schools, or research institutes. Our freeze-substitution system employs a simple Styrofoam box filled with dry ice that is used to maintain the samples at –80 °C for freeze-substitution. The box is placed in a standard refrigerator-freezer unit for gradual warming to −20 °C and embedding in Lowicryl. We use a metal block with holes drilled in it to hold the cryovials of *FS media* upright and to provide a slower rate of temperature change during warming from −80 °C to –20 °C.

We use a homemade UV polymerization chamber for polymerizing blocks of Lowicryl HM20. It consists of an insulated box mounted with two 7-W UV lights. BEEM capsules with samples in liquid resin are held in a wire rack immersed in a temperature-controlled bowl of isopropyl or methyl alcohol. Dry ice is placed in the bottom of the box, and the temperature is maintained at –45 °C by means of a thermocouple-based controller and a heating element wrapped around the bowl.

Commercially available alternatives combine freeze-substitution and UV polymerization capabilities in a single instrument. Such devices include the Leica EM AFS (Leica Microsystems). These instruments offer a versatile and convenient means of achieving controlled, reproducible freeze-substitution and UV polymerization of low-temperature embedding resins. These units have the advantage of offering a wide range of temperatures for initial FS, low temperature fixation, resin infiltration and polymerization, and controlled rates of temperature change throughout the protocol.

2. High-Pressure Freezing and Freeze-Substitution Reagents

The cryoprotectant solution for HPF is 15% dextran (average molecular weight 9.5 kDa, Sigma), 5% BSA in SPP (growth media; see above). We have evaluated a variety of cryoprotectants with respect to the freezing of *Tetrahymena* cells. Consistent and high-quality results are obtained with a mixture of 15% dextran (average molecular weight 9.5 kDa; Sigma) and 5% BSA in culture media. Low MW dextran (9.5–11 kDa) is less viscous at the same concentration than the more commonly used 40 kDa dextran and allows for easier handling of the *Tetrahymena* cells.

The aluminum specimen carriers (hats), Type A and Type B, can be purchased from Technotrade International and are available from several sources. These and many other styles of specimen carriers have been reviewed (McDonald *et al.*, 2007). The freeze-substitution media (*FS media*) for Epon embedding is 2% OsO_4 and 0.1% UA in acetone. It is prepared by placing 12.25 mL of anhydrous acetone in a vial and using 1 mL of the acetone to dissolve the 0.25 g OsO_4 in a glass ampoule (EMS: Electron Microscopy Sciences, Hatfield, PA). The dissolved osmium tetroxide was returned to the vial and placed on dry ice. Repeating the process quickly dissolved all of the OsO_4. Then 0.25 mL of 5% UA (EMS) in methanol (stored at −20 °C) is added to the solution. This *FS media* is kept on dry ice until aliquoted (1 mL/vial) into 1.8 mL cryovials (Nunc), which are stored under liquid nitrogen until needed.

For embedding in Lowicryl HM20, the freeze-substitution media (*FS media*) is 0.25% glutaraldehyde and 0.1% UA in acetone. It is prepared by adding 0.25 mL 10% glutaraldehyde in acetone (EMS) and 0.2 mL of a 5% UA/methanol stock solution to 9.55 mL acetone. The *FS media* is then aliquoted to cryovials and stored as described above.

3. Immuno-Labeling of Thin Sections Instrumentation

Immuno-labeling is done in a covered glass Petri dish lined with moist filter paper and Parafilm. The droplets of blocking solution and antibodies are placed on the Parafilm, and the dish is set on a magnetic stir plate. Adjust the speed of the stirrer to cause very slow rotation of the nickel grids on the droplets. Be sure to use nonmagnetic-self-closing tweezers when handling nickel grids.

4. Immuno-Labeling of Thin Sections Reagents

PBST is 10 mM sodium phosphate, 150 mM sodium chloride and 0.1% Tween-20. The blocking solution is 1% nonfat dry milk powder (w/v) in PBST. The solution is centrifuged at 1500 × g prior to use to remove undissolved solids.

We have had success with various rabbit polyclonal antibodies to GFP prepared by individual investigators. Unfortunately, we have yet to identify a commercially made α-GFP antibody that works reliably for IEM. Goat-anti-rabbit-15 nm gold or 10 nm gold (Ted Pella, Redding, CA) secondary antibodies were diluted 1:20 in blocking solution.

V. Discussion

We have presented imaging techniques for *Tetrahymena* cells focused on the use of fluorescent proteins to tag genes for live-cell light microscopy, and on the preparation of cells for electron microscopy by high-pressure freezing and freeze-substitution. These techniques along with previously published techniques for either light or electron microscopy make for a strong suite of technologies that enable high quality cytology in *Tetrahymena*. We look forward to seeing additional tools and techniques deployed in these cells, such as photoconversion of Dendra tags, the use of SNAP tags, the application of super resolution imaging, and the incorporation of computational modeling to develop predictive models of cellular processes, as future advances to achieve the cutting-edge cell biology research that can be accomplished in *Tetrahymena*.

Acknowledgments

We thank Aaron Turkewitz and Eric Cole for helpful discussions in using *Tetrahymena* cell immobilization, Doug Chalker for discussions of fluorecent protein tagging strategies, and Jacek Gaertig for discussions of fluorescent protein tagging strategies and for information on limiting leakiness from MTT promoters. We thank Eileen O'Toole (Boulder Laboratory for 3D EM of Cells), Janet Meehl, and Christina Clarissa for helpful advice and assistance with electron microscopy and tomography. Our work on *Tetrahymena* has been supported by the NIH (GM74746 to MW), the March of Dimes Foundation (#1-FY07-422 to MW), as well as the Damon Runyan Cancer Research Foundation, Pew Charitable Trust Scholars Program in Biomedical Sciences, and American Cancer Society (ASC-IRG#57-001-50) to CGP.

The authors declare no conflicts of interest.

References

Allen, R. D. (1969). The morphogenesis of basal bodies and accessory structures of the cortex of the ciliated protozoan *Tetrahymena pyriformis*. *J. Cell Biol.* **40**, 716–733.

Aufderheide, K. J. (2008). An overview of techniques for immobilizing and viewing living cells. *Micron* **39**, 71–76.

Boldrin, F., Santovito, G., Formigari, A., Bisharyan, Y., Cassidy-Hanley, D., Clark, T. G., and Piccinni, E. (2008). MTT2, a copper-inducible metallothionein gene from *Tetrahymena thermophila*. *Comp. Biochem. Physiol. C Toxicol. Pharmacol.* **147**, 232–240.

Bright, L. J., Kambesis, N., Nelson, S. B., Jeong, B., and Turkewitz, A. P. (2010). Comprehensive analysis reveals dynamic and evolutionary plasticity of Rab GTPases and membrane traffic in *Tetrahymena thermophila*. *PLoS Genet.* **6**, e1001155.

Busch, C. J., Vogt, A., and Mochizuki, K. (2010). Establishment of a Cre/loxP recombination system for N-terminal epitope tagging of genes in *Tetrahymena*. *BMC Microbiol.* **10**, 191.

Caro, L. G., and Palade, G. E. (1964). Protein synthesis, storage, and discharge in the pancreatic exocrine cell. An autoradiographic study. *J. Cell Biol.* **20**, 473–495.

Cheeseman, I. M., and Desai, A. (2005). A combined approach for the localization and tandem affinity purification of protein complexes from metazoans. *Sci. STKE.* **2005**, l1.

Cole, E. S., Anderson, P. C., Fulton, R. B., Majerus, M. E., Rooney, M. G., Savage, J. M., Chalker, D., Honts, J., Welch, M. E., Wentland, A. L., Zweifel, E., and Beussman, D. J. (2008). A proteomics approach to cloning fenestrin from the nuclear exchange junction of *Tetrahymena*. *J. Eukaryot. Microbiol.* **55**, 245–256.

Cole, E. S., and Stuart, K. R. (2000). Nuclear and cortical histology for brightfield microscopy. *Methods Cell Biol.* **62**, 313–322.

Cole, E. S., Stuart, K. R., Marsh, T. C., Aufderheide, K., and Ringlien, W. (2002). Confocal fluorescence microscopy for *Tetrahymena thermophila*. *Methods Cell Biol.* **70**, 337–359.

Dave, D., Wloga, D., and Gaertig, J. (2009). Manipulating ciliary protein-encoding genes in *Tetrahymena thermophila*. *Methods Cell Biol.* **93**, 1–20.

Dentler, W. (2000). Fixation of *Tetrahymena* cells for electron microscopy. *Methods Cell Biol.* **62**, 323–331.

Gavin, R. H., Hoey, J. G., and Garces, J. A. (2000). Immunoelectron microscopy of *Tetrahymena*. *Methods Cell Biol.* **62**, 333–343.

Giddings Jr., T. H., Meehl, J. B., Pearson, C. G., and Winey, M. (2010). Electron tomography and immuno-labeling of *Tetrahymena thermophila* basal bodies. *Methods Cell Biol.* **96**, 117–141.

Glkey, J. C., and Staehelin, L. A. (1986). Advances in ultrarapid freezing for the preservation of cellular ultrastructure. *J. Electron Microsc. Tech.* **3**, 177–210.

Gurskaya, N. G., Verkhusha, V. V., Shcheglov, A. S., Staroverov, D. B., Chepurnykh, T. V., Fradkov, A. F., Lukyanov, S., and Lukyanov, K. A. (2006). Engineering of a monomeric green-to-red photoactivatable fluorescent protein induced by blue light. *Nat. Biotechnol.* **24**, 461–465.

Jacobs, M. E., DeSouza, L. V., Samaranayake, H., Pearlman, R. E., Siu, K. W., and Klobutcher, L. A. (2006). The *Tetrahymena thermophila* phagosome proteome. *Eukaryot. Cell* **5**, 1990–2000.
Jamieson, J. D., and Palade, G. E. (1967a). Intracellular transport of secretory proteins in the pancreatic exocrine cell: I. Role of the peripheral elements of the Golgi complex. *J. Cell Biol.* **34**, 577–596.
Jamieson, J. D., and Palade, G. E. (1967b). Intracellular transport of secretory proteins in the pancreatic exocrine cell: II. Transport to condensing vacuoles and zymogen granules. *J. Cell Biol.* **34**, 597–615.
Kataoka, K., Schoeberl, U. E., and Mochizuki, K. (2010). Modules for C-terminal epitope tagging of *Tetrahymena* genes. *J. Microbiol. Methods* **82**, 342–346.
Kilburn, C. L., Pearson, C. G., Romijn, E. P., Meehl, J. B., Giddings Jr., T. H., Culver, B. P., Yates 3rd, J. R., and Winey, M. (2007). New *Tetrahymena* basal body protein components identify basal body domain structure. *J. Cell Biol.* **178**, 905–912.
Larsen, J. (1989). The influence of growth phase and culture conditions of *Tetrahymena* on effects of cadmium. *Toxicology* **58**, 211–223.
Loidl, J., and Scherthan, H. (2004). Organization and pairing of meiotic chromosomes in the ciliate *Tetrahymena thermophila*. *J. Cell Sci.* **117**, 5791–5801.
Malone, C. D., Anderson, A. M., Motl, J. A., Rexer, C. H., and Chalker, D. L. (2005). Germ line transcripts are processed by a Dicer-like protein that is essential for developmentally programmed genome rearrangements of *Tetrahymena thermophila*. *Mol. Cell Biol.* **25**, 9151–9164.
Matsuda, A., Shieh, A. W., Chalker, D. L., and Forney, J. D. (2010). The conjugation-specific Die5 protein is required for development of the somatic nucleus in both *Paramecium* and *Tetrahymena*. *Eukaryot. Cell* **9**, 1087–1099.
McDonald, K. (1999). High-pressure freezing for preservation of high resolution fine structure and antigenicity for immunolabeling. *Methods Mol. Biol.* **117**, 77–97.
McDonald, K. L., Morphew, M., Verkade, P., and Muller-Reichert, T. (2007). Recent advances in high-pressure freezing: equipment- and specimen-loading methods. *Methods Mol. Biol.* **369**, 143–173.
Meehl, J. B., Giddings Jr., T. H., and Winey, M. (2009). High pressure freezing, electron microscopy, and immuno-electron microscopy of *Tetrahymena thermophila* basal bodies. *Methods Mol. Biol.* **586**, 227–241.
Meselson, M., and Stahl, F. W. (1958). The replication of DNA in *Escherichia coli*. *Proc. Natl. Acad. Sci. U.S.A.* **44**, 671–682.
Orias, E., Flacks, M., and Satir, B. H. (1983). Isolation and ultrastructural characterization of secretory mutants of *Tetrahymena thermophila*. *J. Cell Sci.* **64**, 49–67.
Pearson, C. G., Giddings Jr., T. H., and Winey, M. (2009a). Basal body components exhibit differential protein dynamics during nascent basal body assembly. *Mol. Biol. Cell* **20**, 904–914.
Pearson, C. G., Osborn, D. P., Giddings Jr., T. H., Beales, P. L., and Winey, M. (2009b). Basal body stability and ciliogenesis requires the conserved component Poc1. *J. Cell Biol.* **187**, 905–920.
Puig, O., Caspary, F., Rigaut, G., Rutz, B., Bouveret, E., Bragado-Nilsson, E., Wilm, M., and Seraphin, B. (2001). The tandem affinity purification (TAP) method: a general procedure of protein complex purification. *Methods* **24**, 218–229.
Salmon, E. D., Leslie, R. J., Saxton, W. M., Karow, M. L., and McIntosh, J. R. (1984a). Spindle microtubule dynamics in sea urchin embryos: analysis using a fluorescein-labeled tubulin and measurements of fluorescence redistribution after laser photobleaching. *J. Cell Biol.* **99**, 2165–2174.
Salmon, E. D., Saxton, W. M., Leslie, R. J., Karow, M. L., and McIntosh, J. R. (1984b). Diffusion coefficient of fluorescein-labeled tubulin in the cytoplasm of embryonic cells of a sea urchin: video image analysis of fluorescence redistribution after photobleaching. *J. Cell Biol.* **99**, 2157–2164.
Shang, Y., Song, X., Bowen, J., Corstanje, R., Gao, Y., Gaertig, J., and Gorovsky, M. A. (2002). A robust inducible-repressible promoter greatly facilitates gene knockouts, conditional expression, and overexpression of homologous and heterologous genes in *Tetrahymena thermophila*. *Proc. Natl. Acad. Sci. U.S.A.* **99**, 3734–3739.
Sprague, B. L., and McNally, J. G. (2005). FRAP analysis of binding: proper and fitting. *Trends Cell Biol.* **15**, 84–91.

Stemm-Wolf, A. J., Morgan, G., Giddings Jr., T. H., White, E. A., Marchione, R., McDonald, H. B., and Winey, M. (2005). Basal body duplication and maintenance require one member of the *Tetrahymena thermophila* centrin gene family. *Mol. Biol. Cell* **16**, 3606–3619.

Stuart, K. R., and Cole, E. S. (2000). Nuclear and cytoskeletal fluorescence microscopy techniques. *Methods Cell Biol.* **62**, 291–311.

Tondravi, M. M., and Yao, M. C. (1986). Transformation of *Tetrahymena thermophila* by microinjection of ribosomal RNA genes. *Proc. Natl. Acad. Sci. U.S.A.* **83**, 4369–4373.

Ueno, H., Gonda, K., Takeda, T., and Numata, O. (2003). Identification of elongation factor-1alpha as a Ca^{2+}/calmodulin-binding protein in *Tetrahymena* cilia. *Cell Motil. Cytoskeleton* **55**, 51–60.

Walczak, C. E., Rizk, R. S., and Shaw, S. L. (2010). The use of fluorescence redistribution after photobleaching for analysis of cellular microtubule dynamics. *Methods Cell Biol.* **97**, 35–52.

Wiley, E. A., Ohba, R., Yao, M. C., and Allis, C. D. (2000). Developmentally regulated rpd3p homolog specific to the transcriptionally active macronucleus of vegetative *Tetrahymena thermophila*. *Mol. Cell Biol.* **20**, 8319–8328.

Wolfe, J., and Colby, R. H. (1981). A method for the immobilization of living *Tetrahymena*. *Exp. Cell Res.* **134**, 313–317.

Yao, M. C., and Yao, C. H. (1991). Transformation of *Tetrahymena* to cycloheximide resistance with a ribosomal protein gene through sequence replacement. *Proc. Natl. Acad. Sci. U.S.A.* **88**, 9493–9497.

Yao, M. C., Yao, C. H., Halasz, L. M., Fuller, P., Rexer, C. H., Wang, S. H., Jain, R., Coyne, R. S., and Chalker, D. L. (2007). Identification of novel chromatin-associated proteins involved in programmed genome rearrangements in *Tetrahymena*. *J. Cell Sci.* **120**, 1978–1989.

CHAPTER 14

Purification of *Tetrahymena* Cytoskeletal Proteins

Jerry E. Honts

Department of Biology, Drake University, 2507 University Avenue, Des Moines, IA, USA

Abstract

Like all eukaryotic cells, *Tetrahymena thermophila* contains a rich array of cytoskeletal proteins, some familiar and some novel. A detailed analysis of the structure, function, and interactions of these proteins requires procedures for purifying the individual protein components. Procedures for the purification of actin and tubulin from *Tetrahymena* are reviewed, followed by a description of a procedure that yields proteins from the epiplasmic layer and associated structures, including the tetrins. Finally, the challenges and opportunities for future advances are assessed.

I. Overview

The elaborate cell surface architecture of ciliates is maintained and propagated by an equally complex membrane-associated cytoskeleton (for more details see Chapter 5 by Wloga and Frankel). This cytoskeletal framework is important for the integrity of the cell as well as the organization and function of cortical organelles.

0091-679X/10 $35.00
DOI 10.1016/B978-0-12-385967-9.00010-4

What makes a ciliated protozoan like *Tetrahymena thermophila* exciting yet challenging to study is the remarkable diversity of cytoskeletal elements present. Even for a single type of cytoskeletal protein, such as kinesin, the predicted number of kinesins exceeds that of any other currently sequenced multicellular organism (Eisen *et al.*, 2006). *Tetrahymena* cells include the staples of eukaryotic cells, actin filaments, and microtubules, but also elements that may be specific to the ciliates, or perhaps more generally, the alveolates (Gould *et al.*, 2011).

Some cytoskeletal proteins were first described in *Tetrahymena*, such as dynein motors in cilia (Gibbons and Rowe, 1965). Over the last several decades, many other cytoskeletal proteins have been identified, some of which appear to be specific to the alveolates in general and *Tetrahymena* in particular (Gould *et al.*, 2011; Honts and Williams, 2003). Purification of the individual proteins has, however, proven challenging. Several key aspects of *Tetrahymena* protein biochemistry make this work challenging: the relative complexity of the mixture, the powerful proteases that are released on cell lysis, and the insolubility of many of the component proteins.

This chapter will describe what is currently known to work in protocols for purifying the cytoskeletal protein components of *Tetrahymena thermophila*. It focuses on the purification of the quantitatively predominant cytoskeletal proteins, potentially yielding them in milligram quantities. It extends or refines some of the previously published methods and finishes by describing some of the challenges and opportunities that lie ahead.

II. Purification of Actin and Actin-Binding Proteins

Actin is a major cytoskeletal constituent of many eukaryotic cells. Like other eukaryotes, *Tetrahymena* contains a diverse family of actin and actin-related proteins (see Chapter 5 for an overview). But actin does not appear to be a quantitatively prominent protein in the cortical cytoskeleton of *Tetrahymena* as it is in the cortex of many other eukaryotic cells. Whereas it is relatively easy to prepare several milligrams of pure actin protein from the budding yeast *Saccharomyces cerevisiae* by means of DNase I-affinity chromatography (Honts *et al.*, 1994), this approach proved problematic in *Tetrahymena*. First, *Tetrahymena* actin does not appear to bind to DNase I, and second, actin is rapidly degraded by proteases upon cell lysis (Hirono *et al.*, 1989).

The successful purification of *Tetrahymena* actin has been described by Hirono *et al.* (1989). The key to the development of this protocol was the preparation of an antibody directed against the N-terminus of *Tetrahymena pyriformis* actin (Act1p). This made it possible to screen chromatographic fractions by immunoblotting with an antibody that was known to be specific to *Tetrahymena pyriformis* actin. Anti-actin antibodies have been used in conjunction with DNase I-affinity chromatography to search for actin in *Tetrahymena*, but the 47-kDa protein identified by this strategy (Mitchell and Zimmerman, 1985) would seem to be different than that purified by Hirono *et al.* (1989).

Briefly, in the procedure described by Hirono *et al.* (1989), an acetone powder from *Tetrahymena pyriformis* was extracted with G-actin buffer and the extract was rapidly clarified. The clarified extract was loaded onto a Q-Sepharose anion exchange resin and rapidly eluted with a salt gradient. The eluted fractions were then subjected to gel filtration chromatography, followed by a second pass over a Q-Sepharose column. The eluted actin was concentrated by ammonium sulfate precipitation and dialyzed against assembly buffer. The net result was a single 43-kDa band resolved in SDS-polyacrylamide gels, with a yield of 1.0 mg of protein from 14 g of acetone powder. The purified protein was capable of forming filaments indistinguishable from muscle actin (Hirono *et al.,* 1989). But they also noted that this actin was unusual in terms of its biochemical properties, such as an inability to bind phalloidin. Presumably this is a reflection of the fact that *Tetrahymena* actin is only 75% identical to highly conserved actins found in fungal and animal cells (Cupples and Pearlman, 1986).

Hirono *et al.* (1989) attribute their success to the rapid processing of the extracted material. Protease inhibitors (notably, the cysteine/serine protease inhibitor leupeptin) were included throughout the procedure, but it would appear that the most important thing was to quickly separate the actin protein from other cellular proteins, especially the proteases.

Although this protocol was developed using *Tetrahymena pyriformis*, it seems reasonable that it would also be effective in purifying actin from *Tetrahymena thermophila* since the amino sequences of actins from these two species are nearly identical (Hirono *et al.,* 1987). Purification of actin from *Tetrahymena thermophila* would benefit from an antibody made specifically against synthetic peptides from *Tetrahymena thermophila* actin, particularly because the N-terminal peptides of *Tetrahymena pyriformis* and *Tetrahymena thermophila* have slightly different amino acid sequences. This protocol (and others in this review) certainly benefits from the increased commercial availability of a wide variety of protease inhibitors, including cocktails of inhibitors specifically targeted against cysteine proteases. Finally, it would be worthwhile to explore alternate protocols for purification of nonmuscle actins which likewise do not depend on DNase I-affinity chromatography (Schafer *et al.*, 1998).

The function of actin within a cell is regulated by its interaction with a diverse set of actin-binding proteins, with the result that different actin filament structures with distinct functions are often formed within a common cytoplasm (Michelot and Drubin, 2011). A number of actin-binding proteins have been purified from *Tetrahymena*: profilin (Edamatsu *et al.*, 1990), translation elongation factor-1 alpha (Kurasawa *et al.*, 1996), fimbrin (Watanabe *et al.*, 1998), p85 (Gonda and Numata, 2002), and ADF/cofilin-like protein (Shiozaki *et al.*, 2009).

At least one type of myosin has also been partially purified from *Tetrahymena* (Garces *et al.*, 1995), but a total of 13 different myosins have been predicted from analysis of the *Tetrahymena thermophila* macronuclear genomic sequence (Eisen *et al.*, 2006), and these fall into three distinct subclasses (Sugita *et al.*, 2011).

III. Purification of Tubulin and Tubulin-Associated Proteins

Given the abundance of microtubular structures in *Tetrahymena*, it would appear to be an ideal source of tubulin for biochemical studies. Tubulin is found in the axoneme of the cilia as well as throughout the cytoplasm of cells, notably within the cortex (see Chapter 5). Procedures for the purification of *Tetrahymena* tubulin have been previously described, based on *in vitro* polymerization of extracted tubulin subunits promoted by axonemal-derived microtubule "seeds" or the addition of taxol (Maekawa and Sakai, 1978; Suprenant *et al.,* 1985). Recently, progress has been made in the preparation of large quantities of very pure tubulin by Lyons-Abbott *et al.* (2010), based on a protocol for the purification of *Leishmania* tubulin (Werbovetz *et al.,* 1999).

In the Lyons-Abbott *et al.* (2010) procedure, *Tetrahymena thermophila* cells were sonicated in the presence of protease inhibitors (notably leupeptin at 25 μg/mL), and the clarified lysate was loaded onto a DEAE-Sepharose Fast Flow column, washed, and then eluted with 0.3 M KCl and 0.75 M glutamate (pH 6.9). Tubulin-containing fractions were pooled and microtubules were assembled by the addition of $MgCl_2$, DMSO, and GTP. After a 30-min incubation at 37 °C, microtubules were pelleted at 50,000 × g. The pellet was rinsed and microtubules were disassembled by sonication and incubation on ice. Another round of centrifugation yielded a tubulin-rich supernatant suitable for biochemical studies. The solution of tubulin αβ heterodimers obtained by this procedure was notably free of contaminating proteins, including high-molecular-weight microtubule-associated proteins (MAPs). The review by Sackett *et al.* (2010) presents recommendations specific to the purification of *Tetrahymena* tubulin.

The preparation of dynein motor proteins from *Tetrahymena* cilia has been described previously (Johnson, 1986). The purification of other microtubule-based motors (kinesins and kinesin-like proteins) and MAPs remain to be described, but these will be aided by the mass spectrometric analysis of proteins that co-sediment with microtubules in pelleting assays. Given the large number of dynein and kinesin proteins predicted to be encoded by the *Tetrahymena thermophila* genome (see Eisen *et al.,* 2006), purification of the individual proteins will likely require some form of molecular labeling such as epitope tagging (Busch *et al.,* 2010; Kataoka *et al.,* 2010).

The purification and characterization of tubulin (as well as the other cytoskeletal proteins mentioned in this review) will also need to take into account the existence of a diverse array of post-translational covalent modifications of the tubulin proteins (Wloga and Gaertig, 2010). These modifications may well affect the extraction and chromatographic purification of these proteins, but these could also be exploited in affinity-based methods that recognize specific types of covalent modifications.

IV. Purification of Proteins from the Cortical Cytoskeleton

Two major methods for the isolation and fractionation of the *Tetrahymena* cytoskeleton have been described (Williams *et al.*, 1979, 1990). The TritonX-100 high-salt (THS) method described by Williams *et al.* (1990) also proved useful in the study of the membrane skeleton of *Euplotes* (Williams *et al.,* 1989a) and *Paramecium* (Williams *et al.,* 1989b).

Difficulties were encountered in scaling up these procedures to purify the high-molecular-weight epiplasmic proteins originally labeled as bands A, B, and C (Williams *et al.,* 1979). Lysates obtained by these methods were often so viscous that it was difficult to pellet the resulting cytoskeletal residues – even by means of ultracentrifugation. An alternate procedure was developed to bypass many of these difficulties, which used a combination of Triton X-100 with 0.15 M KI and 0.05 M $MgCl_2$. Like the THS procedure, the Triton X-100/KI (TKI) method yielded cortical residues depleted of microtubules, but enriched in the high-molecular-weight epiplasmic proteins, the tetrins, and a number of other proteins. This method has been used to analyze the proteomic composition of the membrane skeleton of *Tetrahymena thermophila* (Honts *et al.*, manuscript in preparation).

This TKI procedure has been used successfully with wild-type and mutant strains of *Tetrahymena thermophila* as well as *Tetrahymena pyriformis* strain GL and *Tetrahymena americanus*. Wild-type strains of *Tetrahymena thermophila* (such as SB210) release copious amounts of mucus upon lysis, making the procedure more difficult in the earlier steps (but not impossible). Most recent work has avoided this problem altogether by using the *Tetrahymena thermophila* secretory mutant strain SB281 (Maihle and Satir, 1985).

A. Purification of Epiplasmic Proteins

1. Solutions

2X TKI lysis buffer: 2% Triton X-100, 0.30 M KI, 4 mM EGTA, 100 mM $MgCl_2$, 50 mM PIPES-NaOH, adjusted to pH 6.9.

Wash buffer: 1 mM EGTA, 25 mM PIPES-NaOH, adjusted to pH 6.9.

Alkaline extraction buffer: 30 mM Tris base, 2 mM EGTA (free acid), and pH ~ 9.0. Alternatively, 10 mM Tris–HCl, pH 9.0, supplemented with a 0.5 M EGTA-NaOH, pH 8.0, stock to a final concentration of 2 mM EGTA.

Protease inhibitor stock solutions: 10 mg mL^{-1} leupeptin hemisulfate in H_2O (stored in aliquots at −80 °C); 400 mM *N*-ethylmaleimide in absolute ethanol (stored at −20 °C in dark, warm to room temperature to dissolve crystals before use); 200 mM *o*-phenanthroline in absolute ethanol (stored at −20 °C).

20X polymerization buffer: 3.0 M KCl, 50 mM $CaCl_2$, 1.0 M PIPES-NaOH, adjusted to pH 6.9.

Modified 2X gel sample buffer: 125 mM Tris–HCl, pH 8.8, 4% SDS, 100 mM DTT, 10% glycerol, 2 mM EGTA, and 0.05% Bromophenol Blue. Store at −20 °C.

2. Procedure

1. To a chilled beaker with a stir bar, add 100 mL of ice-cold TKI lysis buffer per ~10 g of *Tetrahymena thermophila* strain SB281 cells (~10 mL of packed

cells). To this add protease inhibitors at 2× their final concentration immediately before use: leupeptin at 20 μg/mL (for smaller preparations). Alternatively, *N*-ethylmaleimide at 2 mM and *o*-phenanthroline at 2 mM were used during methods development for large-scale preparations (>10 g of cells), but only in the lysis buffer; the remaining solutions included leupeptin at 10 μg mL^{-1}.

2. Harvest cells from log-phase culture to get approximately 10 mL of packed cells. Wash these cells with 10 mM Tris–HCl, pH 7.4, to remove as much of the culture media as possible.
3. Quickly but gently resuspend the pelleted washed cells in ice-cold deionized water to a volume equal to the volume of the 2X lysis buffer.
4. While stirring the 2X TKI lysis buffer at moderate speed, quickly pour in the cell suspension and stir for at least 1 min. Check by light microscopy to see that cell lysis has occurred and that greater than 99% of the cells have been extracted to form cytoskeletal "ghosts."
5. Harvest the TKI insoluble residues by means of centrifugation of the lysate at least 10,000 × g for 20 min at 4 °C. Carry out all subsequent steps at 0-4 °C.
6. Remove as much of the supernatant as possible, and with a spatula, gently resuspend the pelleted cytoskeletal residues in the wash buffer, supplemented with leupeptin at 10 μg mL^{-1}. Centrifuge the washed residues at 10,000 × g for 10 min.
7. Resuspend the pelleted residues in 20 mL of alkaline extraction buffer with leupeptin at 10 μg mL^{-1}. Thoroughly homogenize the pellet using (1) multiple passages (10–20 times) through a syringe with an 18 gauge cannula, or (2) sonication until all the cytoskeletal residues have been disrupted, as judged by light microscopy. As much as possible, avoid foaming during homogenization.
8. Transfer the homogenized residues to thick-walled polycarbonate tubes with aluminum caps (Beckman) for ultracentrifugation, and centrifuge at 100,000 × g for 1 h.
9. Carefully open the tubes containing the clarified extract. Remove the supernatant avoiding the pellet as well as the lipid overlayer. It is preferable to sacrifice some material to avoid the lipid overlayer to get a clear extract.
10. To the recovered supernatant, add concentrated magnesium chloride solution to a final concentration of 10 mM. Centrifuge this extract again at 100,000 × g for 1 h.
11. Carefully withdraw and transfer the supernatant to a new tube, avoiding the pellet and any remaining lipid overlayer. To this final clarified extract, add 20X polymerization buffer (1/19 of the volume of the clarified extract).
12. The material is allowed to precipitate overnight on ice. It will often form delicate cobweb-like structure within the tube. Light microscopic observations reveal the presence of a meshwork of microscopic fibers. These can then be harvested by centrifugation at 10,000 × g for 30 min. The pellet can be resolubilized in gel sample buffer and boiled for 2 min. The 2X gel sample buffer is used directly to ensure the complete extraction of larger pellets.

   The resulting material is enriched in three major proteins: epiplasmic band A protein (Epa1p), epiplasmic band C protein (Epc1p), and the 25-kDa

Tetrahymena calcium-binding protein (Tcb2p), as assessed by mass spectrometric analysis of bands excised from gels (manuscript in preparation). It also contains smaller quantities of other proteins, including epiplasmic band B protein (Epb1b).

Notes

The TKI procedure for preparing microtubule-depleted cytoskeletal preparations was briefly described by Honts and Williams (2003). This procedure has proven to be quite robust in that the same set of major proteins (Epa1p, Epc1p, and Tcb2p) is reliably obtained. Variations of this procedure have been explored, such as different starting types of cytoskeletal residues (including those that include microtubules), different buffers with different pH values, and different compositions for the final polymerization buffer. Even with these alterations, the same three proteins were reliably recovered.

The proteins solubilized by the alkaline low-salt extraction buffer remain in solution, even after $MgCl_2$ is added to 10 mM. However, the addition of $CaCl_2$ in excess of chelating capacity of EGTA results in the formation of delicate web like networks (Fig. 1(a)) that can be easily precipitated by low-speed centrifugation (once disturbed they will eventually sediment at the bottom of the tube in which they

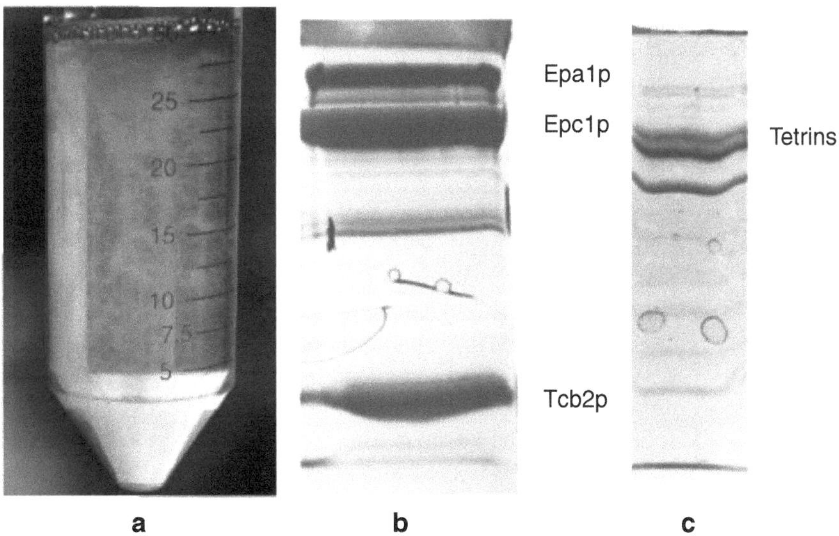

Fig. 1 Purification of Epc1p and tetrin-enriched fractions. (a) Weblike meshwork present in solutions following *in vitro* assembly of Epc1p and associated proteins. (b) SDS-polyacrylamide gel separation of proteins in the Epc1p-enriched precipitate, recovered by low-speed centrifugation after *in vitro* assembly. (c) Gel separation of tetrin polypeptides recovered as a by-product of the Epc1p preparation.

formed). It has been found that it is possible to precipitate the proteins with a polymerization buffer containing 2 mM EGTA, 150 mM KCl, 10 mM $MgCl_2$, and 1 mM $CaCl_2$. Under these conditions, the concentration of free calcium ions can be estimated to be less than 1 μM using the MaxChelator program (Patton *et al.*, 2004)

Precipitates formed in the presence of calcium ions are difficult to resolubilize in nondenaturing buffers following centrifugation. Other precipitants that trigger the formation of a web-like meshwork include ammonium sulfate, polyethylene glycols, and high concentrations of KCl. All of these precipitants yield the same set of three predominant proteins (Fig. 1(b)). The precipitated proteins can be resolubilized by sonication in extraction buffer, but it is preferable to avoid any form of precipitation if the proteins are to be used for *in vitro* assembly assays or loaded onto a column for further chromatographic fractionation.

Preparations have been made from cells from early logarithmic cultures to early stationary phase cultures with the same basic result. It is preferable to harvest cells from mid-logarithmic phase cultures since the concentration of secreted cysteine proteases increases in late logarithmic cultures (Sackett *et al.,* 2010). Regardless, cells should be thoroughly but gently washed with 10 mM Tris, pH 7.4 buffer prior to lysis, to minimize the retention of proteases secreted into the culture media.

A variety of strategies to minimize proteolysis have been employed during the development of this procedure, mostly based on the goal of inhibiting the prevalent cysteine proteases found in *Tetrahymena* cells. In smaller preparations, leupeptin was typically used, at a concentration of 10 μg mL^{-1}. As a more economical alternative for larger scale preparations, 1 mM *N*-ethylmaleimide was used to target the reactive cysteine residue of cysteine proteases released in the lysis step, but it is possible that this method could modify reactive cysteines in cytoskeletal proteins being prepared as well. So far no obvious differences have been observed between proteins prepared in the presence of these two different sets of inhibitors.

One other problem encountered in this procedure is that the alkaline extract contains a significant amount of lipid material that forms an overlayer during the first ultracentrifugation step. Great care must be taken to avoid the bulk of this material when transferring the clarified extract to a clean tube. That way most of the residual lipid material can be removed in the second clarification step (after adding $MgCl_2$ to 10 mM). A small amount of this material can be removed by filtration through glass wool or passage through a 0.2 μm filter – but there are limits to which these methods can clarify the extract. Careful recovery of the clear phase of the first supernatant is the key to getting a clear extract. Likewise, an addition to glycerol or sucrose (5% w/v) to the lysis and wash buffer has been found to eliminate some (but not all) of this material at the beginning of the procedure.

This procedure was been used to analyze the proteins in EPC1-knockout cells (Williams, 2004). Despite the absence of a major protein component of the epiplasmic layer, the cytoskeletal "ghosts" appear relatively normal as judged by light microscopic observations. Western blot analysis has shown that as expected, the Epc1 protein is absent, but two others proteins, Epa1p and Tcb2p remain, as do a number of minor proteins, some of which had been obscured by the abundance of the Epc1 protein present in wild-type preparations.

B. Purification of Tetrin Proteins

A procedure for preparing the nonmicrotubular fibrous components of the oral apparatus, specifically the tetrins, has been described (Honts and Williams, 1990). However, it has been found that the tetrins can be obtained as a byproduct of the method described above, continuing from step 7 in section IV A 2. While it is possible to use the pellet derived from step 8 above for this purpose, it is easier to first perform a lower speed centrifugation before going on to step 8, to obtain a less compacted pellet that can be more easily extracted to obtain a solution enriched in the tetrin proteins.

1. Solutions

Tetrin solubilization buffer: 1.0 M KI, 1 mM EGTA, 10 mM Tris–HCl, adjusted to pH 9.0. Add leupeptin to 10 μg mL^{-1} before use.
Dialysis buffer: 1 mM EGTA, 5 mM Tris–HCl, adjusted to pH 8.0.

2. Procedure

1. Centrifuge the homogenized material from step 7 above at 10,000 × g for 30 min.
2. The low-speed supernatant (which will be very cloudy) can be processed further as above, following on with step 8, to obtain the epiplasmic proteins. The low-speed pellet is resuspended in 20 mL of tetrin solubilization buffer containing 1 M potassium iodide and homogenized using one of the methods described above.
3. This 1.0 M KI extract is centrifuged at 100,000 × g for 1 h. The supernatant is withdrawn and transferred to a new tube, carefully avoiding the pellet and lipid overlayer, as described above.
4. The clarified extract is transferred to a dialysis bag and is dialyzed against 2.0 L of buffer at 4 °C, usually overnight.
5. The dialysate (containing insoluble filamentous material) is then centrifuged at 10,000 × g for 30 min to harvest the tetrin filament proteins. The addition of $MgCl_2$ to 5 mM has been found to enhance the recovery of the tetrin proteins.

Notes

The pelleted material is enriched in homologs of the four tetrin polypeptides previously described by Honts and Williams (1990) for *Tetrahymena pyriformis.* Mass spectrometric analysis has confirmed the presence of the three *Tetrahymena thermophila* tetrin homologs identified by Brimmer and Weber (2000), as well as the one not identified in their study. It, therefore, appears that this preparation (Fig. 1(c)) contains *Tetrahymena thermophila* homologs of the four tetrin proteins first identified in *Tetrahymena pyriformis* (Honts and Williams, 1990).

V. Challenges and Opportunities

The first challenge in the purification of cytoskeletal protein from *Tetrahymena thermophila* is the complexity of the mixture of cytoskeletal proteins present in the cell. Published protocols, and those described in this review, provide a means to fractionate the *Tetrahymena* cytoskeleton into specific sets of proteins: actin, tubulin, epiplasmic proteins, and tetrins. Ongoing work in my laboratory is aimed at developing a comprehensive fractionation scheme to purify all the major cytoskeletal proteins in *Tetrahymena* using a combination of selective extractions and precipitations coupled with chromatographic separations.

It may be of benefit to overexpress the protein of interest under the control of an inducible promoter (Boldrin *et al.*, 2006; Shang *et al.*, 2002; Yu *et al.*, 2011). This will likely be necessary when working with quantitatively minor proteins. Alternatively, heterologous overexpression in bacteria may be an option, if the coding sequence of the *Tetrahymena* gene is redesigned for optimal expression in bacteria. As the cost of gene synthesis continues to drop, this may be a practical strategy for the synthesis of small cytoskeletal proteins or their individual protein domains.

Likewise recombinant DNA methods exist to engineer tagged versions of the individual proteins (Busch *et al.,* 2010; Kataoka *et al.,* 2010). Epitope-tagged versions of these proteins provide an alternative avenue for the purification of proteins that may be difficult to prepare by more traditional means.

Even with the use of cysteine protease specific inhibitors such as leupeptin and E-64, some proteins still seem to be sensitive to proteolytic degradation (although it is possible that some of this represents processing by *in vivo* degradation pathways). In the same way the SB281 secretory mutant strain provides a way to avoid the mucus produced by wild-type cells upon Triton X-100 detergent lysis, the field would benefit from strains that are deficient in protease activity. In budding yeast, the BJ5623 strain (described in Jones, 1991) proved valuable to be the background of cells used to prepare actin and fimbrin proteins (Honts *et al.,* 1994). The work of Herrmann *et al.* (2006) describes a step in that direction.

The methods for the preparation of actin, tubulins, and their accessory proteins provide exciting opportunities to study the function of highly conserved eukaryotic proteins within the context of a complex eukaryotic cell over the course of the cell division cycle or during mating, or to study the cell biology of ciliogenesis, phagocytosis, or intracellular transport.

Insolubility of the membrane skeletal proteins such as Epc1p and tetrins remains a problem. While these proteins can be made soluble (that is, they stay in supernatant after centrifugation at 100,000 × g for 60 min) under dissociating (at high concentrations of KI; in low-ionic-strength buffers) or denaturing conditions (urea or guanidine hydrochloride), the maintenance of these conditions during chromatographic separations has proven difficult. Progress has been made, but much more remains to be done.

Proteomic analysis of the *Tetrahymena thermophila* cytoskeleton (Gould *et al.*, 2011) including various substructures such as the cilia (Smith *et al.*, 2005), basal bodies (Kilburn *et al.*, 2007), and the epiplasm (Honts *et al.*, manuscript in preparation) have revealed a rich array of molecular targets for further study, some familiar (like tubulin) but many poorly characterized–if at all. This situation will change rapidly, especially with rapid advances being made in the molecular tools that can be applied to *Tetrahymena* genes and the proteins that they encode.

Acknowledgments

The author wishes to thank Heather Said, Erica Zbornik, Freddy Witarsa, Jennifer Seidel, and Jodi Gullicksrud for their help in testing and refining the protocols presented above. I am grateful to Joseph Frankel for help suggestions with respect to this manuscript. This work was supported by a faculty research grant provided by Drake University.

References

Boldrin, F., Santovito, G., Gaertig, J., Wloga, D., Cassidy-Hanley, D., Clark, T. G., and Piccinni, E. (2006). Metallothionein gene from *Tetrahymena thermophila* with a copper-inducible-repressible promoter. *Eukaryot. Cell* **5**, 422–425.

Brimmer, A., and Weber, K. (2000). The cDNA sequences of three tetrins, the structural proteins of the *Tetrahymena* oral filaments, show that they are novel cytoskeletal proteins. *Protist* **151**, 171–180.

Busch, C. J., Vogt, A., and Mochizuki, K. (2010). Establishment of a Cre/loxP recombination system for N-terminal epitope tagging of genes in *Tetrahymena*. *BMC Microbiol.* **10**, 191.

Cupples, C. G., and Pearlman, R. E. (1986). Isolation and characterization of the actin gene from *Tetrahymena thermophila*. *Proc. Natl Acad. Sci. U.S.A.* **83**, 5160–5164.

Edamatsu, M., Hirono, M., and Watanabe, Y. (1990). Purification and characterization of *Tetrahymena* profilin. *Biochem. Biophys. Res. Commun.* **170**, 957–962.

Eisen, J. A., Coyne, R. S., Wu, M., Wu, D., Thiagarajan, M., Wortman, J. R., Badger, J. H., Ren, Q., Amedeo, P., Jones, K. M., Tallon, L. J., Delcher, A. L., Salzberg, S. L., Silva, J. C., Haas, B. J., Majoros, W. H., Farzad, M., Carlton, J. M., Smith, R. K. J., Garg, J., Pearlman, R. E., Karrer, K. M., Sun, L., Manning, G., Elde, N. C., Turkewitz, A. P., Asai, D. J., Wilkes, D. E., Wang, Y., Cai, H., Collins, K., Stewart, B. A., Lee, S. R., Wilamowska, K., Weinberg, Z., Ruzzo, W. L., Wloga, D., Gaertig, J., Frankel, J., Tsao, C. C., Gorovsky, M. A., Keeling, P. J., Waller, R. F., Patron, N. J., Cherry, J. M., Stover, N. A., Krieger, C. J., del Toro, C., Ryder, H. F., Williamson, S. C., Barbeau, R. A., Hamilton, E. P., and Orias, E. (2006). Macronuclear genome sequence of the ciliate *Tetrahymena thermophila*, a model eukaryote. *PLoS Biol.* **4**, e286.

Garces, J. A., Hoey, J. G., and Gavin, R. H. (1995). Putative myosin heavy and light chains in *Tetrahymena*: co-localization to the basal body-cage complex and association of the heavy chain with skeletal muscle actin filaments *in vitro*. *J. Cell Sci.* **108**, 869–881.

Gibbons, I. R., and Rowe, A. J. (1965). Dynein: a protein with adenosine triphosphatase activity from cilia. *Science* **149**, 424–426.

Gonda, K., and Numata, O. (2002). p85 binds to G-actin in a Ca(2+)/calmodulin-dependent manner, thus regulating the initiation of cytokinesis in *tetrahymena*. *Biochem. Biophys. Res. Commun.* **292**, 1098–1103.

Gould, S. B., Kraft, L. G., van Dooren, G. G., Goodman, C. D., Ford, K. L., Cassin, A. M., Bacic, A., McFadden, G. I., and Waller, R. F. (2011). Ciliate pellicular proteome identifies novel protein families with characteristic repeat motifs that are common to alveolates. *Mol. Biol. Evol.* **28**, 1319–1331.

Herrmann, L., Erkelenz, M., Aldag, I., Tiedtke, A., and Hartmann, M. W. (2006). Biochemical and molecular characterisation of *Tetrahymena thermophila* extracellular cysteine proteases. *BMC Microbiol.* **6**, 19.

Hirono, M., Endoh, H., Okada, N., Numata, O., and Watanabe, Y. (1987). *Tetrahymena* actin. Cloning and sequencing of the Tetrahymena actin gene and *identification* of its gene product. *J. Mol. Biol.* **194**, 181–192.

Hirono, M., Kumagai, Y., Numata, O., and Watanabe, Y. (1989). Purification of *Tetrahymena* actin reveals some unusual properties. *Proc. Natl. Acad. Sci. U.S.A.* **86**, 75–79.

Honts, J. E., Sandrock, T. S., Brower, S. M., O'Dell, J. L., and Adams, A. E. (1994). Actin mutations that show suppression with fimbrin mutations identify a likely fimbrin-binding site on actin. *J. Cell Biol.* **126**, 413–422.

Honts, J. E., and Williams, N. E. (1990). Tetrins: polypeptides that form bundled filaments in *Tetrahymena*. *J. Cell Sci.* **96**, 293–302.

Honts, J. E., and Williams, N. E. (2003). Novel cytoskeletal proteins in the cortex of *Tetrahymena*. *J. Eukaryot. Microbiol.* **50**, 9–14.

Johnson, K. A. (1986). Preparation and properties of dynein from *Tetrahymena* cilia. *Methods Enzymol.* **134**, 306–317.

Jones, E. W. (1991). Tackling the protease problem in *Saccharomyces cerevisiae*. *Methods Enzymol.* **194**, 428–453.

Kataoka, K., Schoeberl, U. E., and Mochizuki, K. (2010). Modules for C-terminal epitope tagging of *Tetrahymena* genes. *J. Microbiol. Methods* **82**, 342–346.

Kilburn, C. L., Pearson, C. G., Romijn, E. P., Meehl, J. B., Giddings, T. H., Jr, Culver, B. P., Yates, J. R., 3rd, and Winey, M. (2007). New *Tetrahymena* basal body protein components identify basal body domain structure. *J. Cell Biol.* **178**, 905–912.

Kurasawa, Y., Hanyu, K., Watanabe, Y., and Numata, O. (1996). F-actin bundling activity of *Tetrahymena* elongation factor 1 alpha is regulated by Ca^{2+}/calmodulin. *J. Biochem.* **119**, 791–798.

Lyons-Abbott, S., Sackett, D. L., Wloga, D., Gaertig, J., Morgan, R. E., Werbovetz, K. A., and Morrissette, N. S. (2010). alpha-Tubulin mutations alter oryzalin affinity and microtubule assembly properties to confer dinitroaniline resistance. *Eukaryot. Cell* **9**, 1825–1834.

Maekawa, S., and Sakai, H. (1978). Characterization and in vitro polymerization of *Tetrahymena* tubulin. *J. Biochem.* **83**, 1065–1075.

Maihle, N. J., and Satir, B. H. (1985). Protein secretion in *Tetrahymena thermophila*: characterization of the secretory mutant strain SB281. *J. Cell Sci.* **78**, 49–65.

Michelot, A., and Drubin, D. G. (2011). Building distinct actin filament networks in a common cytoplasm. *Curr. Biol.* **21**, R560–R569.

Mitchell, E. J., and Zimmerman, A. M. (1985). Biochemical evidence for the presence of an actin protein in *Tetrahymena pyriformis*. *J. Cell Sci.* **73**, 279–297.

Patton, C., Thompson, S., and Epel, D. (2004). Some precautions in using chelators to buffer metals in biological solutions. *Cell Calcium* **35**, 427–431.

Sackett, D. L., Werbovetz, K. A., and Morrissette, N. S. (2010). Isolating tubulin from nonneural sources. *Methods Cell Biol.* **95**, 17–32.

Schafer, D. A., Jennings, P. B., and Cooper, J. A. (1998). Rapid and efficient purification of actin from nonmuscle sources. *Cell Motil. Cytoskeleton* **39**, 166–171.

Shang, Y., Song, X., Bowen, J., Corstanje, R., Gao, Y., Gaertig, J., and Gorovsky, M. A. (2002). A robust inducible-repressible promoter greatly facilitates gene knockouts, conditional expression, and overexpression of homologous and heterologous genes in *Tetrahymena thermophila*. *Proc. Natl Acad. Sci. U.S.A.* **99**, 3734–3739.

Shiozaki, N., Nakano, K., Takaine, M., Abe, H., and Numata, O. (2009). Usual and unusual biochemical properties of ADF/cofilin-like protein Adf73p in ciliate *Tetrahymena thermophila*. *Biochem. Biophys. Res. Commun.* **390**, 54–59.

Smith, J. C., Northey, J. G., Garg, J., Pearlman, R. E., and Siu, K. W. (2005). Robust method for proteome analysis by MS/MS using an entire translated genome: demonstration on the ciliome of *Tetrahymena thermophila*. *J. Proteome Res.* **4**, 909–919.

Sugita, M., Iwataki, Y., Nakano, K., and Numata, O. (2011). Unique sequences and predicted functions of myosins in *Tetrahymena thermophila*. *Gene* **480**, 10–20.

Suprenant, K. A., Hays, E., LeCluyse, E., and Dentler, W. L. (1985). Multiple forms of tubulin in the cilia and cytoplasm of *Tetrahymena thermophila*. *Proc. Natl. Acad. Sci. U.S.A.* **82**, 6908–6912.

Watanabe, A., Kurasawa, Y., Watanabe, Y., and Numata, O. (1998). A new *Tetrahymena* actin-binding protein is localized in the division furrow. *J. Biochem.* **123**, 607–613.

Werbovetz, K. A., Brendle, J. J., and Sackett, D. L. (1999). Purification, characterization, and drug susceptibility of tubulin from *Leishmania*. *Mol. Biochem. Parasitol.* **98**, 53–65.

Williams, N. E. (2004). The epiplasm gene EPC1 influences cell shape and cortical pattern in *Tetrahymena thermophila*. *J. Eukaryot. Microbiol.* **51**, 201–206.

Williams, N. E., Honts, J. E., and Kaczanowska, J. (1990). The formation of basal body domains in the membrane skeleton of *Tetrahymena*. *Development* **109**, 935–942.

Williams, N. E., Honts, J. E., Lu, Q., Olson, C. L., and Moore, K. C. (1989a). Identification and localization of major cortical proteins in the ciliated protozoan, *Euplotes eurystomus*. *J. Cell Sci* **92**, 433–439.

Williams, N. E., Honts, J. E., and Stuart, K. R. (1989b). Properties of microtubule-free cortical residues isolated from *Paramecium tetraurelia*. *J. Cell Sci.* **92**, 427–432.

Williams, N. E., Vaudaux, P. E., and Skriver, L. (1979). Cytoskeletal proteins of the cell surface in Tetrahymena I. Identification and localization of major proteins. *Exp. Cell Res.* **123**, 311–320.

Wloga, D., and Gaertig, J. (2010). Post-translational modifications of microtubules. *J. Cell Sci.* **123**, 3447–3455.

Yu, T., Barchetta, S., Pucciarelli, S., La Terza, A., and Miceli, C. (2011). A novel robust heat-inducible promoter for heterologous gene expression in *Tetrahymena thermophila*. Protist, http://dx.doi.org/10.1016/j.protis.2011.07.003.

CHAPTER 15

Behavioral Bioassays and Their Uses in *Tetrahymena*

Todd M. Hennessey and Thomas J. Lampert

Department of Biological Sciences, University at Buffalo, Amherst, NY, USA

Abstract

The swimming behaviors of *Tetrahymena* can be used in sensitive behavioral bioassays for estimating the effects of drugs, mutations, and other conditions on the physiological state of the cell. These assays can be used in both forward and reverse genetic approaches to help understand cellular functions from genotype to phenotype.

I. Introduction

One of the goals of modern molecular biology is to understand how the functional expression of a gene product can lead to an observable phenotype in a cell. This is often reflected in the behavioral phenotypes, which they produce. In ciliates such as *Tetrahymena* and *Paramecium*, behavioral bioassays provide additional ways to characterize both primary phenotypes as well as pleiotropic effects because the behavioral repertoire of these cells is quite extensive (Jennings, 1976). Similarly, changes in swimming behaviors can be used to gain insights into the effects and side

0091-679X/10 $35.00
DOI 10.1016/B978-0-12-385967-9.00015-3

effects of drugs, toxins (Lainhart *et al.*, 2009), or other conditions (Hennessey, 1989). Therefore, behavioral bioassays can serve as convenient *in vivo* indicators of normal and altered physiological states.

The swimming behaviors of ciliates like *Tetrahymena* and *Paramecium* have been used in many types of behavioral bioassays to estimate the physiological and biochemical status of both wild-type and mutant cells. Most of these assays originated from the "Genetic Dissection" approach developed in *Paramecium* by Ching Kung (Kung *et al.*, 1975), Mihoko Takahashi (Takahashi, 1979; Takahashi *et al.*, 1980), and others (Bell *et al.*, 2007; Matt *et al.*, 1980) who used behavioral mutants to disrupt sensory transduction pathways. This forward genetics approach has relied heavily on accurate and reproducible behavioral bioassays to identify the key members of such pathways. In these experiments, random mutagenesis was commonly used to generate behavioral mutants with altered swimming behaviors. Genetic, electrophysiological, biochemical, and other approaches followed to identify the change in genotype that led to the altered phenotype of a behavioral mutant. The behavioral bioassays played pivotal roles in these characterizations because these assays were ways to follow the phenotype in genetic crosses and to identify ionic conductances and biochemical factors responsible for these phenotypes. With the advent of the gene knockout procedures initiated by Gaertig *et al.* (1994), a reverse genetics approach can now be used to help understand the functions of identified genes in *Tetrahymena*. It is very important to point out at the outset that these types of behavioral bioassays are valuable as estimates of possible alterations in behavioral mutants but proper physiological and biochemical assays must always follow to verify the basis for that alteration.

The first of these initial behavioral bioassays in *Paramecium* revolved around examining the role of the ciliary voltage-dependent inward Ca^{2+} current in Ca^{2+}-dependent ciliary reversal. Because both *Paramecium* and *Tetrahymena* produce Ca^{2+}-based action potentials in response to sufficient depolarizations (Eckert, 1972; Onimaru *et al.*, 1980), each action potential produces an inward Ca^{2+} current (Satow and Kung, 1979). Since this voltage-dependent inward Ca^{2+} current is localized to the cilia (Dunlap, 1977), each action potential causes a rise in intraciliary Ca^{2+} concentrations. Detergent permeabilization experiments in both *Paramecium* (Kung and Naitoh, 1973; Naitoh and Kaneko, 1972) and *Tetrahymena* (Goodenough, 1983) have shown that when the axonemal Ca^{2+} concentration rises above 10^{-6} M, the direction of ciliary beat reverses. Therefore, backward swimming in ciliates such as *Paramecium* and *Tetrahymena* is due to Ca^{2+}-dependent ciliary reversals (Eckert, 1972).

Behaviorally, action potentials produce backward jerks called avoiding reactions (AR) and stronger or more prolonged stimuli can cause backward swimming (continuous ciliary reversals or CCR) for many seconds. An observed correlation between the duration of backward swimming in a strong depolarizing solution (high K^{+}) and the size of this inward Ca^{2+} current suggested that the duration of backward swimming would be a simpler way to estimate cellular excitability than intracellular electrophysiology (Haga *et al.*, 1984). This provided for faster screening for possible Ca^{2+} channel alterations in both mutants and wild type. However, it was later found that this assay can also be affected by axonemal defects (Hennessey *et al.*, 2002; Hinrichsen *et al.*, 1984) and by changes in the Ca^{2+} removal processes (Evans *et al.*, 1987).

Another prominent swimming behavior relates to changes in forward swimming speed. The general model from *Paramecium* is that hyperpolarizations cause an increase in ciliary beat frequency and increased forward swim speed while subthreshold depolarizations cause decreased ciliary beat frequency and slower forward swim speed (Machemer, 1988). It has been suggested that intracellular cAMP may regulate swim speed (Bonini *et al.*, 1986) but its exact role is not yet clear (Hennessey *et al.*, 1985; Nakaoka and Machemer, 1990). Swim speed assays are also affected by the waveform of ciliary beat (Bonini *et al.*, 1986). Mutants have been described in *Tetrahymena* that have normal beat frequency but swim slowly because of a change in the waveform of ciliary beat (Wood *et al.*, 2007). Therefore, swim speed assays are good first estimates for possible changes in ion channel activities, intracellular Ca^{2+}, cAMP or axonemal function but proper assays must follow to verify the basis for the change.

Responses to chemoattractants can involve changes in swim speeds and changes in the frequency of direction changes. It has been proposed in *Paramecium* that chemoattraction to high (mM) concentrations of chemoattractants requires membrane hyperpolarization and increased swim speed (Valentine *et al.*, 2008). The other contributor to attraction is a decrease in the frequency of spontaneous AR, allowing them to swim in straighter paths toward an attractant (Bell *et al.*, 2007). While some chemoattractants cause hyperpolarization in *Tetrahymena* other attractants have been shown to be effective in micromolar concentrations without changes in either swim speed or membrane potential (Grønlien *et al.*, 2010; Lampert *et al.*, 2011).

In contrast to chemoattractants, low (μM) concentrations of chemorepellents and other stimuli can cause somatic depolarizing receptor potentials in both *Paramecium* and *Tetrahymena* (Hennessey, 2005; Hennessey and Kuruvilla, 1999) and the resultant action potentials trigger avoidance. Similar avoidance behavior has been described in the depolarizing response to anterior stimulation in both *Paramecium* (Eckert *et al.*, 1972; Ogura and Machemer, 1980) and *Tetrahymena* (Onimaru *et al.*, 1980). If a cell bumps into an obstacle, the depolarization can trigger an action potential if this stimulus is strong enough. This causes the cell to swim backward briefly, reorient and regain forward swimming in a new, random direction. Posterior mechanical stimulation causes hyperpolarization and fast forward swimming, allowing these cells to escape from something that approaches it from behind. Thermal avoidance has also been described for *Paramecium* (Hennessey and Nelson, 1979) and *Tetrahymena* (Connolly *et al.*, 1985) and these behaviors also rely on changes in swim speed and AR. Therefore, behavioral bioassays can help in identifying the underlying mechanisms involved in chemoattraction, chemorepulsion, mechanosensory responses, and thermal avoidance in these ciliates.

These behavioral bioassays can be used either as first screens or additional screens. As a first screen, they can be used to gather information about the primary effect of a mutation or some other treatment (like a drug). As second screens, they can help to determine whether or mutant or treatment is due to a specific effect or if it is just a general effect on responsiveness to all stimuli. For example, if a mutant or drug effect is found that eliminates the responses to chemorepellents, it should be checked to make sure that these cells are capable of showing AR by secondary screen assays such as high Ba^{2+}. If they show good AR in Ba^{2+}, it suggests that they are capable of producing action potentials. If these cells do not respond to Ba^{2+}, they should be checked in the detergent SDBS (sodium dodecylbenzenesulfonate) to see if they are able to show ciliary reversals at all. An axonemal mutant might not even show AR in SDBS, so it would not respond to either Ba^{2+} or chemorepellents.

II. Materials and Methods

A. Forward Swim Speed Assays

1. Simple Swim Speed Assay

The simplest forward swimming speed assay is to observe cells swimming under a dissecting microscope with some kind of distance marker, such as a calibrated mark on a slide. For example, if several 1.0 cm marks are made on the underside of a slide, a stopwatch can be started whenever a cell passes over one of the marks. Only cells swimming straight during the entire length of the mark should be included. Swim speed can be expressed as cm/s or mm/s.

2. Use of Negative Geotaxis to Estimate Swim Speed

Indirect estimations of swim speed can be done in either a spectrophotometer or in a glass column. If cells are mixed in a cuvette, the change in OD_{600} over time can be used to estimate swim speed as the cells swim up past the beam due to their negative geotaxis (natural tendency to swim upward). Similarly, cells can be injected through a serum stopper at the bottom of a long glass column and aliquots can be taken off the top at set time periods and counted. A plot of the number of cells reaching the top over time can be used to estimate swim speed. Both of these approaches should be followed by standard swim speed assays because they could also be affected by their ability to show geotaxis. If a mutation or other condition does not affect swim speed but does have an effect in these assays, it might indicate an effect on the mechanisms underlying geotaxis.

3. Analysis of Swim Speed with Digital Movies

The more direct assays of forward swim speed are done best by making digital movies through a dissection microscope. Digital movie cameras can be obtained that are relatively inexpensive and can mount on the eyepiece of any microscope. Microscopes can also be purchased with dedicated digital cameras as part of the microscope. One of the most important concerns is to have a uniform light source, such as diffuse lighting from below the sample. Many light sources that shine on a sample from above produce a beam of light that interacts with the sample at an angle that produces inconsistent lighting across the field. This can make digital analysis more difficult. Also, swim speeds should only be taken from cell paths that are straight and uninterrupted by stops or turns.

A very simple way to measure swim speeds is to put a piece of clear plastic wrap over the monitor and play back the digital movie on the screen. Using a stopwatch and a magic marker, a mark can be made at the start of a swim path at the same time that the stopwatch is started. When the stopwatch is stopped, a final mark is made. If a known distance marker is projected onto the screen, a calibration can be determined to convert observed distance on the screen into real distance on the slide. With the distance and time known, speed can be expressed as mm/s. As before, only straight swimming cells should be included.

4. Digital Image Analysis

The best way to measure swim speed is by measuring the lengths of swim paths from digital videos with the aid of digital imaging software. The ImageJ program works very well for this, and this program is free to download from http://rsbweb.nih.gov/ij. For some digital videos, it may also be necessary to convert them into the right format for Image J. A good program for this is called VitualDub, and it is available as a free download from http://www.virtualdub.org/. For example, a digital movie from a Motic program can be converted to the proper avi format by VirtuaDub (with no audio) and entered into the ImageJ program through file, import, avi. An Avi Reader is also available for this purpose from ImageJ as a plugin. When the file is opened, Image, stacks and Z Project should be selected. With the choice of Min Intensity, swim paths can be displayed as lines, similar to "time-lapse" photography. The ZProjection box lets you chose the length of the swim paths to be analyzed. For example, if the Start slice is 1 and the Stop slice is 300, that means that there are 300 slices (or "stills) making up this video. If you know that the video is 10 s long, going from Start slice 1 to Stop slice 30 will show the first second paths. The length of the paths can be measured by right clicking on Straight line selections (a box on the bottom of the control panel with a straight line on it), selecting Freehand Lines, tracing the length of the path and recording it by going to Analyzing and clicking Measure. The length of the path will be shown in a Results box. This number can be converted to mm by using the same procedures to obtain an image of a calibrated length standard. Knowing the time and length of the path can produce data as mm/s.

The effects of a mutation or other condition on forward swim speeds can be assayed in a number of ways. If cells are incubated in 16 mM K^+ for 1 h, shifting them to 4 mM K^+ will cause them to hyperpolarize and swim faster forward. Forward swim speed can also be increased by some chemoattractants (like 0.1% proteose peptone), incubation for 45 min in 1.0 mM IBMX or addition of 0.5 mM theophylline. Slower forward swimming can be elicited by transfer of cells from 4 to 16 mM

K^+ or addition of many different kinds of depolarizing ions (Machemer, 1989; Saimi and Kung, 1987) or drugs such as inhalation anesthetics (Pope *et al.*, 1978). For example, there are some mutants of *Tetrahymena* that have a slow basal swim speed but they can speed up in response to some stimuli (Wood *et al.*, 2007) while there are others that cannot speed up to normal levels under any condition (Suryavanshi *et al.*, 2010). Such mutants can be used to help understand the regulation of ciliary beat. It is also very important to monitor the pH in test solutions because changes in pH (sometimes caused by the addition of test compounds) can affect swim speeds too.

Although it is usually assumed that an increased swim speed is associated with an increase ciliary beat frequency, this type of analysis should also be followed by high-speed digital analysis of beat frequency and ciliary waveform analysis because some mutants have been shown to have normal ciliary beat frequency but decreased forward swim speed because of an altered ciliary waveform (Wood *et al.*, 2007). Electrophysiological analysis can be used to confirm membrane potential changes (Hennessey and Kuruvilla, 1999) and intracellular Ca^{2+} measurements can detect gross changes in intracellular Ca^{2+} (Nam *et al.*, 2009).

B. Assays to Quantitate Deviations from Normal Forward Swimming

Interruption of forward swimming behavior usually represents membrane depolarization but the extent and duration of that depolarization can be reflected in the response. In response to small depolarizations, the ciliary beat frequency decreases and the cell slows down but stronger depolarizations can elicit repetitive action potentials and ARs (Machemer, 1988).

1. Avoiding Reaction Assay

The common assay for ARs is performed by simply observing cells under a dissecting microscope. In a typical AR assay, single cells are transferred to a test solution with a micropipette or capillary tube (Fig. 1) but a pipetman can also be used to transfer larger amounts of cells. To prepare cells for these assays, they should be washed in the test buffer (typically 50 μM $CaCl_2$, 10 mM MOPS, and pH 7.2 with Tris base) by centrifugation. A sample of cells in culture can be centrifuged in a 100 mL pear-shaped flask at 500 g for 2 min to pellet the cells. This pellet is transferred to 100 mL of this same wash solution and centrifuged again. The final washed pellet is resuspended in the test buffer to the same volume as the initial culture sample. This is left at room temperature for 30 min to several hours for the cells to acclimatize to this solution before testing. In the single cell transfer assay, each individual cell transferred is scored for either a clear deviation from forward swimming (AR) within a few seconds of observation or not. This is done many times and the statistic generated is the percent of cells observed showing AR (mean % AR ± SD). This is often done in at least three blocks of ten cells each so that the mean ± SD can be determined with an $n = 3$ or more. This can also be assayed by mixing cells with a test solution and scoring digital images for obvious AR but this

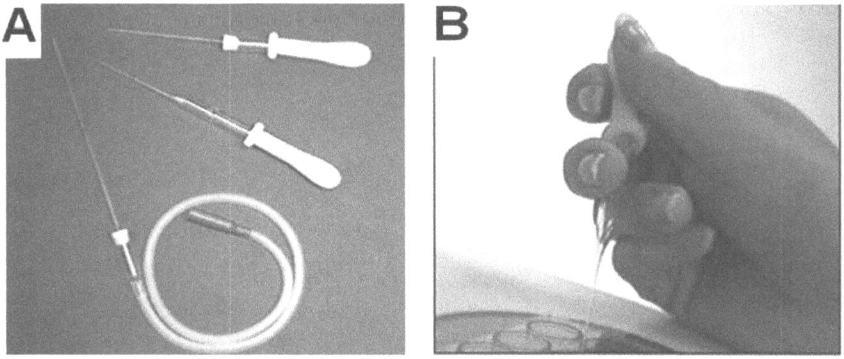

Fig. 1 Various kinds of micropipets can be used to capture individual Tetrahymena for behavioral bioassays. (A) A 10 μL capillary pipet can be used with a long piece of tubing connected to a mouthpiece (lower image). Volumes can be taken up and expelled by gentle pressure. This same capillary holder can be disconnected from the piece of tubing and connected to a pipet bulb (top image) for operation by hand. The image in the middle is a micropipette pulled by heating a 9-inch Pasteur pipet over a flame and pulling it. (B) The most efficient way to hold these micropipets is to rest your hand on a firm surface, move the pipet around with your fingers, and depress the bulb with your thumb.

can miss any initial, transient ARs because there is always a delay due to the movement caused during mixing. Good general stimuli for AR in *Tetrahymena* are low concentrations of Ba^{2+} (less than 0.5 mM), detergents like SDBS (Suryavanshi *et al.*, 2010) and chemorepellents like GTP (Hennessey, 2005). Unlike *Paramecium*, ARs are not seen well in *Tetrahymena* (CU427) in response to Mg^{2+} or Na^{+}. If a mutation or other condition causes a loss of AR in Ba^{2+}, this should be followed by assaying for AR in SDBS. Since SDBS is a detergent, it permeabilizes the cells and by-passes the Ca^{2+} channels to expose the axoneme to high Ca^{2+}. Therefore, cells with decreased excitability (like CU428) show little or no AR in Ba^{2+} but good AR in SDBS while some axonemal mutants, such as KO6 (Hennessey *et al.*, 2002) do not show AR in either Ba^{2+} or SDBS (personal observation).

2. Percent Direction Change Assay

Digital images of swim paths can be used to obtain a general "plus or minus" type of assay to determine whether a cell is swimming forward in a straight path or not. This is can be expressed as the percent of cells showing direction changes (PDC), percent AR or turning frequency. To estimate PDC, an image such as those shown in Figs. 2 and 3 are scored by simply counting the number of straight paths and the number of interrupted paths. Swim paths (1 s duration) are scored as either generally straight forward swimming or deviating at least 17° from linearity. The PDC represents the percentage of cells that deviated from linear swim paths at least once. Therefore, the PDC equals the number of interrupted paths divide by the total number of paths scored. For example, the PDC of Fig. 3(C) is about 13% while

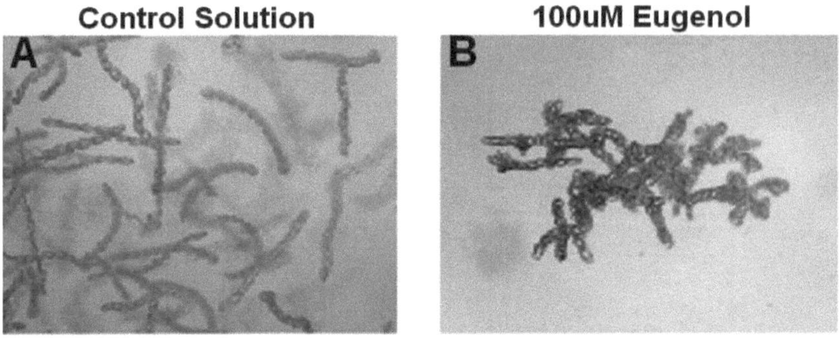

Fig. 2 Swim paths from digital videos and ImageJ. (A) Swim tracks are shown for wild-type cells (CU427) about 10 s after they were added to a control solution. This was a 1-s exposure so each line represent the distance they swam in 1 s. (B) Another sample of these cells was added to a solution containing 100 μM eugenol. This is a chemorepellent that causes repetitive ARs (Rodgers et al., 2008). This was also a 1-s exposure taken about 10 s after the cells were added to the solution.

the PDC in Fig. 2(B) is 100%. Since this is not always easy to measure, many measurements should be done to assure statistical significance.

The swim paths obtained as digital images can also be analyzed for the general shapes of these paths. Since these cells normally swim in a helical path (Machemer, 1988), the path can be seen as either a tight, straight line (see Fig. 3(C)) or as a more obvious helix (Fig. 3(D)). The degree of this helical pitch has been shown in *Paramecium* to change as a result of conditions such as the addition of cGMP (Bonini *et al.*, 1986), chemoattractants (Crenshaw and Edelstein-Keshet, 1993), and during geotaxis (Mogami and Baba, 1998). Deviations from straight swimming can also be seen as whirling in place (Fig. 3(B)), AR (Fig. 2(B)) or other behaviors that change the swim path. As in the case of swim speed assays, it is very important to monitor the pH in test solutions because changes in pH (sometimes caused by the addition of test compounds) can affect all of these assays because both high- and low-pH can cause AR on their own.

Some conditions, such as prolonged exposure to high Ba^{2+} (Schein *et al.*, 1976) or Ni^{2+} (Larsen and Satir, 1991) can slow the forward swimming speed down to the point of immobilization. A Ba^{2+} paralysis solution contains 0.01 mM Na_2HPO_4, 1.0 mM NaH_2PO_4, 2.0 mM Na-citrate, 0.1 mM $CaCl_2$, 10.0 mM $BaCl_2$, and 16.0 mM NaCl (Schein, 1976), and a Ni^{2+} immobilization solution contains 10 mM Ni^{2+}. The immobilizing effects of these ions can be assayed as either the rate of decrease in swim speed over time, time until immobilization of at least 90% of the cells or simply swim speed at a set time after adding the cells to the solution. However, immobilization can also be seen in compounds that cause deciliation. To distinguish immobilization from deciliation, cells can be transferred back into a control solution to see if they regain forward movement. Cells can also be analyzed microscopically for loss of cilia. Deciliated cells can also regenerate the cilia after deciliation after about 4 h (Skriver and Williams, 1980) but dead cells will not.

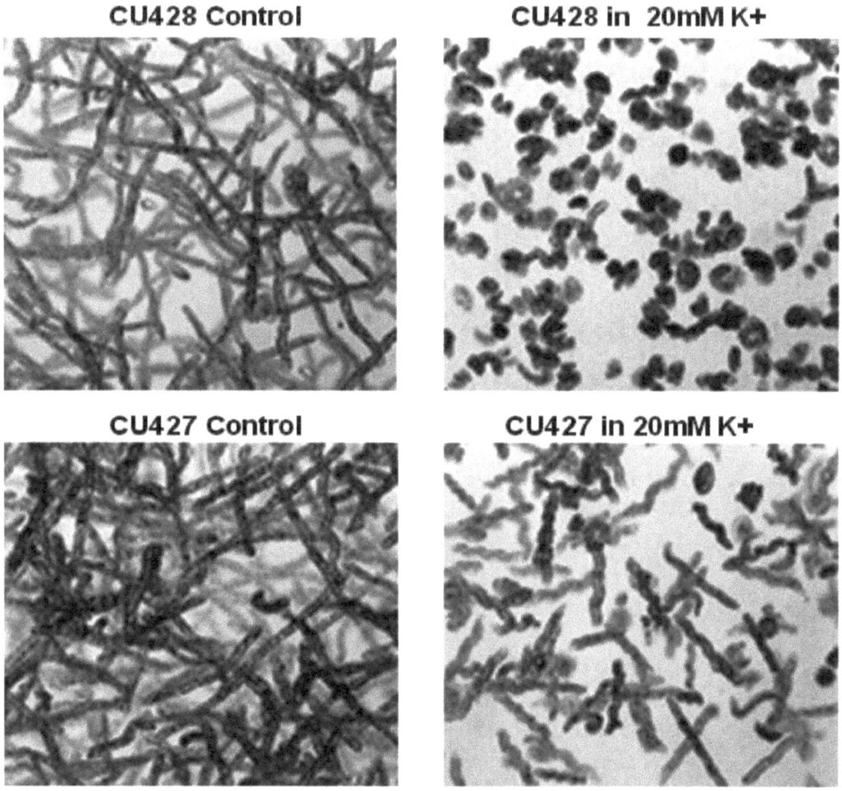

Fig. 3 Effects of high K^+ on swim paths. One second exposures of swim paths showed mostly forward swimming in the wild-type controls (A and C). The length of these paths reflects the swim speeds. Longer paths are faster swimming cells. (B) Addition of CU428 to 20 mM K^+ causes most of the cells to whirl in place. (D) CU427 cells show a different response. The majority of the cells swim backward. Some continue to swim backward for up to 20–30 s. This shows that dramatic differences in swimming behaviors can be even be seen between strains of wild type.

Immobilized or deciliated cells can be distinguished from dead cells by transferring them to a nutritive media. Death is the ultimate behavioral change, and dead cells will not regain the ability to grow in a nutritive media. It is also possible to immobilize cells by crosslinking the cilia with immobilization antigens (Ron *et al.*, 1992). In this case, microscopic analysis can be used to identify the activity of the contractile vacuole (Patterson and Sleigh, 1976). In live cells, contractile vacuole activity can be seen in regular intervals. Dead cells lose this "heartbeat." Trypan blue exclusion and MTT viability assays (Zilberg and Sinai, 2006) can also be used to assay for cell death.

If a depolarization is strong and prolonged, the cell can swim backward for up to 20–30 s due to continuous ciliary reversals. This is generally thought to represent a period of prolonged depolarization and/or high intraciliary Ca^{2+} so it can be affected by changes in ion channel activities (Haga *et al.*, 1984), Ca^{2+} removal systems (Evans *et al.*, 1987) or the Ca^{2+} sensitivity of the axonemal Ca^{2+}-dependent reversal machinery (Hinrichsen *et al.*, 1984). The assays for this can be either simply timing the duration of backward swimming of individuals transferred into a test solution or by measuring the lengths of the backward swim paths in digital images. Since the cells often whirl in place before regaining forward swimming, the duration of CCR can be measured as either the duration of backward swimming only or the total time necessary to regain forward swimming (which includes the whirling time).

C. Chemosensory Assays

There are many kinds of chemoresponse assays in *Tetrahymena* but the three that we have found to work best are a well assay (Fig. 4), the two-phase assay (Koppelhus *et al.*, 1994) (Fig. 5(A)), and the three-way stopcock assay (Van Houten, 1978) (Fig. 5(B)). Using several different assays helps to control for any artifacts that may be assay specific.

1. The Well Assay

A flat slide with an etched or raised circle on it works best but depression slides can also be used. This can be done in two ways. The first is to put 50–100 μL of a test solution on the slide and add a 1.0 μL drop of cells from a pipetman into the middle of the slide. If the cells do not respond to the test solution, they will swim out into the solution. If the solution contains a depolarizing agent like a chemorepellent, the cells will jerk back and forth and not move out from where they were placed on the slide.

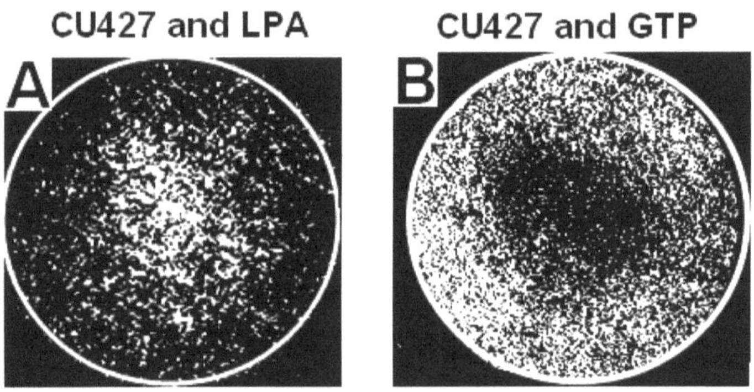

Fig. 4 A bulk assay for chemoresponses of a population of cells. Wild-type (CU427) cells were washed in a control solution and starved for 2 days. About 80 μL of these cells were spread out on a slide with a raised circle on it. (A) One microliter of 1 mM LPA was dropped into the center and a digital picture was taken about 5 min later. Most of the cells were clustered in the middle, showing chemoattraction and chemoaccumulation. (B) One microliter of 10 μM GTP was added to the center of a similar population of cells. The cells spread out and formed a zone of clearing in the middle, indicative of chemorepulsion. Both pictures were digitally enhanced in ImageJ to add contrast.

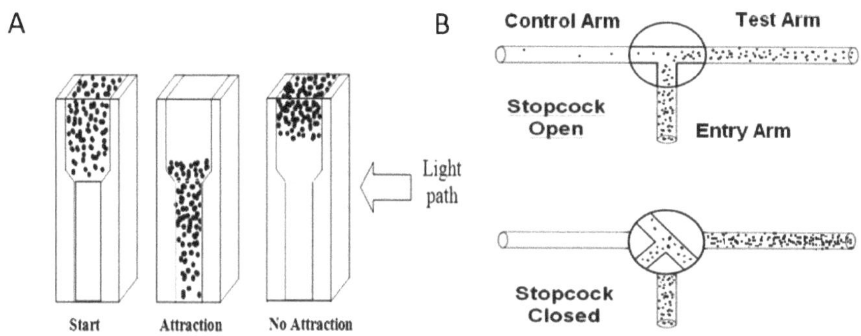

Fig. 5 Wild-type (CU427) cells were washed in control solution and starved for 2 days. (A) At the start of the two-phase assay, 1.5 mL of cells are layered on top of a 1.0-mL solution containing the test compound with 30 mg/mL nycodenz added to stabilize these two phases. This is placed in a spectrophotometer. Since the light path is near the interface of these two solutions, attraction will bring the cells through the light path and cause an increase in OD_{600}. If there is no attraction, the cells rise to the top of the cuvette by negative geotaxis. (B) For the three-way stopcock assay, similarly starved cells are added to the entry arm with a test solution in the test arm and control. When the stopcock is open, the cells move into it and make a choice to turn one way or the other. After 30 min, the stopcock is closed and the cells removed from each arm for counting after staining them with Lugol's stain (2.0 g KI and 1.0 g I_2 in 100 mL water). The index of chemotaxis (Iche) is determined by dividing the number of cells in the test arm by the total number of cells in the test and control arms together. An index of chemotaxis greater than 0.5 indicates chemoattraction toward the solution in test arm. A number of 0.5 indicates no preference for either side. A number of less than 0.5 indicates repulsion.

An example of this is shown in Fig. 2(B). The second way is to add the test solution to the cells. Examples of this are shown in Fig. 4.

2. Two-Phase Assay

A common way to assay chemoresponses in *Tetrahymena* is with the two-phase spectrophotometric assay (Koppelhus *et al.*, 1994). Cells must be starved in the test buffer for at least 2 days for optimal responses to the chemoattractants. To set up the assay, the test compound (like LPA for example) is added to 10 mM Tris buffer with 3% Nycodenz (w/v) added, and this mixture is placed in bottom of cuvette. A Pasteur pipette is used to carefully place 1.5 mL of cells (in the same Tris buffer) on top, creating upper layer. Care must be taken to not disturb the interphase. The cuvette is placed in the spectrophotometer, and the OD_{600} is monitored continuously for 30 min with the data printed in 60 s intervals. Chemoattraction is shown by a significant increase in the OD_{600} within the first 10–15 min.

3. The Three-Way Stopcock Assay

This is the most commonly used chemoresponse assay in *Paramecium* (Van Houten, 1978). Cells are washed in a control solution (typically 10 mM Tris,

50 μM Ca^{2+} buffered to pH 7.2 with MOPS) and put in the entry arm with the three-way stopcock closed. The stopcock itself contains the control solution, as does the control arm. The test arm contains the same solution with the compound to be tested added. To start the assay, the stopcock is opened and it is closed after 30 min to end the assay. Cells are removed from the control and test arms and counted by eye after staining with iodine (Lugol's stain). The Index of Chemotaxis (Iche) is determined by dividing the number of cells in the test arm by the total number of cells in the test and control arms. If this number is greater than 0.5, it indicates chemoattraction but if it is lower than 0.5 it indicates avoidance or chemorepulsion. If the number is near 0.5, then no chemoresponse was shown to that compound. An Index of Motility (Im) can also be obtained by dividing the number of cells in the test arm by the total number of cells in the entire assay. To test for drug effects, the cells can be pre-incubated in the drug for 30 min (or longer if necessary) and the drug would be included in all arms of the assay.

4. Chemosensory Adaptation

Chemosensory adaptation is the loss of responsiveness to a chemoeffector as a function of time of exposure to that compound. Chemosensory adaptation has been described in *Paramecium* (Kim *et al.*, 1997) and *Tetrahymena* (Kim *et al.*, 1999). To assay for chemosensory adaptation, cells can be incubated for a set time (usually 30 min) in an effective concentration of the compound but the chemoeffector must be washed away before retesting. If the de-adaptation time is long enough (at least 2 min), adapted cells can be added to an excess of a wash solution and quickly centrifuged. The pellet would be the washed cells. When these washed cells are tested in the original chemoeffector, they will not respond if they have adapted. De-adaptation can be assayed by leaving the cells in the wash solution and assaying them over time for return of the response. Cross-adaptation studies can be done by adapting cells to one chemoeffector and testing for their responses to others (Kim *et al.*, 1999).

D. Other Behavioral Assays

1. Pattern Formation

Starved cells show different kinds of patterns in shallow cultures (Fig. 6(A)). To see these patterns, cultures can be observed over time either as a 50 mL cultures in a 250 mL flask or as 10 mL cultures in 60 mm × 15 mm petri dishes. They usually start showing these patterns after 3 or more days.

2. Mating Assays

Mating behavior can be used to determine whether a mutant or condition affects the ability of cells to express a mating type, pair with a different mating type, fertilize

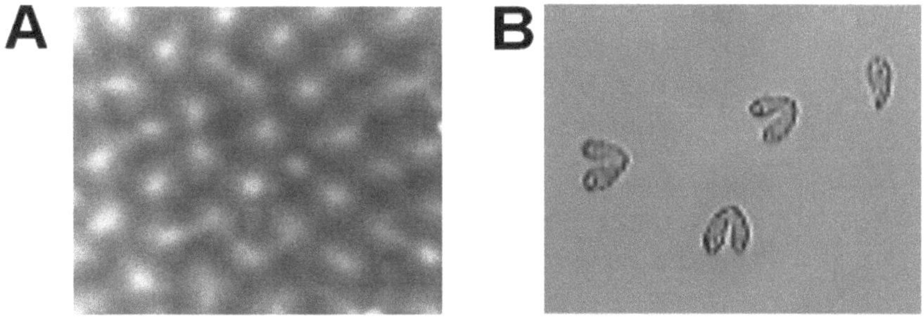

Fig. 6. Pattern formation and mating pairs. (A) A 10.0 mL culture of wild-type CU427 was grown for 2 days in a small (60 mm × 15 mm) Petri dish and a digital image was recorded with a black background. The white areas are high-cell densities, and the back areas have lower cell densities. This view is from an area of the dish that was about 2.2 cm wide and about 1.9 cm high. (B) Mating pairs. This shows mating pairs that are CU427 and CU428 about 2 h after mixing these opposite mating types. This shows three pairs and one unmated individual (on the right).

properly, and to express many other mating phenotypes (Bruns and Cassidy-Hanley, 1999; Hamilton and Orias, 1999). For our simple mating assay, we are only asking whether or not they can form pairs and if fertilization can occur between the pairs. To set cells up for mating, we chose two different mating types and grow them at 30 °C in 250 mL flasks with 50 mL of proteose peptone media in each for 2 days. After 2 days of growth, we count an aliquot of cells to determine the number of cells/mL. The cells are collected by centrifugation in sterile pear-shaped flask for 2 min at 500 g but the volume of cells is adjusted so that roughly the same number of cells are obtained in both pellets. These pellets are then added 100 mL of sterile 10 mM Tris (pH 7.4) with a sterile 9-inch Pasteur pipet. After a second centrifugation, the pellet is added to 50 mL of sterile 10 mM Tris in a 500 mL sterile flask. They are then left at 30 °C overnight. In the morning, the two cultures are mixed and kept at 30 °C. The appearance of pairs (Fig. 6(B)) should start to be seen after about one and half hours, although it may take up to 4 h. The percent of cell mating over time can be determined by taking out an aliquot, staining them at set times with Lugol's stain and counting the number of pairs. Since cells can often pair and come apart without fertilizing, a test for fertilization should also be done. This can be done by assaying for a micronuclear drug resistance marker (such as 5-methylpurine or cycloheximide) which should only be expressed if fertilization has occurred (Bruns and Cassidy-Hanley, 1999; Hamilton and Orias, 1999).

3. Growth and Viability Assays

Culture samples will be diluted, stained with iodine (Lugol's solution) and counted to determine the number of cells/mL. This will be plotted over a period of 4 to 5 days to determine the initial growth rates and the maximal cell densities. Viability can be assayed by trypan blue exclusion. Cells placed in 0.2% (w/v) trypan blue will exclude the dye if they are alive but trypan blue can enter dead cells and turn them blue. The redox indicator MTT has also been used as a viability assay (Zilberg and Sinai, 2006).

4. Other Assays

Additional behavioral assays include and mucocysts discharge as an assay for exocytosis (Turkewitz *et al.*, 1999), india ink uptake as an assay for endocytosis (Tiedtke *et al.*, 1988) and the rate of contractile vacuole activity as an assay for osmoregulation (Patterson and Sleigh, 1976).

III. Discussion and Summary

Since behavioral bioassays can be used to estimate possible physiological changes, they can be used to give a "behavioral physical" to new mutants or cells in a new condition (like a new drug). An initial look at growth, basal and stimulated swim speed, and ability to show AR can provide insights into the physiological state of the cell. For example, cells with a slow basal swim speed could have changes in either resting membrane ion conductances, axonemal function, or the uncharacterized mechanisms governing ciliary beat frequency and waveform. This could be followed with electrophysiological analysis (Hennessey and Kuruvilla, 1999), assays for axonemal function in detergent permeabilized cells (Goodenough, 1983), high-speed digital image analysis of ciliary beat frequency and waveform (Wood and Hennessey, 2007), axonemal sliding assays (Holwill and Satir, 1990), and many other biochemical and molecular approaches. Since most of these assays are quite involved, the simpler behavioral bioassays can be used to decide which assays to proceed to next.

One of the major problems is to define what is "wild-type" behavior. For example, there are dramatic behavioral differences between two classical wild type used in mating assays, CU427 and CU428. As shown in Fig. 3, CU427 are more responsive to depolarizing stimuli than CU428. Therefore, the results of behavioral bioassays in one strain may not be comparable to another. There might also be effects of mutations and drugs that are strain-specific, making comparisons of conclusions in the literature confusing.

Day-to-day consistency is often a problem because many of these behavioral bioassays are very sensitive to relatively minor changes. If an observation is made about behavioral changes in a new mutant, new drug, or other novel condition, it is necessary to show that this observation is reproducible every time that a new culture is grown and analyzed. Controls for the growth phase, pH of the test solution and other factors must be taken into account. Even seemingly trivial concerns such as the size of the opening of the pipet, the effects of mechanical stimulation during centrifugation and pipetting, number of cells/mL in the assay and room temperature must be carefully considered. Also, comparisons to other strains and species can often be misleading so keeping the controls within the same strain can be advantageous.

Other stimuli have also been described that could be adapted to new behavioral bioassays. While responses to changes in temperatures have been well studied in *Paramecium* (Hennessey and Nelson, 1979; Nakaoka *et al.*, 1987; Tawada and Oosawa, 1972), this has not been studied as extensively in *Tetrahymena.* The AR assay could be modified to be done on a temperature-controlled slide to characterize the %AR at different temperatures but the classical thermal accumulation and thermal avoidance types of assays are difficult in *Tetrahymena* because their small size makes them too susceptible to movement by convection currents (personal observations). Although *Tetrahymena* show good electrophysiological responses to mechanical stimulation (Onimaru *et al.*, 1980), no behavioral bioassays have been described to be used in the study of this type of sensory response. Bioassays have also been used to study geotaxis in *Tetrahymena* (Noever *et al.*, 1994), making this another area for future studies.

Many of the types of behavioral bioassays described above could also be performed by computer analyses of digital movies as shown by Clark and Nelson (1991) with *Paramecium*. There are also commercially available computer programs available to do this with *Tetrahymena* but the cost is often prohibitive. These types of procedures could provide greater statistical strength and less chance of experimental bias than the current methods.

References

Bell, W. E., Preston, R. R., Yano, J., and Van Houten, J. L. (2007). Genetic dissection of attractant-induced conductances in *Paramecium. J. Exp. Biol.* **210**, 357–365.

Bonini, N. M., Gustin, M. C., and Nelson, D. L. (1986). Regulation of ciliary motility by membrane potential in *Paramecium*: a role for cyclic AMP. *Cell Motil Cytoskeleton* **6**(3), 256–272.

Bruns, P. J., and Cassidy-Hanley, D. (1999). Methods for genetic analysis. *Methods Cell Biol.* **62**, 229–240.

Clark, K. D., and Nelson, D. L. (1991). An automated assay for quantifying the swimming behavior of *Paramecium* and its use to study cation responses. *Cell Motil. Cytoskeleton* **19**, 91–98.

Connolly, J. G., Brown, I. D., Lee, A. G., and Kerkut, G. A. (1985). Temperature-dependent changes in the swimming behaviour of *Tetrahymena pyriformis* – NT1 and their interrelationships with electrophysiology and the state of membrane lipids. *Comp. Biochem. Physiol.* **81**(2), 303–310.

Crenshaw, H. C., and Edelstein-Keshet, L. (1993). Orientation by helical motion – II. Changing the direction of the axis of motion. *Bull. Math. Biol.* **55**(1) 213–130.

Dunlap, K. (1977). Localization of calcium channels in *Paramecium caudatum*. *J Physiol.* **271**(1), 119–133.

Eckert, R. (1972). Bioelectric control of ciliary activity. *Science* **176**, 473–481.

Eckert, R., Naioth, Y., and Friedman, K. (1972). Sensory mechanisms in *Paramecium*: I. Two components of the electrical response to mechanical stimulation of the anterior surface. *J. Exp. Biol.* **56**, 683–694.

Evans, T. C., Hennessey, T. M., and Nelson, D. L. (1987). Electrophysiological evidence suggests a defective Ca^{2+} control mechanism in a new *Paramecium* mutant. *J. Membr. Biol.* **98**(3), 275–283.

Gaertig, J., Gu, L., Hai, B., and Gorovsky, M. A. (1994). High frequency vector mediated. transformation and gene replacement in *Tetrahymena*. *Nucleic Acids Res.* **22**, 5391–5398.

Goodenough, U. W. (1983). Motile detergent-extracted cells of *Tetrahymena* and *Chlamydomonas*. *J. Cell Biol.* **96**(6), 1610–1621.

Grønlien, H. K., Rønnevig, A. K., Hagen, B., and Sand, O. (2010). Chemo-accumulation without changes in membrane potential in the microstome form of the ciliate *Tetrahymena vorax*. *J. Exp. Biol.* **213**(23), 3980–3987.

Haga, N., Forte, M., Ramanathan, R., Hennessey, T., Takahashi, M., and Kung, C. (1984). Characterization and purification of a soluble protein controlling Ca-channel activity in *Paramecium*. *Cell* **39**(1), 71–78.

Hamilton, E. P., and Orias, E. (1999). Genetic crosses: setting up crosses, testing progeny and isolating phenotypic assortants. *Methods Cell Biol.* **62**, 219–228.

Hennessey, T. M. (2005). Responses of the ciliates *Tetrahymena* and *Paramecium* to external ATP and GTP. *Purinergic Signalling* **1**, 101–110.

Hennessey, T. M. (1989). Ion currents of *Paramecium*: effects of mutations and drugs. *In* "Evolution of the First Nervous Systems," (P. A. V. Anderson, ed.), pp. 215–235. Plenum Press, New York.

Hennessey, T. M., Kim, D. Y., Oberski, D. J., Hard, R., Rankin, S. A., and Pennock, D. G. (2002). Inner arm dynein 1 is essential for Ca^{++}-dependent ciliary reversals in *Tetrahymena thermophila*. *Cell Motil. Cytoskeleton.* **53**(4), 281–288.

Hennessey, T. M., and Kuruvilla, H. G. (1999). Electrophysiology of Tetrahymena. *In* "Methods in Cell Biology," (D. J. Asai, and J. D. Forney, eds.), Vol. 62, pp. 363–376. Academic Press, New York.

Hennessey, T., Machemer, H., and Nelson, D. L. (1985). Injected cyclic AMP increases ciliary beat frequency in conjunction with membrane hyperpolarization. *Eur. J. Cell Biol.* **36**(2), 153–156.

Hennessey, T. M., and Nelson, D. L. (1979). Thermosensory behavior in *Paramecium* tetraurelia: a quantitative assay and some factors that influence thermal avoidance. *J. Gen. Microbiol.* **112**, 337–347.

Hinrichsen, R. D., Saimi, Y., Hennessey, T., and Kung, C. (1984). Mutants in *Paramecium tetraurelia* defective in their axonemal response to calcium. *Cell Motil.* **4**(4), 283–295.

Holwill, M. E., and Satir, P. (1990). A physical model of microtubule sliding in ciliary axonemes. *Biophys J.* **58**(4), 905–917.

Jennings, H. S. (1976). *Behavior of the Lower Organisms.* Indiana University Press, Bloomington and London.

Kim, M. Y., Kuruvilla, H. G., and Hennessey, T. M. (1997). Chemosensory adaptation in Paramecium involves changes in both repellent binding and the consequent receptor potentials. *J. Comp. Biochem. Physiol.* **118**(3), 589–597.

Kim, M. Y., Kuruvilla, H. G., Ragu, S., and Hennessey, T. M. (1999). ATP reception and. chemosensory adaptation in *Tetrahymena thermophila*. *J. Exp. Biol.* **202**(4), 407–416.

Koppelhus, U., Hellung-Larsen, P., and Leick, V. (1994). An improved quantitative assay for chemokinesis in *Tetrahymena*. *Biol. Bull.* **187**, 8–15.

Kung, C., Chang, S. Y., Satow, Y., Houten, J. V., and Hansma, H. (1975). Genetic dissection of behavior in Paramecium. *Science.* **188**(4191), 898–904.

Kung, C., and Naitoh, Y. (1973). Calcium-induced ciliary reversal in the extracted models of "Pawn," a behavioral mutant of Paramecium. *Science.* **179**, 195–196.

Lainhart, W., Stolfa, G., and Koudelka, G. B. (2009). Shiga toxin as a bacterial defense against a eukaryotic predator, Tetrahymena thermophila. *J. Bact.* **191**(16), 5116–5122.

Lampert, T. J., Coleman, K. D., and Hennessey, T. M. (2011). Chemoattraction to lysophosphatidic acid does not require a change in membrane potential in *Tetrahymena thermophila*. *Cell Biol. Int.* **35**(5), 519–528.

Larsen, J., and Satir, P. (1991). Analysis of Ni^{2+}-induced arrest of *Paramecium* axonemes. *J. Cell Sci.* **99**, 33–40.

Machemer, H. (1988). Motor control of cilia. *In* "Paramecium," (H. -D. Gortz, ed.), pp. 216–233. Springer-Verlag, Berlin.

Machemer, H. (1989). Cellular behavior modulated by ions: electrophysiological implications. *J. Protozool.* **36**, 463–487.

Matt, H., Plattner, H., Reichel, K., Lefort-Tran., &., and Beisson, J. (1980). Genetic dissection of the final exocytosis steps in *Paramecium tetraurelia* cells: trigger analysis. *J. Cell Sci.* **46**, 41–60.

Mogami, Y., and Baba, S. A. (1998). Super-helix model: a physiological model for gravitaxis of *Paramecia*. *Adv. Space Res.* **21**(8/9), 1291–1300.

Naitoh, Y., and Kaneko, H. (1972). Reactivated Triton-extracted models of *Paramecium*: modification of ciliary movements by calcium ions. *Science* **176**, 523–524.

Nakaoka, Y., Kurotani, T., and Itoh, H. (1987). Ionic mechanism of thermoreception in *Paramecium*. *J. Exp. Biol.* **127**, 95–103.

Nakaoka, Y., and Machemer, H. (1990). Effects of cyclic nucleotides and intracellular Ca on voltage-activated ciliary beating in *Paramecium*. *J. Comp. Physiol. A* **166**, 401–406.

Nam, S. -W., Kim, S. -T., Lee, K. -M., Kim, S., Kou, S., Lim, J., Hwang, H., Joo, M., Jeong, B., Yoo, S., and Park, S. (2009). *N*-Methyl-D-aspartate receptor-mediated chemotaxis and Ca^{2+} signaling in *Tetrahymena pyriformis*. *Protist* **160**, 331–342.

Noever, D. A., Cronise, R., and Matsos, H. C. (1994). Preferred negative geotactic orientation in mobile cells: *Tetrahymena* results. *Biophys J.* **67**, 2090–2095.

Ogura, A., and Machemer, H. (1980). Distribution of mechanoreceptor channels in the Paramecium surface membrane. *J. Comp. Physiol.* **135**, 233–242.

Onimaru, H., Ohki, K., Nozawa, Y., and Naitoh, Y. (1980). Electrical properties of *Tetrahymena*, a suitable tool for studies on membrane excitation. *Proc. Japan Acad.* **56**, 538–543.

Patterson, D. J., and Sleigh, M. A. (1976). Behavior of the contractile vacuole of *Tetrahymena pyriformis W*: a redescription with comments on the terminology. *J Protozool.* **23**(3), 410–417.

Pope, W. D. B., Jones, A. J., Halsey, M. J., and Numm, J. F. (1978). Pressure enhancement of the depressant effect of halothane on cilial beat. *Canad. Anaesth. Soc. J.* **25**(4), 319–322.

Rodgers, L. R., Markle, K., and Hennessey, T. M. (2008). Responses of the ciliates *Tetrahymena* and *Paramecium* to vertebrate odorants and tastants. *J. Euk. Microbiol.* **55**(1), 27–33.

Ron, A., Williams, N. E., and Doerder, F. P. (1992). The immobilization antigens of *Tetrahymena thermophila* are glycoproteins. *J. Protozool.* **39**(4), 508–510.

Saimi, Y., and Kung, C. (1987). Behavioral genetics of *Paramecium*. *Ann. Rev. Genetics* **21**, 47–65.

Satow, Y., and Kung, C. (1979). Voltage sensitive Ca-channels and the transient inward current in *Paramecium tetraurelia*. *J. Exp. Biol.* **78**, 149–161.

Schein, S. J. (1976). Nonbehavioral selection for pawns, mutants of *Paramecium Aurelia*. with decreased excitability. *Genetics.* **84**(3), 453–468.

Schein, S. J., Bennett, M. V. L., and Katz, G. M. (1976). Altered calcium conductance in Pawns, behavioural mutants of *Paramecium aurelia*. *J. Exp. Biol.* **65**, 600–714.

Skriver, L., and Williams, N. E. (1980). Regeneration of cilia in starved *Tetrahymena thermophila* involves induced synthesis of ciliary proteins but not synthesis of membrane lipids. *Biochem J.* **188** (3), 695–704.

Suryavanshi, S., Bernard Eddé, B., Fox, L., Guerrero, S., Griffin, P., Hard, R., Hennessey, T., Kabi, A., Malison, D., Pennock, D., Sale, W., Wloga, D., and Gaertig, J. (2010). Tubulin glutamylation regulates the motility of cilia. *Curr. Biol.* **20**, 435–440.

Takahashi, M. (1979). Behavioral mutants in *Paramecium caudatum*. *Genetics.* **91**(3), 393–408.

Takahashi, M., Onimaru, H., and Naitoh, Y. (1980). A mutant of *Tetrahymena* with a non-excitable membrane. *Proc Jpn Acad.* **56**, 585.

Tawada, K., and Oosawa, F. (1972). Responses of *Paramecium* to temperature change. *J. Euk. Microbiol.* **19**(1), 53–57.

Tiedtke, A., Hunseler, P., Elorin-Christensen, J., and Florin-Christensen, M. (1988). Exocytosis, endocytosis and membrane recycling in *Tetrahymena thermophila*. *J. Cell Sci.* **89**, 515–520.

Turkewitz, A. P., Chilcoat, N. D., Haddad, A., and Verbsky, J. W. (1999). Regulated secretion in *Tetrahymena*. *Methods Cell Biol.* **62**, 347–360.

Valentine, M., Yano, J., and Van Houten, J. (2008). Chemosensory transduction in *Paramecium*. *Jpn. J. Protozool.* **41**, 1–7.

Van Houten, J. L. (1978). Two mechanisms of chemotaxis in *Paramecium*. *J. Comp. Physiol.* **127**, 167–174.

Wood, C. R., Hard, R., and Hennessey, T. M. (2007). Targeted gene disruption of dynein heavy chain 7 of *Tetrahymena thermophila* results in altered ciliary waveform and reduced swim speed. *J. Cell Sci.* **120** (17), 3075–3085.

Zilberg, D., and Sinai, T. (2006). Optimization and validation of a colorimetric assay for *Tetrahymena* sp. *Survival. Res. Microbiol.* **157**(4), 355–359.

CHAPTER 16

Tetrahymena in the Classroom

Joshua J. Smith*, Emily A. Wiley† and Donna M. Cassidy-Hanley‡

*Department of Biomedical Sciences, Missouri State University, Springfield, MO 65897, USA

†W.M. Keck Science Department, Claremont McKenna, Pitzer, and Scripps Colleges, Claremont, CA 91711-5916, USA

‡Department of Microbiology and Immunology, College of Veterinary Medicine, Cornell University, Ithaca, New York, USA

Abstract

Tetrahymena has been a useful model in basic research in part due to the fact it is easy to grow in culture and exhibits a range of complex processes, all within a single cell. For these same reasons *Tetrahymena* has shown enormous potential as a teaching tool for fundamental principles of biology at multiple science education levels that can be integrated into K-12 classrooms and undergraduate and graduate college laboratory courses. These *Tetrahymena*-based teaching modules are inquiry-based experiences that are also effective at teaching scientific concepts, retaining students in science, and exciting students about the scientific process. Two learning communities have been developed that utilize *Tetrahymena*-based teaching modules. Advancing Secondary Science Education with *Tetrahymena* (ASSET) and the Ciliate Genomics Consortium (CGC) have developed modules for K-12 students and college-level curriculums, respectively. These modules range from addressing topics in ecology, taxonomy, and environmental toxicity to more advanced concepts in biochemistry,

0091-679X/10 $35.00
DOI 10.1016/B978-0-12-385967-9.00016-5

proteomics, bioinformatics, cell biology, and molecular biology. An overview of the current modules and their learning outcomes are discussed, as are assessment, dissemination, and sustainability strategies for K-12 and college-level curriculum.

I. Introduction/Rationale

The last decade has seen the development of *Tetrahymena thermophila* as a remarkable tool for both teaching and research. *Tetrahymena* combines the simplicity of an easily manipulated single-cell system with a structural and functional complexity comparable to that of higher metazoans. Its genetics, development, cell physiology, biochemistry, and ultrastructure are well characterized (for overviews of the system, see Elliott, 1973; Gall, 1986; Hill, 1972; Nanney, 1980), and an array of biochemical, physiological, genetic, and molecular techniques have been developed to facilitate use of the *Tetrahymena* system in basic research (Asai and Forney, 2000; chapters in this volume). As a result, *Tetrahymena* has become a well-published model for research in fundamental areas of cell biology, including telomere assembly, histone modification, programmed genome rearrangement, membrane trafficking, endocytosis, exocytosis, tubulin structure, cilia (cytoskeletal) dynamics, among others (e.g., see reviews by Chalker and Yao, 2011; Mochizuki, 2010; Turkewitz *et al.*, 2002; Turkewitz, 2004). For many of the same reasons that *Tetrahymena* has proven useful in basic research, it has enormous potential as a teaching tool. Its complex behaviors are easily manipulated in small-volume cultures, and provide a convenient window into many of the fundamental principles of biology. Practical aspects of working with *Tetrahymena* make it ideal for the classroom. Most importantly, *Tetrahymena's* free-living, non-pathogenic lifestyle makes it completely safe to work with. In addition, *Tetrahymena* exhibits a wide range of biological characteristics and activities that are suitable for investigation of fundamental biological concepts at multiple educational levels, from K-12 classrooms to undergraduate college courses. Integrating the fruits of research with *Tetrahymena* into the science curriculum at the K-12 and college levels provides students with access to new hands-on approaches to important scientific ideas, and benefits researchers by increasing public awareness of the relevance and importance of basic research.

Efforts to integrate *Tetrahymena* into the biology curriculum at the high school and college level have been greatly aided by several key tools, starting with organization of the genome sequence into a searchable Wiki that is easily accessed and used by students at all levels, even those just learning to navigate genome databases (http://www.ciliate.org). Gene expression data is also available in an easy-to-use format at the Tetrahymena Functional Genomics Database (http://tfgd.ihb.ac.cn/). Molecular genetic tools facilitating the straightforward engineering of epitope-tagged proteins for localization and protein complex analyses have recently been developed for college-level courses. In addition, *Tetrahymena*-based teaching modules incorporating bioinformatics are being developed for high school use. The development of genomic and proteomic based teaching tools is greatly facilitated by access to the user-friendly database interfaces provided by the *Tetrahymena* research community. In addition to bioinformatics resources, the use of *Tetrahymena* as a teaching tool is enhanced by the availability of a variety of strains exhibiting diverse genotypes and phenotypes (see Ch. 8 in this volume, and http://tetrahymena.vet.cornell.edu/) and by the ongoing development of new *Tetrahymena* research techniques that can be modified for classroom use.

At the college level, many teaching modules using *Tetrahymena* have been independently developed by faculty at a variety of institutions. A workshop called "Ciliates in the Classroom," held as part of the semi-annual Ciliate Molecular Biology research conference, was initiated over 15 years ago as a place where college teaching faculty share their ideas, protocols, and resources, and collaborate on larger teaching initiatives. In parallel, national recommendations for effective biology teaching reform evolved from federal and private organizations, including the American Association for the Advancement of Science (AAAS), National Research Council (NRC), Howard Hughes Medical Institute (HHMI), Project Kaleidoscope (PKAL), and others (American Society for Cell Biology Education and Committee, 1992; Brewer and Smith, 2011; Fairweather, 2008; NRC, 2003, 2009, 2011; Project Kaleidoscope, 1991). These tout inquiry-based experiences for students as effective ways of both teaching scientific concepts and retaining students in science. Members of the ciliate community saw an opportunity to address these recommendations by developing ciliate-based research activities into scalable modules for integration into college classrooms at multiple levels. Initiated through the workshop discussions, a fairly extensive set of modules now exist, which engage students in the excitement of original research on a variety of biological questions, while simultaneously teaching central concepts in cell and molecular biology, biochemistry, ecology, and taxonomy.

Recommendations for transforming biology teaching to retain students in science included the building of learning communities, which can help inspire and support students with a variety of backgrounds in science (AAAS, 2011; NRC, 2009). In particular, learning communities are supportive of under-represented students such as women and racial minorities (Jones *et al.*, 2010; Treisman, 1992). *Tetrahymena* biologists have recently capitalized on this idea by involving undergraduate students directly in the functional annotation of *Tetrahymena* genes through the research modules described below. Faculty and students engaging these modules are part of the Ciliate Genomics Consortium (CGC), which was initiated in 2006 with funding from NSF. Participating students can immediately disseminate and share their results with the broader ciliate community through a database for unpublished results (http://tet.jsd.claremont.edu/) that is linked to the *Tetrahymena* Genome Database. Through this mechanism, students are immediately connected with and part of the larger research community. Assessment has shown that this experience has validated their ability to make real scientific contributions and has empowered them to immediately pursue further research experiences (Wiley, unpublished).

Tetrahymena has also been featured in teaching exercises that can be used in middle and high school classrooms (e.g., Bozzone, 2000; http://www.ascb.org/newsfiles/exercises_cell_bio.pdf), and has been utilized in high school science fairs (e.g., http://www.all-science-fair-projects.com/science_fair_projects_encyclopedia/

Tetrahymena_thermophila; Intel 2011 Science Fair; http://www.societyforscience.org/document.doc?id=295) and other independent student projects that encourage more in-depth student involvement with scientific concepts, and stimulate enthusiasm and commitment beyond the standard science classroom format. Currently, K-12 educational use of *Tetrahymena* is being expanded under the auspices of an NIH SEPA–funded program (ASSET: Advancing Secondary Science Education with Tetrahymena; http://tetrahymenaasset.vet.cornell.edu/), specifically designed to develop *Tetrahymena*-based laboratory modules targeted to middle and high school students.

In this chapter we outline the classroom research modules that are available at multiple levels, and their learning outcomes that align with the most recent recommendations for transforming biology education. Additionally, we discuss assessment outcomes and future plans to improve sustainability and strengthen interactions between students in the research community and faculty at a variety of institutions.

II. Materials and Methods

A. *Tetrahymena* in the College curriculum

The following classroom research modules have common learning outcomes:

1. develop facility with designing experiments; improve understanding of controls;
2. improve data analysis, interpretation, and presentation skills;
3. learn effective scientific communication and general collaboration (teamwork) skills;
4. improve scientific writing skills;
5. deal with ambiguity through engaging original questions; and
6. aid in the structural and functional annotation of the *Tetrahymena* genome.

All of the following modules and instructor guides are available on the Ciliate Genomics Consortium (CGC) website (http://tet.jsd.claremont.edu)

1. Current Modules

Bioinformatics Modules (requires 3–6 h of class time)

These modules guide students through identification of a gene(s) of interest that belongs to a particular gene family, or that may be a homolog of a specific gene in another organism. Students analyze gene sequence for features such as splice sites, translation start, and stop codons, and analyze alignments of expressed sequence tags (ESTs) to evaluate the gene structure predictions. Students predict protein features through a variety of free online programs (Fig. 1A). Students can compare protein sequences and construct phylogenetic trees (Fig. 1B). The bioinformatics modules may be used independently or as precursors to other modules.

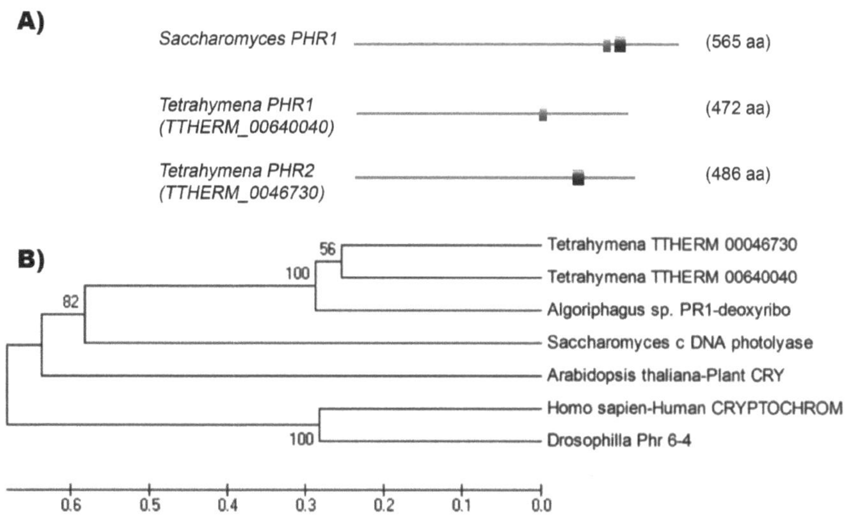

Fig. 1 Bioinformatics results. (A) Schematic of functional domains in Photolyase (PHR1) obtained using ExPASy PROSITE (http://prosite.expasy.org/). *Saccharomyces* contains both photolyase signature domains 1 (Light Gray) and 2 (Dark Gray) while the *Tetrahymena* homologs PHR1 and PHR2 contain only one photolyase signature domain. *Tetrahymena* PHR1 has signature domain 1 (Light Gray) while *Tetrahymena* PHR2 has signature domain 2 (Dark Gray). (B) UPGMA phylogenetic tree of various photolyase-related proteins from various organisms. These trees help predict the degree of conservation in the protein and to what organisms it is most closely related (*data generated by Scott Kelsey and Archana Shrestha in BMS658: Recombinant DNA techniques, Biomedical Sciences, Missouri State University*). Phylogenetic tree analysis was performed using ClustalW (http://www.genome.jp/tools/clustalw/) and Mega 5.0 (http://www.megasoftware.net/). (For color version of this figure, the reader is referred to the web version of this book.)

Students learn to retrieve gene information from various databases, the difference between genomic and coding sequences, gene structure (start and stop codons, introns, and exons), the concept of ESTs, the use of gene translation tools, organism-specific codon usage, BLAST searching, sequence alignments, concepts relating to functional domains, and evolutionary sequence conservation.

Gene Structure Determination (5 × 4-h laboratory periods)

The coding sequences of all putative *Tetrahymena* genes have been computationally predicted. These predictions now require experimental testing. In this module, students work in teams to identify introns, exons, and the 5' and 3' termini of gene transcripts and coding sequences by PCR.

Students learn standard molecular techniques (primer design, PCR, agarose gel electrophoresis, genomic DNA isolation, RNA isolation, cDNA synthesis, nucleic acid quantification), the difference between genomic and coding sequences, introns versus exons, basic gene structure, the concept of 3' and 5' untranslated regions, effective data presentation, and the use of gene graphics software.

Gene Expression Analysis (5 × 4-h laboratory periods)

In this module students assess the relative amount of expression of a gene of interest throughout different stages in the *Tetrahymena* life cycle. This takes advantage of processes and physiological changes that are synchronized through conjugation. Gene expression is evaluated through assessing the relative production of gene transcripts at different time points in the life cycle by reverse transcriptase PCR and agarose gel electrophoresis (Fig. 2A). Alternatively, expression can be measured using quantitative real-time PCR to graph the relative levels of expression under various treatment conditions (Fig. 2B-D). Results can be compared to microarray-

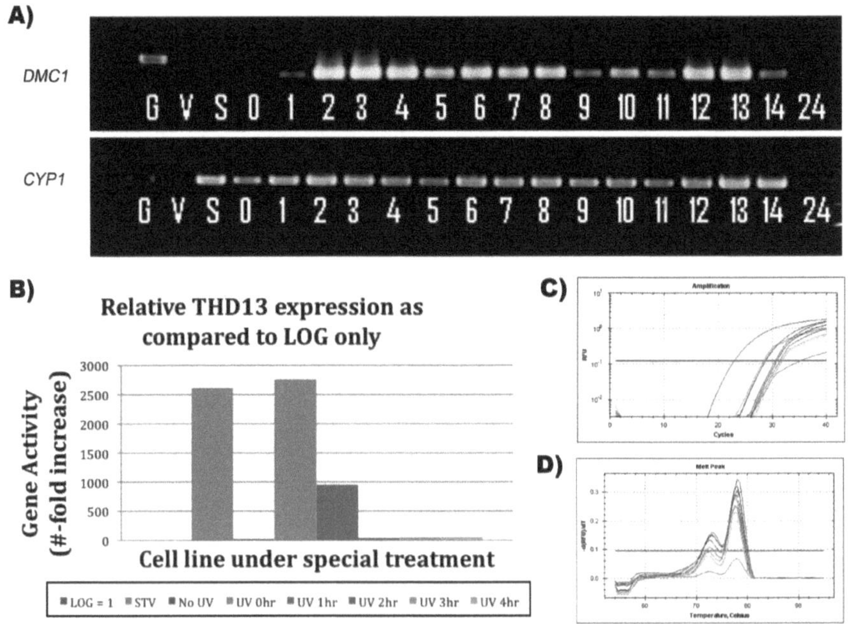

Fig. 2 Gene expression analysis by reverse transcriptase PCR (RT-PCR). (A) Semi-quantitative RT-PCR of *DMC1* and *CYP1* (control gene) during conjugation (0–14 and 24 h post mixing), vegetative growth (V), starvation for 18 h (S), and genomic DNA control (G). Large changes in expression can be visualized on the gel. Here *DMC1* is highly expressed at 2–4 h and 12–13 h during conjugation (*data generated by Emily Gallichotte, independent research, Keck Science Department, Pitzer Colleges).* (B) Relative gene expression of *THD13* (sirtuin-like deacetylase) determined using quantitative real-time RT-PCR. Data was obtained by dividing the *THD13* signal by that for *HHP1* and the numbers were set relative to the value for the logarithmically growing sample (LOG) being equal to 1.0. Various conditions can be analyzed such as starvation (STV) and treatment with 100 J/m^2 ultraviolet light and allowing recovery after the treatment (UV 0 h–UV 4 h). (C) PCR amplification graph from the real-time PCR machine illustrates the amount of amplification that occurs for each condition. This raw data is saved in a excel file to yield the data that was used to generate the normalized relative *THD13* expression in panel B. (D) Melt peak graph from the real-time PCR machine shows the purity of the products in each reaction. Large peak represents the cDNA product expected while the small peak represents some genomic DNA contamination (*data in panels B-D generated by Christopher Reynolds in BMS558: Recombinant DNA Techniques, Biomedical Sciences, Missouri State University).* (See color plate.)

based expression profiles published on Tetrahymena Genome Expression Database (TGED; http://tged.ihb.ac.cn/; Miao *et al.*, 2009)

Students learn standard molecular techniques (genomic DNA isolation, RNA isolation, cDNA synthesis, PCR, primer design, agarose gel electrophoresis, nucleic acid quantification), cytological analysis by fluorescence microscopy, effective data presentation, and use of graphics software.

Protein Localization (6 × 4-h laboratory periods)

In this module students use PCR to clone a gene to fuse with a fluorescent protein (GFP, RFP, YFP, or CFP). This can be done with conventional restriction enzyme digestion and subsequent ligation or through the use of Gateway technology and LR recombinase methods (adapted from Invitrogen technology). The tagged gene (under the cadmium-inducible *MTT1* promoter) is then transformed into *Tetrahymena* through either electroporation or use of biolistic particle bombardment procedures. Positive clones obtained through drug selection are induced with $CdCl_2$ to express the tagged protein, and fluorescence microscopy is used to determine localization of the protein in the cells (Fig. 3). Results for protein localization to

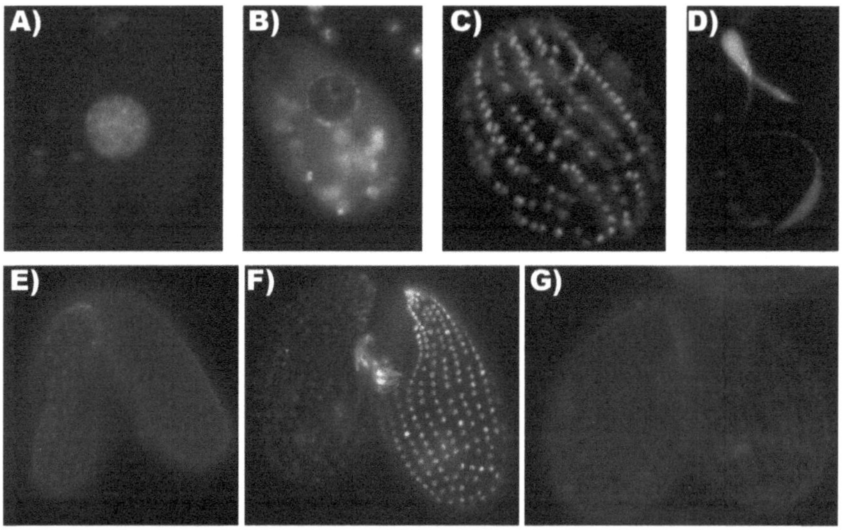

Fig. 3 Localization of GFP and RFP tagged genes. (A) GFP-HHP1 localized to chromatin bodies in the macronucleus (*data generated by Katerina Yale, thesis research, Keck Science Department, Claremont McKenna College*). (B) GFP-THD14 localized to the nucleoli around the edge of the macronucleus (*data generated by Allen Chen, thesis research, Keck Science Department, Claremont McKenna College*). (C) GFP-THD14 localized to the mitochondria in *Tetrahymena* (*data generated by Kristin Slade, Keck Science Department, Claremont Colleges*). (D) GFP-DMC1 localized to the micronucleus in meiotic prophase (crescent phase) during conjugation at 3 h (*data generated by Emily Gallichotte, thesis research, Keck Science Department, Pitzer College*). (E) GFP-KDA3 and RFP-TCBP in conjugating *Tetrahymena*. (F) GFP-KDA3 and RFP-TetrinA in conjugating *Tetrahymena*. (G) GFP-KDC2 and RFP-TCBP in conjugating *Tetrahymena* (*data in panels E-G generated by students in BIOL 3492: Laboratory Experiments with Eukaryotic Microbes, Biology Department, Washington University*). (See color plate.)

various organelles and structures in *Tetrahymena* improves the functional annotation for that protein.

Students learn standard molecular techniques (PCR, agarose gel electrophoresis, restriction enzyme digestion, cloning), *Tetrahymena* transformation techniques, fluorescence microscopy, data collection and analysis, cellular organelles and structures, preparation of figures, and data presentation.

Protein Interaction through Immunoprecipitation and Mass Spectrometry (6 x 4-h laboratory periods)

In this module students use PCR to clone a gene with an epitope tag (2HA, FLAG, HIS_6, etc.) fused to it. This can be done with conventional restriction enzyme digestion and subsequent ligation or through the use of Gateway technology and LR recombinase methods (adapted from Invitrogen technology). The tagged gene constructs (under control of the cadmium-inducible MTT1 promoter) are then transformed into *Tetrahymena* through either electroporation or use of biolistic particle bombardment procedures. After drug selection of positive clones they are induced with $CdCl_2$, protein extracts isolated, and immunoblot analysis performed to determine expression of the tagged protein. Extracts are then used in an immunoprecipitation assay to isolate proteins that interact with the tagged protein via mass spectrometry (Fig. 4). Results can help annotate the protein complexes and their possible functions in the cell.

Students learn standard molecular techniques (PCR, agarose gel electrophoresis, restriction enzyme digestion, cloning), *Tetrahymena* transformation techniques, protein isolation techniques, SDS-PAGE, western blot analysis, Coomassie staining, data collection and analysis, preparation of figures, and presentation of data.

2. Assessment

Assessment of College Modules

The research modules have been integrated into the laboratory component of various courses at different levels, including upper division molecular biology courses with laboratory, a research-based laboratory course for students in their second year, and a first-year honors life sciences course. Assessment data have been collected for over 4 years using attitudinal surveys, confidence surveys, written reports and oral presentations, empirical observations, tracking student placement into research positions, Student Assessment of Learning Gains (SALG; Seymour *et al.*, 2000), and published metrics for evaluation of student research experiences such as the Survey of Undergraduate Research Experiences (SURE; Lopatto, 2008).

These various assessments show significant gains in the following areas: (1) increased confidence with each step of the scientific process: generating hypotheses, experimental design and execution, data analysis, data presentation, scientific writing, oral presentation of results, and keeping scientific records; (2) enhanced student effort, investment in the project, and sense of making valuable scientific

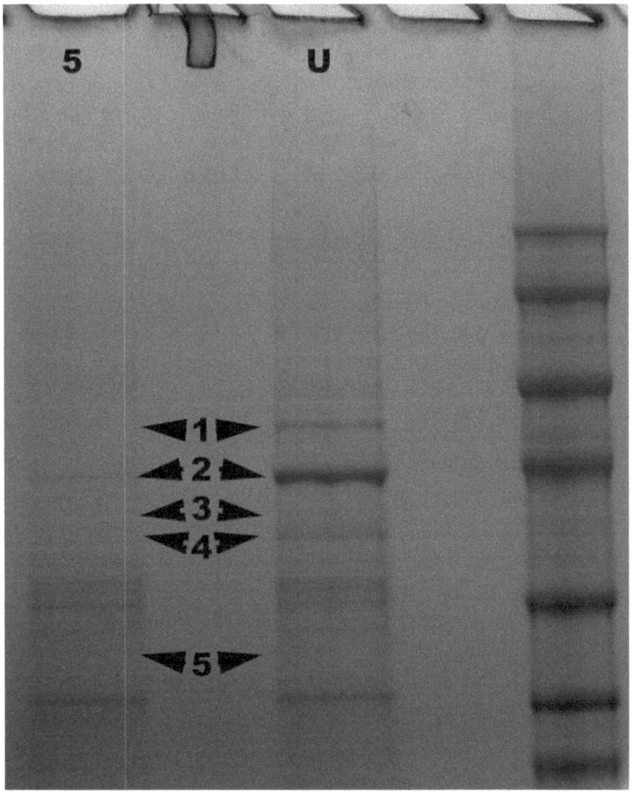

Fig. 4 Coomassie-stained SDS-PAGE of proteins interacting with FLAGHis$_6$-URM1. Protein extract from an untagged strain, CU522 (5), and a strain containing a FLAGHis6-URM1 epitope tag (U) were used in an immunoprecipitation with Anti-FLAG agarose beads. The beads were boiled with a non-denaturing loading buffer and the entire sample was loaded on a 4–18% SDS-PAGE. The gel was Coomassie stained and five bands were excised and sent for analysis by mass spectrometry (Doug Beussman, Chemistry Department, St. Olaf College). Bands in the URM1 immunoprecipitation lane yielded potential interacting proteins in all but band 5 and no proteins were identified in the control lane (5) by mass spectrometry *(data generated by Allie Maltzman and Kyle Cottrell in BMS558: Recombinant DNA Techniques, Biomedical Sciences, Missouri State University).*

contributions; (3) increased interest in pursuing further research experiences and considering research science as a career option; and (4) a sense of belonging to a larger research community and peer cohort in science.

B. *Tetrahymena* in K-12 Curriculum

A number of teaching modules, primarily aimed at middle and high school students, are being developed by the NIH-funded ASSET program at Cornell University. Each module is designed to stimulate active, inquiry-driven learning of key biological concepts using living *Tetrahymena*, a eukaryotic organism that encompasses many of the biological attributes of metazoans but does not engender any of the conflicting emotions and reactions often invoked by the use of multicellular organisms in the classroom. Since it is critical that each module be effective in addressing relevant core concepts and be presented in a format that is both teacher and student friendly, each module is developed with input from active middle and high school teachers, tested multiple times in pilot classrooms, and evaluated at each step by an independent external evaluator. The ASSET *Tetrahymena*-based teaching modules incorporate many of the ideas expressed as part of the new framework for K-12 science education recently put forth by the National Research Council's Committee on Conceptual Framework for the New K-12 Science Education Standards (National Research Council 2011). Each of the three dimensions broadly outlined in the above plan are addressed. First, the modules allow students to actively engage in the practice of science, from asking questions to carrying out investigations to analyzing and interpreting data. All utilize a hands-on approach designed to help stimulate student curiosity, interest, and motivation. Second, the modules help to identify concepts that are fundamental to developing a coherent, broad-based view of science by directly examining basic biological interactions such as cause-and-effect relationships, structure and function, and evolutionary change. Third, the modules directly address many core ideas that are fundamental to an understanding of the life sciences, ranging from structures and processes in organisms to ecology, heredity, and evolution.

Because of the ease and flexibility with which *Tetrahymena* can be utilized in the development of educational tools, *Tetrahymena*-based modules can be used to address a wide range of educational goals, addressing both scientific practice and biological content. For example, the ASSET modules address scientific practice and the scientific method by fostering student observation and formulation of questions in a meaningful and testable way, by helping students understand experimental design and the importance of variables and accurate and meaningful measurements, and by encouraging students to critically analyze data and compare experimental outcomes. The biological content of the ASSET modules is wide-ranging, addressing issues from evolution to predator–prey interactions to bioinformatics to ecology and toxicology. Examples of the ASSET modules currently available or under development are briefly described below. As testing and evaluation are completed, each module is made available on the ASSET website (http://tetrahymenaasset.vet.cornell.edu/). These modules are presented to encourage use among educators at all levels, and to serve as models for future development of new, innovative teaching modules utilizing *Tetrahymena's* unique biological attributes.

I. Current Modules

Micro-Evolution

The micro-evolution lab provides a rare opportunity for students to address evolution and natural selection with a hands-on experiment that can be completed in less than 2 weeks. The experiment utilizes *Pseudomonas fluorescens*, a common, nonpathogenic saprophyte that colonizes soil, water, and plant surface environments and *Tetrahymena thermophila* to demonstrate diversifying selection in response to predator–prey interaction. The basic laboratory can be expanded to compare resource competition and predation as driving forces behind adaptive radiation. Students observe predator-driven real-time evolution in a micro-environment in about a week. In the presence of a *Tetrahymena* predator, clear phenotypic changes in bacterial growth pattern and niche formation are observed in liquid culture, and related changes in bacterial colony formation on agar plates are easily distinguished. Colony formation is dependent on what niche bacteria occupy in the liquid culture, for example, surface biofilm compared to bottom dwellers.

Students learn mechanisms of evolution; population dynamics; diversity of organisms; mutation; differences between prokaryotic and eukaryotic cells; interdependence in nature; and continuity and change.

Phagocytosis

The phagocytosis laboratory uses *Tetrahymena* to investigate the processes of ingestion, phagocytosis, and vacuole formation in cells, and the effects of various factors on these physiological processes. Students monitor vacuole development during feeding using digital cameras to record the data, learning microscopy and data collection and analysis as an integral part of the exercise. Students can look at feeding preferences, including live versus dead food, big versus small food, organic versus inorganic material. The lab can be combined with a consideration of mutational effects by incorporating the use of mutants unable to eat for various reasons, for example, mutants that fail to form a mouth at certain temperatures, or mutants with swimming defects that are unable to efficiently filter food from the water. The laboratory can also be used to address other issues such as pollution and toxicology, and can be combined with laboratories addressing the effects of cigarette smoke and alcohol.

Students learn cellular energetics; chemistry of life; energy transfer; structural similarity between single cell and multicellular organisms; and relationship of structure to function.

Cannibalism and Interspecific Predation

The cannibalism and interspecific predation laboratory utilizes both *T. thermophila* and *T. vorax*. Morphogenetic change is addressed by observing the *T. vorax* shift from microstome to macrostome form. As part of the laboratory, stomatin, the transformation inducing substance produced by *T. thermophila*, is isolated using simple techniques that can easily be carried out in high school classrooms. The change in morphology is obvious even under relatively low magnification. The laboratory can be used to address chemical induction of morphological changes, evolutionary differences among different species within the same genus, and, since the macrostome forms consume their own species as well as (preferably) other

Tetrahymena species, inter- and intra-specific predation. The use of inexpensive digital cameras to record morphological changes and predation events adds enormously to student interest and enthusiasm for this exercise.

Students learn about diversity of *Tetrahymena*; relationship of structure to function; mechanisms of evolution; population dynamics; growth and survival of organisms; diversity and adaptation of organisms; inter-organismal relationships; and species variation.

Mutation

The mutation laboratory utilizes *Tetrahymena* strains carrying temperature-sensitive mutations in genes involved in cell division. At permissive temperatures, the cells are normal, but when shifted to the restrictive temperature, they exhibit easily observed and characterized changes in morphology. This is a clear and simple demonstration of mutations, how temperature-sensitive mutations work, and the effects of permissive and restrictive temperature effects in a safe, simple system. The module can be expanded to look at the long- and short-term effects of exposure to restrictive temperatures and recovery parameters, and can be combined with other modules to stimulate student development of independent inquiry projects. For more advanced classes, these mutants can also be used as a basis for genetic investigation of gene dominance and gene complementation.

Students learn concepts related to mutation; gene regulation; complementation, and dominance.

Mating

The mating laboratory addresses issues associated with sex and reproduction from an evolutionary perspective, using *Tetrahymena* as a model system for examining sexual and asexual reproduction. The laboratory takes advantage of the fact that, under appropriate conditions, *Tetrahymena* can propagate either vegetatively or sexually, addressing the relative impact of genetics, growth rate, and population size on reproductive advantage. Students design an experiment to identify the mating type of unknown clones of *Tetrahymena* by testing with clones of known mating type, based on the lack of self-mating within clones of the same mating type. The effects of various environmental conditions on mating behavior can also be examined, and the module can be expanded to include a discussion of chemotaxis and the role of surface proteins in cell-to-cell communication.

Students learn concepts related to population dynamics; gene regulation; science as a process; reproduction; and heredity.

Pattern Formation

The pattern formation laboratory deals with spontaneous pattern formation in standing cultures of *Tetrahymena*. The honeycomb patterns formed in standing cultures of *Tetrahymena* have been linked to convection, chemotaxis, and cell-cell interaction, making this module not only useful as an independent biology laboratory but also potentially interesting as a cross-platform vehicle for linking biology, math, and physics. Since various conditions affect spontaneous pattern formation and reaggregation, including media, cell density, and environmental conditions, the laboratory lends itself to independent student inquiry and experimental design.

Students learn science as a process; cell–cell interactions; and cell response to environmental stimuli.

Toxicology

The toxicology laboratory examines the response of *Tetrahymena* cells to a variety of basic substances, looking at both lethal toxicity (LD_{50}) and chronic sub-lethal toxicity. Students can use different concentrations of common substances suggested in the module (for example, shampoo, detergents, toothpaste, Windex) to determine lethal and sub-lethal dosage, or, under teacher supervision, design their own experiment to look at common substances of interest to them. The laboratory offers an opportunity for students at all levels to begin to address experimental design, the effect of experimental variables, and potential long-term environmental impact of even commonly used items.

Students learn about environmental impact on growth and survival; ecosystems and relationships between organisms.

Effects of Cigarette Smoke

The cigarette laboratory is an offshoot of the Toxicology lab that looks directly at the effects of cigarette smoke on overall cell viability, motility, and behavior, and relates the observed behaviors to ciliary activity. Cigarette smoke is bubbled into alcohol or water, and the resultant extract is added to cell cultures and the effects observed both directly and using digital image capture. Methods are presented to allow students to analyze digital images of treated and control cells by tracking cell swimming patterns using freely available software (ImageJ, http://rsbweb.nih.gov/ij/). A comparison between *Tetrahymena* cilia and cilia present in human lungs is presented. The simplicity of the laboratory design allows students considerable freedom to design experiments, and ask and answer their own questions.

Students learn concepts related to motility; cell response to environmental stimuli; and science as a process.

Effects of Alcohol

The alcohol laboratory is another offshoot of the Toxicology laboratory that looks directly at the effects of alcohol (beer) on overall cell viability, motility, and behavior. The effects of various concentrations of beer and the non-alcoholic equivalent on *Tetrahymena* viability and behavior are analyzed using digital image capture and cell tracking software. Students are encouraged to discuss the biology behind the observed behaviors and to design their own experiments using the cells and methodologies provided.

Students learn concepts related to motility; environmental impact on growth and survival; cell response to environmental stimuli; and science as a process.

Chemosensory Response

The chemosensory response laboratory addresses cell response to a variety of substances, from avoidance/attraction response to modification of ciliary beat, and presents optional sections to allow teachers to tailor module use to specific classroom levels. Middle school students examine the response of the cells to simple substances such as herbs, spices, or citrus peel, some of which are provided with the kit, or they can design their own experiments to test the avoidance/attraction response to a variety of substances of interest to them. High school students can examine more sophisticated chemical responses, for example to GTP. This laboratory has been adapted for use in classes ranging from $4^{th}/5^{th}$ grade through AP biology.

Students learn cell responses to environmental stimuli; and interorganismal relationships.

Osmolarity

The osmolarity laboratory provides a clear, easy technique for identifying and quantifying cell response to changes in osmotic conditions. Contractile vacuoles in *Tetrahymena* are large and visible even with the fairly low-quality microscopes generally available in high school biology labs, and changes are readily captured for detailed analysis using the single frame, time lapse, and movie options available with low-cost digital cameras. *Tetrahymena* contractile vacuoles gather and expel water in periodic fashion, with the rhythm and rate of contraction dependent on environmental factors. The laboratory addresses membrane permeability and osmoregulation in a free-swimming cell under a variety of conditions. The exercise can be varied according to class level, from simply observing changes to collecting and graphing data on rates of contraction under different conditions to student-designed experiments altering the cell environment in specific ways and recording and analyzing the results.

Students learn concepts related to water and chemistry of life; physiological regulation; relationship of structure to function; homeostasis; feedback mechanisms; and structural similarity between single cell and multicellular organisms.

Field Research

The field research module combines field collection, cell culture, molecular biology (in the form of DNA isolation, PCR, DNA sequencing), and bioinformatics within one extended laboratory. It involves field collection of local *Tetrahymena* species, isolation of DNA from wild-caught cells or unknown *Tetrahymena* strains (isolated from known locations) provided by the *Tetrahymena* Stock Center, PCR amplification of fragments for sequencing and bioinformatic analysis, bar coding, and examination of genetic diversity. Students have an opportunity to publish their results on the ASSET website, where information regarding all the *Tetrahymena* species identified around the country will be made available. Sequence data will permit them to identify the *Tetrahymena* species collected and place it on a phylogenetic tree. Ultimately, the information collected by students performing this module will be fed into a larger database that will provide an overview of genetic variation among natural populations of *Tetrahymena thermophila*. Students will be able to track this data as it is collected from remote sites. The laboratory gives students an opportunity to be part of an ongoing research project and strengthens their understanding of the scientific process.

Students learn science as a process; evolution; continuity and change; interdependence in nature; molecular genetics; diversity of organisms; and species variation.

Microscopic Life Around Us

The microscopic life around us module is a field laboratory that allows students to experience the abundant microbial life in nearby aqueous natural habitats. Students are introduced to the diversity of organisms that exist in something as seemingly simple as a drop of water, using basic keys to identify some of the more common inhabitants. Students can also collect non-aqueous material such as hay or twigs, and examine flora and fauna that emerges when the objects are placed in water or various media. A more sophisticated version of the laboratory involves comparing the abundance and variety of life forms from various water sources, and relating differences to possible environmental factors. For example, samples from ponds treated to eliminate algae or temporary run-off ponds from agricultural sites might be compared to relatively pristine ponds or lakes, or the variation in standing versus rapidly running water examined. Students are encouraged to be creative in examining the world around them. The laboratory is designed to foster student interest and involvement in scientific inquiry.

Students learn about the diversity of life; differences between prokaryotic and eukaryotic cells; interdependence in nature; ecosystems and relationships between organisms; and species variation.

Growth and Population Dynamics

The growth and population dynamics laboratory is purposely multifaceted. It can be used to simply address questions of cell growth, including lag, log, and stationary phases in microbial cultures, but can easily be expanded to encompass questions of population density and sustainability relative to environmental resources. The laboratory can be used to look at issues including population growth in response to food abundance and shortage and the effects of environmental change (temperature, salinity, water quality, introduction of competitors, etc) on population growth and maintenance. The laboratory exploits *Tetrahymena's* natural lag, log, stationary, crash growth cycle, allowing students to create conditions that examine specific effects on each phase of the cycle, for example, by input of additional nutrients, removal of nutrients, environmental changes, and changes in population density. Students are encouraged to design their own experiments asking fundamental questions about growth, population, and sustainability.

Students learn about population dynamics; science as a process; and growth and survival of organisms.

Cilia: Growth and Regeneration

The cilia growth and regeneration lab offers a unique opportunity to manipulate an important cell organelle without killing the cell. Ciliary loss can be non-lethally induced in *Tetrahymena* by several methods, and cells are capable of regenerating their cilia in a few hours. However, in the mean time they are immobile, subject to increased predation, and unable to feed. The laboratory provides an experimental approach to examining organelle growth by chemically inducing cells to shed their cilia, and monitoring regeneration by the resumption of swimming behavior monitored by time-lapse photography or direct observation. The effects of external conditions on regeneration are easily monitored as well.

Students learn concepts related to motility; regeneration; and cell response to environmental stimuli.

Exocytosis and Secretion

The exocytosis and secretion laboratory deals with the induction of stimulus-dependent secretion, a topic well studied in *Tetrahymena*. This laboratory can also be linked to evolution of defense mechanisms in lower eukaryotes. Students induce secretion of a gelatinous capsule around the whole cell (visually similar to the jelly capsule surrounding frog eggs) in response to Alcian blue. Cell reaction is captured by microscope mounted digital cameras using real-time and time lapse exposures. Survival of the cells and ability to secrete a second capsule in response to stimulation are explored. Classroom modifications include examining the role of various stimulants and calcium concentration in the secretory process. Students are encouraged to formulate their own questions and design experiments to answer them, for example, what effect media or temperature or cell density has on secretory behavior.

Students learn about subcellular organization; defense mechanisms; relationship of structure to function; science as a process; interorganismal relationships; and survival mechanisms.

2. Assessment

Assessment of K-12 modules.

Evaluation plays a critical part in the assessment of any educational tool. For example, a full evaluation plan, compiled by an external evaluator to guide project development and assessment, is an ongoing part of the ASSET program, and is presented as a model for evaluation of future *Tetrahymena*-based educational materials. The results obtained so far serve as a significant indication of the overall usefulness of *Tetrahymena* as a teaching tool. Each ASSET module undergoes both formative and summative evaluations. Formative evaluation includes working directly with those involved in developing and testing the modules to insure that the exercises are teacher and student friendly, self-explanatory, scientifically accurate, and pedagogically sound. Testing and evaluation is carried out both in the classroom and in teacher workshops in an iterative process, with multiple rounds of revision. Teachers provide feedback to satisfy the first three criteria, and scientific and pedagogical advisors provide feedback to satisfy the final two criteria. Summative evaluation is used to determine overall impact. Teacher outcomes are determined by successful implementation of curriculum and modules in the classroom, an increase in content knowledge, an increase in the comfort level of teaching science, and an increase in use of hands-on science laboratory modules. Student educational outcomes are based on an assessment of increase in content knowledge, increase in science process skill development, and increase in understanding of scientific research. Direct classroom observation by ASSET staff and external evaluators (based on Horizon Research Observation protocol: http://www.horizon-research.com/instruments/hri_instrument.php?inst_id=14), pre- and post-testing of students in classrooms using the modules, and pre- and post-testing of teachers taking part in ASSET workshops are all used in evaluating the program. Thus far, use of the modules has uniformly been associated with positive gains in teacher and student outcomes, supporting the use of *Tetrahymena* in the development of teacher and student-friendly laboratory exercises. Additionally, all evaluations and comments from teachers using the modules indicate that *Tetrahymena* provides a unique teaching tool that can be used in the classroom to excite students about the scientific enterprise and increase their interest in science overall.

III. Discussion

The education initiatives discussed in this chapter are inherently sustainable in their design. With an estimated ~25,000 genes in the *Tetrahymena* genome, investigating the function of each through the modules developed for college-level curriculum will provide opportunity for many years of novel contributions from students. In addition, as gene knockout cell lines are created by the community, they may be tested for phenotypes using the middle- and high school level modules for novel investigations for years into the future. To enhance dissemination of student results, the Ciliate Genomics Consortium is doubling efforts to produce a user-friendly Wiki database for student-generated results. Plans to link this database with the Tetrahymena Genome Database Wiki are in progress. The CGC aims to regularly offer workshops to train faculty to use the research modules in their classrooms. The consortium is also seeking funding to support module implementation in a variety of classrooms, as they cost more than typical laboratory exercises.

As new molecular tools are developed for use with *Tetrahymena,* additional teaching modules will be developed in parallel. College-level modules will be published on the Ciliate Genomics Consortium for Education website (http://tet.jsd.claremont.edu/), and K-12 modules will be published on the ASSET website (http://tetrahymenaasset.vet.cornell.edu/). Some of the modules currently in development include chromatin immunoprecipitation (ChIP), differential centrifugation of organelles, and transcription regulation through a luciferase assay.

Tetrahymena has the potential to become an intrinsic part of K-12 education, but to fulfill that promise, the use of *Tetrahymena*-based teaching modules must be expanded, and a sustainable model for the continued use of these and future teaching modules must be developed. One approach that both increases the use of *Tetrahymena* in K-12 classrooms and promotes sustainability is the development of a local expert teacher program such as the one currently being developed by ASSET. Under this program, regionally dispersed expert teachers drawn from various K-12 outreach programs act as focal points for the local use of the ASSET *Tetrahymena*-based teaching modules. The teachers attend an NIH SEPA–funded intensive 3-day workshop at Cornell University that familiarizes them with *Tetrahymena* biology and the standard methods needed to use *Tetrahymena* in the classroom. The workshop also provides intensive hands-on experience using the teaching modules, and suggestions for optimal methods for sharing ASSET modules with other teachers. Participants in the expert teacher program commit to providing a minimum of two training sessions on the use of *Tetrahymena* in the classroom for other teachers in their district. This approach creates a core group of knowledgeable teachers willing to share their *Tetrahymena* expertise with other teachers, expands interest in the use of *Tetrahymena* in the classroom among active teachers in disperse geographical locations, and informs other educational outreach programs about possible collaborative interactions involving *Tetrahymena.* An interactive website dedicated to K-12 educational use of *Tetrahymena* provides discussion groups and a blog for further support of participating teachers. Webinars provide an inexpensive mechanism for providing addition information on new or revised modules. Sustainability is further enhanced by the *Tetrahymena* Stock Center, which provides a dependable source of cells for the program. Although the initial workshops are somewhat costly, the subsequent potential for expanded access to the *Tetrahymena* modules using local teacher resources makes the overall program relatively cost-efficient. Training local teachers in the use of *Tetrahymena* modules also provides a comparatively inexpensive mechanism for increasing and sustaining use of these educational tools in underfunded schools serving at-risk students. Providing support for workshop training for one expert teacher can ultimately facilitate the use of *Tetrahymena* in an entire school district, and impact large numbers of students.

Acknowledgments

The ASSET (Advancing Secondary Science Education with *Tetrahymena*) program is funded by NIH NCRR SEPA grant 5 R25 RR025126-03. The *Tetrahymena* Stock Center is supported by NIH grant # P40 RR019688-01A1. DCH is supported by NIH NCRR SEPA grant 5 R25 RR025126-03 and by NIH grant # P40 RR019688-01A1. The Ciliate Genomics Consortium is funded by NSF CAREER grant #0545560 to EAW. DCH is a founding member of Tetragenetics, Inc., Ithaca, NY, a company developing and utilizing *Tetrahymena* as a platform for the manufacture of biotechnological products. Neither this connection, nor any financial, personal, or professional interest of any of the author, has influenced this paper.

References

American Society for Cell Biology Education Committee. (1992). Exercises in Cell Biology for the Undergraduate Laboratory. http://www.ascb.org/newsfiles/exercises_cell_bio.pdf.

Asai, D. L., and Forney, J. D. (2000). *Tetrahymena thermophila*. *In* "Methods in Cell Biology", p. 580. Academic Press, San Diego.

Bozzone, D. M. (2000). Investigating phagocytosis in *Tetrahymena*: an experimental system suitable for introductory & advanced instruction. *The American Biology Teacher* **62**(2), 136.

Brewer, C. A., and Smith, D. (2011). American Association for the Advancement of Science (AAAS), eds. Vision and Change In Undergraduate Biology: A Call to Action. http://www.visionandchange.org/finalreport.

Chalker, D., and Yao, M. C. (2011). DNA elimination in ciliates: transposon domestication and genome surveillance. *Annu. Rev. Genet* **45**, 227–246.

Elliott, A. M. (1973). *Biology of Tetrahymena*. Dowden, Hutchinson and Ross, Stroudsburg, PA, pp. 508.

Fairweather, J. (2008). Linking evidence and promising practices in science, technology, engineering, and mathematics (STEM) undergraduate education: A status report for the National Academies National Research Council Board of Science Education. http://www7.nationalacademies.org/bose/Fairweather_CommissionedPaper.

Gall, J. G. (1986). *Molecular Biology of the Ciliated Protozoa*. Academic Press, New York, NY, pp. 252.

Hill, D. L. (1972). *The Biochemistry and Physiology of Tetrahymena*. Academic Press, New York & London, pp. 230.

Jones, M. T., Barlow, A., and Villarejo, M. (2010). The importance of undergraduate research for minority persistence and achievement in biology. *J. Higher Educ.* **81**, 82–115.

Lopatto, D. (2008). Exploring the benefits of undergraduate research: the SURE survey. *In* "Creating Effective Undergraduate Research Programs in Science," (R. Taraban, and R. L. Blanton, eds.), pp. 112–132. Teachers College Press, New York.

Miao, W., Xiong, J., Bowen, J., Wang, W., Liu, Y., Braguinets, O., Grigull, J., Pearlman, R. E., Orias, E., and Gorovsky, M. A. (2009). Microarray analyses of gene expression during the *Tetrahymena thermophila* life cycle. *PLoS ONE* **4**(2), e4429.

Mochizuki, K. (2010). RNA-directed epigenetic regulation of DNA rearrangements. *Essays Biochem.* **48** (1), 89–100.

Nanney, D. L. (1980). *Experimental Ciliatology: An Introduction to Genetic and Developmental Analysis in Ciliates*. John Wiley & Sons, New York, NY, pp. 304.

National Research Council (NRC)(2003). *Bio 2010: Transforming Undergraduate Education for Future Research Biologists*. National Academies Press, Washington, DC.

National Research Council (NRC)(2009). *A New Biology for the 21st Century: Ensuring the United States Leads the Coming Biology Revolution*. National Academies Press, Washington, DC.

National Research Council (2011). *A Framework for K-12 Science Education: Practices, Crosscutting Concepts, and Core Ideas. Committee on a Conceptual Framework for New K-12 Science Education Standards. Board on Science Education, Division of Behavioral and Social Sciences and Education.* The National Academies Press, Washington, DC.

Project Kaleidoscope(1991). *What Works: Building Natural Science Communities.* Project Kaleidoscope, Washington, DC.

Seymour, E., Wiese, D., Hunter, A., and Daffinrud, S. M. (2000). Creating a Better Mousetrap: On-line Student Assessment of their Learning Gains. Paper presentation at the National Meeting of the American Chemical Society, San Francisco, CA.

Treisman, U. (1992). Studying students studying calculus: a look at the lives of minority mathematics students in college. *College Math. J.* **23**, 272–362.

Turkewitz, A. P., Orias, E., and Kapler, G. (2002). Functional genomics: the coming of age for *Tetrahymena thermophila. Trends Genet.* **18**, 35–40.

Turkewitz, A. P. (2004). Out with a Bang! Tetrahymena as a Model System to Study Secretory Granule Biogenesis. *Traffic* **5**, 63–68.

INDEX

E

F

U

V

VOLUMES IN SERIES

Founding Series Editor
DAVID M. PRESCOTT

Volume 1 (1964)
Methods in Cell Physiology
Edited by David M. Prescott

Volume 2 (1966)
Methods in Cell Physiology
Edited by David M. Prescott

Volume 3 (1968)
Methods in Cell Physiology
Edited by David M. Prescott

Volume 4 (1970)
Methods in Cell Physiology
Edited by David M. Prescott

Volume 5 (1972)
Methods in Cell Physiology
Edited by David M. Prescott

Volume 6 (1973)
Methods in Cell Physiology
Edited by David M. Prescott

Volume 7 (1973)
Methods in Cell Biology
Edited by David M. Prescott

Volume 8 (1974)
Methods in Cell Biology
Edited by David M. Prescott

Volume 9 (1975)
Methods in Cell Biology
Edited by David M. Prescott

Volume 10 (1975)
Methods in Cell Biology
Edited by David M. Prescott

Advisory Board Chairman
KEITH R. PORTER

Volume 21B (1980)
Normal Human Tissue and Cell Culture, Part B: Endocrine, Urogenital, and Gastrointestinal Systems
Edited by Curtis C. Harris, Benjamin F. Trump, and Gray D. Stoner

Volume 22 (1981)
Three-Dimensional Ultrastructure in Biology
Edited by James N. Turner

Volume 23 (1981)
Basic Mechanisms of Cellular Secretion
Edited by Arthur R. Hand and Constance Oliver

Volume 24 (1982)
The Cytoskeleton, Part A: Cytoskeletal Proteins, Isolation and Characterization
Edited by Leslie Wilson

Volume 25 (1982)
The Cytoskeleton, Part B: Biological Systems and In Vitro Models
Edited by Leslie Wilson

Volume 26 (1982)
Prenatal Diagnosis: Cell Biological Approaches
Edited by Samuel A. Latt and Gretchen J. Darlington

Series Editor
LESLIE WILSON

Volume 27 (1986)
Echinoderm Gametes and Embryos
Edited by Thomas E. Schroeder

Volume 28 (1987)
Dictyostelium discoideum: Molecular Approaches to Cell Biology
Edited by James A. Spudich

Volume 29 (1989)
Fluorescence Microscopy of Living Cells in Culture, Part A: Fluorescent Analogs, Labeling Cells, and Basic Microscopy
Edited by Yu-Li Wang and D. Lansing Taylor

Volume 30 (1989)
Fluorescence Microscopy of Living Cells in Culture, Part B: Quantitative Fluorescence Microscopy—Imaging and Spectroscopy
Edited by D. Lansing Taylor and Yu-Li Wang

Volume 31 (1989)
Vesicular Transport, Part A
Edited by Alan M. Tartakoff

Volume 32 (1989)
Vesicular Transport, Part B
Edited by Alan M. Tartakoff

Volume 33 (1990)
Flow Cytometry
Edited by Zbigniew Darzynkiewicz and Harry A. Crissman

Volume 34 (1991)
Vectorial Transport of Proteins into and across Membranes
Edited by Alan M. Tartakoff

Selected from Volumes 31, 32, and 34 (1991)
Laboratory Methods for Vesicular and Vectorial Transport
Edited by Alan M. Tartakoff

Volume 35 (1991)
Functional Organization of the Nucleus: A Laboratory Guide
Edited by Barbara A. Hamkalo and Sarah C. R. Elgin

Volume 36 (1991)
***Xenopus laevis:* Practical Uses in Cell and Molecular Biology**
Edited by Brian K. Kay and H. Benjamin Peng

Series Editors
LESLIE WILSON AND PAUL MATSUDAIRA

Volume 37 (1993)
Antibodies in Cell Biology
Edited by David J. Asai

Volume 38 (1993)
Cell Biological Applications of Confocal Microscopy
Edited by Brian Matsumoto

Volume 39 (1993)
Motility Assays for Motor Proteins
Edited by Jonathan M. Scholey

Volume 40 (1994)
A Practical Guide to the Study of Calcium in Living Cells
Edited by Richard Nuccitelli

Volume 41 (1994)
Flow Cytometry, Second Edition, Part A
Edited by Zbigniew Darzynkiewicz, J. Paul Robinson, and Harry A. Crissman

Volume 67 (2001)
Centrosomes and Spindle Pole Bodies
Edited by Robert E. Palazzo and Trisha N. Davis

Volume 68 (2002)
Atomic Force Microscopy in Cell Biology
Edited by Bhanu P. Jena and J. K. Heinrich Hörber

Volume 69 (2002)
Methods in Cell–Matrix Adhesion
Edited by Josephine C. Adams

Volume 70 (2002)
Cell Biological Applications of Confocal Microscopy
Edited by Brian Matsumoto

Volume 71 (2003)
Neurons: Methods and Applications for Cell Biologist
Edited by Peter J. Hollenbeck and James R. Bamburg

Volume 72 (2003)
Digital Microscopy: A Second Edition of Video Microscopy
Edited by Greenfield Sluder and David E. Wolf

Volume 73 (2003)
Cumulative Index

Volume 74 (2004)
Development of Sea Urchins, Ascidians, and Other Invertebrate Deuterostomes: Experimental Approaches
Edited by Charles A. Ettensohn, Gary M. Wessel, and Gregory A. Wray

Volume 75 (2004)
Cytometry, 4th Edition: New Developments
Edited by Zbigniew Darzynkiewicz, Mario Roederer, and Hans Tanke

Volume 76 (2004)
The Zebrafish: Cellular and Developmental Biology
Edited by H. William Detrich, III, Monte Westerfield, and Leonard I. Zon

Volume 77 (2004)
The Zebrafish: Genetics, Genomics, and Informatics
Edited by William H. Detrich, III, Monte Westerfield, and Leonard I. Zon

Volume 78 (2004)
Intermediate Filament Cytoskeleton
Edited by M. Bishr Omary and Pierre A. Coulombe

Volume 91 (2009)
Cilia: Structure and Motility
Edited by Stephen M. King and Gregory J. Pazour

Volume 92 (2009)
Cilia: Motors and Regulation
Edited by Stephen M. King and Gregory J. Pazour

Volume 93 (2009)
Cilia: Model Organisms and Intraflagellar Transport
Edited by Stephen M. King and Gregory J. Pazour

Volume 94 (2009)
Primary Cilia
Edited by Roger D. Sloboda

Volume 95 (2010)
Microtubules, in vitro
Edited by Leslie Wilson and John J. Correia

Volume 96 (2010)
Electron Microscopy of Model Systems
Edited by Thomas Müeller-Reichert

Volume 97 (2010)
Microtubules: In Vivo
Edited by Lynne Cassimeris and Phong Tran

Volume 98 (2010)
Nuclear Mechanics & Genome Regulation
Edited by G.V. Shivashankar

Volume 99 (2010)
Calcium in Living Cells
Edited by Michael Whitaker

Volume 100 (2010)
The Zebrafish: Cellular and Developmental Biology, Part A
Edited by: H. William Detrich III, Monte Westerfield and Leonard I. Zon

Volume 101 (2011)
The Zebrafish: Cellular and Developmental Biology, Part B
Edited by: H. William Detrich III, Monte Westerfield and Leonard I. Zon

Volume 102 (2011)
Recent Advances in Cytometry, Part A: Instrumentation, Methods
Edited by Zbigniew Darzynkiewicz, Elena Holden, Alberto Orfao, William Telford and Donald Wlodkowic

Volume 103 (2011)
Recent Advances in Cytometry, Part B: Advances in Applications
Edited by Zbigniew Darzynkiewicz, Elena Holden, Alberto Orfao, Alberto Orfao and Donald Wlodkowic

Volume 104 (2011)
The Zebrafish: Genetics, Genomics and Informatics 3rd Edition
Edited by H. William Detrich III, Monte Westerfield, and Leonard I. Zon

Volume 105 (2011)
The Zebrafish: Disease Models and Chemical Screens 3rd Edition
Edited by H. William Detrich III, Monte Westerfield, and Leonard I. Zon

Volume 106 (2011)
Caenorhabditis elegans: Molecular Genetics and Development 2nd Edition
Edited by Joel H. Rothman and Andrew Singson

Volume 107 (2011)
Caenorhabditis elegans: Cell Biology and Physiology 2nd Edition
Edited by Joel H. Rothman and Andrew Singson

Volume 108 (2012)
Lipids
Edited by Gilbert Di Paolo and Markus R Wenk

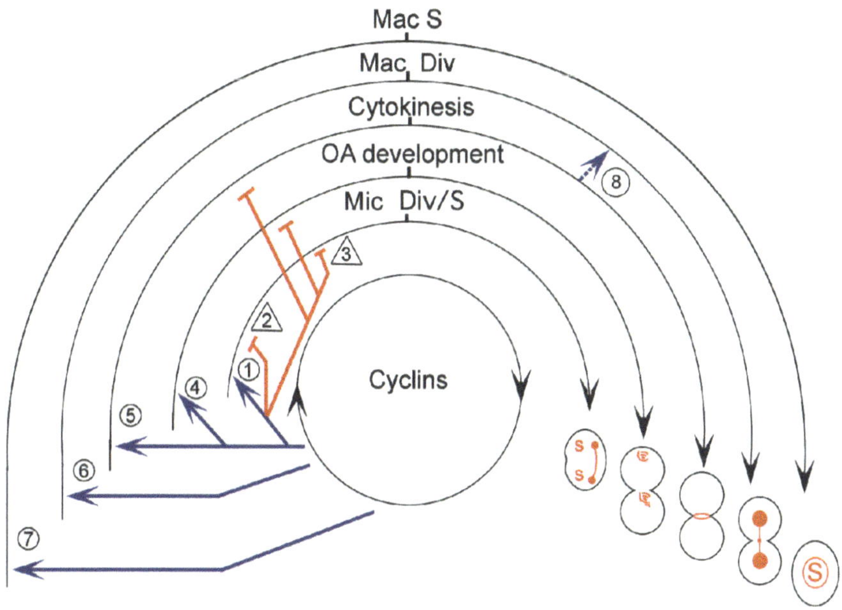

Plate 1 (Figure 7.3 on page 183 of this volume).

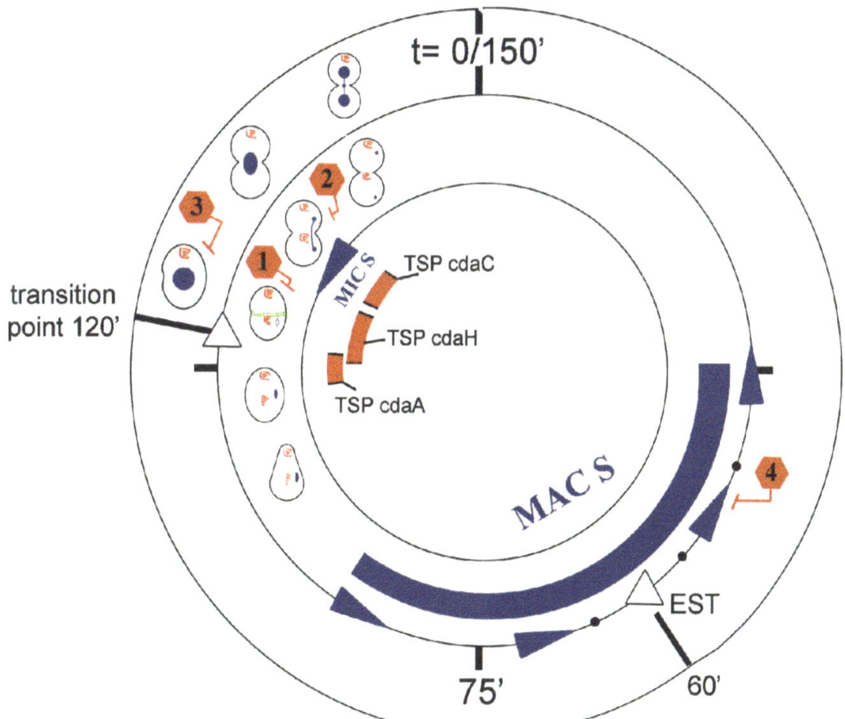

Plate 2 (Figure 7.5 on page 186 of this volume).

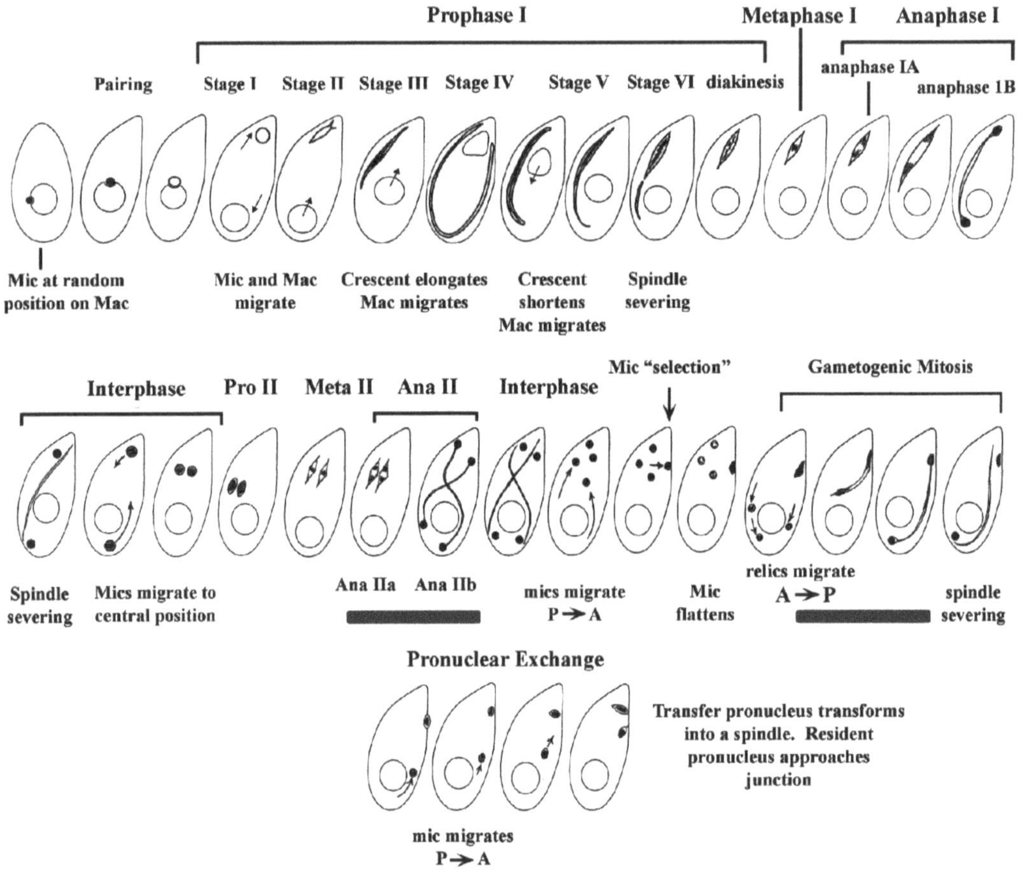

Plate 3 (Figure 7.6 on page 199 of this volume).

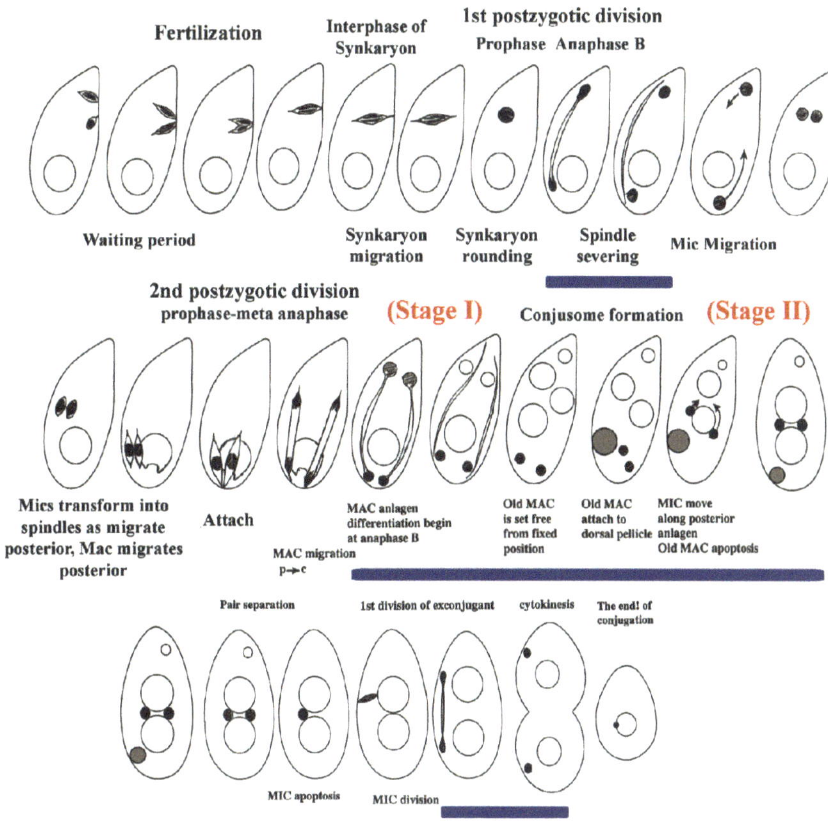

Plate 4 (Figure 7.7 on page 200 of this volume).

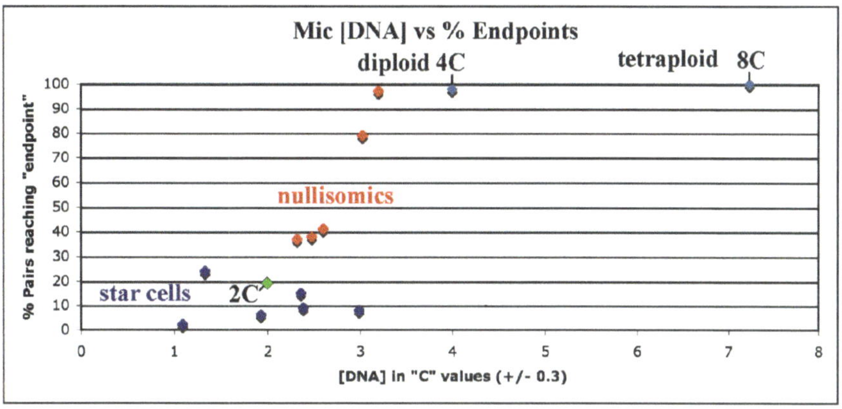

Plate 5 (Figure 7.8 on page 209 of this volume).

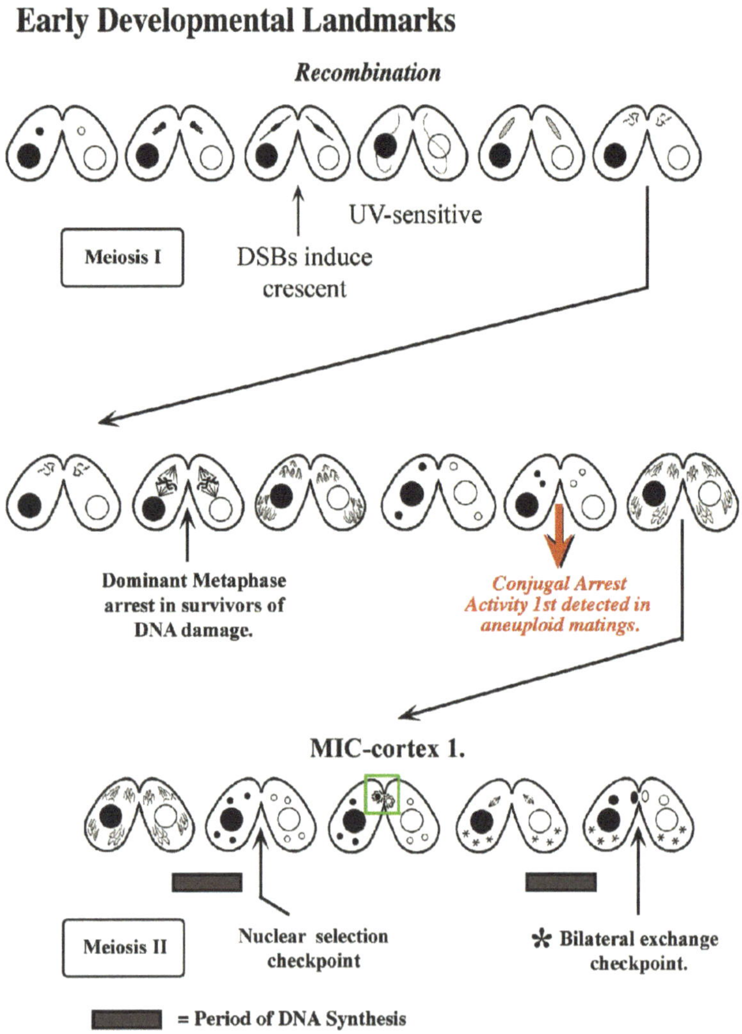

Plate 6 (Figure 7.9 on page 226 of this volume).

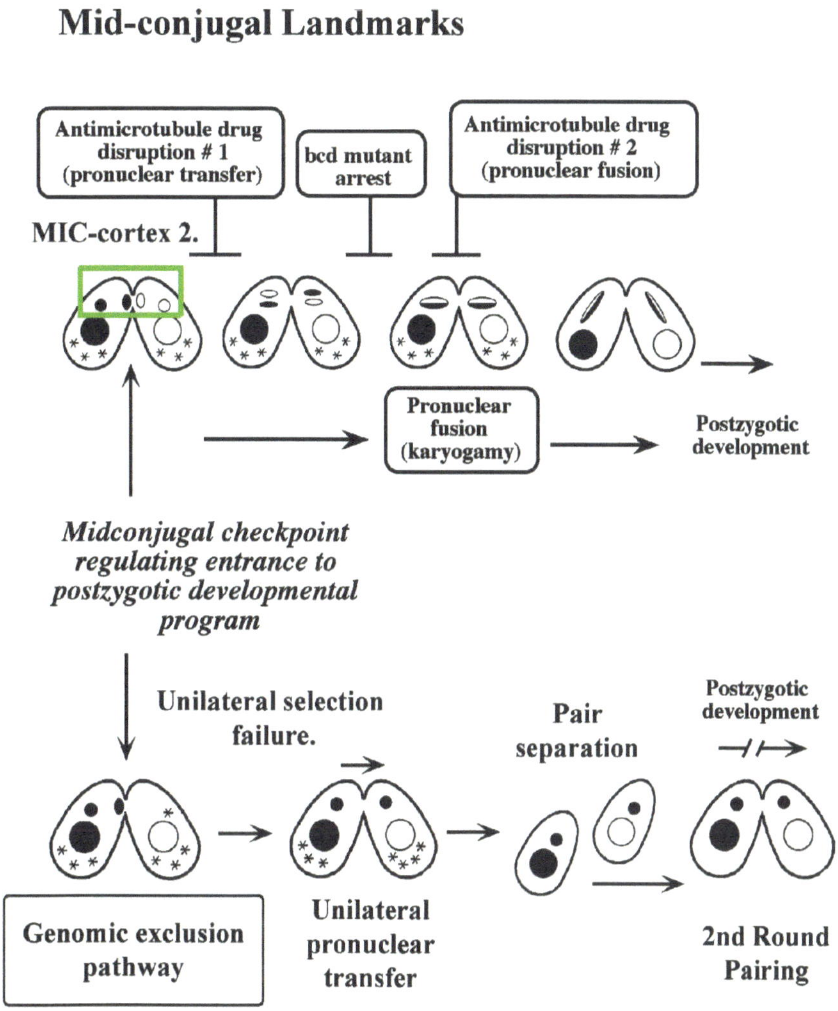

Plate 7 (Figure 7.10 on page 227 of this volume).

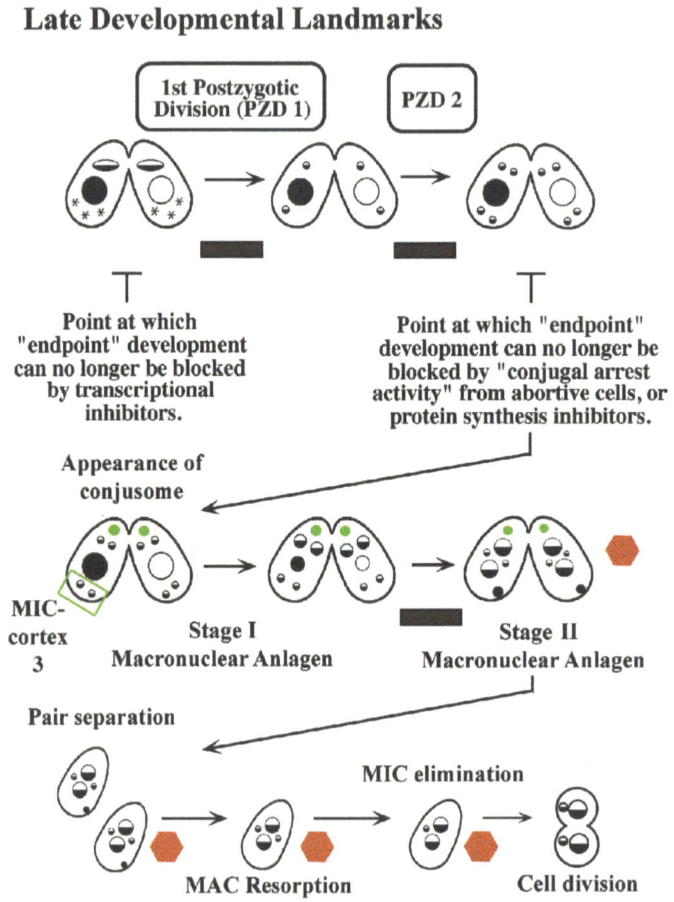

Plate 8 (Figure 7.11 on page 228 of this volume).

Plate 9 (Figure 9.3 on page 288 of this volume).

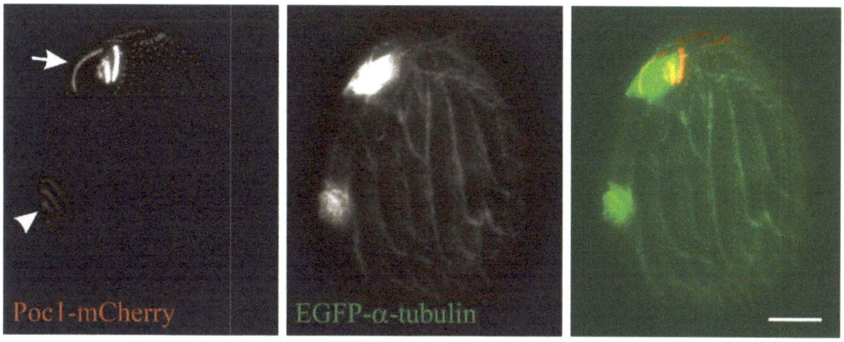

Plate 10 (Figure 13.1 on page 360 of this volume).

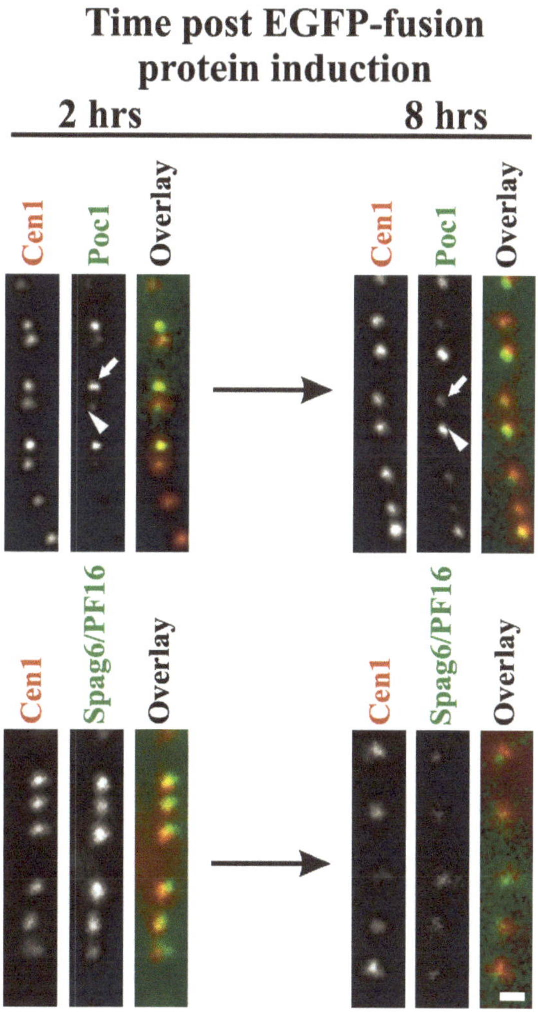

Plate 11 (Figure 13.2 on page 363 of this volume).

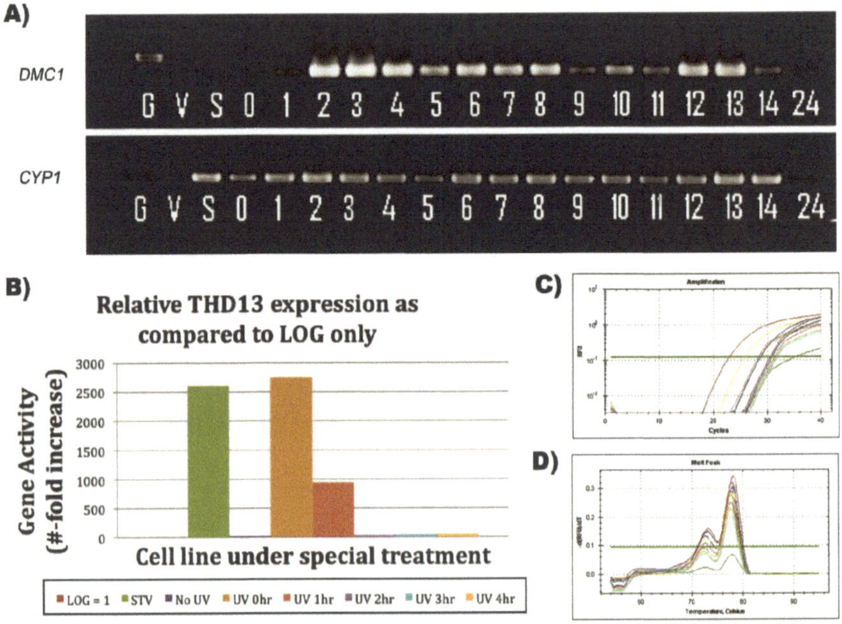

Plate 12 (Figure 16.2 on page 416 of this volume).

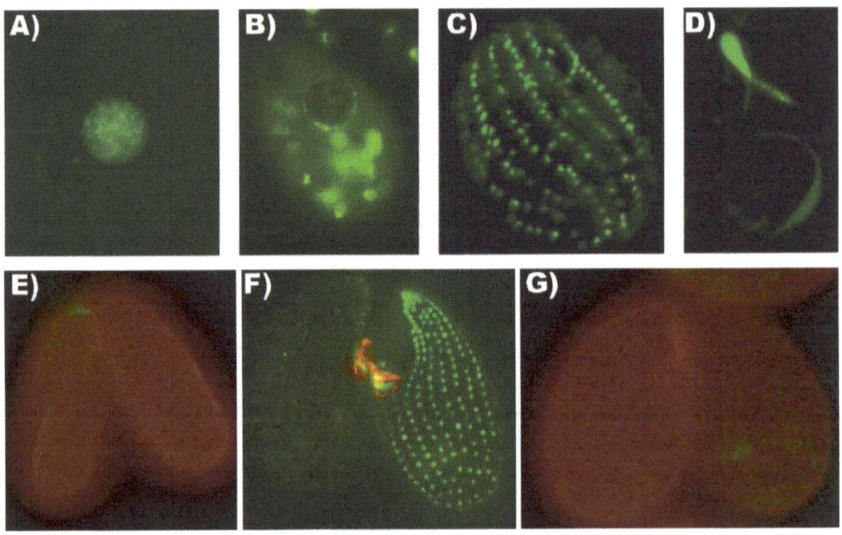

Plate 13 (Figure 16.3 on page 417 of this volume).

FSC
www.fsc.org
MIX
Papier | Fördert gute Waldnutzung
FSC® C083411

Kolibri 360 GmbH
Ferdinand-Jühlke-Straße 7
99095 Erfurt, Deutschland
produktsicherheit@kolibri360.de